Heinz von Foerster Conference on Emergent Quantum Mechanics 2011

(EmerQuM11)

AA008879

Journal of Physics: Conference Series Volume 361

Vienna, Austria
11-13 November 2011

ISBN: 978-1-62276-778-6
ISSN: 1742-6588

Printed from e-media with permission by:

Curran Associates, Inc.
57 Morehouse Lane
Red Hook, NY 12571

Some format issues inherent in the e-media version may also appear in this print version.

Printed by Curran Associates, Inc. (2013)

For permission requests, please contact the Institute of Physics
at the address below.

Institute of Physics
Dirac House, Temple Back
Bristol BS1 6BE UK

Phone: 44 1 17 929 7481
Fax: 44 1 17 920 0979

techtracking@iop.org

Additional copies of this publication are available from:

Curran Associates, Inc.
57 Morehouse Lane
Red Hook, NY 12571 USA
Phone: 845-758-0400
Fax: 845-758-2634
Email: curran@proceedings.com
Web: www.proceedings.com

Papers

Plenary Lectures

Posters

Preface

These proceedings comprise the plenary lectures and poster contributions of the 'Heinz von Foerster Conference 2011' on *Emergent Quantum Mechanics* (EmerQuM11), which was held at the University of Vienna, 11–13 November 2011.

With the 5th International Heinz von Foerster Conference convened at the occasion of von Foerster's 100th birthday, the organizers opted for a twin conference to take place at the Large and Small Ceremonial Halls of the University's main building, respectively. The overall topic was chosen as 'Self-Organization and Emergence', a topic to which von Foerster was an early contributor. While the first conference ('Self-Organization and Emergence in Nature and Society') addressed a more general audience, the second one ('Emergent Quantum Mechanics') was intended as a specialist meeting with a contemporary topic that could both serve as an illustration of von Foerster's intellectual heritage and, more generally, point towards future directions in physics. We thus intended to bring together many of those physicists who are interested in or are working on attempts to understand quantum mechanics as emerging from a suitable classical (or, more generally, deeper level) physics.

EmerQuM11 was organized by the Austrian Institute for Nonlinear Studies (AINS), with essential support from the Wiener Institute for Social Science Documentation and Methodology (WISDOM), the Department of Contemporary History at the University of Vienna, and the Heinz von Foerster-Gesellschaft. There were a number of individuals who contributed to the smooth course of our meeting and whom I would like to sincerely thank: Christian Bischof, Thomas Elze, Marianne Ertl, Gertrud Hafner, Werner Korn, Angelika Krawanja, Florian Krug and his team, Sonja Lang, Albert Müller, Ilse Müller, Irene Müller, Karl Müller, Armin Reautschnig, Marion Schirrmacher, Anton Staudinger, Roman Zlabinger, and, last but not least, my AINS colleagues Siegfried Fussy, Herbert Schwabl and Johannes Mesa Pascasio, the latter in particular for his invaluable technical help with these proceedings.

Funds made available by the Federal Ministry of Science and Research (BMWF), the City of Vienna MA7 Science Funding, the Faculty of Historical and Cultural Studies, Blaha Office Furniture, and Padma AG Zurich are gratefully acknowledged.

As for the nature of the search for a 'deeper level' foundation of quantum mechanics, a first difficulty already arises with respect to the question: *Where do we start?* One may look for quite different points of departure, such as an encompassing theory of quantum gravity. Or one may find arguments for the necessity to base one's approach at least on a relativistic formulation of the problem. Or one may discard searching for general principles for the time being, and develop an explicit physical model first. And so on. In fact, this is actually what is happening today in different research programs for emergent quantum mechanics, a fact which is also reflected in the rich variety of approaches presented at our meeting. This may be considered a very welcome situation, reminding us of Heinz von Foerster's dictum: 'Act always so as to increase the number of choices.' However, some may view this variety also as a drawback: There is not (yet?) a single, definite alternative theory that would challenge orthodox positions, for example, by providing different experimental predictions. However, the prevailing orthodoxy has shown throughout the 20th century to the present day, that a too restrictive attitude towards theoretical alternatives can lead to almost a standstill in coping with the serious shortcomings and contradictions of present-day physics. Many of us remember famous quantum physicists repeating in an almost mantra-like fashion that quantum theory, or experimental evidence, 'excludes hidden variables as a possibility', along with a reference to some or other newly found 'impossibility proof'. Yet we also recall John Bell's famous counter-statement: 'What is proved by impossibility proofs is lack of imagination.' In this sense, therefore, our task is to make use of the variety of the different approaches offered, for it is in scrutinizing each of them that a chance for further progress and understanding may emerge.

The papers collected in these proceedings essentially follow the order of the plenary talks during the conference program, with the addition of the poster presentations. Prior to the contributions to EmerQuM11, the very first paper of these proceedings presents the opening lecture by Yves Couder who addressed both twin conferences with his talk on wave-particle duality in a classical system. (Although he was not able to participate personally, the contribution of Robert Carroll, a member of the academic advisory board, is included here. Regrettably, the lectures by John Bush, Marek Czachor, Mark Everitt, Felix Finster, and Lee Smolin could not be included, partly for copyright reasons.) Finally, I would like to thank Sarah Toms and Graham Douglas and their team at IOP Publishing (Bristol) for their friendly advice and help during the preparation of these proceedings.

Vienna, April 2012

Gerhard Grössing

Journal of Physics: Conference Series **361** (2012) 011002 doi:10.1088/1742-6596/361/1/011002

Peer review statement

All papers published in this volume of *Journal of Physics: Conference Series* have been peer reviewed through processes administered by the proceedings Editors. Reviews were conducted by expert referees to the professional and scientific standards expected of a proceedings journal published by IOP Publishing.

Probabilities and trajectories in a classical wave-particle duality

Y. Couder

Matière et Systèmes Complexes, Université Paris Diderot, CNRS - UMR 7057, Bâtiment Condorcet, 10 rue Alice Domon et Léonie Duquet, 75013 Paris, France

E-mail: couder@physique.ens.fr

E. Fort

Institut Langevin, ESPCI ParisTech and Université Paris Diderot, CNRS UMR 7587, 10 rue Vauquelin, 75231 Paris Cedex 05, France

E-mail: emmanuel.fort@paris7.jussieu.fr

Abstract. Several recent experiments were devoted to walkers, structures that associate a droplet bouncing on a vibrated liquid with the surface waves it excites. They reveal that a form of wave-particle duality exists in this classical system with the emergence of quantum-like behaviours. Here we revisit the single particle diffraction experiment and show the coexistence of two waves. The measured probability distributions are ruled by the diffraction of a quantum-like probability wave. But the observation of a single walker reveals that the droplet is driven by a pilot wave of different spatial structure that determines its trajectory in real space. The existence of two waves of these types had been proposed by de Broglie in his "double solution" model of quantum mechanics. A difference with the latter is that the pilot-wave is, in our experiment, endowed with a "path memory". When intrusive measurements are performed, this memory effect induces transient chaotic individual trajectories that generate the resulting statistical behaviour.

1. Introduction

At the origin of this work is the discovery that it is possible, at classical scale, to obtain a steady regime of propagation for a structure associating a localized particle and a wave. The particle is a droplet bouncing on the vibrated surface of a liquid and the wave is the surface wave generated by the bouncing. It is known that when a liquid bath is vertically vibrated at a frequency f_0 with an acceleration amplitude larger than an onset value, the whole surface becomes unstable and a wave pattern of frequency $f_0/2$ forms spontaneously due to parametric forcing. This is the Faraday instability [1]. The dynamical coupling of the droplet to the wave occurs immediately below this threshold. For such a forcing and when the droplet has the correct size, its bouncing becomes sub-harmonic so that the droplet acts as an efficient local exciter of Faraday waves. In this regime the bouncing droplet and the wave it generates become phase-locked and the whole object, the drop and its associated wave, becomes spontaneously propagative [2,3]. In spite of the system being dissipative, this self-propagative structure is sustained. The bath oscillation provides energy both to the drop to maintain its bouncing and to the waves through

their parametric forcing.
We called these structures "walkers" and various experiments were undertaken to understand how the particle and the wave could have a common motion on the 2D surface of the bath. They revealed that some behaviours thought to be strictly reserved to quantum duality showed up at macroscopic scale in the dynamics of these objects. Single particle diffraction and interference were observed [4] as well as a form of tunneling effect [5]. Furthermore, when an applied transverse force gives a walker a circular motion, it was found that its possible orbits were quantized and an analogy existed with the quantum Landau orbits in a magnetic field [6]. All these results were shown to emerge from the spatio-temporal non-locality of the walker due to what we called its "wave-mediated path-memory" which is described in detail in references [6, 7]. At each collision with the interface the droplet emits a circular travelling capillary wave that triggers the formation of a localized mode of Faraday standing waves of wavelength λ_F. Because of the proximity of the instability threshold, these waves are sustained during "a memory time" τ that becomes large near the Faraday instability threshold [7]. As a result the wave field is the linear superposition of the successive Faraday waves emitted by past bounces. Its complex interference structure thus contains a memory of the recent trajectory. Furthermore, since the travelling waves move faster than the drop, the wave field also contains information about the obstacles that lie ahead. Hence, two non-local effects exist in the wave-field driving the motion of the droplet: the past bounces influence directly the present (direct propulsion) and the trajectory is perturbed by scattered waves from distant obstacles in a kind of echo-location effect. This interplay between the droplet motion and its associated wave field makes it a macroscopic implementation of a pilot-wave dynamics.
Such a system is reminiscent of the early de Broglie models for quantum systems [8-11]. In the present article we will limit ourselves to revisiting our diffraction experiment [4] in the terms used by de Broglie in his pilot-wave and double solution model [10, 11]. This will lead us to a discussion of the disordered trajectories generated by intrusive measurements and the resulting probabilistic behaviours.

2. The experiment and its results

In the experiment reported in ref. [4] we had investigated the diffraction of walkers as they pass through one slit. The reader can refer to this letter for the experimental set-up as well as for the precise description of the results. Here we will revisit these results by thought experiments in which the observational means are restricted. We can thus artificially introduce the measurement problem central to quantum physics. With this aim we first present the statistical results that can be obtained with reduced means.

2.1. The blind statistical results

The thought experiment is sketched in Fig. 1. We assume that we have a source of walkers and that the only information we can get about a walker is the position of the droplet *that we can only obtain by destroying it.* A single droplet trajectory cannot be tracked and the surface waves are not visible. We thus have to resort to statistical measurements. Between a source of walkers and a detector screen, we place a wall opened with a single slit of width L. The source is located sufficiently far so that the walkers it emits have a normal incidence ($\alpha_i = \pi/2$) on the wall. Their flux is homogeneous and they impinge with the same probability in all regions of the wall and the slit. The source is weak so that successive walkers cross the slit one at a time. On the screen placed at a distance behind the slit the impacts of the successive droplets accumulate. The experiment shows that a diffraction pattern builds-up. The resulting histogram (Fig.1) of droplet deviations α_d is well fitted by the modulus of the amplitude of diffraction of a plane

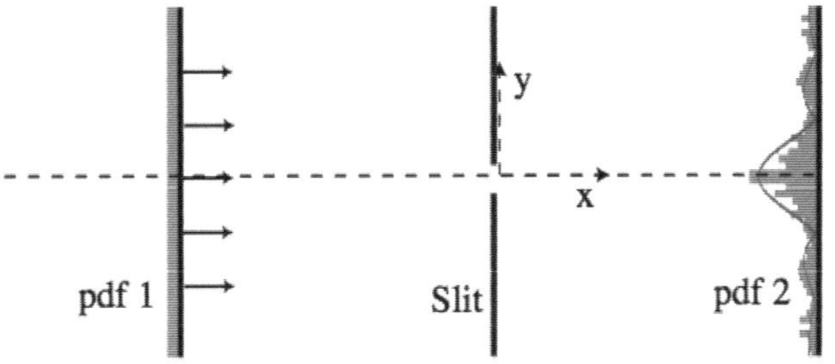

Figure 1. Sketch of the diffraction experiment with reduced means of observation. If only a detection of the locus of impact of the walkers on a screen is possible, the only experimental result is that the probability distribution function for the deviations is given by the diffraction of a plane wave of wavelength λ_F passing through the slit of width L.

wave of wavelength λ_F passing through the aperture of width L:

$$f(\alpha_d) = \frac{A \sin(\pi L \sin \alpha_d / \lambda_F)}{\pi L \sin \alpha / \lambda_F} \tag{1}$$

This means that the probability distribution of the deviations of a droplet is given by the diffraction of a plane wave [4]. This result is similar to what would be obtained with electrons or photons except that the distribution would then be given by the square of the wave amplitude.

2.2. *Diffraction as an intrusive measurement*

Passing through the slit can be considered as a means by which we measure a droplet's position in the y direction. Before crossing the slit, due to the homogeneity of the flow in the semi-infinite medium, a given droplet has an equal probability to be anywhere in the y direction and its velocity component V_y is zero. After having passed through the slit we know its position: it is located at the abscissa $y = 0$ with a precision $\delta y \sim \pm L/2$. This information has been obtained at the cost of a loss of definition on the transverse component V_y of its velocity. An uncertainty δV_y has appeared. As in quantum mechanics, this uncertainty has been generated by the intrusion of the measurement itself. The measured probability distribution functions, both before the slit and after it, are similar to those of a plane wave of wavelength λ_F. This wave is a probability wave: it plays a similar role to that of the wave function in quantum mechanics.

2.3. *The non-intrusive observation*

The elementary structures of our experiment are bouncing droplets and their associated surface waves. Measurements involving exclusively the interactions of this type of objects would certainly be intrusive. However in this system non-intrusive observations can be obtained with visual observation. By switching on the light, we can observe with the naked eye either the droplet only, or both the droplet and its wave field. The droplet alone appears as a point-like particle. Fig. 2 shows its recorded trajectories during two successive crossings of the slit. The trajectories are observed to be complex in the slit region. In spite of rather similar initial conditions they deviate differently.

The origin of such individual trajectories is revealed only when we observe both the particle and the fluid interface. As deduced from the blind experiments, the association of the drop with a wave of wavelength λ_F is responsible for the diffraction. However, the observed wave field is very different from the wave that determines the probabilities. No common plane wave is

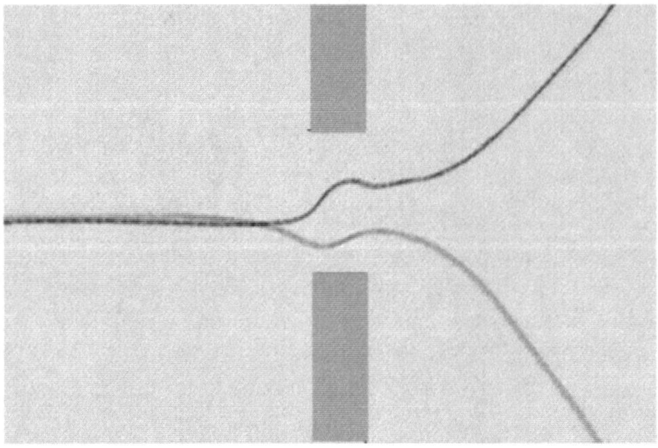

Figure 2. Two measured trajectories of single droplets passing through the slit. The similar initial conditions generate very different trajectories.

observed before the slit crossing and no common diffracted wave is observed behind it. Instead, the droplet is seen to be at the centre of a wave packet with a complex and constantly evolving structure (Fig. 3). The nature of this wave packet is now well-understood and described in Fort *et al.*[6].

The topography of the liquid surface is determined by the superposition of circular standing waves centred on the positions of the past bounces of the droplet. The global wave-field thus has an interference structure that contains a memory of the particle's recent trajectory. This is what we have called [6,7] the "wave-mediated path-memory" of the walker. The horizontal motion of the droplet is driven iteratively, bounce after bounce, by its coupling to this wave. At each collision with the interface the drop undergoes a damping due to viscous friction but it is simultaneously given a momentum increment by its shock with the slanted oscillating surface. The direction of the kick results from the local slope of the interface at the point of impact (see Fig. 4). The resulting velocity of the drop is thus :

$$m\frac{d\mathbf{r_i}}{dt} \propto \nabla\zeta(\mathbf{r_i}, \mathbf{t_i}) \tag{2}$$

where m is the mass of the droplet, $\mathbf{r_i}$ its position in the plane at the collision of order i and $\zeta(\mathbf{r_i}, \mathbf{t_i})$ the height of the liquid interface in $\mathbf{r_i}$ at time t_i.
This path memory model retains all the dynamical behaviour of the experimental walker. Its numerical implementation is described in detail in ref. [6,7]. It accounts for the spontaneous motion of the droplet as well as for the structure of the resulting pilot-wave field. It was used in ref. [4] to investigate the diffraction of walkers. Figure 5 shows the computed trajectories of individual walkers impinging on the slit and coupled to waves that have a large path memory. The individual trajectories (Fig. 5) confirm the main characteristics observed in the experiment.
- The droplets are deviated long before they reach the slit. Since the propagative front of the wave field moves faster than the droplet, it has an echo-location function. The standing Faraday waves that it generates contain information on the shape of the boundaries.
- The complexity of the trajectories in the region of the slit leads to their divergence. Droplets having very similar initial trajectories far from the slit can undergo different deviations. This

Journal of Physics: Conference Series **361** (2012) 012001 doi:10.1088/1742-6596/361/1/012001

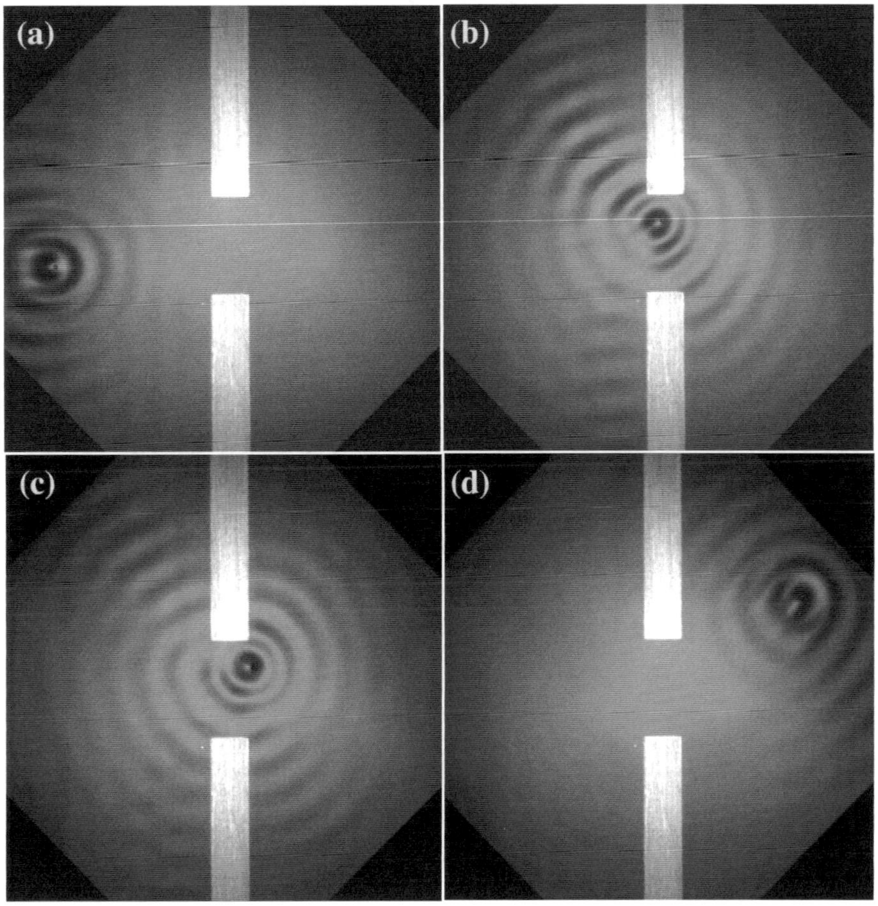

Figure 3. Four photographs of a walker as it passes through the slit. Note the complexity of the pilot wave field when the drop is in the slit region (b and c).

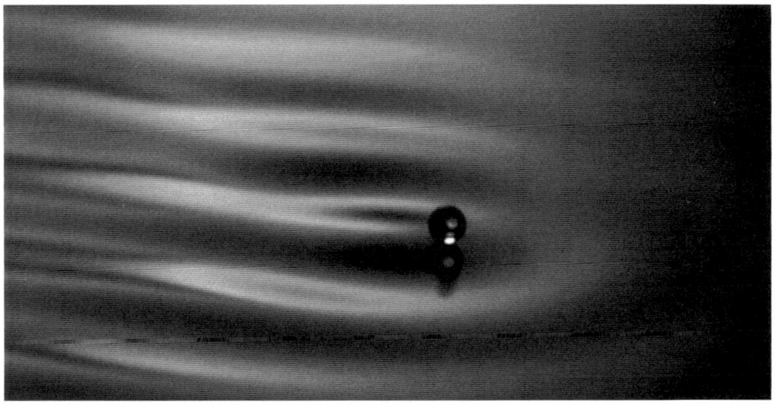

Figure 4. Close-up photograph of the collision of the droplet with the locally slanted interface due to the wave.

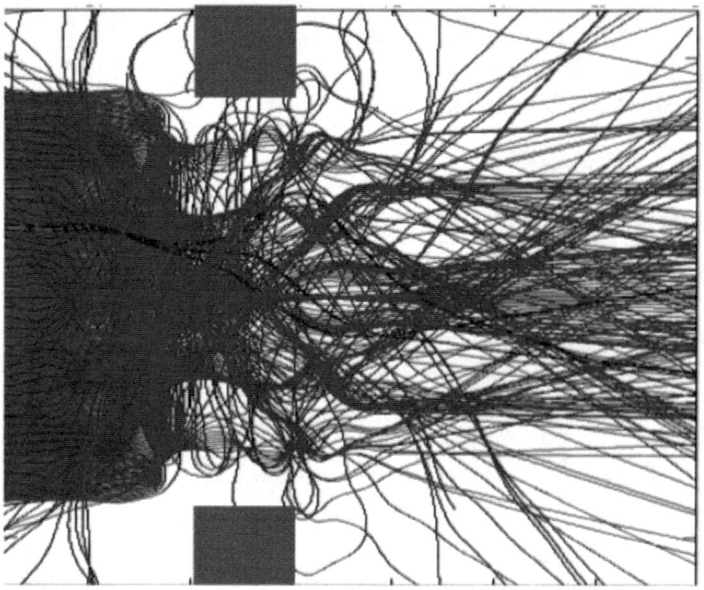

Figure 5. The simulation of the motion of a hundred walkers impinging on a slit. Note that, as in the experiment, some trajectories cross the symmetry axis of the apparatus.

appears as the source of the observed uncertainty.
- Some of the trajectories cross the symmetry line. A drop impinging on the left of the slit can be deviated to the right.
- Finally, in the numerical simulation, as in the experiment, a diffraction pattern of a plane wave emerges from the statistics of the deviations of a large number of independent realizations [4].

3. The relation with de Broglie pilot wave models

Our experiment is characterized by the coexistence of two waves. One is an individual physical wave excited by the drop that has a pilot-wave role during the trajectory in space and time. The second is not visible at any particular time but determines the probabilities of the measurements. This appears very close to the hypothesis of a double solution put forward by de Broglie with the aim of restoring determinism in quantum mechanics.
It is useful here to summarize very briefly the historical evolution of the pilot wave models of quantum mechanics usually gathered as "de Broglie-Bohm model". This latter terminology turns out to be misleading since the theoretical approaches of these two authors, though inter-related, are very different from one another.

3.1. de Broglie's pilot wave model

In his initial work in 1925, de Broglie [8] proposed that a particle of rest mass m_0 (such as an electron) has an oscillation at a frequency $\nu_0 = m_0c^2/h$. This singularity is surrounded by a standing wave-field of wavelength $\lambda_{dB} = h/mV$ formed by the superposition of an emitted wave radiated away (retarded wave) and a wave converging towards the particle (advanced wave). In his model the particle is guided by the wave, its velocity being linked to the wave by

$$m\frac{d\mathbf{x_i}}{\mathbf{dt}} \propto \nabla_\mathbf{i}\mathbf{S} \tag{3}$$

where S is the phase of the guiding wave.
Such an association of the particle to a wave field was meant to introduce the possibility of non-newtonian behaviours. This was summarized by de Broglie in the following terms [10]: *"the particle's motion ... guided by the wave... would not follow the laws of classical mechanics according to which the particle is subject only to the action of forces which act on it along its trajectory, and does not suffer any repercussion from the existence of obstacles which may be situated far away outside its trajectory. In my (...) conception, on the contrary, the movement of the singularity should experience the influence of all the obstacles which hinder the propagation of the wave with which it is connected. This circumstance would explain the existence of the phenomena of interference and diffraction".*
Inspired by this work, Schrödinger initially introduced his equation while seeking de Broglie's wave. It was only later that Born demonstrated that the Schrödinger equation was related to the probability of presence. de Broglie then suggested in 1927 [9] that in all quantum systems two wave fields have to be considered:
- The first wave-field (that he called **u**) is generated by the particle. It is a real wave that has a singularity at its core. (In later versions he proposed that the core region could be strongly non-linear). For de Broglie this wave-particle structure had well-defined trajectories in the physical space. The particle could be submitted to forces but its association to the wave also gave it a non-local sensitivity to the environment and a non-newtonian behaviour.
- The second wave-field is ψ the linear and smooth solution of the Schrödinger equation that is related to the statistical density of probability of presence.

3.2. Bohm's pilot wave model
Much later, in 1952, David Bohm [12] revisited the idea of a pilot wave and investigated the situation where the wave solution of the Schrödinger equation would be the pilot wave. This idea had been briefly considered by de Broglie himself but then renounced [13]. Bohm used the Madelung transformation of the Schrödinger equation to obtain the following equation of motion:

$$m\frac{d\mathbf{x_i}}{dt} \propto \nabla_\mathbf{i}\mathbf{S}_\psi \tag{4}$$

where is the phase of the Schrödinger wave function. This equation is similar to eq.3, but deceptively so, since these two equations are related to two different wave fields (**u** and ψ, respectively). In the Bohmian model where the wave function is decoupled from the trajectory itself, a slightly modified Newtonian mechanics can be recovered by addition of a "quantum potential".
The Bohmian trajectories, being computed from Schrödinger's equation, remain the trajectories of probability densities, rather than those of individual particles. This is the reason for which de Broglie wrote shortly after Bohm's publication [10] : *"... David Bohm took up the pilot-wave theory again. His work is very interesting in many ways (...) But since Bohm's theory regards the wave ψ as a physical reality, it seems to me to be unacceptable in its present form".* This led de Broglie to restate [11] his "double solution" hypothesis where a wave field **u** is associated to a given particle and guides its motion. However, little progress was done along this line. The **u** waves remained undetected and it was unclear how they could generate the random behaviours responsible for the statistics. For these reasons, research in this domain appeared to have stalled. By contrast, the recent developments of the Bohmian approach have permitted the computation of trajectories in such situations like the interference and diffraction of particle beams [14]. In

a recent work [15], statistical trajectories that could be related to the Bohmian trajectories have been observed experimentally. They are smoother than those shown here in Figure 2 and 5. Another difference is that the Bohmian trajectories do not cross the symmetry axis of the system. Those passing on the left (right) of the slits are always deviated to the left (right). This can be seen as a characteristic difference between the Bohmian trajectory that concerns a probability density and the individual trajectory of a single particle.

4. Discussion and conclusion

The results about diffraction that we have summarized demonstrate that there exist in our macroscopic system two waves of the type imagined by de Broglie for quantum mechanics. One of these waves is a pilot wave: the droplet is guided by the local gradient of its phase (Eq. 2) as proposed by de Broglie (Eq.3) for particles. The other wave is a Schrödinger-like probability wave. In this experiment the generation of the probabilistic behaviours can be observed directly, shedding light on the relation between these two waves. Because of its macroscopic scale, our experiment lends itself to performing either intrusive or non-intrusive measurements. The non-intrusive observation of the phenomena associated with an intrusive measurement reveals how it generates a chaotic burst in which both the pilot wave and the droplet trajectory are strongly disturbed [16]. In, e.g., the diffraction experiment, there is a feedback effect. As the wave field becomes more complex (because of the reflected waves) the trajectory bends. In reverse, as the trajectory bends the complexity of the generated wave field increases. However, whatever its complexity, the wave field still has to satisfy the external boundary conditions. There is a statistical sampling of the environment through the propagation of a single walker so that statistical properties are present within the pilot wave. This is why the characteristic interference pattern of the slit shows up in the statistics. More generally, the eigenmodes of the configuration of a detector will contribute to the pilot wave configurations. For this reason a signature of these modes is present in the probability distribution of the particle. In our experiment this effect is entirely related to the path memory effect, a characteristic absent from de Broglie's models. If this idea was transferred to the quantum world where all measurements are intrusive, such an effect could be responsible for both the uncertainty principle and the quantum projection effect. This would require that a wave-mediated path-memory gives a form of spatio-temporal non-locality to quantum particles. To our knowledge this type of hypothesis has not been investigated yet.

Acknowledgements During our exploration of this system we worked in teams on the various aspects of the dynamics of walkers. The present state of our understanding results from the path-memory of these collaborations with Arezki Boudaoud, Antonin Eddi, Mathieu Labousse, Marc Miskin, Frédéric Moisy, Julien Moukhtar, Stéphane Perrard, Suzie Protière, Maurice Rossi and Eric Sultan. This work was sponsored by the Agence Nationale de la Recherche and by the Institut Universitaire de France.

References

[1] Faraday M 1831 *Philos. Trans. R. Soc. London* **52** 299-340, On the forms and states of fluids on vibrating elastic surfaces.
[2] Couder Y., Protière S., Fort E., Boudaoud A. 2005 *Nature* **437** 208, Walking and orbiting droplets.
[3] Protière S., Y Couder, and Boudaoud A. 2006 *J. Fluid Mech.* **554** 85-108, A particle-wave association on a fluid interface.
[4] Couder Y. and Fort E. 2006 *Phys. Rev. Lett.* **97** 154101, Single-particle diffraction and interference at a macroscopic scale.
[5] Eddi A., Fort E., Moisy F. and Couder Y. 2009 *Phys. Rev. Lett.* **102** 240401-4, Unpredictable tunneling of a classical wave-particle association.
[6] Fort E., Eddi A., Boudaoud A., Moukhtar J., and Couder Y. 2010 *PNAS* **107** 17515-17520, Path memory induced quantization of classical orbits.

[7] Eddi A., Sultan E., Moukhtar J., Fort E., Rossi M. and Couder Y. 2011 *J. Fluid Mech.* **674** 433-464, Information stored in Faraday waves: the origin of a path memory.
[8] de Broglie L. 1925 *Annales de Physique* **10-III** 22-128, Recherches sur la théorie des quanta. English translation (Kracklauer A.F. 2004), online at Fondation Louis de Broglie, aflb.ensmp.fr/.
[9] de Broglie L. 1927 *Journal de Physique et le radium* **6-5** 225-241, La mécanique ondulatoire et la structure atomique de la matière et du rayonnement.
[10] de Broglie L. 1953 in *Scientific Papers Presented to Max Born on his retirement from the Tait Chair of Natural Philosophy in the University of Edinburgh"* (Oliver and Boyd 1953), English translation (Dileep Karanth 2010) online: http://arxiv.org/abs/1005.4534v1.
[11] de Broglie L. 1959 *Le journal de physique et le radium* **20** 963-979, L'interprétation de la mécanique ondulatoire.
[12] Bohm D. 1952 *Phys. Rev* **85** 166-179 and *Phys. Rev* **85** 180-193, A suggested interpretation of quantum theory in terms of hidden variables. Part 1 and part 2.
[13] At the Solvay conference, de Broglie had admitted that the Schrdinger waves could be the "pilot wave" of the particle. As stated in reference [10] he was later to regret what he considered as a concession.
[14] Philippidis C., Dewdney C. and Hiley B. J. 1979 *Nuovo Cimento B* **52** 15-28, Quantum interference and the quantum potential.
[15] Kocsis S., Braverman B., Ravets S., Stevens M. J., Mirin R., P. Shalm L. K. and Steinberg A.M. 2011 *Science* **332** 1179-1173, Observing the average trajectories of single photons in a two-slit interferometer.
[16] We can note that the chaos we described here has no direct relation to the quantum chaos usually investigated. Most works of this type are devoted to finding a relation between the trajectories of classical particles in, e.g., chaotic stadium-shaped billiards and the behaviour of the Schrödinger waves in the same geometries. These works are thus downstream from Schrödinger probabilities, but do not address their production.

Quantum Theory as an Emergent Phenomenon: Foundations and Phenomenology

Stephen L. Adler
Institute for Advanced Study, Einstein Drive, Princeton, NJ 08540, USA
E-mail: adler@ias.edu

Abstract. I review the proposal made in my 2004 book [1], that quantum theory is an emergent theory arising from a deeper level of dynamics. The dynamics at this deeper level is taken to be an extension of classical dynamics to non-commuting matrix variables, with cyclic permutation inside a trace used as the basic calculational tool. With plausible assumptions, quantum theory is shown to emerge as the statistical thermodynamics of this underlying theory, with the canonical commutation-anticommutation relations derived from a generalized equipartition theorem. Brownian motion corrections to this thermodynamics are argued to lead to state vector reduction and to the probabilistic interpretation of quantum theory, making contact with phenomenological proposals [2, 3] for stochastic modifications to Schrödinger dynamics.

1. Introduction

Quantum mechanics is our most successful physical theory, but its foundational issues have been under debate for over eighty years. There are two possibilities:

- *Quantum theory is exact*, with (perhaps) a need for reinterpretation at the foundational level. Popular interpretations are "Many worlds", "Bohmian mechanics", "Histories", and "Quantum theory as information". Experimentally, they are indistinguishable, and I will have nothing further to say about them.
- *Quantum theory is not exact*, but is a very accurate asymptotic approximation to a deeper level theory – a pre-quantum dynamics. This is the idea developed in my 2004 book [1], and forms the basis for this talk.

Motivations for considering the possibility that quantum theory is only approximate include:

- *Historical.* "Exact" theories have usually turned out to be approximations accurate only in a certain domain, such as Newtonian mechanics and thermodynamics.
- *The riddle of canonical quantization.* The standard approach to creating a quantum theory is to write down a classical theory and then to "quantize" it by replacing the Poisson bracket by $-i/\hbar$ times the corresponding commutator. But this raises some questions. (i) If quantum theory is more fundamental, why can't it be obtained directly from an operator theory? (ii) What is the origin of the Planck constant $\hbar$? In my opinion, canonical quantization looks like an algorithm to invert the classical limit of quantum theory, and not a fundamental principle.

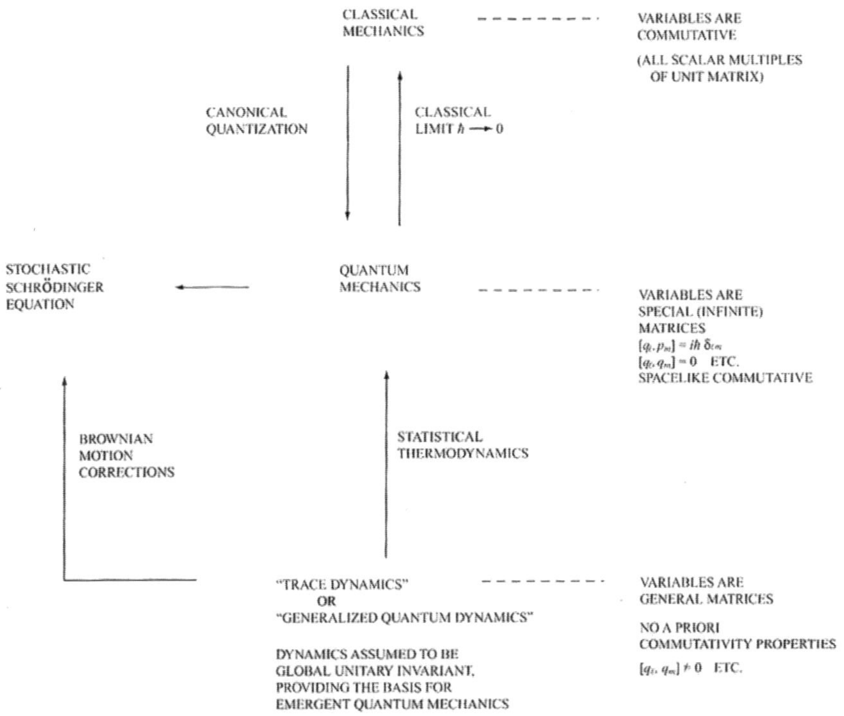

Figure 1. Diagrammatic relations between the theories discussed here: classical mechanics, quantum mechanics and quantum field theory, trace dynamics, and stochastically modified Schrödinger picture quantum mechanics.

- *Measurement problem.* Quantum mechanics is a linear theory, that is, the state vector at time t is related to the state vector at time 0 by the linear relation $|\psi(t)\rangle = U(t)|\psi(0)\rangle$. On the other hand, "measurements" involve a nonlinear stochastic "state vector reduction" induced by interaction with a "classical" apparatus. This raises profound questions: Where do the probabilities come from? Is there a single dynamical framework that incorporates both the linear Schrödinger evolution part of quantum theory, and the nonlinear, stochastic state vector reduction part?

The proposal for a pre-quantum theory developed in [1] and described in this talk is illustrated schematically in Fig. 1.

2. Overview of Trace Dynamics

"Trace Dynamics" is a noncommutative generalization of classical Lagrangian and Hamiltonian dynamics. To illustrate how it works for bosonic variables, let $\{q_r\}$ be a set of noncommuting coordinates, and $\{\dot{q}_r\}$ (with the dot denoting $\partial/\partial t$) their time derivatives. Let $L = L[\{q_r\}, \{\dot{q}_r\}]$ be a polynomial in its arguments; since the arguments do not commute, the derivative $\delta L/\delta q_r$ is not defined. Now let us use *cyclic invariance* of the trace: define the *trace Lagrangian* $\mathbf{L}$ by $\mathbf{L} = \mathrm{Tr}L$. Let us vary to form $\delta\mathbf{L}$, and reorder cyclically so that all δq_r and $\delta\dot{q}_r$ stand on the right. This gives the *definition*

$$\delta \mathbf{L} = \mathrm{Tr} \sum_r \left(\frac{\delta \mathbf{L}}{\delta q_r} \delta q_r + \frac{\delta \mathbf{L}}{\delta \dot{q}_r} \delta \dot{q}_r \right) \quad , \tag{1}$$

which defines the derivatives of the trace Lagrangian $\delta \mathbf{L}/\delta q_r$ and $\delta \mathbf{L}/\delta \dot{q}_r$. One can now show that requiring a stationary trace action

$$0 = \delta \mathbf{S} = \delta \int_{-\infty}^{\infty} dt \mathbf{L} \tag{2}$$

directly gives the *operator* Euler-Lagrange equations

$$0 = \frac{\delta \mathbf{L}}{\delta q_r} - \frac{d}{dt} \frac{\delta \mathbf{L}}{\delta \dot{q}_r} \quad . \tag{3}$$

2.1. Conserved quantities

Since the commutation relations of the q_rs are arbitrary, we have obtained a dynamics that is more general than quantum mechanics. To recover quantum mechanics, we consider the equilibrium statistical mechanics of this dynamics. This is obtained by noting that there are three *generic conserved quantities* in trace dynamics.

- Trace Hamiltonian: Let us define the canonical momentum conjugate to q_r by $p_r = \delta \mathbf{L}/\delta \dot{q}_r$, and introduce the trace Hamiltonian $\mathbf{H}$ defined by

$$\mathbf{H} = \mathrm{Tr} \sum_{\mathbf{r}} \mathbf{p_r} \dot{\mathbf{q}}_{\mathbf{r}} - \mathbf{L} \quad . \tag{4}$$

  Then $\delta \mathbf{H}/\delta q_r = -\dot{p}_r$ and $\delta \mathbf{H}/\delta p_r = \dot{q}_r$, and a simple calculation gives

$$\begin{aligned} \frac{d\mathbf{H}}{dt} &= \mathrm{Tr} \sum_r \left(\frac{\delta \mathbf{H}}{\delta q_r} \dot{q}_r + \frac{\delta \mathbf{H}}{\delta p_r} \dot{p}_r \right) \\ &= \mathrm{Tr} \sum_r (-\dot{p}_r \dot{q}_r + \dot{q}_r \dot{p}_r) = 0 \quad , \end{aligned} \tag{5}$$

  where in the first line we have used the definition of the trace derivative, and in the second line we have used cyclic invariance under the trace to cancel the two terms.

- Operator $\tilde{C}$: My graduate student Andrew Millard first noted, in the context of Weyl-ordered Hamiltonians, that the operator $\tilde{C} = \sum_r [q_r, p_r]$ is conserved, and this turns out to generalize. Suppose that $\mathbf{H}$ is *global unitary invariant*, that is, it involves no noncommutative constants in its definition. Then $\tilde{C}$ as defined above is conserved, and if one introduces Grassmann variables to represent fermionic degrees of freedom, the operator

$$\tilde{C} = \sum_{r,B} [q_r, p_r] - \sum_{r,F} \{q_r, p_r\} \quad , \tag{6}$$

  with the subscripts B, F indicating respectively sums over the bosonic and fermionic degrees of freedom, is the conserved Noether charge operator corresponding to the global unitary invariance of $\mathbf{H}$, that is $d\tilde{C}/dt = 0$. With standard adjointness assignments for the dynamical variables (see [1]) the operator $\tilde{C}$ is anti-self-adjoint.

- Phase space measure: Define the operator phase space measure

$$d\mu \equiv \prod_{r,m,n,A} d\langle m|q_r|n\rangle^A d\langle m|p_r|n\rangle^A \quad , \tag{7}$$

where A indexes the real valued components. Then one can show that $d\mu$ is invariant under the canonical transformation with generator $\mathbf{G}$,

$$\delta p_r = -\delta \mathbf{G}/\delta q_r \ , \quad \delta q_r = \delta \mathbf{G}/\delta p_r \quad , \tag{8}$$

of which the trace Hamiltonian dynamics is a special case. This means that trace dynamics has a generalized Liouville theorem: operator phase space volumes are conserved under Hamiltonian flow. This permits the use of statistical mechanical concepts, since if one postulates a uniform a priori phase space distribution, this uniformity is preserved in time under the Hamiltonian dynamics.

2.2. *Canonical ensemble*

We now introduce the *canonical ensemble* for trace dynamics, which is the phase space distribution that maximizes the entropy subject to the constraints that $\mathbf{H}$ and $\tilde{C}$ have specified average values, and is given by

$$\rho = Z^{-1} \exp(-\tau \mathbf{H} - \mathrm{Tr}\tilde{\lambda}\tilde{C}) \quad . \tag{9}$$

The normalization condition $\int d\mu\rho = 1$ fixes the partition function Z to be

$$Z = \int d\mu \exp(-\tau \mathbf{H} - \mathrm{Tr}\tilde{\lambda}\tilde{C}) \quad . \tag{10}$$

The ensemble parameters are the real number τ, which has the dimensions of inverse mass, and the anti-self-adjoint matrix $\tilde{\lambda}$. Letting $\langle ...\rangle_{\mathrm{AV}}$ denote the average over the canonical ensemble, we can write the average of $\tilde{C}$ in the canonical form

$$\begin{aligned} \langle \tilde{C}\rangle_{\mathrm{AV}} =& i_{\mathrm{eff}} D \quad , \\ i_{\mathrm{eff}}^2 = -1 \ , \quad i_{\mathrm{eff}}^\dagger =& - i_{\mathrm{eff}} \ , \quad [i_{\mathrm{eff}}, D] = 0 \quad . \end{aligned} \tag{11}$$

The simplest case, which we shall assume, is that no direction in the underlying operator space is preferred to any other, so that D is a multiple of the unit matrix, which with some foresight we write as

$$D = \hbar 1 \quad . \tag{12}$$

Since $\langle \tilde{C}\rangle_{\mathrm{AV}}$ is a function of τ and $\tilde{\lambda}$, with this assumption we have $\tilde{\lambda} = i_{\mathrm{eff}}\lambda$, with lambda a real number.

2.3. *Unitary fixing*

Because $\tilde{\lambda} = i_{\mathrm{eff}}\lambda$, the canonical ensemble is invariant under unitary transformations U_{eff} for which $U_{\mathrm{eff}}^\dagger i_{\mathrm{eff}} U_{\mathrm{eff}} = i_{\mathrm{eff}}$, or equivalently, $[U_{\mathrm{eff}}, i_{\mathrm{eff}}] = 0$. This follows immediately from cyclic invariance of the trace,

$$\mathrm{Tr}\tilde{\lambda} U_{\mathrm{eff}}^\dagger \tilde{C} U_{\mathrm{eff}} = \mathrm{Tr} U_{\mathrm{eff}} \tilde{\lambda} U_{\mathrm{eff}}^\dagger \tilde{C} = \mathrm{Tr}\tilde{\lambda}\tilde{C} \quad . \tag{13}$$

Thus the canonical ensemble only partially breaks the global unitary invariance, and this means that we will average too much if we use $\int d\mu\rho$ for forming ensemble averages. So we introduce a restricted measure

$$d\mu' = d\mu|_{\mathrm{OVERALL}\ U_{\mathrm{eff}}\ \mathrm{FROZEN}} \quad , \tag{14}$$

and will use this measure for forming thermodynamic averages.

With the preceding remarks in mind, it is useful to introduce the *effective complex part* of any operator x, which can be any function of the canonical variables. We define x_{eff} to be the part of x that commutes with i_{eff}; this is given by

$$x_{\mathrm{eff}} = \frac{1}{2}(x - i_{\mathrm{eff}} x i_{\mathrm{eff}}) \quad , \tag{15}$$

as verified by noting that

$$i_{\mathrm{eff}} x_{\mathrm{eff}} = \frac{1}{2}(i_{\mathrm{eff}} x + x i_{\mathrm{eff}}) = x_{\mathrm{eff}} i_{\mathrm{eff}} \quad . \tag{16}$$

2.4. Physical observables

Physical observables in trace dynamics are trace quantities, which are independent of the unitary fixing since

$$\mathrm{Tr}\mathcal{O} = \mathrm{Tr}U_{\mathrm{eff}}^{\dagger}\mathcal{O}U_{\mathrm{eff}} \quad . \tag{17}$$

This suffices, since transition probabilities can be written as traces,

$$|\langle\alpha|\mathcal{O}|\beta\rangle|^2 = \langle\alpha|\mathcal{O}|\beta\rangle\langle\beta|\mathcal{O}^{\dagger}|\alpha\rangle = \mathrm{Tr}P_{\alpha}\mathcal{O}P_{\beta}\mathcal{O}^{\dagger} \quad , \tag{18}$$

with P_{α} the projection operator

$$P_{\alpha} = |\alpha\rangle\langle\alpha| = \frac{1}{2\pi i}\oint_{\lambda_{\alpha}} dz \frac{1}{z - \mathcal{O}_{\alpha}} \quad , \tag{19}$$

where the contour integral over z circles λ_{α}, the eigenvalue of the operator $\mathcal{O}_{\alpha}$ on the state $|\alpha\rangle$.

3. How Quantum Field Theory Emerges

We now set up a correspondence

$$\frac{\int d\mu' \rho x_{1\,\mathrm{eff}} \ldots x_{n\,\mathrm{eff}}}{\int d\mu' \rho} \leftrightarrow x_{1\,\mathrm{eff}} \ldots x_{n\,\mathrm{eff}} \quad , \tag{20}$$

where on the right the xs are operators in a quantum field theory with the role of i played by i_{eff}. One can derive *equipartition theorems* or *Ward identities* for the averages on the left of Eq. (20). When τ is very small (so that it is the inverse of a very large mass – the Planck mass?), and when $\tilde{C}$ is replaced by its ensemble average $\langle\tilde{C}\rangle_{\mathrm{AV}} = i_{\mathrm{eff}}\hbar$, these Ward identities have the structure of quantum mechanics, so that inside ensemble averages we have

$$\begin{aligned}
[q_{r\,\mathrm{eff}}, p_{s\,\mathrm{eff}}] =& i\hbar\delta_{rs} + O(\tau) \quad , \\
[q_{r\,\mathrm{eff}}, q_{s\,\mathrm{eff}}] =& 0 + O(\tau) \quad , \\
[p_{r\,\mathrm{eff}}, p_{s\,\mathrm{eff}}] =& 0 + O(\tau) \quad , \\
\dot{x}_{\mathrm{eff}} =& \frac{i_{\mathrm{eff}}}{\hbar}[H_{\mathrm{eff}}, x_{\mathrm{eff}}] + O(\tau) \quad .
\end{aligned} \tag{21}$$

Thus, although the initial dynamical degrees of freedom have general commutation relations, the emergent theory has the usual local commutation relations of quantum theory when r is a spatial box label.

4. Sketch of General Ward Identity Derivation

We now give a brief sketch of how the Ward identities are derived. For general operator $\mathcal{O}$, let

$$\frac{\int d\mu' \rho \mathcal{O}}{\int d\mu' \rho} \equiv \langle \mathcal{O} \rangle_{\mathrm{AV}} \tag{22}$$

be the average defined with respect to the unitary fixed phase space measure. Taking $\mathcal{O} = \{\tilde{C}, i_{\mathrm{eff}}\}W$, we proceed as follows:

- Use translation invariance of the measure $d\mu'$ to write $\int d\mu' \delta[\rho\mathcal{O}] = 0$, where δ is a variation with respect to a variable not frozen by the unitary fixing.
- Neglect the τ term coming from $\delta\rho$: this is an assumption of decoupling of the low energy from the high energy regime.
- Replace $\tilde{C} \to \langle \tilde{C} \rangle_{\mathrm{AV}}$ in integrands.
- Make various choices for W as follows: (i) Taking W proportional to a canonical q or p gives the canonical algebra inside $\langle\rangle_{\mathrm{AV}}$. (ii) Taking W proportional to the operator Hamiltonian H gives the Heisenberg equation of motion inside $\langle\rangle_{\mathrm{AV}}$. (iii) Taking W proportional to a self adjoint generator G gives a unitary canonical transformation inside $\langle\rangle_{\mathrm{AV}}$.
- One can include sources in ρ so that $\langle\rangle_{\mathrm{AV}}$ has source terms that can be varied.
- Some remarks on this derivation are in order:
  - The $O(\tilde{\lambda})$ term in the Ward identities vanishes because $\{i_{\mathrm{eff}}, [\tilde{\lambda}, x]\} \propto [\tilde{\lambda}, x_{\mathrm{eff}}] = 0$.
  - There are conditions for the neglect of the $O(\tau)$ terms in the Ward identities. The first [1] is that it suffices for $\dot{x}_{\mathrm{eff}}$ and $\tilde{C}_{\mathrm{eff}}$ to have disjoint support on operator phase space.
  - A second condition for neglect of the $O(\tau)$ terms is that boson–fermion balance is required. Suppose the theory has only bosons, then the effective canonical algebra with $O(\tau)$ terms neglected reads (omitting subs "eff" on q, p; this is justified for a second type of Ward identity derived in [1])

$$\langle [q_s, p_r] \rangle_{\mathrm{AV}} = i_{\mathrm{eff}} \hbar \delta_{rs} \quad . \tag{23}$$

Setting $r = s$ and summing gives then

$$\langle \sum_r [q_r, p_r] \rangle_{\mathrm{AV}} = \langle \tilde{C}_{\mathrm{AV}} \rangle = i_{\mathrm{eff}} \hbar = i_{\mathrm{eff}} \hbar \sum_r 1 = i_{\mathrm{eff}} \hbar N_B \quad , \tag{24}$$

with N_B the number of bosonic degrees of freedom, which is a contradiction. When both bosons and fermions are present, this is changed to

$$\langle \sum_{r,B} [q_r, p_r] - \sum_{r,F} \{q_r, p_r\} \rangle_{\mathrm{AV}} = \langle \tilde{C}_{\mathrm{AV}} \rangle = i_{\mathrm{eff}} \hbar \quad ; \tag{25}$$

the left hand side is now $i_{\mathrm{eff}} \hbar (N_B - N_F) + \sum O(\tau)$, and so there is no contradiction in neglecting the $O(\tau)$ terms if $N_B = N_F$.

5. Why there are Brownian Motion Corrections

Let us write

$$\tilde{C} = \langle \tilde{C} \rangle_{\mathrm{AV}} + \Delta C = i_{\mathrm{eff}} \hbar + \Delta \tilde{C} \quad , \tag{26}$$

with $\Delta \tilde{C}$ a rapidly fluctuating quantity in operator phase space. Then the combination appearing in the Ward identities becomes

$$\frac{1}{2}\{\tilde{C}, i_{\mathrm{eff}}\} = -\hbar + \frac{1}{2}\{\Delta \tilde{C}, i_{\mathrm{eff}}\} = -\hbar - \hbar(\mathcal{K} + \mathcal{N}) \quad , \tag{27}$$

with $\mathcal{K}$ a fluctuating c-number, and with $\mathcal{N}$ a fluctuating matrix, with an operator analog. An appropriate Ansatz for the $\mathcal{N}$ term then corresponds to the mass-proportional form of the "continuous spontaneous localization" (CSL [2]) stochastic Schrödinger equation. Thus we have plausibility arguments, but *not* a derivation, of the CSL model.

As a simple illustration of a stochastic Schrödinger equation, consider a pointer with center of mass q. (This is the leading small displacement approximation to the Ghirardi-Rimini-Weber and CSL models.) The stochastic Schrödinger equation is

$$d|\psi\rangle = \frac{-i}{\hbar}H|\psi\rangle - \frac{\eta}{2}(q - \langle q\rangle)^2|\psi\rangle dt + \sqrt{\eta}(q - \langle q\rangle)|\psi\rangle dW_t \quad , \tag{28}$$

with $\langle q\rangle = \langle\psi|q|\psi\rangle$ the expectation value of q in the state $|\psi\rangle$, and with dW_t a Brownian noise. We note that:

- Because $|\psi\rangle$ appears in $\langle q\rangle$, this is a *nonlinear* stochastic differential equation.
- This equation can be proved to give state vector reduction on position eigenstates with *Born rule probabilities.*

6. Summary: Trace Dynamics as Pre-Quantum Mechanics

- Thermodynamics, via equipartition theorems (Ward identities) implies the unitary evolution of quantum mechanics in the form of the Heisenberg equation of motion, which can be transformed into the Schrödinger equation.
- Brownian motion corrections, via CSL phenomenology gives the probability interpretation and the Born rule of the reduction postulate of quantum theory.

Appendix A. Proof of Conservation of $\tilde{C}$

Let us assume global unitary invariance of the trace Hamiltonian, so that

$$\mathbf{H}[\{U^\dagger q_r U\}, \{U^\dagger p_r U\}] = \mathbf{H}[\{q_r\}, \{p_r\}] \quad , \tag{A.1}$$

with $U = e^\Lambda$, $U^\dagger = e^{-\Lambda}$, $\Lambda = -\Lambda^\dagger$. Expanding to first order in Λ, this gives

$$\mathbf{H}[\{q_r - [\Lambda, q_r]\}, \{p_r - [\Lambda, p_r]\}] = \mathbf{H}[\{q_r\}, \{p_r\}] \quad , \tag{A.2}$$

so using the definition of the trace derivative we get

$$\mathrm{Tr}\sum_r \left(-\frac{\delta \mathbf{H}}{\delta q_r}[\Lambda, q_r] - \frac{\delta \mathbf{H}}{\delta p_r}[\Lambda, p_r] \right) = 0 \quad . \tag{A.3}$$

Since Λ is arbitrary, this implies the operator relation

$$\begin{aligned} 0 =& \sum_r \left([\frac{\delta \mathbf{H}}{\delta q_r}, q_r] + [\frac{\delta \mathbf{H}}{\delta p_r}, p_r] \right) \\ =& \sum_r ([-\dot{p}_r, q_r] + [\dot{q}_r, p_r]) = \sum_r ([q_r, \dot{p}_r] + [\dot{q}_r, p_r]) = \dot{\tilde{C}} \quad , \end{aligned} \tag{A.4}$$

with $\tilde{C} = \sum_r [q_r, p_r]$. When fermions (Grassmann odd q_r, p_r) are included, the same derivation gives conservation of $\tilde{C} = \sum_{r,B}[q_r, p_r] - \sum_{r,F}\{q_r, p_r\}$.

References

[1] S. L. Adler, *Quantum Theory as an Emergent Phenomenon: The Statistical Mechanics of Matrix Models as the Precursor of Quantum Field Theory*, Cambridge University Press (2004). An initial draft appeared as arXiv: hep-th/0206120. An update to Chapter 6 appeared as arXiv: hep-th/0510120, J. Phys. A **39**, 1397-1404 (2006).

[2] A. Bassi and G. C. Ghirardi, "Dynamical Reduction Models", Phys. Rep. **379**, 257-428 (2003).

[3] S. L. Adler, "Lower and Upper Bounds on CSL Parameters from Latent Image Formation and IGM Heating", arXiv: 0605072, J. Phys. A **40**, 2935-2958 (2007), Erratum–ibid. A **40**, 13501 (2007).

Emergence: Key physical issues for deeper philosophical inquiries

B. L. Hu

Maryland Center for Fundamental Physics and Joint Quantum Institute,
University of Maryland, College Park, Maryland 20742, USA and
Institute for Advanced Study and Department of Physics,
Hong Kong University of Science and Technology, Hong Kong, China.

E-mail: `blhu@umd.edu`

Abstract. In this essay I begin with a sketch of three senses of emergence and a suggestive view on the emergence of time and the direction of time. After trying to identify what issues representative philosophers view as important in their inquires into emergence phenomena in physics I make several observations pertaining to the concepts, methodology and mechanisms required to understand emergence and describe a platform for its investigation. I then identify some key physical issues which I feel need be better appreciated by the philosophers in this pursuit. I end with some comments on the issue of coarse-graining and persistence structures.

1. Introduction

In this talk I present some latest thoughts on three inter-related subjects:

1) Emergence: After describing three different senses of emergence, I point out that effective field theory (EFT) or renormalization group (RG) is a suitable, maybe even necessary, but not sufficient set of conceptual means for describing emergence. EFT or RG [A1] may suggest how different physics manifest at different scales, but we also need to identify the mechanisms or processes whereby different levels of structures and the laws governing them, including the symmetry principles, emerge. That depends on deeper interplay of collectivity, complexity, stochasticity and self-organization.

2) Emergent Gravity: There are at least two intimately related veins in viewing gravity as emergent: a) "General Relativity as Hydrodynamics?" (This viewpoint which can be traced back to Sakharov (1968) [1] was first pronounced in this theme in [2]. Other major proponents of this view are Volovik [3] and Wen [4].)– in the sense that gravity is an effective theory valid only at the long wavelength, low energy limit of some underlying theory (quantum gravity [5]) for the microscopic structures of spacetime and matter [A2][A3]. b) Gravity as Thermodynamics , where such a view is often shaped by considering the effects of an event horizon on the quantum fluctuations of a field, first shown by Bekenstein [10] (see also work of Hawking [11], Unruh [12]), which underlies what is known today as the holography principle [13]. This view is represented by the works of Jacobson [14], Padmanabhan [15] and Verlinde [16] [A4].

3) Gravity and Thermodynamics: Since both gravity and thermodynamics are classical theories of macroscopic structures, if a deep connection exists, we should be able to see their direct relation at this level, without relying on arguments invoking the microscopic structure of matter (quantum fluctuations). This I first posed as a challenge in [A5], one which physicists in the 19c in principle may be able to resolve. If we can meet this challenge we may see the simpler and deeper connection between gravity and thermodynamics without invoking quantum mechanics. If we fail we will perhaps see more clearly the essential role of quantum physics in explaining gravity and the necessary implication that a) either

the macroscopic world is fundamentally quantum. (For viewing spacetime as a condensate, see, e.g, [18]) or b) quantum mechanics is also emergent from a deeper structure [19], a representation of stochastic processes [20], or as a form of organizational rules like statistical mechanics [21].

[A1] For a recent meeting on this topic, see, e.g.,
http://www.perimeterinstitute.ca/Events/ Emergence-and-Effective-Field-Theories/ Schedule/
[A2] For references to writings of the major proponents see, e.g., [6, 7].
[A3] It is easier and more natural for string theorists to view gravity as emergent. See, e.g., [8, 9]
[A4] For references to earlier work and a critique, see, e.g., [17]
[A5] B. L. Hu, "*Gravity and Thermodynamics: What exactly do we want?*" Invited talk at ESF Exploratory Workshop: Gravity and Thermodynamics, SISSA Sept 8, 2011. Further discussion can be found in [22].

****

The above is the Abstract of my talk. In writing up my report, rather than a sketchy summary of all the points raised above, I thought perhaps it is more useful to select one topic and go deeper. Of these, since *emergence* is the overriding issue receiving increasing attention in physics and beyond, in particular, philosophy, I will focus on this subject matter. In keeping with the selected focus of this paper I will leave my Point 1) in the Abstract as is, pending future elaboration, Readers interested in the contents of the discussions in that meeting I would refer them to the recent work of Goldenfeld on the side of the physicists and of Bain from the perspective of a philosopher conversant in relevant physics problems. Point 2) above has been discussed in recent meetings and the reader can find more details and other authors' related work in References given there.

On Point 3) about the challenge I posed: "Can we deduce a relation between Gravity and Thermodynamics without invoking any quantum consideration?" either one succeeds or one fails to provide such a direct link without relying on quantum arguments, the implications for theoretical physics are equally significant. If one can meet this challenge , that thermodynamics rules can be used to understand or even derive gravity without appeal to the microphysical constituents of matter or spacetime and the physical laws governing them, it would be a very important step forward. If on the contrary this bridge between gravity and thermodynamics requires quantum mechanics to build, as is assumed from Bekenstein, Hawking onward, then it says something about the role of quantum mechanics in the macroscopic realm, particularly important for spacetimes in this respect.

This paper is organized as follows: In Sec. 2 I give a sketch of the three senses of emergence [1] which await further development. In Sec. 3 I present some notes on essays written by philosophers trying to understand emergence in the physics context for the purpose of identifying what issues these philosophers view as important in their inquires. In Sec. 4 I make several observations pertaining to the concepts, methodology and mechanisms required to understand emergence and describe a platform for its investigation, namely, nonequilibrium statistical mechanics. In Sec. 5 I try to identify some key physical issues which I feel need be better appreciated by the philosophers who wish to deepen their inquiries and bring themselves closer to practising scientists pursuing research topics bearing on these issue. I then describe how we approach the issue of coarse-graining and persistence structures.

[1] These are thoughts I have toyed with on and off in the past two decades since the 1991 Huelva Workshop on The Physical Origin of Time-Asymmetry and the 1993 Santa Fe Institute Workshop on 'Fluctuations and Forms'.

2. Emergence: Processes and Mechanisms

In this section I jot down some notes on the different senses of emergence which underlie different scientific disciplines, as well as a view of emergent time. These are merely sketches of ideas for ideas, most likely not new, but which I find useful as nucleus for gathering amorphous threads of thoughts for further analysis.

2.1. Three Senses of Emergence

1. Emergence in the sense of *difference in manifestations*– Role of coarse-graining:

* Different manifestations at different levels of structures, hierarchical in form, and corresponding interactions.

* Requires the identification of the range and precision of measurement, thus interfaces necessarily with an observer's probing ability and observation range. Effective field theory is of this nature.

* Stability of emergent structures depends on the degree of coarse-graining

* Robustness of emergent structures against the variation of coarse-graining.

(Vaguely in philosopher's language, the first two bullets correspond to novelty, and the latter two correspond to autonomy. More in the next two sections.)

This aspect of emergence is sensitive to the conditions for the appearance of meta-stable structures. The key issues are embodied in studies of the emergence of quasi-classical domains [28] in environment-induced decoherence [24, 25] and decoherent or consistent histories [29, 30] formulation of quantum mechanics. Although historically these studies focused more on the quantum aspects (transition to classicality), less on the statistical mechanical aspects. See, however, e.g., Sec. 5 of [31] and [32].

2. Emergence in the sense of *conditioning or control*– Role of stochasticity:

* Selection by the system's environment through their interactions. For quantum open systems and decoherence studies this would correspond loosely to the functionality of the "pointer basis" [27]. A particular set of bases is 'selected' by the way a system interacts with its environment. It is more sensitive to the *type* of coupling than the *magnitude* of interaction.

* Adaptation of a biological system to its environment is familiar in evolutionary theory. The emergence of new species is the outcome of both genetic (intrinsic) variation and environmental (extrinsic) selection, the former arising from mutation which is a stochastic variable. This is similar to the environment-induced decoherence of a quantum system although the stochastic element there is from the noise and fluctuations in an environment and the quantum to classical transition is largely not a stochastic process. Note even if a reduced density matrix is almost diagonal, there is still the question of how its diagonal components, the classical probability or the weighing functions, are distributed, which influences the outcome of emergent entities and behavior.

3. Emergence in the sense of *dynamics ('updates', not time evolution)*:

Here I have in mind models from

* Cellula automata (CA) [33]: Simple conjunctive rules between neighboring elements when reiterated in steps can lead to very different structures – many will terminate but some will keep evolving for a long 'time'. There is no time, thus no dynamics in terms of time evolution, simply sequencing. Those which survive are emergent structures since they possess both the qualities of being novel and autonomous. But note that in this sense, given the conjunctive rules of the cells, one can predict exactly the outcome of 'evolution' (thus not in the biological sense 2 above) without any sense of stochasticity involved, even though the outcome may appear irregular or seemingly unpredictable.

* Growth and form in driven diffusive systems [34], self-organization: Different forms emerge arising from the interplay of stochasticity and nonlinearly. Role of nonlinearity in self-

organization is quite familiar. Role of stochasticity (noise and fluctuations) in the genesis of new forms ushered in a new field of soft matter physics in the 80-90s. I want to add that the effect of memory is also important. This calls for the importance of non-Markovian processes in emergent phenomena. For example: every snowflake has its own distinct identity, being a fine record of the density, humidity, temperature as the condensate traverses the different layers of water vapor, growing into a flake. Here, non-Markovian or memory-laden processes, or histories (usually connoted as with memories), do not have to refer to dynamics in the sense of time-evolution. It could also be understood in the sense of 'sequencing' referred to above in the example of cellula automata.

2.2. Sense of Time in Emergence

The essence or relevance of time is in its ordering function. I think it is conceptually clearer to consider three distinct elements: a) configurations in each step as basic entities, b) sequencing or iteration of steps (I prefer not to use the word 'update', since date is a marking of time.). An example of a): energetics of phase transition a la free energy density functional F (T) describes critical phenomena, not critical dynamics. It is a one parameter (temperature T) family of curves, where the minima signify the existence of meta-stable states and one can discuss the probability of tunneling from one to another in this landscape (assuming multi-dimensional space), but there is no dynamics involved. Using CA as an example of b), the conjunction rules produce one configuration in one step, then another after the second step, which make up a sequence. There are sequences which terminate after relatively few iterations, others sustain many iterations and assume particular forms. There is still no sense of time in the sense of dynamics yet.

The third ingredient c) is key. In simple cases like CA, iteration in increasing steps can be viewed as time, as 'update' conjures. But it is more challenging to understand time in complex systems where they can interact with various environments. Using biological systems as example, replication continues the species and provides the stability. This is like the sustaining sequences in CA – those long-living sequences as measured by the number of iterations. Sufficient stability is necessary for a notion of time to emerge. Mutations in the genes of a system provides the species a wider sampling pool to adapt to a changing environment. But this stochastic element also introduces a source of instability. Thus even though the sequencing function could facilitate a notion of micro-time this stochasticity and instability at the cellula level undermines the macro-time as we experience it which requires some sustained unidirectional development.

Note at this stage there is no 'arrow of time' yet. Survival of the species is the precondition for the notion of 'direction of time' to enter. Fluctuation triggers instability in the system in varying degrees: the adverse ones could bring about extinction of the species, the 'beneficial' ones enable the species to better adapt and continue to evolve. A sense of 'direction of time' or a sense of (forward) history comes only when there is sufficient 'success' as measured by the species' ability to survive, thrive and improve on itself. The word 'advanced' in advanced civilization connotes a sense of positive direction of evolution. Thus it not only reflects the passive conditioning by the environment but also allows for some degrees of regulation, control and adjustment by the system itself. This amounts to biasing a random process.

Therefore, in summary, sustained sequencing with some degree of stability is the precondition for the notion of time to emerge and the direction of time emerges only with positive feedback in the sustained sequencing of biased random processes.

The emergence of time and the sense of direction of time are explained above by using the paradigms of evolutionary biology. Placing these issues concepts and techniques in nonequilibrium statistical mechanics including open systems dynamics and stochastic processes enter in essential ways. the physics context, especially invoking nonequilibrium statistical mechanics which will be discussed in a later section. Suffice it here to say that time asymmetry in many physical processes is influenced by many factors: the way one stipulates the boundary conditions and initial states, the time scale of observation in comparison with the dynamical time scale, how one decides what the relevant variables are and how they are separated from the irrelevant ones, how the irrelevant variables are coarse-grained, and what assumptions one makes and what limits one takes in shaping a macroscopic picture from one's imperfect knowledge of the underlying microscopic structure and dynamics. More discussions follow in Sec. 4, 5.

3. Philosophical Studies

In this section I gather some notes in a naive attempt to extract the wisdom from philosophers trying to understand emergence in the physics context. I begin with the rudimentary exercise of finding out what the philosophers interpret the physicists (by these two communities, I mean by necessity a selected subset sampling) in their use of words like "constructivism" "protectorate" and the philosophers' terminology "microphysicalism" "supervenience" and key words used by both communities, such as "novelty" "autonomy" which characterize emergence. Of course these exercises are not just to clarify the terminology, but for the meaning behind these words. What I find in this preliminary exercise is that the philosophers are not unexpectedly mostly interested in issues of their own, in metaphysics, in the philosophy of the mind, etc. From the questions they raised I also see some ambiguity (or not clear enough definitions) in the physicists' use of words like "fundamental", "reductionism", which need be clarified. The focusses of philosophers of science with a physics background or those who are attuned to physicists' language or way of thinking enabled us to see what issues they consider as important to bring back to their own community. At the same time I also see the philosophers' choice of focus is somewhat partial, or skewed, missing some key issues which physicists consider as important in understanding emergent phenomena. This is likely a result of the scarcity of explanations by the practising physicists of their technical findings in the light of emergence phenomena, better yet, in philosophical language. This aspect I'll spell out more explicitly in the last section.

3.1. Sources and Tracks

I shall first identify a few sources of literature on both sides so readers can see how my readings could be limited or biased and identify what I have missed out clearly.

Classic papers in physics: Anderson [35], Laughlin and Pines [36] on the one side and Weinberg [37] on the other.

Philosophers with physics training or attuned to physical issues; Bain, Wilson.

Physicists (at PI wksp) who addressed key issues on emergence: Goldenfeld, Kadanoff

Issues of emergence exemplified in quantum decoherence: Stamp

Philosophers (at PI wksp) who raised key issues on emergence: Batterman, Morrison

Useful introduction: Mainwood PhD thesis(2006), with metaphysical emphasis.

3.2. Terminology, Contents, Meaning and Issues

Physicists usage: **'Fundamental', Reductionism, Constructivism, "protectorate"** [36]
Emergent entities, properties, principles (Weinberg)
Philosopher's usage: **microphysicalism, supervenience**

Defining properties of emergence largely agreed-upon by philosophers:
Novelty, Autonomy(reductive, predictive, causal and/or explanatory)

A description rather than a definition is probably best to illustrate what novelty refers to: *'Instead, at each level of complexity entirely new properties appear'* (Anderson [35], p. 393).

"When you put enough elementary units together, you get something that is more than the sum of these units. A substance made of a great number of molecules, for instance, has properties such as pressure and temperature that no one molecule possesses. It may be a solid or a liquid or a gas, although no single molecule is solid or liquid or gas." (Wheeler and Ford [47] p.341, quoted by Mainwood [40] p.18)

Reductionism– reduction, prediction; derivable, deducible:

"In most philosophical discussions the concept of emergence is intimately related to the following notions: antireductionism, unpredictability, and novelty. Emergence in these contexts is also typically associated with parts and wholes. The idea being that a phenomenon is emergent if its behavior is not reducible to some sort of sum of the behaviors of its parts, if its behavior is not predictable given full knowledge of the behaviors of its parts, and if it is somehow new – most typically this is taken to mean that emergent phenomenon displays causal powers not displayed by any of its parts.1 In addition to irreducibility, unpredictability, and novelty, it is often asserted that emergent phenomena are inexplicable – they defy explanation in terms of the behaviors of their components. And, as "explanation" is typically understood to be explanation by a particular theory, this means that the behavior of the emergent whole is not fully explained by the theory that governs the behavior of the component parts." (Batterman [41] p.1-2)

"Almost all conceptions of emergence characterise novelty by appealing to some conception of predictability or derivability, questioning whether it is possible to obtain one set of properties from another by some mechanical process of logical derivation. Kim's 1995 definition is typical from one concerned with the philosophy of mind: 'a property of a complex system is said to be 'emergent' just in case, although it arises out of the properties and relations characterizing simpler constituents, it is neither predictable from, nor reducible to, these lower-level characteristics.' The thought is that if it is possible to derive systemic properties from those of their components, these properties cannot be truly novel – for they were "there all along" in the parts, and it was merely a matter of careful analysis to make this fact manifest. This leads naturally to a distinction between this 'prediction' or 'reduction' being one that could be carried out in practice, and one that is possible only in principle." (Mainwood [40] p. 27) Thus the distinction between **ontological** (in principle) vs **epistemological** (in practice) **emergence**.

Reductionism- Constructivism

"The workings of our minds and bodies, and of all the animate or inanimate matter of which we have any detailed knowledge, are assumed to be controlled by the same set of fundamental laws, which except under certain extreme conditions we feel we know pretty well." (Anderson [35] p. 393)

... the material composition of organisms is exactly the same as that found in the inorganic world. Further ... none of the events and processes encountered in the world of living organism is in any conflict with the physico-chemical phenomena at the level of atoms and molecules. (Mainwood [40] p. 16, citing Mayr's "constitutive reductionism' in the context of biology [48] p.59)

"... the reductionist hypothesis does not by any means imply a constructionist one: The ability to reduce everything to simple fundamental laws does not imply the ability to start from those laws and reconstruct the universe.

"The constructionist hypothesis breaks down when confronted with the twin difficulties of scale and complexity. (Anderson [35] p. 393)

Microphysicalism and the 'New Emergentists"

What Anderson and Weinberg refer to as 'reductionism' is at odds with the metaphysicians' usage, who call it 'microphysicalism'.

In metaphysics, the claim endorsed by Anderson and Mayr is usually called microphysicalism. [M]icrophysicalism ... is the doctrine that actually (but not necessarily) everything non-microphysical is composed out of microphysical entities and is governed by microphysical laws. (Pettit [49])

Pettit also brings out the two components which constitute microphysicalism. The first is that everything in the empirical world – every particular in that world – is composed in some sense out of subatomic materials. And second is that everything that happens in the empirical world happens, ultimately, under the controlling influence of subatomic forces and laws. (Pettit [50] p.342-3)

Mainwood refers to proponents and adherents of these tenets as the New Emergentists.

On the issue of "in practice" versus "in principle" mentioned above, *the New Emergentists are accused of illegitimately citing evidence for epistemological emergence (a practical difficulty in deriving the properties of a large system) as support for claims of ontological emergence (there exist emergent properties that are "novel" in a metaphysically important sense).* (Mainwood)

Supervenience – a key concept philosophers use for qualifying emergence.

The most promising approach to capturing the 'determination' inherent in property physicalism is through a relation of supervenience. Take a set of objects O, let their physical properties form a set P_A, and all their properties (both physical and ostensibly non-physical) form a set P_B. To capture physicalism via supervenience, we demand that the extensions of elements of P_B are fixed by specifying the extensions of the elements of P_A. The simplest way of accommodating any such n-adic relations amongst objects o_1, o_2, o_3 is to treat it as a one-place property of the larger object composed of all of these. So we need to close the set of objects O under the operation of taking ordered n-tuples.

A little more formally: the properties P_B supervene on P_A with respect to a set of objects O, on which both are defined, iff any two objects in O that match with respect to all properties in P_A, also match with respect to all properties in P_B. The central claim of property physicalism can then be expressed very simply: all properties supervene on physical properties. (Mainwood [40] p. 22)

Using these technical terms in philosophy, Mainwood summarizes, *"... systemic properties are novel, if and only if it is practically impossible to derive them from the microphysical properties mentioned in microphysical supervenience."* ([40] p. 30)

Three proposals of novelty were examined by Mainwood ([40] Sec. 1.5): *1) a failure of inter-theoretic reduction; 2) an impossibility of deducing the systemic properties from the properties of the parts; or 3) a failure of mereological supervenience.* The approaches appeal to three entirely separate distinctions between the properties of parts and wholes; thus according to Mainwood, there are three entirely different sets of criteria for emergence. We leave it for the interested reader to follow his treatise but offer some insights on these issues in a later section.

3.3. Key words and concepts used by physicists

Fundamental:

"The modern theory of critical phenomena has interesting implications for our understanding of what constitutes "fundamental" physics. For many important problems, a fundamental under- standing of the physics involved does not necessarily lie in the science of the smallest available time or length scale. The extreme insensitivity of the hydrodynamics of fluids to the precise physics at high frequencies and short distances is highlighted when we remember that the Navier-Stokes equations were derived in the early nineteenth century, at a time when even the discrete atomistic nature of matter was in doubt. ([43] p.3 quoted by Batterman [41]]

Protectorate:

According to Laughlin and Pines (LP) [36] .. *a (quantum) protectorate, (is) "a stable state of matter whose generic low-energy properties are determined by a higher organizing principle and nothing else."*

An example of LP's "higher organizing principle", used also by Anderson earlier, is spontaneous symmetry breaking. *".. one does not need to prove the existence of sound in a solid, for it follows from the existence of elastic moduli at long length scales, which in turn follows from the spontaneous breaking of translational and rotation symmetry characteristic of the crystalline state."* What is more telling is what follows, *"Conversely, one therefore learns little about the atomic structure of a crystalline solid by measuring its acoustics."*

Morrison [46]) *".. referred to them as 'theoretical principles' in order to capture the idea that there is a dynamical process associated with these principles responsible for producing certain kinds of behavior."*

In the eyes of the theoretical physicists, 'theoretical' is too general a term to identify these principles, and can become nondescriptive. These principles provide not only how the micro-parts are organized into the meso-'whole', and here we are referring two adjacent rungs in a hierarchy, but also the dynamical process which they come into being. This would correspond to the specification of – how to choose or constitute – the collective variables and how to describe their state and dynamics, the collective behavior, which is a important and nontrivial issue.

Parts and Whole relation:

I agree with these summarily statements by Morrison below and by Batterman next:

"Not only do we need to reorient our thinking about the role of natural kinds as a method for differentiating essential features of matter, but new ways of thinking about the part/whole relations involved in defining and describing emergent phenomena are required."

"And perhaps most importantly, a re-evaluation of reductionist strategies for defining the relationship between different theoretical levels is crucial for making sense of emergence in physics. In connection with the "levels" approach characteristic of this picture of emergence, it is important to note the differences between emergence and the fact that phenomena at different scales may obey different fundamental laws." ([46] p.886)

3.4. Overall Assessment

"A number of philosophers of science have imposed these philosophical conceptions upon physical theory in an attempt to address the issue of emergence in physics. While I believe there are some benefits to this methodological approach, on the whole I think it is better to turn the process on its head. We should look to physics and to "emergent relations" between physical theories to get a better idea about what the nature of emergence really is. Trying to impose a conceptual framework designed primarily to deal with the problem of the mental's relation to the physical is by and large unhelpful." (Betterman p.2)

I think the same can be said about the relation of science and philosophy of science in general. If one wants to not just talk about the philosophy of a subject X (the layman's use of the word "philosophize on"), but to provide philosophical insights to the key issues of X, in terms of asking better questions and seeing clearer directions, it is imperative that one should first have the technical mastery of X. E.g., one must have lived a life to talk about the philosophy of life.

4. Understanding emergent phenomena: Some observations and a useful platform

In this section I wish to make some observations on how philosophers can better communicate with, and of greater help to physicists, e.g., sorting out the complexity of issues in problems physicist wrestle with and probing deeper into specific key issues in these problems. Three aspects:

1) **Conceptual Constructs**: It is not enough to rely on a reductionist conceptual approach, such as what Mainard related to about Russell's impact on philosophers, that all mathematical statements can be reduced to set theory. Many physical phenomena defy simple or monolithic descriptions, some may even appear illogical (in reasoning, not in its intrinsic consistency). A new physical theory will become more logical and its conceptual construct more rational (think special relativity and quantum physics) only after it has been proposed, examined and checked against observations and experiments, not before. Physicists create models to capture the essence of the novel phenomena and they make approximations to test out characteristic behavior in different parameter regimes. What they are wrestling with is physical reality in whatever shape and form it be, raw and unruly. Physicists approach truth from reality. This process may be aided by abstractions of theories with some senses of completeness and perfection but the final judgment come from experiments. Here we see the difference in attitude stemming from approaching truth in the epistemic rather than the ontological sense. For emergent phenomena, rather than assuming a linear progression of logical deduction or inference we need a conceptual

scheme which can explain the emergence of a hierarchical construct with a tower of levels of structure, the collective variables most suitable for each level, the laws governing them, and even more demanding, the inter-level meta-constituents and their activities.

2) **Methodology and Vehicles**: It seems that there is still heavy reliance on the method of close scrutiny of the internal logical consistency of an argument – down to every sentence written or word used, to the extent that a slight detected crack in the logic of argument immediately spells the demise of the whole theory and the triumph of the protagonist. This golden example set up by the classic philosophers before the advent of modern science is unfortunately dated and inadequate when applied to physical issues, because reality often defies logic at least in the epistemic level, in contrast to mathematical constructs, especially when the challenge is to come up with a new paradigm to deal with new issues such as emergence under study here. This methodology is not wrong, it works for disciplines based on pure logic. The mathematical tools physicists use to derive equations are constructed with rigorous logic. But there is a divide, often unnoticed or belittled by non-physicists, between casting physical ideas to mathematical models and inferring physical meanings from mathematical results. It is just wrongly applied to arguments, and to the extent arguments cannot be represented rigorous by mathematical concepts, this approach can easily lead one to drawing sweeping or overgeneralized statements or/and missing the key issues.

3) **Mechanisms and Processes** For the study of emergence in physics, a modest way which produces reliable and helpful results for philosophical inquiries is to examine existing clear-cut examples with valid physical theories at both the micro (basic or originating) and macro (effective or emergent) levels, examine the underlying physical issues, extract the commonalities and differences, identify notions, terminology and concepts which need further clarification. Examples are thermodynamics via kinetic theory, hydrodynamics via molecular dynamics, collective excitations in (atomic-based) condensed matter systems; quantum Hall effect [39], chiral-dynamics in QCD, dimensional reduction in Kaluza-Klein theories. In terms of constructing paradigms for emergence from a thorough understanding of the specifics, in addition to the modeling and approximation schemes physicist use all the time, here , what is particularly relevant are the mechanisms and processes whereby a new phenomena emerges at a particular collective (or derived) level from a more basic (or elementary) level. Symmetry breaking mentioned by Anderson has been cited repeatedly by philosophers. There are more, which can be gleaned from the well-established examples mentioned above. We will dwell on a few of these in the following.

A **useful platform** in physics to explore emergence is nonequilibrium statistical mechanics. It consists not just of the formalisms or theorems for physical systems out of equilibrium, not just finding real-time causal equations of motion for the evolution of such systems, but more importantly, for our purposes here, its conceptual constructs in the foundational issues, the physical meaning they convey, but also the methodology adapted to specific physical setups, and the different mechanisms and processes. They all prove to be essential for the explanation of the emergence of forms and structures, even new symmetry principles and constituting laws. Just as equilibrium statistical mechanics, in particular Gibbs' ensemble theory, has acted as the base for condensed matter physics for almost a century, nonequilibrium statistical mechanics is the underpinning of disciplines from soft matter physics to chemical-biological systems to sociological- collective behavior to financial-economic modeling. It is in these complex systems where emergent phenomena manifest most vividly and strikingly.

The key issues of nonequilibrium statistical mechanics, which includes stochastic mechanics, open system dynamics, etc, are: (i) collectivity, coarse-graining, correlations - coherence; (ii) noise, fluctuations, stochasticity; (iii) nonlinearity, nonMarkovianity, nonlocality. They all have a bearing on emergence in varying degrees. We can discuss only one issue here from this list, that of coarse-graining and persistent structures ("protectorates"). A description of various physical systems with reference to how certain information in the system is kept, neglected, lost or degraded is facilitated by the distinction between closed, open and effectively open systems. This aspect which underlies the thematic material of the next section is discussed before (see, e.g, the last sections in [31], excerpts in Chapter 1 of [51]. We learn that even the very first step in defining an open system or an effectively open system but otherwise closed system, is non-trivial. In fact the judicious choice of an appropriate set of collective variables already provides half of the answer to the issues of emergence. (E.g., witness how simple yet powerful the Laughlin wave function is to capture the fractional quantum Hall effect.) The remaining half comes from deriving the dynamics of these collective variables in describing the physical phenomena at this level of structure and interactions. (E.g., thermodynamics in terms of temperature, chemical potential, entropy, enthalpy)

5. Key physical issues worthy of closer attention by philosophers: 3Cs and 3Ds

The 3Cs are "Collectivity, Coarse-graining, Correlation- Coherence". The 3Ds are: Details, Details, Details [2] – minute details in the study of emergence could be important. We feel that in a new field which lacks a well-defined paradigm it is safer to keep all the details in mind before drawing general conclusions than concocting some general principles from pure reasons to explain all the details. [3]

We shall focus on the issue of coarse-graining and persistence structure here. There is no space to include the second set of issues, namely, correlation, and for quantum systems, quantum coherence and entanglement, These are important issues for understanding how one level of structure connects to, or yields / reduces to another. They appear especially acutely for strongly correlated systems or systems with memories because they make the choice of coarse-graining measures particularly difficult. From the projection operator viewpoint, this corresponds to situations where, confronted with the integro-differential equation for one system (containing the dynamics of the other subsystem), one can find no easily justifiable or implementable approximations. Some discussions of this point can be found in [53] which addresses these aspects of macroscopic quantum phenomena with references.

5.1. Elimination of degrees of freedom (DOF) and emergence in effective field theory (EFT)

To set the stage for our discussions I quote the Abstract of Bain [38] who used two examples in physics [39] to illustrate some issues of emergence in effective field theory (EFT) (see, e.g., [44]). As stated in the beginning, the more specific a physical theory or model a philosopher can relate to, the more helpful it is to physicists who want to see how related philosophical issues are being dealt with. The issue of the 'elimination of degrees of freedom' raised therein by Bain and by Wilson [45] related by Bain are relevant to our theses discussed in the next subsection, namely, on the robustness of coarse-graining measures and persistent structures.

Bain [38] suggests that *EFTs satisfy the following disiderata for a notion of emergence:*

(i) Emergence should involve microphysicalism, in the sense that the emergent system should ultimately be composed of microphysical systems that comprise the fundamental system and that obey the fundamental system's laws. (ii) Emergence should involve novelty, in the sense that the properties of the emergent system should not be deducible from the properties of the fundamental system.

[2] I'm not referring to what is contained in the title of [42], but to the details in a physical phenomenon and in a physical theory trying to capture such phenomena.

[3] A section meant to illustrate this point with the processes of environment-induced quantum decoherence (See, e.g., [26, 27]) in the emergence of the classical world is omitted here due to space limitation. Even when a process or mechanism can be and has been identified, different parameter regimes of the factors involved (e.g., low temperature, super-Ohmic spectral density) can yield very different outcomes, see, e.g., [52]. In the decoherent histories approach the existence of quasi-classical domains [28] is, contrary to what was initially conjured, a very complex issue. See, e.g, [32].

These disiderata are underwritten in an EFT by the elimination of degrees of freedom in its construction. Thus the properties of a system described by an effective Langrangian density Leff can be said to emerge from a fundamental system described by a high energy Langrangian density L in the following sense: (a) High-energy degrees of freedom are integrated out of L. This secures microphysicalism insofar as it entails that the degrees of freedom of L_{eff} *are exactly the low-energy degrees of freedom of L. (b)* L_{eff} *is expanded in a local operator expansion (either to guarantee locality or as a means of approximating the functional integral in (a)). The result is dynamically distinct from L, and this secures novelty in the sense of the failure of lawlike deducibility from L of the properties described by* L_{eff}*.*

According to Mainwood ([40] pp. 107, 116), the mechanisms Anderson and Laughlin and Pines identify as underwriting (i) and (ii) are spontaneous symmetry breaking and universality . Neither of these is generally applicable to EFTs, as the Quantum Hall liquid example indicates.

Elimination of DOF *Mainwood ([40] p. 284) further suggests that a nontrivial notion of emergence requires the specification of a physical mechanism to underwrite (i) and (ii), and it might seem that the elimination of degrees of freedom in an EFT is a formal, as opposed to a physical, mechanism.*

As related to by Bain, *Wilson [45] similarly identifies the elimination of degrees of freedom (DOF) as an essential characteristic of a notion of emergence. For Wilson, DOF elimination plays two roles. First it secures the physical acceptability of an emergent entity by securing the lawlike deducibility of the entity's behavior from its composing parts (p. 295), and such physical acceptability partially underwrites physicalism. Second, according to Wilson, DOF elimination entails that an emergent entity is characterized by different law-governed properties and behavior than those of its composing parts, and this suggests that the former cannot be reduced to the latter (p. 301). This failure of ontological reduction might charitably be associated with a notion of novelty (although Wilson's explicit goal is simply to establish peaceful coexistence between physicalism and non-reductivism).*

Bain's comment: *"This might suggest similarity with the above account of emergence in EFTs. However, the type of DOF elimination involved in the construction of an EFT is distinct from Wilson's notion in two major respects. First, DOF elimination in an EFT is typically characterized by a failure of lawlike deducibility: The lawlike behavior of entities described by an EFT cannot, in general, be deduced from the lawlike behavior of the entities described by its high-energy theory. This failure, I suggested above, is what underwrites a notion of novelty. Second, in the presence of such failure, physicalism is preserved, insofar as, in DOF elimination in an EFT, the degrees of freedom of the EFT are exactly the low-energy degrees of freedom of the high-energy theory."*

Our comments: Emergence of a new (macro) level of structure does not come simply by the "elimination" of (micro) degrees of freedom. This emerged level of structure usually comes from a new set of collective variables which bears no relation to the variables of the sub-level. An easy example is again thermodynamics, where one important collective variable is the temperature T. The DOF of the micro theory are the coordinates and momenta (q_i, p_i) of N (N_A Avagadro's number 10^{23}) molecules. Elimination of this huge number of DOF to a few does not lead one any closer to the single collective variable T. It is in fact the wrong way to think and to go. Temperature is related to the kinetic energy of the molecules, not from the elimination of these variables. The judicious choice, or more often than not, the creative construction of a set of collective variables from a given set of micro-variables is required and usually poses the most critical challenge.

The following two subsections excerpted from [31] reflect the complexity of this issue.

5.2. Measures of Coarse-graining

Coarse graining in the most general sense refers to some information lost, removed, or degraded from a system. It could come about because these information is *inaccessible to us*, due to the limited accuracy in our observation or measurement. A drastic example is Planck scale physics, the details of which are mostly lost (hard to retrieve) because the world we live in today is an ultra-low energy construct. For this one needs to invoke ideas like effective field theory [44]. Even when information is fully accessible to us in principle, in practice one may only be interested in some aspects of the system. *We choose to ignore* certain variables such as ignoring the higher order correlations in Boltzmann's kinetic theory, or ignoring the phase information in a quantum system by imposing a random phase approximation. We do this by 'integrating over' or "projecting out' these 'irrelevant' variables.

Here are some examples of coarse-graining in action. We start with the familiar Boltzmann theory: implementation of the molecular chaos assumption (i.e., the 2 particle distribution function $f_2 = f_1 f_1$ can be expressed schematically as a product of two 1-particle distribution functions f_1) entails performing a coarse-graining in the collision integral of space over the range of interaction and of time over the duration of a collision.

Note also that coarse-graining is a necessary but not sufficient condition for entropy generation. It does not always produce a dissipative system. Truncation of the BBGKY hierarchy leads to a closed subsystem composed of n- particle correlation functions whose dynamical equations are unitary. (An example mentioned before is the Vlasov equation describing particle interaction via long range forces.) In quantum field theory equations derived from a finite-loop effective action are also unitary – at one loop the effect of the quantum field on the particles manifests through the renormalize masses and charges (to be exact, the equations of motion derived from a finite-loop effective action are unitary if none of the relevant correlation functions are 'slaved' [4] for ℓ loops, one must keep the first $\ell + 1$th order correlations, otherwise dissipation in the sense defined above sets in - dissipation is absent only in very specific situations, such as a free theory or equilibrium initial conditions). That is perhaps why (if one limits one's attention to loop expansions) statistical mechanical concepts rarely came to the fore, until one starts asking questions of a distinct nature, such as how dissipative dynamics appears in an otherwise unitary system, and the origin and nature of noise in quantum field theory. A causal condition need be introduced to render the dynamics of the subsystem irreversible. This opens up another important theme: effective field theory viewed in the open system framework (See [54] for an example of how noise from the higher energy sector can serve as a measure of the validity of a low energy effective field theory.)

5.3. persistent structures in the physical world – 'protectorates'

We have seen from the above discussions that the appearance of irreversibility is often traced to the initial condition being special in some sense. The dynamics of the system and how it interacts with its environment also enter in determining whether the system exhibits mixing or dissipative behavior. For the sake of highlighting the contrast we could broadly divide the processes into two classes depending on how sensitive they are to the initial conditions versus the dynamics. One can say that the first class is *a priori* determined by the initial conditions, the other is *a posteriori* rather insensitive to the initial conditions. Of the examples we have seen, the first group includes divergent trajectories in molecular (micro) dynamics, Landau damping, vacuum particle creation, the second class includes gas (macro) or fluid dynamics, diffusion, particle creation with interaction, decoherence. Appearance of dissipation is accompanied by a degradation of information via coarse graining (such as the molecular chaos assumption in kinetic theory, restriction to one-particle distribution in particle creation with interaction, 'integrating out' some class of histories in decoherence).

Many perceived phenomena in the observable physical world, including the phenomenon of time-asymmetry, can be understood in the open-system viewpoint via the approximations introduced to the objective microscopic world by a macroscopic observer.[25, 29, 56]

We have discussed the procedures such as the system-environment split and the coarse-graining of the environment which can bring about these results. However, a set of more important and challenging issues remain largely unexplored, i.e., under what conditions the outcomes become less subjective and less sensitive to these procedures. These procedures provide one

[4] See [51] Chapter 6, Sec. 3 and Chapter 9 Sec. 2.3 for a discussion of this concept.

with a viable prescription to get certain general qualitative results, but are still not specific and robust enough to explain how and why the variety of observed phenomena in the physical world arise and stay in their particular ways. To address these issues one should ask a different set of questions:

1) By what criteria are the system variables chosen? – collectivity and hierarchy of structure and interactions

In a model problem, one picks out the system variables – be it the Brownian particle or the minisuperspace variables – by fiat. One defines one's system in a particular way because one wants to calculate the properties of that particular system. But in the real world, certain variables distinguish themselves from others because they possess a relatively well-defined, stable, and meaningful set of properties for which the observer can carry out measurements and derive meaningful results. Its meaningfulness is defined by the range of validity or degree of precision or the level of relevance to what the observer chooses to extract information from. In this sense, it clearly carries a certain degree of subjectivity– not in the sense of arbitrariness in the exercise of free will of the observer, but in the specification of the parameters of observation and measurement. For example, a thermodynamic variable like temperature provide excellent description of systems close to equilibrium; in other regimes one needs to describe the system in terms of kinetic-theoretical or molecular-dynamical variables where temperature carries no direct meaning.

The level of relevance which defines one's system changes with the level of structure of matter and the relative importance of the forces at work at that level. The improvement of the Weinberg-Salam model with W, Z intermediate bosons over the Fermi model of four-point interactions is what is needed in probing a deeper level of interaction and structure which puts the electromagnetic and weak forces on the same footing. Therefore, one needs to explore the rules for the formation of such relatively distinct and stable levels, before one can sensibly define one's system (and the environment) to carry out meaningful inquiries of a statistical nature.

What is interesting here is that these levels of structures and interactions come in approximate hierarchical order (thus, e.g., one doesn't need QCD to calculate the rotational spectrum of a nucleus, and the manifold picture of spacetime will hopefully provide most of what we need in the post-Planckian era). One needs both some knowledge of the hierarchy of interactions and the way effective theories emerge from 'integrating out' variables at very different energy scales in the hierarchical structure (e.g., ordinary gravity plus particle theory regarded as a low energy effective higher-dimension or Kaluza-Klein theory) The first part involves fundamental constituents and interactions and the second part the application of statistical methods. One should also keep in mind that what is viewed as fundamental at one level can be a composite or statistical mixture at a finer level. There are system-environment separation schemes which are designed to accommodate or reflect these more intricate structures, from the mean field-fluctuation field split to the multiple source or nPI formalism (see [51] Chapter 6) for the description of the dynamics of correlations and fluctuations. The validity of these approximations depends quite sensitively on where exactly one wants to probe in between any two levels of structure. Statistical properties of the system such as the appearance of dissipative effects and the associated irreversibility character of the dynamics in an open system certainly depend on this separation.

2) How does the behavior of the subsystem depend on coarse-graining? – sensitivity and variability of coarse-graining, stability and robustness of emergent structure

Does there exist a common asymptotic regime as the result of including successively higher order iterations in the same coarse-graining routine? This measures the sensitivity of the end result to a particular kind of coarse-graining. How well can different kinds of coarse-graining measure produce and preserve the same result? This is measured by its variability. Based on these properties of coarse-graining, one can discuss the relative stability of the behavior of the resultant open system after a sequence of coarse-grainings within the same routine, and its robustness with respect to changes to slightly different coarse-graining routines.

Let us illustrate this problem with some simple examples. When we present a microscopic derivation of the transport coefficients (viscosity, heat conductivity, etc) in kinetic theory via the system-environment separation scheme, we usually get the same correct answer independent of the way the environment is chosen or coarse-grained. Why? It turns out that this is the case only if we operate in the linear-response regime. (See [57]). The linear coupling between the system and the environment makes this dependence simple. This is something we usually take for granted, but has some deeper meaning. For nonlinear coupling, the above problem becomes nontrivial. Another aspect of this problem can be brought out [58, 59] by comparing these two levels of structure and interaction, e.g., the hydrodynamic regime and the kinetic regime. Construct the relevant entropy from the one-particle classical distribution function f_1, that gives us the kinetic theory entropy S_{kt} which is simply $-kH_B$, where H_B is the Boltzmann's H-function. Now compare it with the hydrodynamic entropy function S_{hd} given in terms of the hydrodynamic variables (in this case, the number and energy density), one sees that $S_{hd} > S_{kt}$. A simple physical argument for this result is that the information contained in the correlations amongst the particles are not included in the hydrodynamic approximation. Even within the kinetic theory regime there exist intermediate stages described by suitably chosen variables [59]. The entropy functions constructed therefrom will reflect how much fine-grained information is lost. In this sense S_{hd} is a maximum in the sequence of different coarse-graining procedures. In the terminology we introduced above, by comparison with the other regimes, the hydrodynamic regime is more robust in its structure and interactions with respect to varying levels of coarse-graining. One way to account for this is, as we know, the hydrodynamic variables enter in the description of systems in equilibrium and they obey conservation laws [60, 61, 62]. Further coarse-grainings on these systems is expected to produce the same results, i.e., the hydrodynamic regime is a limit point of sorts after the action from a sequence of coarse-grainings. Therefore, a kind of 'maximal entropy principle' with respect to variability of coarse-graining is one way where thermodynamically robust systems can be located.

While including successively higher orders of the same coarse-graining measure usually gives rise to quantitative differences (if there is a convergent result, that is, but this condition is not guaranteed, especially if a phase transition intervenes), coarse-graining of a different nature will in general result in very different behavior in the dynamics of the open system. Let us look further at the relation of variability of coarse-graining and robustness of structure.

Sometimes the stability of a system with respect to the variability of coarse-graining is an implicit criterion behind the proper identification of a system. For example, Boltzmann's equation governing the one-particle distribution function which gives a very adequate depiction of the physical world is, as we have seen, only the lowest order equation in an infinite (BBGKY) hierarchy. If coarse-graining is by the order of the hierarchy – e.g., if the second and higher order correlations are ignored, then one can calculate without ambiguity the error introduced by such a truncation. The dynamics of the open system which includes dissipation effects and irreversible behavior will not change drastically if one uses a different (say more fine-grained) procedure, such as retaining the fourth order correlations (if the series converges, which is a non trivial issue (see, e.g., [63]). Consider now a different approximation: For a binary gas of large mass discrepancy, if one considers the system as the heavy mass particles, ignore their mutual interactions and coarse-grain the effect of the light molecules on the heavy ones, the system now behaves like a Brownian particle motion described by a Fokker-Planck equation. We get a qualitatively very different result in the behavior of the system.

In general the variability of different coarse-grainings in producing a qualitatively similar result is higher (more variations allowed) when the system one works with is closer to a stable level in the interaction range or in the hierarchical order of structure of matter. The result is more sensitive to different coarse-graining measures if it is far away from a stable structure, usually falling in between two stable levels. Only robust systems survive in nature and carry definite meaning in terms of their persistent structure and systematic evolution. This is where the relation of coarse-graining and persistent structures enters.

So far we have only discussed the activity around one level of robust structure. To investigate the domain lying in-between two levels of structures (e.g., between nucleons and quark-gluons) one needs to first know the basic constituents and interactions of the two levels. This brings back our consideration of levels of structures above. Studies in the properties of coarse-graining can provide a useful guide to venture into the often nebulous and elusive area between the two levels and extract meaningful results pertaining to the collective behavior of the underlying structure. But one probably cannot gain new information about the fine structure and the new interactions from the old just by these statistical measures. (cf. the old bootstrapping idea in particle physics versus the quark model).

Acknowledgments

I thank Professor Grossing for his invitation and hospitality. I wish to thank Professor Hua-Tung Nieh, Director of the Institute for Advanced Study at Tsing Hua University, Beijing, China for his warm welcome during my visit in April-May 2011, Professor Zheng-Yu Weng of IAS-THU for showing me how in his high-T_c superconductivity theory emergence of collective variables and their dynamics comes about and Professor Lu Yu of the Institute of Physics, Chinese Academy of Natural Sciences for a discussion on the proper Chinese technical terminology for emergence. This kind of non mission-driven, non utilitarian work addressing purely intellectual issues is not expected to be supported by any U.S. grant agency.

References

[1] A. D. Sakharov, "Vacuum Quantum Fluctuations in Curved Space and the Theory of Gravitation" Doklady Akad. Nauk S. S. R. 177, 70-71 (1967) [Sov. Phys. - Doklady 12, 1040-1041 (1968)].
[2] B. L. Hu, "GENERAL RELATIVITY AS GEOMETRO-HYDRODYNAMICS" Invited talk at the Second Sakharov International Conference Lebedev Physical Institute, May, 1996. [gr-qc/9607070].
[3] G. E. Volovik, *The Universe in a Helium Droplet* (Clarendon Press 2003). "Fermi-point scenario for emergent gravity" in Proceedings of conference "From Quantum to Emergent Gravity: Theory and Phenomenology" PoS(QG-Ph)043 (2007).
[4] Xiao-Gang Wen, *Quantum Field Theory of Many-Body Systems* (Oxford University Press 2004). Michael Levin, Xiao-Gang Wen "Fermions, strings, and gauge fields in lattice spin models" Phys. Rev. B67 (2003) 245316, B71, 045110 (2005).
[5] Daniele Oriti (ed), *Approaches to quantum gravity* (Cambridge University Press 2009)
[6] B. L. Hu, "EMERGENT / QUANTUM GRAVITY: Macro/Micro Structures of Spacetime", J. Phys. Conf. Ser. 174 (2009) 012015 [arXiv:0903.0878]
[7] B. L. Hu, "Emergent Gravity: Conceptual Development and New Challenges" Review for IJMPD (2013)
[8] N. Seiberg, "Emergent Spacetime" Rapporteur talk at the 23rd Solvay Conference in Physics, December, 2005. [arXiv:hep-th/0601234]
[9] Gary T. Horowitz, Joseph Polchinski, "Gauge/gravity duality" in [5].
[10] J.D. Bekenstein, Black holes and entropy, Phys. Rev. D 7, 2333 (1973).
[11] S.W. Hawking, Particle creation by black holes, Commun. Math. Phys. 43, 199 (1975).
[12] W.G. Unruh, Notes on black-hole evaporation, Phys. Rev. D 14, 870 (1976).
[13] G. 't Hooft, "Dimensional reduction in quantum gravity," arXiv:gr-qc/9310026, in Abdus Salam Festschrift: A Collection of Talks (World Scientific, Singapore, 1993). L. Susskind, "The World as a hologram," J. Math. Phys. 36 (1995) 6377 -6396 [arXiv:hep-th/9409089].
[14] T. Jacobson, Phys. Rev. Lett. 75, 1260 (1995). C. Eling, R. Guedens, T. Jacobson, Phys. Rev. Lett. 96, 121301 (2006).
[15] T. Padmanabhan, Rep. Prog. Phys. 73 (2010) 046901 "Thermodynamical Aspects of Gravity: New insights"

[arXiv:0911.5004]. T. Padmanabhan , Aseem Paranjape, Entropy of Null Surfaces and Dynamics of Spacetime Phys. Rev. D75 064004, (2007).
[16] E. P. Verlinde, "On the Origin of Gravity and the Laws of Newton" [arXiv:1001.0785]
[17] B. L. Hu, "Gravity and Nonequilibrium Thermodynamics of Classical Matter" Int. J. Mod. Phys. D 20:697-716 (2011) [arXiv:1010.5837]
[18] B. L. Hu, "Can Spacetime be a Condensate?" Int. J. Theor. Phys. 44 (2005) 1785 [gr-qc/0503067]; The Universe as an Ultimate Macroscopic Quantum Phenomenon, *Invited talk at the Quantum Physics of Nature (QUPON) Conference* Vienna, Austria, 2005
[19] S. A. Adler, *Quantum Theory as an Emergent Phenomenon: The Statistical Mechanics of Matrix Models as the Precursor of Quantum Field Theory* (Cambridge University Press, 2009)
[20] E. Nelson, Talk in this meeting.
[21] G. 't Hooft, Talk in this meeting
[22] B. L. Hu, "Gravity and Nonequilibrium Thermodynamics of Quantum Fields" Invited Talk at DICE2012, September 2012, Tuscany, Italy. Proceedings in J. Phys. Conf. Ser. (2013)
[23] R Balescu *Equilibrium and Nonequilibrium Statistical Mechanics* (New York: John Wiley & Sons 1975)
[24] W H Zurek *Phys. Today* **44** 36 (1991)
[25] D. Giulini, C. Kiefer, E. Joos, J. Kupsch, I.O. Stamatescu and H.D. Zeh, *Decoherence and the Appearance of a Classical World in Quantum Theory* 2nd ed. (Berlin: Springer-Verlag 2003)
[26] J.P. Paz and W. Zurek, "Environment-induced decoherence and the transition from quantum to classical", lectures at the 72nd Les Houches Summer School "Coherent Matter Waves" (1999)
[27] W.H. Zurek, Decoherence, einselection, and the quantum origins of the classical, Rev. Mod. Phys. 75, 715-C775 (2003).
[28] J.B. Hartle and M. Gell-Mann, Classical equations for quantum systems, Phys. Rev. D 47, 3345 (1993).
[29] R. Omn'es, *The Interpretation of Quantum Mechanics* (Princeton University Press, Princeton, NJ, 1994).
[30] R. B. Griffiths, *Consistent Quantum Theory* (Cambridge University Press 2003)
[31] B. L. Hu, in *The Physical Origin of Time-Asymmetry*, collection of essays from the meeting in Huelva, Spain, 1991 ed. J.J. Halliwell, J. Perez-Mercader and W.H. Zurek (Cambridge University Press, Cambridge, 1994) [gr-qc/9302021]
[32] F. Dowker and A. Kent, On the consistent histories approach to quantum mechanics, J. Stat. Phys. 82, 1575 (1996).
[33] Stephen Wolfram, *Cellular Automata And Complexity: Collected Papers* Westview Press (1994)
[34] See, e.g., B. Schmittmann and R. K. P. Zia, "Statistical Mechanics of Driven Diffusive Systems", in *Phase Transitions and Critical Phenomena*, Vol. 17, pp. 1-220 eds. C. Domb and J. L. Lebowitz (Academic, N.Y., 1995)
[35] P. Anderson, (1972) 'More is Different', Science 177: 393-396.
[36] R. Laughlin, and D. Pines (2000) 'The Theory of Everything', Proceedings of the National Academy of Sciences 97: 28-31.
[37] S. Weinberg, "Newtonianism, reductionism and the art of congressional testimony". Nature, 330:433-437 (1987)
[38] J. Bain, 'Effective Field Theories', in Batterman, B. (ed.) The Oxford Handbook of Philosophy of Physics, Oxford University Press (2011).
[39] J. Bain, 'Condensed Matter Physics and the Nature of Spacetime', in Dieks, D. (ed.) The Ontology of Spacetime, Vol. 2, Elsevier Press, pp. 301-329 (2008) .
[40] P. Mainwood, 'Is More Different? Emergent Properties in Physics', unpublished dissertation, University of Oxford (2006).
[41] R. W. Batterman, "Emergence, Singularities, and Symmetry Breaking" Foundations of Physics, October, 2010.
[42] R. W. Batterman, *The Devil in the Details: Asymptotic Reasoning in Explanation, Reduction, and Emergence*, Oxford University Press (New York), 2002
[43] David R. Nelson, *Defects and Geometry in Condensed Matter Physics* (Cambridge University Press, 2002)
[44] J. Polchinski,'Effective Field Theory and the Fermi Surface', in J. Harvey and J. Polchinski (eds.), Proceedings of 1992 Theoretical Advanced Studies Institute in Elementary Particle Physics. World Scientific: Singapore (1993) . [arXiv: hepth/92110046]
[45] J. Wilson, 'Non-reductive Physicalism and Degrees of Freedom', British Journal for Philosophy of Science 61: 279-311 (2010) .
[46] M. Morrison, "Emergence, Reduction, and Theoretical Principles: Rethinking Fundamentalism", Philosophy of Science, 73: 876-887 (2006)
[47] J. A. Wheeler and K. Ford. *Geons, Black Holes and Quantum Foam: A Life in Physics* (W. W. Norton, 1998).

[48] E. Mayr, *The Growth of Biological Thought.* Harvard University Press, 1982.
[49] P. Pettit, "Microphysicalism without contingent micro-macro laws". Analysis, 55: 253 7, 1994.
[50] P. Pettit, "The Common Mind. An essay on Psychology, Society and Politics". Oxford University Press, 1996.
[51] E. Calzetta and B. L. Hu, *Nonequilibrium Quantum Field Theory*, (Cambridge University Press, 2008)
[52] J.P. Paz, S. Habib and W.H. Zurek, Reduction of the wave packet: Preferred observable and decoherence time scale, Phys. Rev. D 47, 488 (1993).
[53] C. H. Chou, B. L. Hu and Y. Subasi, "Macroscopic Quantum Phenomena from the Correlation, Coupling and Criticality Perspectives" J. Phys. Conf. Ser. 330, 012003 (2011) [arXiv:1107.3008v1]
[54] E. Calzetta and B. L. Hu, Phys. Rev. D55, 3536-3551 (1997).
[55] R. Omn‘es, *The Interpretation of Quantum Mechanics* (Princeton University Press, Princeton, NJ, 1994).
[56] A. Peres, *Quantum Theory: Concepts and Methods* (Kluwer, Dordrecht, 1993).
[57] R. Feynman and F. Vernon, The theory of a general quantum system interacting with a linear dissipative system, Ann. Phys. (N.Y.) 24, 118 (1963).
[58] R. Balian and M. Veneroni, Incomplete descriptions, relevant information, and entropy production in collision processes, Ann. Phys. (N.Y.) 174, 229-244 (1987).
[59] H. Spohn 1991 *Large Scale Dynamics of Interacting Particles* (Berlin: Springer-Verlag)
[60] J.B. Hartle, R. Laflamme, and D. Marolf, Conservation laws in the quantum mechanics of closed systems, Phys. Rev. D 51, 7007 (1995).
[61] T. Brun and J.J. Halliwell, Decoherence of hydrodynamic histories: A simple spin model, Phys. Rev. D 54, 2899 (1996).
[62] J.J. Halliwell, Decoherent histories and hydrodynamic equations, Phys. Rev. D 58, 105015 (1998).
[63] R. Dorfman, Some recent development in the kinetic theory, in H.J. Raveché (editor) Perspectives in Statistical Physics: M.S. Green Memorial Volume (North Holland, Amsterdam, 1981).

Four questions for quantum-classical hybrid theory [1]

Hans-Thomas Elze

Dipartimento di Fisica "Enrico Fermi", Largo Pontecorvo 3, I-56127 Pisa, Italia

E-mail: elze@df.unipi.it

Abstract. Four questions are discussed which may be addressed to any proposal of a quantum-classical hybrid theory which concerns the direct coupling of classical and quantum mechanical degrees of freedom. In particular, we consider the formulation of hybrid dynamics presented recently in Ref. [1]. This linear theory differs from the nonlinear ensemble theory of Hall and Reginatto, but shares with it to fulfil all consistency requirements discussed in the literature, while earlier attempts failed. – Here, we additionally ask: Does the theory allow for superposition, separable, and entangled states originating in the quantum mechanical sector? Does it allow for "Free Will", as introduced, in this context, by Diósi [2]? Is it local? Does it provide hints for the emergence of quantum mechanics from dynamics beneath?

PACS numbers: 03.65.Ca, 03.65.Ta

1. Introduction

The hypothetical direct coupling of quantum mechanical and classical degrees of freedom – *"hybrid dynamics"* – departs from quantum mechanics. It has been researched at length for practical as well as theoretical reasons. In particular, the standard Copenhagen interpretation entails the unresolved measurement problem which, together with the fact that quantum mechanics needs interpretation, in order to be operationally well defined, may indicate that it deserves amendments. It has been recognized early on that a theory which *dynamically* bridges the quantum-classical divide should have an impact on the measurement problem [3] as well as on attempts to describe consistently the interaction between quantum matter and classical spacetime [4].

Numerous works have appeared, in order to formulate hybrid dynamics in a satisfactory way. Generally, they showed deficiencies in one or another respect. Which has led to various no-go theorems in view of a list of desirable properties or consistency requirements, see, for example, Refs. [5, 6]:

- Conservation of energy.
- Conservation and positivity of probability.
- Separability of quantum and classical subsystems in the absence of their interaction, recovering the correct quantum and classical equations of motion.
- Consistent definitions of states and observables; existence of a Lie bracket structure on the algebra of observables that suitably generalizes Poisson and commutator brackets.

[1] Main parts of author's talk at *"Heinz von Foerster Congress / Emergent Quantum Mechanics"* (Vienna, Nov. 2011) are in Ref. [1]; questions are considered here which were raised afterwards or in related correspondence.

- Existence of canonical transformations generated by the observables; invariance of the classical sector under canonical transformations performed on the quantum sector only and *vice versa*.
- Existence of generalized Ehrenfest relations (*i.e.* the correspondence limit) which, for bilinearly coupled classical and quantum oscillators, are to assume the form of the classical equations of motion ("Peres-Terno benchmark" test [7]).

These issues have been reviewed in recent works by Hall and Reginatto where they have introduced the first theory of hybrid dynamics that conforms with the points listed above [8, 9, 10]. Their ensemble theory is based on configuration space, which requires a certain nonlinearity of the action functional from which it is derived and entails effects that might allow to falsify this proposal experimentally.

We have proposed an alternative theory of hybrid dynamics based on notions of phase space [1]. This is partly motivated by work on related topics of general linear dynamics and classical path integrals [11, 12] and extends work by Heslot, who demonstrated that quantum mechanics can entirely be rephrased in the language and formalism of classical analytical mechanics [13]. Introducing unified notions of states on phase space, observables, canonical transformations, and a generalized quantum-classical Poisson bracket, this has led to an intrinsically linear hybrid theory, which fulfils the above consistency requirements.

Objects that somehow reside between classical and quantum mechanics have been described recently also in a statistical theory, based on very different premises than the hybrid theories considered here [14]. It would be interesting to uncover any relation, if it exists.

In any case, it must be emphasized that it is also of great practical importance to better understand quantum-classical hybrids appearing in quantum mechanical approximation schemes. These typically address many-body systems or interacting fields, which are naturally separable into quantum and classical subsystems, for example, representing fast and slow degrees of freedom (keywords: Born-Oppenheimer approximation, mesoscopic systems, "semiclassical quantum gravity"); for references see Ref. [1].

Furthermore, concerning the hypothetical emergence of quantum mechanics from some coarse-grained deterministic dynamics (see, for example, Refs. [15, 16, 17] with numerous references to earlier work), the quantum-classical backreaction problem seems to appear in a new form, namely regarding the interplay of fluctuations among underlying deterministic and emergent quantum mechanical degrees of freedom. Which can be rephrased succinctly as the question: *"Can quantum mechanics be seeded?"*

Besides constructing the quantum-classical hybrid formalism and showing how it conforms with the above consistency requirements, we discussed the possibility to have classical-environment induced decoherence, quantum-classical backreaction, a deviation from the Hall-Reginatto proposal in presence of translation symmetry, and closure of the algebra of hybrid observables [1].

Presently, we add four questions to accompany the above consistency requirements:

- Does the hybrid theory allow for the superposition principle, for separable and for entangled states originating in the quantum mechanical sector?
- Does it allow for "Free Will", a notion introduced, in this context, by Diósi [2]?
- Does it describe a local interaction of classical and quantum mechanical subsystems which evolve in spacetime?
- Does it provide hints concerning the study of emergence of quantum mechanics from deterministic dynamics beneath?

which may be addressed to any model intended to describe quantum-classical hybrids.

The remainder of the paper is organized as follows. In Section 2., we collect some of the results of Ref. [1], which will be useful in the following. In Sections 3. to 6., we discuss, if not answer, the questions just listed with respect to the hybrid theory proposed in Ref. [1], followed by concluding remarks in Section 7.

2. Linear quantum-classical hybrid dynamics – a summary

In the following two subsections, we will present some important results drawn from the exposition of classical Hamiltonian mechanics and its generalization incorporating quantum mechanics by Heslot [13]. This will form the starting point of our brief review of the hypothetical direct coupling between quantum and classical degrees of freedom, as developed in Ref. [1]; the reader familiar with the earlier derivations may directly pass to Section 3.

2.1. Classical mechanics

The evolution of a *classical* object is described with respect to its $2n$-dimensional phase space, which is identified as its *state space.* A real-valued regular function on the state space defines an *observable*, *i.e.*, a differentiable function on this smooth manifold.

Darboux's theorem shows that there always exist (local) systems of so-called *canonical coordinates*, commonly denoted by (x_k, p_k), $k = 1, \dots, n$, such that the *Poisson bracket* of any pair of observables f, g assumes the standard form [18]:

$$\{f, g\} = \sum_k \Big(\frac{\partial f}{\partial_{x_k}} \frac{\partial g}{\partial_{p_k}} - \frac{\partial f}{\partial_{p_k}} \frac{\partial g}{\partial_{x_k}} \Big) \; . \tag{1}$$

This is consistent with $\{x_k, p_l\} = \delta_{kl}$, $\{x_k, x_l\} = \{p_k, p_l\} = 0$, $k, l = 1, \dots, n$, and reflects the bilinearity, antisymmetry, derivation-like product formula, and Jacobi identity which define a Lie bracket operation, $f, g \to \{f, g\}$, mapping a pair of observables to an observable.

Compatibility with the Poisson bracket structure restricts general transformations $\mathcal{G}$ of the state space to so-called *canonical transformations* which do not change the physical properties of the object under study; e.g., a translation, a rotation, a change of inertial frame, or evolution in time. Such $\mathcal{G}$ induces a change of an observable, $f \to \mathcal{G}(f)$, and is an automorphism of the state space compatible with its Poisson bracket structure. Due to the Lie group structure of the set of canonical transformations, it is sufficient to consider infinitesimal transformations generated by the elements of the corresponding Lie algebra. Then, an *infinitesimal transformation* $\mathcal{G}$ is *canonical*, if and only if for any observable f the map $f \to \mathcal{G}(f)$ is given by $f \to f' = f + \{f, g\}\delta\alpha$, with some observable g, the so-called *generator* of $\mathcal{G}$, and $\delta\alpha$ an infinitesimal real parameter.

Thus, for the canonical coordinates, in particular, an infinitesimal canonical transformation amounts to:

$$x_k \;\to\; x'_k = x_k + \frac{\partial g}{\partial p_k}\delta\alpha \;, \tag{2}$$

$$p_k \;\to\; p'_k = p_k - \frac{\partial g}{\partial x_k}\delta\alpha \;, \tag{3}$$

employing the Poisson bracket given in Eq. (1).

This analysis shows the fundamental relation between observables and generators of infinitesimal canonical transformations in classical Hamiltonian mechanics.

2.2. Quantum mechanics

An important achievement of Heslot's work is the realization that the previous analysis can be generalized and applied to quantum mechanics; this concerns the dynamical aspects as well as the notions of states, canonical transformations, and observables.

To begin with, we recall that the Schrödinger equation and its adjoint can be derived from a well known action principle [1]. Which shows immediately that they have the character of Hamiltonian equations of motion.

To this must be added the *normalization condition* for the state vector $|\Psi\rangle$:

$$\mathcal{C} := \langle\Psi(t)|\Psi(t)\rangle \stackrel{!}{=} \text{constant} \equiv 1 \ , \tag{4}$$

which is an essential ingredient of the probability interpretation associated with state vectors. – Adding here that state vectors that differ by an *unphysical constant phase* are to be identified, we are reminded that the *quantum mechanical state space* is formed by the rays of an underlying Hilbert space, *i.e.*, forming a complex projective space.

2.2.1. The oscillator representation Quantum mechanical evolution can be described by a unitary transformation, $|\Psi(t)\rangle = \hat{U}(t-t_0)|\Psi(t_0)\rangle$, with $U(t-t_0) = \exp[-i\hat{H}(t-t_0)]$, which formally solves the Schrödinger equation. It follows immediately that a stationary state, characterized by $\hat{H}|\phi_i\rangle = E_i|\phi_i\rangle$, with a real energy eigenvalue E_i, performs a simple harmonic motion, *i.e.*, $|\psi_i(t)\rangle = \exp[-iE_i(t-t_0)]|\psi_i(t_0)\rangle \equiv \exp[-iE_i(t-t_0)]|\phi_i\rangle$. Henceforth, we assume a denumerable set of eigenstates of the Hamilton operator.

With the Hamiltonian character of the underlying equation(s) of motion, the harmonic motion suggests to introduce what we called *oscillator representation.* We consider the expansion of any state vector with respect to a complete orthonormal basis, $\{|\Phi_i\rangle\}$:

$$|\Psi\rangle = \sum_i |\Phi_i\rangle(X_i + iP_i)/\sqrt{2} \ , \tag{5}$$

where the generally time dependent expansion coefficients are written in terms of real and imaginary parts, X_i, P_i. Employing this expansion, allows to evaluate what will serve as the *Hamiltonian function, i.e.*, $\mathcal{H} := \langle\Psi|\hat{H}|\Psi\rangle$:

$$\mathcal{H} = \frac{1}{2}\sum_{i,j}\langle\Phi_i|\hat{H}|\Phi_j\rangle(X_i - iP_i)(X_j + iP_j) =: \mathcal{H}(X_i, P_i) \ . \tag{6}$$

Choosing especially the set of energy eigenstates, $\{|\phi_i\rangle\}$, as basis, we obtain:

$$\mathcal{H}(X_i, P_i) = \sum_i \frac{E_i}{2}({P_i}^2 + {X_i}^2) \ , \tag{7}$$

hence the name *oscillator representation.* The reasoning leading to this result indicates that (X_i, P_i) can play the role of *canonical coordinates* in the description of a quantum mechanical object and its evolution with respect to the state space.

This interpretation is substantiated by the fact that the Schrödinger equation is recovered by evaluating $|\dot{\Psi}\rangle = \sum_i |\Phi_i\rangle(\dot{X}_i + i\dot{P}_i)/\sqrt{2}$ according to Hamilton's equations of motion, using the Hamiltonian function of Eq. (6) or (7). – Furthermore, the *constraint* of Eq. (4) becomes:

$$\mathcal{C}(X_i, P_i) = \frac{1}{2}\sum_i({X_i}^2 + {P_i}^2) \stackrel{!}{=} 1 \ . \tag{8}$$

Thus, the vector with components given by the canonical coordinates (X_i, P_i), $i = 1, \ldots, N$, is constrained to the surface of a $2N$-dimensional sphere with radius $\sqrt{2}$. This constraint presents a major difference to classical Hamiltonian mechanics.

Similarly as in Subsection 2.1., it is natural to introduce also here a *Poisson bracket* for any two *observables* on the *spherically compactified state space, i.e.* real-valued regular functions F, G of the coordinates (X_i, P_i):

$$\{F, G\} = \sum_i \Big(\frac{\partial F}{\partial X_i}\frac{\partial G}{\partial P_i} - \frac{\partial F}{\partial P_i}\frac{\partial G}{\partial X_i} \Big) \ . \tag{9}$$

Then, the Hamiltonian acts as the generator of time evolution of any observable O:

$$\frac{\mathrm{d}O}{\mathrm{d}t} = \partial_t O + \{O, \mathcal{H}\} \ , \tag{10}$$

and one verifies that the constraint, Eq. (8), is conserved under the Hamiltonian flow:

$$\frac{\mathrm{d}\mathcal{C}}{\mathrm{d}t} = \{\mathcal{C}, \mathcal{H}\} = 0 \ . \tag{11}$$

It remains to demonstrate the *compatibility* of the notion of *observable* introduced here – as in classical mechanics – with the one adopted in quantum mechanics. This concerns, in particular, the implementation of *canonical transformations* and the role of observables as their generators.

2.2.2. Canonical transformations and quantum observables The Hamiltonian function has been introduced as observable in the Eq. (6) which provides a direct relation to the corresponding quantum observable, namely the expectation value of the self-adjoint Hamilton operator. This is indicative of the general structure to be discussed now.

Referring to Refs. [1, 13] for all details, we summarize here the main points:

• A) *Compatibility of unitary transformations and Poisson structure.* – The canonical transformations discussed in Section 2.1. represent automorphisms of the classical state space which are compatible with the Poisson brackets. In quantum mechanics automorphisms of the Hilbert space are implemented by unitary transformations. This implies a transformation of the canonical coordinates here, *i.e.*, of the expansion coefficients (X_i, P_i) introduced in Eq. (5). Analysing this, one finds that the fundamental Poisson brackets remain invariant under unitary transformations. Thus, *unitary transformations on Hilbert space are canonical transformations on the (X, P) state space.*

• B) *Self-adjoint operators as observables.* – Any infinitesimal unitary transformation $\hat{U}$ can be generated by a self-adjoint operator $\hat{G}$, such that:

$$\hat{U} = 1 - i\hat{G}\delta\alpha \ , \tag{12}$$

which will lead to the quantum mechanical relation between an observable and a self-adjoint operator. In fact, straightforward calculation shows that in the present case we have:

$$X_i \ \rightarrow \ X_i' = X_i + \frac{\partial\langle\Psi|\hat{G}|\Psi\rangle}{\partial P_i}\delta\alpha \ , \tag{13}$$

$$P_i \ \rightarrow \ P_i' = P_i - \frac{\partial\langle\Psi|\hat{G}|\Psi\rangle}{\partial X_i}\delta\alpha \ . \tag{14}$$

Then, the relation between an observable G, defined in analogy to Section 2.1., and a self-adjoint operator $\hat{G}$ can be inferred from Eqs. (13)–(14) within a few steps:

$$G(X_i, P_i) = \langle\Psi|\hat{G}|\Psi\rangle \ , \tag{15}$$

by comparison with the classical result. We find that a *real-valued regular function G of the state is an observable, if and only if there exists a self-adjoint operator $\hat{G}$ such that Eq. (15) holds.* This implies that *all quantum observables are quadratic forms* in the X_i's and P_i's, which are essentially fewer than in the corresponding classical case; see, however, Subsection 3.2 for the generalization necessitated by interacting quantum-classical hybrids.

• C) *Commutators as Poisson brackets.* – Relation (15) between observables and self-adjoint operators together with the Poisson bracket (9) allow to demonstrate the important result:

$$\{F,G\} = \langle\Psi|\frac{1}{i}[\hat{F},\hat{G}]|\Psi\rangle \quad , \tag{16}$$

with both sides of the equality considered as functions of the variables X_i, P_i and with the commutator defined as usual. This shows that the *commutator is a Poisson bracket with respect to the (X,P) state space* and relates the algebra of observables, in the sense of the classical construction of Section 2.1., to the algebra of self-adjoint operators in quantum mechanics.

• D) *Normalization, phase arbitrariness, and admissible observables.* – Coming back to the normalization condition $\langle\Psi|\Psi\rangle \stackrel{!}{=} 1$, which compactifies the state space, cf. the constraint Eq. (8), it must be preserved under infinitesimal canonical transformations, since it belongs to the structural characteristics of the state space. This leads to a constraint on admissible observables, which turns out to be compatible with the requirement that observables G are invariant under an infinitesimal phase transformation $|\Psi\rangle \rightarrow |\Psi\rangle\cdot\exp(i\delta\theta)$, with constant $\delta\theta$. Conversely, assuming this *phase invariance* of observables, we recover that Hilbert space vectors differing by a constant phase are indistinguishable and represent the same physical state.

We note that any observable G with an expansion as in Eq. (6) automatically satisfies the invariance requirements of item D). Explicit calculation shows:

$$\{\mathcal{C},G\} = \sum_j \Big(\frac{\partial G}{\partial P_j}X_j - \frac{\partial G}{\partial X_j}P_j\Big) = 0 \quad , \tag{17}$$

assuming that:

$$G(P_i,X_i) := \langle\Psi|\hat{G}|\Psi\rangle = \frac{1}{2}\sum_{i,j} G_{ij}(X_i - iP_i)(X_j + iP_j) \quad , \tag{18}$$

and where $G_{ij} := \langle\Phi_i|\hat{G}|\Phi_j\rangle = G^*_{ji}$, for a self-adjoint operator $\hat{G}$.

In conclusion, quantum mechanics shares with classical mechanics an even dimensional state space, a Poisson structure, and a related algebra of observables. Yet it differs essentially by a restricted set of observables and the requirements of phase invariance and normalization, which compactify the underlying Hilbert space to the complex projective space formed by its rays.

2.3. Quantum-classical Poisson bracket and separability

The result of aligning classical and quantum mechanics in the way summarized above suggests to introduce a *generalized Poisson bracket* for observables defined on the Cartesian product state space of CL (classical) *and* QM (quantum mechanical) sectors of a hybrid:

$$\{A,B\}_\times := \{A,B\}_{\mathrm{CL}} + \{A,B\}_{\mathrm{QM}} \tag{19}$$

$$:= \sum_k \Big(\frac{\partial A}{\partial_{x_k}}\frac{\partial B}{\partial_{p_k}} - \frac{\partial A}{\partial_{p_k}}\frac{\partial B}{\partial_{x_k}}\Big) + \sum_i \Big(\frac{\partial A}{\partial_{X_i}}\frac{\partial B}{\partial_{P_i}} - \frac{\partial A}{\partial_{P_i}}\frac{\partial B}{\partial_{X_i}}\Big) \quad , \tag{20}$$

for any two observables A, B. It is bilinear and antisymmetric, leads to a derivation-like product formula and obeys the Jacobi identity.

Let us say *an observable "belongs" to the CL (QM) sector, if it is constant with respect to the canonical coordinates of the QM (CL) sector.* – Then, the generalized Poisson bracket has the additional important properties:

- It reduces to the Poisson brackets introduced in Eqs. (1) and (9), respectively, for pairs of observables that belong *either* to the CL *or* the QM sector.
- It reduces to the appropriate one of the former brackets, if one of the observables belongs only to either one of the two sectors.
- It reflects the *separability* of CL and QM sectors, since $\{A, B\}_\times = 0$, if A and B belong to different sectors.

The physical relevance of separability is this: *If a canonical transformation is performed on the QM (CL) sector only, then all observables that belong to the CL (QM) sector remain invariant.*

3. Superposition, separable and entangled states from the QM sector?

Whether our description of quantum-classical hybrids incorporates the possibility of superposition, separable, and entangled states originating in the QM sector can be answered right away in the affirmative. This follows from the fact that it represents quantum mechanics exactly, in the absence of a hybrid coupling, as we have shown [1]. This continues to hold, as long as the dynamical variables pertaining to the CL sector can be considered as external parameters for the QM sector. However, further comment is necessary when the QM-CL hybrid is fully dynamical and we will come to this in due course.

3.1. Definition and interpretation of hybrid states

It may be useful to look from a different angle at the question how different kinds of quantum states enter the description of hybrids. – We recall that the hybrid density ρ has been introduced in Ref. [1] as the expectation with respect to a given state vector of a self-adjoint, positive semi-definite, trace normalized density operator $\hat{\rho}$ [1]:

$$\rho(x_k, p_k; X_i, P_i) := \langle\Psi|\hat{\rho}(x_k, p_k)|\Psi\rangle = \frac{1}{2}\sum_{i,j}\rho_{ij}(x_k, p_k)(X_i - iP_i)(X_j + iP_j) \ , \tag{21}$$

using the oscillator expansion, Eq. (5), and $\rho_{ij}(x_k, p_k) := \langle\Phi_i|\hat{\rho}(x_k, p_k)|\Phi_j\rangle = \rho^*_{ji}(x_k, p_k)$. It describes a *quantum-classical hybrid ensemble* by a real-valued, positive semi-definite, normalized, and possibly time dependent regular function, the probability distribution ρ, on the Cartesian product state space canonically coordinated by $2(n+N)$-tuples $(x_k, p_k; X_i, P_i)$; the variables x_k, p_k, $k = 1, \dots, n$ and X_i, P_i, $i = 1, \dots, N$ are reserved for the CL and QM sector, respectively.

Furthermore, expanding $\hat{\rho}$ in terms of its eigenstates, $\hat{\rho} = \sum_j w_j|j\rangle\langle j|$, one obtains:

$$\rho(x_k, p_k; X_i, P_i) = \sum_j w_j(x_k, p_k)\mathrm{Tr}(|\Psi\rangle\langle\Psi|j\rangle\langle j|) \tag{22}$$

$$= \sum_j w_j(x_k, p_k)|\langle j|\Psi\rangle|^2 \ , \tag{23}$$

with $0 \le w_j \le 1$ and $\sum_j \int \Pi_l(\mathrm{d}x_l \mathrm{d}p_l) w_j(x_k, p_k) = 1$. – This suggests that $\rho(x_k, p_k; X_i, P_i)$, when properly normalized, is the *probability density to find in the hybrid ensemble the QM state* $|\Psi\rangle$, parametrized by X_i, P_i through Eq. (5), *together with the CL state* given by a point in phase space, specified by the coordinates (x_k, p_k).

Clearly, this interpretation does not depend on whether $\hat{\rho}$ stands for any *pure state*, which might be a coherent superposition of other pure states in the Hilbert space of the QM sector, or for a *mixed state*, i.e., a probabilistic superposition of pure states.

Note, in particular, how a probabilistic mixture of two densities, $\rho := p\rho^{(1)} + (1-p)\rho^{(2)}$, with $0 \leq p \leq 1$, can be represented according to Eq. (23):

$$\rho(x_k, p_k; X_i, P_i) = \sum_j \Big(p w_j(x_k, p_k)^{(1)} + (1-p) w_j(x_k, p_k)^{(2)} \Big) |\langle j|\Psi\rangle|^2 \ , \tag{24}$$

which is consistent with the interpretation given. – It must be emphasized that our definition of the density, Eq. (21), is not suitable to ask questions like *"What is the probability to find in the hybrid ensemble a given QM mixed state?"*. It does not contain sufficient information for such, more general purposes.

However, the observations made allow us to address the cases of *separable and entangled states* relevant for a composite quantum system as part of a hybrid. For simplicity, we treat a bi-partite QM sector; the generalization for multi-partite systems follows by induction. – Generally, two *QM systems* A *and* B *are defined to be separable*, if their state can be prepared as a statistical mixture of tensor product states [19]:

$$\hat{\rho}^{(\mathrm{AB})} = \sum_l w_l \hat{\rho}_l^{(\mathrm{A})} \otimes \hat{\rho}_l^{(\mathrm{B})} \ , \quad \text{with} \ \sum_l w_l = 1 \ , \quad w_l \geq 0 \ . \tag{25}$$

If this is not the case, A and B are said to be in an *entangled state*. If A and B are separable, then there exist at most classical but no quantum correlations between them. For a pure state, this is equivalent to the statement that it is separable if and only if it is of the form $|\Psi^{(\mathrm{AB})}\rangle = |\Psi^{(\mathrm{A})}\rangle \otimes |\Psi^{(\mathrm{B})}\rangle$; here, the sum in Eq. (25) has only one nonvanishing term.

Considering *separable states*, generally, we can adapt Eqs. (22)–(23) to the case at hand by expanding each one of the self-adjoint factors $\hat{\rho}_l^{(\mathrm{A})}$ and $\hat{\rho}_l^{(\mathrm{B})}$ in the definition, Eq. (25), in terms of its eigenstates, e.g., $\hat{\rho}_l^{(\mathrm{A})} = \sum_{j_l} w_{j_l}^{(\mathrm{A})} |\mathrm{A}; j_l\rangle\langle \mathrm{A}; j_l|$. Inserting these expansions, the Eq. (23) is replaced by:

$$\begin{aligned} \rho(x_k, p_k; X_i, P_i) &= \sum_{l, j_l, j_l'} w_l(x_k, p_k) w_{j_l}^{(\mathrm{A})}(x_k, p_k) w_{j_l'}^{(\mathrm{B})}(x_k, p_k) |\langle j_l, j_l'|\Psi\rangle|^2 &(26)\\ &= \sum_{j, j'} w_j^{(\mathrm{A})}(x_k, p_k) w_{j'}^{(\mathrm{B})}(x_k, p_k) |\langle j, j'|\Psi\rangle|^2 \ , &(27) \end{aligned}$$

where $|j_l, j_l'\rangle := |A; j_l\rangle \otimes |B; j_l'\rangle$, and the second equality holds for an overall pure state with only one term contributing to the sum over l, hence l-dependence is suppressed; note that each one of the weights w has the meaning of a probability. The interpretation here is an obvious generalization of the one following Eq. (23).

Next, we distinguish pure and mixed entangled states of the QM composite system. – The case of *pure entangled states* can be easily fit into the above considerations. To the bi-partite (finite-dimensional) Hilbert space, in which a pure state has a given tensor product structure, corresponds a factorized algebra of QM observables. It has been shown recently that this situation is unitarily equivalent to one where the Hilbert space is unstructured, provided the algebra of observables is suitably rearranged (*"Tailored Observables Theorem"*) [20]. Thus, the tensor product structure can be effectively eliminated and the entanglement properties shifted to the observables, and vice versa. With an unstructured Hilbert space, we are back to the situation discussed in the context of Eqs. (22)–(23). In this way, in our formulation, pure entangled states from the QM sector can naturally occur in a QM-CL hybrid.

Mixed states present a more subtle situation. This is immediately obvious if one looks at the limiting case of a *totally mixed state*, described by $\hat{\rho} = N^{-1}\sum_{i=1}^{N}|\Phi_i\rangle\langle\Phi_i| \equiv N^{-1}\mathbf{1}_N$, where $\{\Phi_i\}$ stands for any complete orthonormal basis of the N-dimensional Hilbert space of the composite. By definition, cf. Eq. (25), this state is separable for all (factorized) representations, and thus can be handled as before. By continuity, this holds for states in a sufficiently small neighbourhood of the totally mixed state. However, it turns out that the possibility (or not) to switch between entangled and separable forms of the state, by tailoring the algebra of observables, which we needed for our above arguments, still cannot be ascertained for mixed states in all generality. Criteria to assess this situation are presently under investigation, see Ref. [21], with references to earlier work therein. In conclusion, for *mixed entangled states* an intuitive interpretation of the hybrid density, along the lines sketched above, is still missing. Yet we are confident that this will work out eventually.

3.2. Evolution of hybrid densities

Turning to dynamics, we proposed a straightforward generalization of Hamiltonian mechanics (see also Section 4.), which we employ here in the form of a *Liouville equation*, in order to describe the evolution of hybrid ensembles [1]. This is based on Liouville's theorem and the generalized Poisson bracket defined in Eqs. (19)–(20) and assumes the compact form:

$$-\partial_t\rho = \{\rho, \mathcal{H}_\Sigma\}_\times \ , \tag{28}$$

with $\mathcal{H}_\Sigma \equiv \mathcal{H}_\Sigma(x_k, p_k; X_i, P_i)$ and:

$$\mathcal{H}_\Sigma := \mathcal{H}_{\rm CL}(x_k, p_k) + \mathcal{H}_{\rm QM}(X_i, P_i) + \mathcal{I}(x_k, p_k; X_i, P_i) \ , \tag{29}$$

which defines the relevant *Hamiltonian function*, including a hybrid interaction. We require $\mathcal{H}_\Sigma$ to be an *observable*, last not least to have a meaningful notion of conserved energy.

An important advantage of Hamiltonian flow and a general property of the Liouville equation in this context is [18]:

- The normalization and positivity of the probability density ρ are conserved in the presence of a genuine hybrid interaction. Therefore, the interpretation of ρ as a probability density remains valid.

However, the simple relation between ρ as a function of the QM "phase space" variables X_i, P_i and an expectation of a self-adjoint density operator $\hat{\rho}$, as employed so far, does not continue to hold generally under hybrid evolution. As pointed out in Section 5.4 of Ref. [1], the oscillator expansion of observables, such as in the second of Eqs. (21), has to be replaced by a more general form, appropriate for what we named *almost-classical observables*, cf. below.

This comes about, since the "classical part" of the bracket, $\{A, B\}_{\rm CL}$, can generate terms which do *not* qualify as observable with respect to the QM sector; here we assume that A and B are both hybrid observables, as defined before. Such terms are of the form:

$$\begin{aligned}
&\frac{1}{4}\sum_{i,i',j,j'}\{A_{ij}, B_{i'j'}\}_{\rm CL}(X_i - iP_i)(X_j + iP_j)(X_{i'} - iP_{i'})(X_{j'} + iP_{j'}) \\
&= \sum_{i,i',j,j'}\langle\Psi|\Phi_i\rangle\langle\Psi|\Phi_{i'}\rangle\{A_{ij}, B_{i'j'}\}_{\rm CL}\langle\Phi_j|\Psi\rangle\langle\Phi_{j'}|\Psi\rangle \ ,
\end{aligned} \tag{30}$$

where we used the oscillator expansion, Eq. (5), and:

$$\{A_{ij}, B_{i'j'}\}_{\rm CL} = \sum_k\Big(\frac{\partial A_{ij}}{\partial_{x_k}}\frac{\partial B_{i'j'}}{\partial_{p_k}} - \frac{\partial A_{ij}}{\partial_{p_k}}\frac{\partial B_{i'j'}}{\partial_{x_k}}\Big) \ , \tag{31}$$

since, for example, $A \equiv A(x_k, p_k; X_i, P_i) = \sum_{i,j} A_{ij}(x_k, p_k)(X_i - iP_i)(X_j + iP_j)$, cf. Eq. (18).

In this way, evolution of hybrid observables, of the density ρ in particular, can induce a *structural change*: while continuing to be CL observables, they do not remain QM observables (quadratic forms in X_i's and P_i's). They fall outside of the product algebra generated by the observables to which we confined ourselves in Section 2. Therefore, if we wish to maintain the formal consistency of our scheme, we have to assume:

- The algebra of hybrid observables is closed under the QM-CL Poisson bracket operation, implemented by $\{\ ,\ \}_\times$.

Thus, the product algebra of CL and QM observables is replaced by its larger closure. This amounts to a *physical hypothesis*, as we shall discuss in the remainder of this section.

Referring to the phase space coordinates (X_i, P_i), we define an *almost-classical observable* as a real-valued regular function of pairs of factors like $(X_i - iP_i)(X_j + iP_j)$, such as in the left-hand side of Eq. (30), subject to the constraint: $\mathcal{C}(X_i, P_i) = \frac{1}{2}\sum_i (X_i^{\ 2} + P_i^{\ 2}) \stackrel{!}{=} 1$.

This normalization constraint, cf. Eq. (8), is preserved under the evolution, since $\{\mathcal{C}, \mathcal{H}_\Sigma\}_\times = 0$, in the presence of QM-CL hybrid interaction. Furthermore, consistently with the closure of the enlarged algebra of observables, we find:

$$\{\mathcal{C}(X_i, P_i), \mathcal{G}(x_k, p_k; X_i, P_i)\}_\times = \{\mathcal{C}(X_i, P_i), \mathcal{G}(x_k, p_k; X_i, P_i)\}_{\rm QM} = 0 \ , \qquad (32)$$

where $\mathcal{G}(x_k, p_k; X_i, P_i)$ stands for any almost-classical observable, including QM observables as a special case, of course. This follows, since the explicit form of the generalized observables, as on the left-hand side of Eq. (30), or generalizations including additional pairs of factors like $(X_i - iP_i)(X_j + iP_j)$, leads to a sum of terms, each one vanishing as if it stemmed from a QM observable, which commute with the constraint represented by $\mathcal{C}$ under the Poisson bracket.

According to this definition, members of the complete algebra of hybrid observables, generally, are *classical* with respect to coordinates (x_k, p_k) and *almost-classical* with respect to coordinates (X_i, P_i), which can be restated as:

- QM observables (quadratic forms in phase space coordinates) form a subset of almost-classical observables which, in turn, form a subset of classical observables (real-valued regular functions of phase space coordinates), cf. Section 2.

Physical consequences of this enlarged "classical × almost-classical algebra" for interacting QM-CL hybrids are illustrated by the following *Gedankenexperiment*.

Consider a quantum together with a classical object subject to a transient hybrid interaction. As long as the hybrid interaction is ineffective, both objects evolve independently according to Schrödinger's and Hamilton's equations, respectively. However, once they form an interacting hybrid, the corresponding phase space density changes from a factorized form, in absence of any initial correlation, to become an almost-classical/classical hybrid observable. Most likely, the density maintains such a mixed character, even when the hybrid interaction ceases.

This outcome contradicts naive expectation that quantum and classical objects evolve separately in quantum and classical ways, when they no longer interact. – Two possibilities come to mind. Either persistence of the almost-classical/classical character is a *physical effect* accompanying QM-CL hybrids, if they exist. Or our description needs to be augmented with a *reduction mechanism* by which evolving observables return to standard QM or CL form, following a hybrid interaction. Both possibilities seem quite interesting in their own right.

Presently, we take the enlargement of the QM algebra of observables, induced by a hybrid interaction, as **"first hint"** that features of QM-CL hybrids might be relevant for an understanding of *how QM emerges*. One would like to understand how a large algebra of classical observables (functions on phase space) is reduced, possibly via almost-classical observables at an intermediary stage, to a smaller QM algebra (linear operators on Hilbert space) for an object that becomes "quantized".

4. Does hybrid dynamics allow for "Free Will"?

In order to address this question, we recall the generalized Hamiltonian equations of motion derived earlier [1]. They reflect the dynamical structure underlying hybrid evolution, according to our proposal, and correspond in detail to the Liouville equation of the previous section.

We consider hybrid systems described by a generic classical *Hamiltonian function* and a quantum mechanical *Hamiltonian operator*, respectively:

$$\mathcal{H}_{\mathrm{CL}} := \sum_k \frac{p_k^2}{2} + v(x_l) \; , \tag{33}$$

$$\hat{H}_{\mathrm{QM}} := \frac{\hat{P}^2}{2} + V(\hat{X}) \; , \tag{34}$$

where $v(x_l) \equiv v(x_1, \dots, x_n)$ and V denote relevant potentials and all masses are set equal to one, for simplicity. Furthermore, there is a *self-adjoint hybrid interaction operator* $\hat{I}(x_k, p_k; \hat{X}, \hat{P})$, invoking symmetrical (Weyl) ordering of the noncommuting operators $\hat{X}$ and $\hat{P}$. By Eq. (15), this gives rise to the following *Hamiltonian function* $\mathcal{H}_\Sigma$:

$$\begin{aligned}\mathcal{H}_\Sigma = \sum_k \frac{p_k^2}{2} + v(x_l) + \langle\Psi|\Big(\frac{\hat{P}^2}{2} + V(\hat{x})\Big)|\Psi\rangle + \langle\Psi|\hat{I}(x_k, p_k; \hat{X}, \hat{P})|\Psi\rangle \\ =: \mathcal{H}_{\mathrm{CL}}(x_k, p_k) + \mathcal{H}_{\mathrm{QM}}(X_i, P_i) + \mathcal{I}(x_k, p_k; X_i, P_i) \; ,\end{aligned} \tag{35}$$

when evaluated in a pure state $|\Psi\rangle$, invoking the oscillator representation of Eq. (5). With these definitions in place, one finds the equations of motion by the rules of Hamiltonian dynamics.

The equations of motion for the CL *observables* x_k, p_k are:

$$\dot{x}_k = \{x_k, \mathcal{H}_\Sigma\}_\times = p_k + \partial_{p_k}\mathcal{I}(x_k, p_k; X_i, P_i) \; , \tag{36}$$

$$\dot{p}_k = \{p_k, \mathcal{H}_\Sigma\}_\times = -\partial_{x_k} v(x_l) - \partial_{x_k}\mathcal{I}(x_k, p_k; X_i, P_i) \; . \tag{37}$$

Similarly, we obtain for the QM *variables* X_i, P_i, which are *not observables*:

$$\dot{X}_i = \{X_i, \mathcal{H}_\Sigma\}_\times = \partial_{P_i}\mathcal{H}_{\mathrm{QM}}(X_j, P_j) + \partial_{P_i}\mathcal{I}(x_k, p_k; X_j, P_j) \tag{38}$$

$$= E_i P_i + \partial_{P_i}\mathcal{I}(x_k, p_k; X_j, P_j) \; , \tag{39}$$

$$\dot{P}_i = \{P_i, \mathcal{H}_\Sigma\}_\times = -\partial_{X_i}\mathcal{H}_{\mathrm{QM}}(X_j, P_j) - \partial_{X_i}\mathcal{I}(x_k, p_k; X_j, P_j) \tag{40}$$

$$= -E_i X_i - \partial_{X_i}\mathcal{I}(x_k, p_k; X_j, P_j) \; , \tag{41}$$

where Eqs. (39) and (41) follow, if the oscillator expansion is performed with respect to the stationary states of $\hat{H}_{\mathrm{QM}}$, cf. Eqs. (6)–(7).

Notably, the Eqs. (36), (37) together with Eqs. (38), (40), or together with Eqs. (39), (41), form a *closed set* of $2(n+N)$ deterministic equations, where n denotes the number of CL degrees of freedom and N the dimension of the QM Hilbert space (assumed denumerable, if not finite). Earlier we contrasted this exact result with *generalized Ehrenfest equations* obtained for the hybrid model [1], which we will employ in Section 5.

Following Diósi [2], someone might ask "Do I have the freedom (*"Free Will"*) to measure one of the CL observables, say z, and, conditioned on its value, to perturb the subsequent hybrid evolution in such a way that the following reasonable properties are guaranteed?":

- i) the variable z is a smooth real function of time;

- ii) the CL and QM variables, as well as the density ρ (with its statistical interpretation), coexist and depend on each other;
- iii) the perturbation at time t_0 affects the QM-CL hybrid in a causal way, i.e., it has no effect at earlier times $t < t_0$.

The classical variable is called *tangible*, in this case.

In the present hybrid theory the CL observables pertain to a classical object with its own dynamics, in the absence of interaction with a quantum object; they are *not* thought to derive somehow from measurements performed on this or an auxiliary quantum object, unlike in all models discussed by Diósi. So measuring z could be the result of a coupling to another CL object, which serves as *apparatus*.

In this case, it seems clear that property i) holds, provided the perturbation applied to the hybrid, in the form of some $\delta_z(t)$, added to z for $t \geq t_0$, is sufficiently smooth; the effect on the equations of motion, Eqs. (36)–(41), is simply to change initial conditions of the subsequent evolution. Similarly, properties ii) and iii) hold as a property of the Hamiltonian dynamics considered; consequently, this applies for the density ρ, evolving according to the equivalent Liouville equation, Eq. (28), as we discussed in Subsection 3.2.

We conclude that there is "Free Will" and classical observables are *tangible* in our theory.

5. Does locality remain intact in hybrids?

Let us consider the Ehrenfest relations for hybrids [1]. Here we look at the case of bilinearly coupled oscillators, which yields a very simple version of these equations and still illustrates the issue of locality. – The following discussion applies as well to the general equations of motion of Section 4, as we shall see shortly.

Earlier we obtained the closed set of equations determining the CL observables x_k, p_k and QM *coordinate and momentum observables* X, P, defined by:

$$X(X_i, P_i) := \langle\Psi|\hat{X}|\Psi\rangle \quad , \quad P(X_i, P_i) := \langle\Psi|\hat{P}|\Psi\rangle \quad , \tag{42}$$

following our construction, cf. Section 2., and assuming here a *set of* CL oscillators coupled bilinearly to *one* QM oscillator, as defined by:

$$\mathcal{H}_{\mathrm{CL}} := \sum_k \Big(\tfrac{1}{2m_k}p_k^2 + \tfrac{m_k\omega_k^2}{2}x_k^2\Big) \quad , \tag{43}$$

$$\hat{H}_{\mathrm{QM}} := \tfrac{1}{2M}\hat{P}^2 + \tfrac{M\Omega^2}{2}\hat{X}^2 \quad , \tag{44}$$

$$\hat{I} := \hat{X}\sum_k \lambda_k x_k \quad . \tag{45}$$

The constants m_k, M, ω_k, Ω, and λ_k denote masses, frequencies, and couplings, respectively. – For this hybrid system, the general equations of motion and Ehrenfest relations indeed reduce to a simple *closed* set of equations:

$$\dot{x}_k = \tfrac{1}{m_k}p_k \quad , \tag{46}$$

$$\dot{p}_k = -m_k\omega_k^2 x_k - \lambda_k X \quad , \tag{47}$$

$$\dot{X} = \tfrac{1}{M}P \quad , \tag{48}$$

$$\dot{P} = -M\Omega^2 X - \sum_k \lambda_k x_k \quad . \tag{49}$$

We observe that the *backreaction* of QM on CL subsystem appears, *as if* the CL subsystem was coupled to another CL oscillator. For a more general discussion of backreaction of QM on CL subsystems, see Ref. [1], in particular, concerning the role of quantum fluctuations.

However, as defined above, the QM observables are expectations and we have explicitly, with $\hat{X}|q\rangle = q|q\rangle$, for example:

$$\begin{aligned} X(X_i, P_i) &= \int \mathrm{d}q'\mathrm{d}q \, \langle\Psi|q'\rangle\langle q'|\hat{X}|q\rangle\langle q|\Psi\rangle = \int \mathrm{d}q \; q|\Psi(q)|^2 \\ &= \frac{1}{2}\sum_{i,j}(X_i - iP_i)(X_j + iP_j) \int \mathrm{d}q \, \langle\Phi_i|q\rangle q\langle q|\Phi_j\rangle \;, \end{aligned} \tag{50}$$

or, more generally, for hybrid interactions that include typical position measurement interactions:

$$\mathcal{I}(x_k, p_k; X_i, P_i) = \frac{1}{2}\sum_{i,j}(X_i - iP_i)(X_j + iP_j) \int \mathrm{d}q \; \Phi_i^*(q)\Phi_j(q)\mathcal{I}(x_k, p_k; q) \;, \tag{51}$$

with $\{|\Phi_j\rangle\}$ representing any complete orthonormal set of states. According to quantum theory, if the QM subsystem of the hybrid is evolving in spacetime, the integration over the variable q is understood as an *integration over all space*. Thus, the coupling in Eqs. (35) or (45) is generally *highly nonlocal* indeed.

This nonlocality has been held as a deficiency against mean-field semiclassical theories and the present theory, as discussed so far, shares this problem.

However, several remarks are in order here. – First of all, we may introduce a localizing factor and redefine the hybrid interaction:

$$\begin{aligned} \mathcal{I}(x_k, p_k; X_i, P_i) &:= \langle\Psi|\hat{\mathcal{I}}(x_k, p_k; \hat{X})\delta(\hat{X} - x_k\hat{\mathbf{1}})|\Psi\rangle && (52) \\ &= \frac{1}{2}\sum_{i,j}(X_i - iP_i)(X_j + iP_j)\Phi_i^*(x_k)\Phi_j(x_k)\mathcal{I}(x_k, p_k; x_k) \;. && (53) \end{aligned}$$

For the bilinear interaction of Eq. (45), for example, we obtain here by localization:

$$\begin{aligned} \mathcal{I}(x_k; X_i, P_i) &= \sum_k \lambda_k x_k \cdot x_k|\Psi(x_k)|^2 \\ &= \frac{1}{2}\sum_{i,j,k} \lambda_k {x_k}^2\Phi_i^*(x_k)\Phi_j(x_k)(X_i - iP_i)(X_j + iP_j) \;, \end{aligned} \tag{54}$$

where the couplings λ_k include a suitable dimensional factor, as compared to before.

This definition of the hybrid interaction is consistent with our formal framework and solves the locality issue. It describes the interaction in terms of the CL coordinates, however, weighted by appropriate wave function factors. The latter reflect that the QM subsystem is generally not localized. Of course, more general forms of the localizing factor are possible and can be extended to field theories as well.

By localizing the hybrid interaction in this way, we *assume ad hoc* that the QM subsystem is probed by the CL subsystem at its location.

Due to the nonlinear features of the localization, the generalized Ehrenfest relations obtain additional terms, as compared to our earlier results [1], which we will not explore here.

Instead, we would like to draw attention to an aspect of *locality of hybrid dynamics*, in particular of the general equations of motion, Eqs. (36)–(41). This will turn out to be independent of whether localization, as discussed above, is present or not.

Suppose a physicist lacking any knowledge of quantum mechanics were presented with these equations (plus the normalization constraint, Eq. (8)). – *We know* that these equations present independent CL and QM sectors, in the absence of a hybrid interaction. – However, he/she would naturally interpret them to describe the dynamics of a composite CL object, with part of its phase space compactified (due to the constraint). Looking at it this way, *he/she finds a perfectly local dynamics*. In fact, *our* knowledge of nonlocal features can be traced to the definition of the canonical coordinates and momenta X_i, P_i, introduced by the oscillator representation, Eq. (5), since: $X_i/\sqrt{2} = \mathrm{Re}\int dq\ \Phi_i^*(q)\Psi(q)$ and $P_i/\sqrt{2} = \mathrm{Im}\int dq\ \Phi_i^*(q)\Psi(q)$. Therefore, spatially nonlocal (and probabilistic) features enter only by reference to the QM wave function.

We tentatively conclude that the spatial interpretation of QM phenomena is, in some sense to be better understood, of a secondary character and take this as **"second hint"** that features of QM-CL hybrids might be relevant for an understanding of *how QM emerges.*

6. Are there hints for the emergence of quantum mechanics?

Our presentation of QM-CL hybrid dynamics has been based on suitably re-presenting the QM sector in the generalized analytical mechanics framework incorporating quantum mechanics, which was outlined by Heslot [13]. This has allowed us, in particular, to generate the hybrid evolution by a generalized Poisson bracket and to have a unified QM-CL description of states and observables.

As we discussed in Sections 3. and 5., we have found at least two hints that our present description may have already something to say about dynamics beyond quantum, classical, and QM-CL hybrid mechanics, indicating structures that lead us beneath quantum mechanics.

A caveat is in order here. Dynamics from which quantum mechanics could emerge by some coarse-graining process for large scales will not necessarily have the Hamiltonian structure that we invoked. It could be of a much more general, less structured kind, such as based on discrete deterministic automata [15].

However, the present work could help to identify deviations from quantum mechanics, which seemingly works perfectly on the scales accessible to present-day experiments. For this purpose, the Hamiltonian dynamics here can serve as a model, in order to parametrize such deviations.

It has been repeatedly pointed out by V.I. Man'ko and collaborators that classical states may differ widely from what could be obtained as the "$\hbar \to 0$" limit of quantum mechanical ones. Furthermore, they show that all states can be classified by their "tomograms" as *either* CL *or* QM, CL *and* QM, and *neither* CL *nor* QM [22, 23]. Yet, in order to explain these *"Man'ko classes of states"*, some unknown dynamics seems missing.

In this context, we find the **"first hint"** of Section 3. very interesting, namely that a consistent hybrid description forces an enlarged algebra of observables upon us, as compared to the quantum mechanical one. More specifically, the QM observables obtain through a *shrinking of the algebra of observables*, via intermediary almost-classical observables, from the largest algebra of classical observables. This seems to complement and underline the findings by Man'ko in our approach, where all states are represented in a suitable phase space. We have also emphasized the *spherical compactification of QM state space*, which is absent classically.

It would be most interesting to find underlying dynamical reasons for such *structural change* that occurs when a CL object interacts with a QM object. This might help to explain how quantum mechanics emerges in a deterministic world.

Furthermore, a **"second hint"**, concerning the relevance of QM-CL hybrids for an understanding of how QM emerges, is coming from Section 5. There, we have seen that the hybrid dynamics might appear as a perfectly local deterministic scheme, if it were not for seemingly added-on interpretation of the QM sector with the help of a wave function defined on its to-be configuration space. Let us assume that spatial notions are somehow related to how gravity enters the picture. Then, leaving the latter aside at this time, one might devise a rather general dynamical framework, possibly incorporating dissipation in the form of an attractor mechanism, already alluded to in Refs. [24, 25, 26, 27], which drives classical through hybrid to quantum mechanics.

7. Concluding remarks

We have discussed four questions in these notes which can be addressed to any proposal of a quantum-classical hybrid theory, in addition to the consistency requirements listed in the introduction, which, in turn, have been widely studied by others. Presently, we considered these questions in the context of the hybrid theory of Ref. [1], which meets all consistency requirements and can be seen as a generalization of proposals incorporating mean field theory.

We emphasize that here the classical sector is not necessarily meant to present an approximation for some of the quantum mechanical degrees of freedom in a fully quantum mechanical multi-partite system. Rather, we have continued to study here, as well as earlier, whether such a hybrid model as ours can stand formally on its own and meet precisely all the consistency requirements posed.

Thus, the questions raised at present are meant to further illuminate the contents and limitations of our or similar models.

The *first question* essentially asks, whether the proposed formalism is general enough to incorporate all possible quantum states, including superpositions, entangled or separable pure and mixed states. Which we have answered positively, with the exception of mixed entangled states, where the necessary tailoring of observables has not yet been achieved in quantum mechanics proper in all generality, cf. Ref. [21]. We have interpreted (Section 3.) the encountered necessity of an enlarged algebra of almost-classical observables as a **"first hint"** that hybrid dynamics might offer a glimpse beyond quantum mechanics.

The *second question* concerns the control of a hybrid system by external classical means, obviously a question of great practical importance. Indeed, our proposal allows for this and for "Free Will", in particular, in the sense defined by Diósi [2].

The *third question*, whether locality is not violated by the proposed hybrid dynamics, as in other proposals related to mean field theory, has been discussed in some detail. Only if we invoke an ad hoc localization of the relevant interactions, this seems to be guaranteed. On the other hand, leaving out the interpretation of the quantum mechanical sector related to a wave function on configuration space, we have argued that here we may have an interesting indication of dynamics – a **"second hint"** – that transcends the usual quantum mechanical framework.

In Section 5., we outlined what might be taken forward from quantum-classical hybrid dynamics tackling the emergence of quantum mechanics from a deterministic theory beneath, which comprises the *fourth question*. Based on the formal developments of Ref. [1] and further discussion here, we seem to get closer to be able to study these issues in some detailed models.

Acknowledgments

It is a pleasure to thank L. Diósi, P. Hajicek, M.J.W. Hall, and M. Reginatto for asking related questions and G. Grössing and co-organizers for inviting me to the *"Heinz von Foerster Congress / Emergent Quantum Mechanics"*. [2]

References

[1] Elze H-T 2011 Linear dynamics of quantum-classical hybrids *Preprint* arXiv:1111.2276
[2] Diósi L 2011 Classical-quantum coexistence: a 'Free Will Test' *Preprint* arXix:1202.2472 (this volume)

[2] After completion of this work, we have noticed Ref. [28], where issues of Section 3 are discussed in different ways. A stringent requirement of *statistical consistency* is introduced which is, however, based on the assumption that its validity extends from QM to quantum-classical hybrid dynamics.

[3] Sherry T N and Sudarshan E C G 1978 Interaction between classical and quantum systems: A new approach to quantum measurement. I. *Phys. Rev.* D **18** 4580; *do.* II. 1979 *Phys. Rev.* D **20** 857
[4] Boucher W and Traschen J 1988 Semiclassical physics and quantum fluctuations *Phys. Rev.* D **37** 3522
[5] Caro J and Salcedo L L 1999 Impediments to mixing classical and quantum dynamics *Phys. Rev.* A **60** 842
[6] Diósi L, Gisin N and Strunz W T 2000 Quantum approach to coupling classical and quantum dynamics *Phys. Rev.* A **61** 022108
[7] Peres A and Terno D R 2001 Hybrid classical-quantum dynamics *Phys. Rev.* A **63** 022101
[8] Hall M J W and Reginatto M 2005 Interacting classical and quantum ensembles *Phys. Rev.* A **72** 062109
[9] Hall M J W 2008 Consistent classical and quantum mixed dynamics *Phys. Rev.* A **78** 042104
[10] Reginatto M and Hall M J W 2009 Quantum-classical interactions and measurement: a consistent description using statistical ensembles on configuration space *J. Phys.: Conf. Ser.* **174** 012038 (*Preprint* arXiv:0905.2948)
[11] Elze H-T, Gambarotta G and Vallone F 2011 General linear dynamics – quantum, classical or hybrid *J. Phys.: Conf. Ser.* **306** 012010 (*Preprint* arXiv:1103.3589)
[12] Elze H-T, Gambarotta G and Vallone F 2011 A path integral for classical dynamics, entanglement, and Jaynes-Cummings model at the quantum-classical divide *Int. J. Qu. Inf. (IJQI)* **9** Suppl. **1** 203-224 (*Preprint* arXiv:1006.1569)
[13] Heslot A 1985 Quantum mechanics as a classical theory *Phys. Rev.* D **31** 1341
[14] Wetterich C 2009 Zwitters: particles between quantum and classical *Preprint* arXiv:0911.1261
[15] 't Hooft G 2010 Classical cellular automata and quantum field theory *Int. J. Mod. Phys.* A **25** No. 23 4385-4396; *do.* 2009 *Preprint* arXiv:0908.3408
[16] Elze H-T 2009 Does quantum mechanics tell an atomistic spacetime? *J. Phys.: Conf. Ser.* **174** 012009 (*Preprint* arXiv:0906.1101); *do.* 2006 Deterministic models of quantum fields *J. Phys.: Conf. Ser.* **33** 399 (*Preprint* arXiv:gr-qc/0512016)
[17] Adler S L 2005 *Quantum Mechanics as an Emergent Phenomenon* (Cambridge, UK: Cambridge U. Press)
[18] Arnold V I 1978 *Mathematical Methods of Classical Mechanics*, Graduate Texts in Mathematics, Vol. 60 (Berlin: Springer-Verlag), Part III
[19] Diósi L 2007 *A Short Course in Quantum Information Theory* Lect. Notes Phys. **713** (Berlin, Heidelberg: Springer)
[20] Harshman N L and Ranade K S 2011 Observables can be tailored to make any pure state entangled (or not) *Phys. Rev.* A **84** 012303 (*Preprint* arXiv:1102.0955)
[21] Thirring W, Bertlmann R A, Köhler P and Narnhofer H 2011 Entanglement or separability: The choice of how to factorize the algebra of the density matrix *Preprint* arXiv:1106.3047
[22] Man'ko O V and Man'ko V I 2004 Classical mechanics is not the $\hbar \to 0$ limit of quantum mechanics *J. Russ. Laser Res.* **25**(5) 477 (*Preprint* arXiv:quant-ph/0407183)
[23] Chernega V N and Man'ko V I 2011 System with classical and quantum subsystems in tomographic probability representation *Preprint*
[24] Blasone M, Jizba P and Vitiello G 2001 Dissipation and quantization *Phys. Lett.* **A287** 205 (*Preprint* arXiv:hep-th/0007138)
[25] 't Hooft G 2007 Emergent quantum mechanics and emergent symmetries *AIP Conf. Proc.* **957** 154-163 (*Preprint* arXiv:0707.4568)
[26] Elze H-T 2008 Note on the existence theorem in "Emergent quantum mechanics and emergent symmetries" *J. Phys. A.: Math. Theor.* **41** 304020 (*Preprint* arXiv:0710.2765); *do.* 2009 The attractor and the quantum states *Int. J. Qu. Inf. (IJQI)* **7** 83 (*Preprint* arXiv:0806.3408)
[27] Blasone M, Jizba P, Scardigli F and Vitiello G 2009 Dissipation and quantization for composite systems *Phys. Lett.* A **373** 4106 (*Preprint* arXiv: 0905.4078)
[28] Salcedo L L 2012 Statistical consistency of quantum-classical hybrids *Phys. Rev.* A **85** 022127 (*Preprint* arXiv:1201.4237)

A study of the Lorentz-Dirac equation in complex space-time for clues to emergent quantum mechanics

Mark Davidson
Spectel Research Corp, Palo Alto, CA USA
E-mail: mdavid@spectelresearch.com

Abstract. A hypothetical equation of motion is proposed for Kerr-Newman particles. Obtained by analytic continuation of the Lorentz-Dirac equation into complex space-time, the goal is to study the implications of this theory for emergent quantum mechanics. A new class of "runaway" solutions is found which bear similarity to the quantum phenomenon of Zitterbewegung. Other solutions incorporating external forces are also presented. The electromagnetic fields generated by these motions are studied. It is found that the retarded times are multi-sheeted functions of the field points. As a consequence, the solutions are not unique. A cloaking mechanism is found which inhibits electromagnetic radiation by combining the fields from several Riemann sheets for the retarded time. These results reinforce to some extent, but also pose additional conceptual questions for the idea that Kerr-Newman solutions provide insight into elementary particles and into emergent quantum mechanics.

1. Introduction

In the general relativity literature Kerr-Newman (KN) models for elementary particles have been studied [1, 2, 3, 4, 5, 6]. Their gyromagnetic g factor is exactly 2, the same as with the Dirac equation [3]. Complex manifold techniques have proved useful [7, 8, 9]. Emergent quantum mechanics at the Planck scale is now an active research area[10, 11, 12, 13, 14]. A dynamical theory for the KN particle could yield clues for this program [15]. The author has argued that this emergence might be cosmological in origin [16, 17]. The Lorentz-Dirac equation (LDE) is typically derived for exactly point-like classical charged particles. We study here a complexified version of it, and find new solutions which are similar to Zitterbewegung [18, 19, 20, 15].

2. A brief overview of the Kerr-Newman solution and complex Minkowski space

The KN particle is modeled as a point charge in a complex Minkowski space (CM^4) [21, 22, 8, 9, 1, 2, 4, 6]. Gravity is ignorable for elementary particle scale parameters [2]. The Riemann-Silberstein complex vector field is used $\mathbf{W} = \mathbf{E} + i\mathbf{B}$, or a complex Faraday tensor is introduced [22] $W^{\mu\nu} = F^{\mu\nu} + i * F^{\mu\nu}$, $*W = -iW$, where F is the Faraday tensor and $*F$ is its dual. W is termed anti self-dual. The real part gives the physical Faraday tensor when evaluated on the real hyperspace $F^{\mu\nu}_{phys} = Re\, W^{\mu\nu}$. The static Kerr-Newman particle is modelled as a point charge located at a point in complex 3-space at $\mathbf{z_0} = \mathbf{x_0} + i\mathbf{b}$, where $\mathbf{x_0}$ and $\mathbf{b}$ are real 3-vectors. One introduces a complex Coulomb potential $\Phi = q/\sqrt{(\mathbf{z} - \mathbf{z_0})^2}$. Φ is holomorphic in each spatial coordinate and double-sheeted. It follows that $\nabla^2\Phi = 0$ on the real hyperspace so long as $\mathbf{b} \neq 0$. The prescription for making physical sense out of this complex displacement is to take the Riemann-Silberstein vector as given by the analytic continuation of the electric field to complex values. Amazingly, this recipe yields the correct electromagnetic field of the Kerr-Newman solution $\mathbf{W} = -\nabla\Phi$. As there is no charge on the real hyperspace, this leads to the so-called "sourceless" Kerr-Newman solution. The covariant version of this formula starts with a calculation of the analytic continuation of the real Faraday tensor to its value at the complex source point $\mathbf{z}_0$, $F^{\mu\nu} = \partial^{[\mu}\Phi\delta^{0\nu]}$, where δ here denotes the Kronecker delta. This tensor will not be anti-self-dual automatically. In order to calculate the tensor W we must project out the anti self-dual part. Then the real part will give the physical Faraday tensor. The metric tensor is found by solving the equations of general relativity by assuming that the metric can be expressed in Kerr-Schild coordinates in the form $g_{\mu\nu} = \eta_{\mu\nu} - 2Hk_\mu k_\nu$ where k is a null vector field, H a scalar field, and η is the Minkowski metric. Introducing a discontinuity on the Riemann cut leads to charges as sources for the fields [23]. When considering values for charge, mass, and angular momentum (q, m, and j) of elementary particles, the gravitational metric is well-approximated by a Minkowski metric except very near to the ring singularity [2]. They include the case where the mass is negative. The total electromagnetic energy stored in the static electric and magnetic fields of a Kerr-Newman solution is infinite [4]. Despite this, the mass is finite, and is arbitrary and unrelated to the electromagnetic energy. Consequently, we cannot interpret these solutions as purely electromagnetic particles. The Kerr-Newman solutions have event horizons whenever the mass is sufficiently large such that (in Planck units) $m^2 > (j/m)^2 + q^2$ [2, 3]. For the case of elementary particles, event horizons would not be present, and consequently the ring singularity would be naked.

3. The Lorentz-Dirac equation (LDE) and Runaway solutions

The LDE is [24, 25, 26, 27, 28] $m_0 a^\mu = F^\mu_{ext} + m_0\tau_0\left(\dot{a}^\mu + a^\lambda a_\lambda v^\mu\right)$ (metric signature is (+,-,-,-)). The notation and conventions are the same as [15] where F^μ_{ext} is any external force and $\tau_0 = \frac{2q_0^2}{3m_0c^3}$. The vast literature on this equation is reasonably in agreement on two points. First, if we consider exactly point particles then the general consensus is that the LDE is correct in the sense that straightforward electromagnetic analysis leads to it. Second, this equation has one of two possible unphysical properties - either it has runaway solutions, or it has preacceleration. There is a tendency to replace it in practical calculations by an approximate equation developed first by Landau and Lifshitz [29, 30] which is suitable for non-point-like quasi-particles and free of pathologies. Since the Kerr-Newman solutions are predicated on the existence of an exactly point-like singularity in complex space-time, we are forced to consider the LDE and not the quasi-particle equation for describing them. The term $m_0\tau_0\dot{a}$ is called the Schott term. Its origin can be understood from the following equation for the total momentum of the charged particle including electromagnetic momentum [31]. $p^\mu = mv^\mu - \frac{2}{3}e^2a^\mu$. The term $m_0\tau_0 a^2 v$ is the force caused by emitted radiation. In the method developed in this paper, the runaway solutions lead to interesting behaviour, and therefore we study them in detail. The most famous runaway solution is [27] for motion in the 0-3 plane, $v^\mu = (cosh(Ae^{\tau/\tau_0} + B),\ 0,\ 0,\ sinh(Ae^{\tau/\tau_0} + B))^\mu$, where A and B are constants. Call these type I runaway solutions. Here the focus will be on another type of runaway solution. Look for a solution which has vanishing acceleration squared in the absence of force, $a^\lambda a_\lambda = 0$, then the LDE becomes, $a^\mu = \tau_0\dot{a}^\mu$. The solution is therefore

$$a^\mu(\tau) = e^{\tau/\tau_0}a^\mu(0),\ a^\lambda(0)a_\lambda(0) = 0 \tag{1}$$

$$v^\mu(\tau) = \tau_0 e^{\tau/\tau_0}a^\mu(0) + u^\mu,\ z^\mu(\tau) = (\tau_0)^2\, e^{\tau/\tau_0}a^\mu(0) + u^\mu\tau + \Omega^\mu \tag{2}$$

where Ω^u and u^μ are constant 4-vectors. In complex space-time, there are new solutions of this type, and these shall be the focus of this paper. These solutions shall be referred to as type II. Note that the Laplace transform in τ of a type II solution has a simple pole whereas the type

I solutions have an infinite number of poles. So the type II solutions are considerably simpler. Runaway solutions have been dismissed as unphysical for over 100 years and they have posed a profound challenge for classical electromagnetism. Note that both type I and II solutions are entire functions of τ if it is considered as a complex variable. They are also entire functions of the constants of the motion $a^\mu(0)$, Ω^μ, and u^μ.

4. Analytic continuation of free particle equation and solutions

We let all variables become complex. Considering only τ, the position of the particle becomes not a curve but rather a two dimensional surface embedded in complex Minkowski space or CM^4. This is because, as a complex variable, τ has both a real and imaginary part. We shall require that the particle moves along a curve in complex space-time parametrized by some real variable s so that the curve of the particle in complex Minkowski space is given by $z^\mu(\tau(s))$. The analytic continuation of the LDE and its solutions to complex τ are uniquely determined by their real hyperspace forms. We could interpret this as meaning that any analytic function $\tau(s)$ yields a possible solution for the charged particle's motion due to self-forces while moving in complex space-time. Despite this non-uniqueness, one particular analytic continuation will be studied in detail here because of its similarity to Zitterbewegung. Starting with all real values for the various initial parameters, set $\tau(s) = s$ for s real and ranging from $-\infty$ to $+\infty$ and then analytically continue this curve to the curve $\tau(s) = is$ along the imaginary τ axis. An analytic transformation which accomplishes this is the following continuous function as λ vary from 0 to 1. $\tau(\lambda, s) = s + (i-1)\lambda s$, $\lambda \in [0,1]$. After the continuation in τ is performed, the resulting real part of each coordinate is interpreted as the physical position of the particle in physical (ie. ordinary 4D) Minkowski space. Additional analytic continuations of the initial values Ω^μ, u^μ, $a(0)^\mu$ will also be made in order to obtain a physically sensible solution. The result is $\tau = is$ and consequently the LDE becomes

$$-m_0\frac{d^2z^\mu}{ds^2} = F^\mu_{ext} + im_0\tau_0\left(\frac{d^3z^\mu}{ds^3} - \left(\frac{d^2z}{ds^2}\right)^2\frac{dz^\mu}{ds}\right) \tag{3}$$

We wish to interpret s as a new scaled proper time for the motion of the particle in the real hyperspace. Notice that then the inertial mass has the wrong sign. The Kerr-Newman metric has both positive and negative mass solutions though, so we therefore choose m_0 to be negative. We have then, in the absence of external force and defining τ_0 positive

$$m\frac{d^2z^\mu}{ds^2} = im\tau_0\left(\frac{d^3z^\mu}{ds^3} - \left(\frac{d^2z}{ds^2}\right)^2\frac{dz^\mu}{ds}\right), \quad m = -m_0 > 0 \tag{4}$$

Considering the type II runaway solutions we have

$$a_s^\mu(s) = d^2z^\mu/ds^2 = e^{-is/\tau_0}a_s^\mu(0), \quad a_s^\lambda(0)a_{s\lambda}(0) = 0 \tag{5}$$

$$v_s^\mu(s) = dz^\mu/ds = i\tau_0 e^{-is/\tau_0}a_s^\mu(0) + V^\mu \tag{6}$$

$$z_s^\mu(s) = -\tau_0^2 e^{-is/\tau_0}a_s^\mu(0) + V^\mu s + \Omega^\mu \tag{7}$$

where s varies from $-\infty$ to ∞, and V and Ω are constant 4-vectors, with V real and time-like.

5. Circular rotating solution for free particles

Consider analytic continuation of the constant 4-vector $a_s(0)$ to null vectors of the form $a_\pm(0) = \delta_a(0, 1, \pm i, 0)$ which have opposite chirality. Although δ_a can be a complex constant, by choice of the zero point for s we can absorb its phase, and so we take it to be real and positive. We let Ω^μ have a small imaginary part to give it the properties of a spinning particle as dictated by the Kerr-Newman static solution $\Omega^\mu = x_0^\mu + ib^\mu$ where b^μ is real and space-like, and in the "average rest frame" where $V^\mu = (\gamma, 0, 0, 0)$, $b^\mu = (0, \overrightarrow{b})$, and γ is a real constant. In this example we have chosen to take $\vec{b}$ along the 3-direction to maintain symmetry. It might be oriented in other directions too, but these shall not be considered here. The magnitude of $\vec{b}$ is the angular momentum divided by mass for the Kerr-Newman particle. For an electron, it would be $\hbar/2m_e$. We shall allow values of $\gamma \neq 1$ as this allows us to interpret the real part of z_s as a true particle trajectory with proper time given by s. Note that $v_s \cdot v_s = \gamma^2 \neq 1$, so this is a departure from ordinary classical mechanics. However, note also that with this complex acceleration, there is no rest frame for the particle which can be reached from the Laboratory frame with a real-valued Lorentz transformation. Writing out the real part (with δ_a real and positive)

$$Re(z_\pm^\mu) = -\tau_0^2\delta_a(0,\, cos(s/\tau_0),\, \pm sin(s/\tau_0),\, 0)^\mu + V^\mu s + x_0^\mu \tag{8}$$

and we have for the proper velocity of the real position

$$\frac{d}{ds}Re(z_\pm^\mu) = -\tau_0\delta_a(0,\, -sin(s/\tau_0),\, \pm cos(s/\tau_0),\, 0)^\mu + V^\mu \tag{9}$$

and this must satisfy $\frac{d}{ds}Re(z_\pm^\mu)\frac{d}{ds}Re(z_{\pm\mu}) = \gamma^2 - (\tau_0\delta_a)^2 = 1$ which is required if this is to be interpreted as a real time-like particle, so that s is the proper time for the real projection of the particles motion. Therefore we must have $|\delta_a| = \sqrt{\gamma^2 - 1}/\tau_0$. It also follows for these solutions that

$$\left(\frac{d}{ds}z_\pm^\mu\right)a_{\pm\mu} = 0,\ \left(\frac{d}{ds}Re(z_\pm^\mu)\right)Re(a_{\pm\mu}) = 0,\ and\ (dz/ds)^2 = V^\mu V_\mu = \gamma^2 \tag{10}$$

In order to fix γ, we identify the oscillation with Zitterbewegung as described by the Dirac equation [32, 18, 20, 33, 19] which satisfies $\omega_z = 2m_{obs}/\hbar$ where m_{obs} is the positive physical mass of the observed particle, and the one to be used in the Dirac equation. Laboratory time is given by $t = Re\ z^0 = V^0 s = \gamma s$. Expressing the oscillation in terms of this time, equating it with the Zitterbewegung angular frequency, and considering that special relativity requires that the mechanical energy of the rotating point mass be $m_{obs} = \gamma m$ yields $\gamma^2 = \frac{3\hbar}{4q^2}$ as shown in [15]. Introducing the fine structure constant $\alpha = \frac{e^2}{\hbar}$ we have for a particle with charge e, $\gamma = \sqrt{\frac{3}{4\alpha}} \approx 10.1379$. We can interpret the solution as a particle with an internal clock which has twice the frequency as mandated by quantum mechanics - the de Broglie clock [34] which is the same as the result for the Dirac equation. In the average rest frame $Re\ z^\mu$ describes a circular orbit of radius $\tau_0^2\delta_a$ and angular frequency ω_z. This is the projection onto the real hyperspace that is interpreted as the actual motion of the particle in physical space-time. The laboratory speed, determined by the real part of z is $\tau_0^2 \mid \delta_a \mid \omega$. If we accept the usual arguments regarding Zitterbewegung, then this would be the speed of light. This is only approximately true in our case. We find rather $\delta_a = \sqrt{\gamma^2 - 1}/\tau_0$ so that the radius of oscillation is then approximately half of the (reduced) Compton wavelength $\lambda_C = 2\gamma\tau_0$ and $r = \frac{\lambda_C\sqrt{\gamma^2-1}}{2\gamma}$, so that the speed is given by $r\omega_z = \sqrt{\gamma^2 - 1}/\gamma \approx .995c$. In most models for Zitterbewegung the speed is taken to be exactly c and the radius is given by [18, 33, 19] $r_S = \frac{\hbar}{2m_{obs}} = \gamma\tau_0$. This value, originally due to Schroedinger, agrees with our radius to better than 1%. The angular momentum due to this circular motion alone may be calculated to be $L = \frac{\hbar}{2}\frac{\gamma^2-1}{\gamma^2}$ which is slightly smaller than the spin of the electron by about 1%. The electromagnetic fields' angular momentum might make up the deficit. Experimental proof of Zitterbewegung comes from the Darwin term [35] in hyperfine splitting of atomic lines in spectroscopy, and more directly by observing resonant behaviour in electron channeling [36].

6. Liénard-Wiechert potentials and the fields

The Liénard-Wiechert potentials provide a procedure for calculating the electromagnetic fields produced by a particle. In ordinary electromagnetism they are [25] $A^\mu(x) = \frac{qv^\mu(\tau)}{v(\tau)\cdot(x-z(\tau))}\ |_{\tau=\tau_r}$. Note that $\frac{dz^\mu(\tau)/d\tau}{dz(\tau)/d\tau\cdot(x-z(\tau))} = \frac{dz^\mu(\tau(s))/ds}{dz(\tau(s))/ds\cdot(x-z(\tau(s)))}$, and where τ_r is the proper time which satisfies the null root equation $(x-z(\tau))\cdot(x-z(\tau)) = 0$ or $(x-z(\tau(s)))\cdot(x-z(\tau(s))) = 0$. After analytic continuation the coordinates are expressed in terms of the new variable s and the complex valued potential A is used to calculate a complex Faraday tensor F and its dual $*F$ from which the anti self-dual tensor W can be calculated from. If the particle trajectory is time-like in the (real valued) case of standard electromagnetic theory, then there are exactly two solutions to the null root equation, one in the past (retarded time) and one in the future (advanced time). When we analytically continue these equations into the complex plane, the null root equation becomes for complex z and s and real x, $(x-z(s))^\mu(x-z(s))_\mu = 0$. This condition in the complex case is fundamentally different since both the real and imaginary parts must vanish. We are considering here the situation where z is parametrized along a curve by a real parameter s, and the null root equation will quite likely have no solutions at all for real s. In general, the null root equation will have solutions which are not on the path of the particle. Nevertheless, these solutions are obtained by analytic continuation from ordinary solutions, and so they are valid ones. This is an important - indeed profound - difference as compared to the usual case. The set of complex solutions $\{s_i\}$ can be quite large, even infinite. Some subset will be causal, but even this subset can be infinite. Why shouldn't we include a superposition of several different roots at the same time? The solutions for the electromagnetic field are not unique and may be characterized by the number of roots included in the solution, by the particular values of these roots, and by their respective weights in the sum. For single-root solutions we have, letting $v^\mu = dz^\mu/ds$, $A^\mu(x) = \frac{qv_s^\mu(s)}{v_s(s)\cdot(x-z(s))}$, $s \in \{s_i\}$. If there are more than one retarded solution, then there is no reason to exclude multiple root solutions. For an N root solution we have $A^\mu(x) = \sum_{j=1}^N w_j \frac{qv_s^\mu(s_j)}{v_s(s_j)\cdot(x-z(s_j))}$, $s_j \in \{s_i\}$, $\sum_{j=1}^N w_j = 1$. These 4-potentials are in general complex. The rule for calculating the complex Faraday tensor from them is $W^{\mu\nu} = \partial^{[\mu}A^{\nu]} + i * \partial^{[\mu}A^{\nu]}$, and the physical Faraday tensor on the real hyperspace x is the real part of this $F_{physical} = Re\, W$. The weight factors w_j are arbitrary except that we would expect that they sum to one, and we may want to include only causal retarded times in the sum. For the time being we shall assume that these weighting factors are constants, but later on we shall consider the possibility that they are dependent on position. These multi-root solutions would look to an observer capable of measuring the electromagnetic fields produced by them like a particle made up of multiple constituents, but they are really analytic continuations associated with a single particle trajectory. So we see we have a whole family of possible solutions. Already in the literature there is evidence for this non-uniqueness. It is exploited in Wheeler-Feynman electrodynamics [37] by allowing a sum of advanced and retarded potentials to give solutions to Maxwell's equation, although this is a fairly trivial case in comparison. The author has wondered whether these multi-root solutions might be related to the quark model for hadrons. They could look like a particle made up of multiple constituent point particles if probed electromagnetically. But these "particles" might be hard to separate because they are related to a single particle trajectory in the complex space, with their positions linked together by the requirement that they must lie on different Riemann sheets of a single root equation, perhaps providing a mechanism for quark confinement. We can think of this analytic continuation by imagining we start with real-space Liénard-Wiechert potential for a single root, and write it in the following way: $A^\mu(x) = \sum_j w_j \frac{qv^\mu(\tau_r)}{v(\tau_r)\cdot(x-z(\tau_r))}$ (We haven't done anything yet, since $\sum w_j = 1$). But now we analytically continue each term in the sum to a different Riemann sheet, and thus obtain a different value for the retarded time for each term in the sum. This is a generalization of the usual analytic continuation for analytic functions in mathematics. The result is a field which is produced by complex source point charges at a countable set of different complex valued retarded times. The maximum number of source points is the number of Riemann sheets of the analytic function which describes the coordinates for the particle.

7. Calculation of retarded proper times for the type II runaway solutions

The Null root equation that must be satisfied is, for a real field point x^μ, $\left(x + \tau_0^2 e^{-is/\tau_0} a_s(0) - Vs - \Omega\right)^2 = 0$. The solution set $\{s_j\}$ will be implicit functions of x, $a_s(0)$, u, and Ω, and in general will not be real and consequently $z(s_j)$ will not lie on the particle's actual trajectory in the complex space. This is a consequence of relying on analytic continuation to guide us into the complex space. One might be tempted to reject this approach because of this development. We shall adopt a more liberal exploratory philosophy and accept this possibility in order to learn the consequences it leads to. We may rewrite this equation in the form

$$(s - C_A)(s - C_R) = De^{-is/\tau_0} \tag{11}$$

$$C_A = \left(+2x^0\gamma + \sqrt{(2x^0\gamma)^2 - 4\gamma^2 R^2}\right)/2\gamma^2,\ C_R = \left(+2x^0\gamma - \sqrt{(2x^0\gamma)^2 - 4\gamma^2 R^2}\right)/2\gamma^2 \tag{12}$$

$$D_\pm = -2\tau_0^2 a_\pm(0)\cdot x/\gamma^2 = 2\tau_0^2(\delta_a r_\perp e^{\pm i\varphi})/\gamma^2 \tag{13}$$

$$R^2 = (x-\Omega)^2,\ \Omega = (0,0,0,ib) \tag{14}$$

It's not possible to solve (11) in closed form, unless one uses a proposed generalization of the Lambert function [38]. We can proceed numerically. Start with C_A and C_R set to zero, so that the equation then becomes: $s^2 = De^{-is/\tau_0}$. This can be solved in terms of Lambert functions [39].

$$s = -2i\tau_0 W_n(\pm i\sqrt{D}/(2\tau_0)),\ n \in \mathbb{Z} \tag{15}$$

Where W_n is the nth branch of the Lambert function. There are an infinite number of Riemann sheets, and an infinite number of roots. Starting with this solution, we can analytically continue (11) to the correct values for C_A and C_R using numerical means. Thus we can conclude that there are likely to be an infinite number of solutions, unless the analytic continuation results in infinitely many of these different starting approximations coalescing to the same value. Numerical simulation of this analytic continuation suggests that this does not happen, and they indicate the existence of an infinite number of solutions. These solutions correspond to different Riemann sheets of the solution s when viewed as a function of x.

7.1. Calculation of the roots for the asymptotic field limit to study radiation
We assume that $|\mathbf{z}| \ll |\mathbf{x}|$ and therefore we make the standard far-field approximation with x^μ real

$$(x - z(s))^2 = 0 \approx \left(t - z^0\right)^2 - (\mathbf{x^2} - 2\mathbf{x} \cdot \mathbf{z}),\ \mathbf{x} = r\hat{\mathbf{r}} \tag{16}$$

$$t - z^0 = t - \gamma s \approx \pm(r - \hat{\mathbf{r}} \cdot \mathbf{z}) = \pm\left(r + \tau_0^2(\mathbf{a}_s(0) \cdot \mathbf{x})e^{-is/\tau_0}/r - \hat{\mathbf{r}} \cdot \mathbf{\Omega}\right) \tag{17}$$

For the retarded solution $s\uparrow$we choose the + sign and obtain

$$\gamma s\uparrow - (t - r + \hat{\mathbf{r}} \cdot \mathbf{\Omega}) + \tau_0^2(\mathbf{a_s}(0) \cdot \hat{\mathbf{r}})e^{-is\uparrow/\tau_0} = 0 \tag{18}$$

$$A \equiv (t - r + \hat{\mathbf{r}} \cdot \mathbf{\Omega}),\ B \equiv \tau_0^2(\mathbf{a_s}(0) \cdot \hat{\mathbf{r}}),\ \gamma s\uparrow - A + Be^{-is\uparrow/\tau_0} = 0 \tag{19}$$

This equation can be solved for $s\uparrow$ again in terms of Lambert functions

$$s\uparrow = A/\gamma - i\tau_0 W_n\left(-\frac{iBe^{-iA/\gamma\tau_0}}{\gamma\tau_0}\right),\ n \in \mathbb{Z} \tag{20}$$

$$s\uparrow_n = (t - r + \hat{\mathbf{r}} \cdot \mathbf{\Omega})/\gamma - i\tau_0 W_n\left(-\frac{i\tau_0^2(\mathbf{a_s}(0) \cdot \hat{\mathbf{r}})e^{-i(t-r+\hat{\mathbf{r}}\cdot\mathbf{\Omega})/\gamma\tau_0}}{\gamma\tau_0}\right) \tag{21}$$

There are an infinite number of solutions as n is an arbitrary integer, and there are advanced solutions as well [15]. These null source points all appear to radiate [15], and so does any finite superposition of them. We view it as unlikely that even an infinite number of roots with fixed weighting factors could suppress radiation.

8. Suppressing or Cloaking radiation with weighting factors which are functions of x

In order to suppress the radiation, we consider allowing the weighting factors w_i to be functions of the field point x. So long as these functions are analytic (or at least holomorphic) in the coordinates, the resulting fields will be too. It is convenient write $A^\mu(x) = A^\mu_{Coul}(x) + A^\mu_{Rad}(x)$

$$A^\mu_{Coul}(x) = \sum_{j=1}^{N} w_j(x)\frac{q\gamma\delta_0^\mu}{v(s_j(x)) \cdot (x - z(s_j(x)))},\ \sum_{j=1}^{N} w_j(x) = 1 \tag{22}$$

$$A^\mu_{Rad}(x) = \sum_{j=1}^{N} w_j(x)\frac{qi\tau_0 e^{-is_j(x)/\tau_0}a^\mu_\pm(0)}{v(s_j(x)) \cdot (x - z(s_j(x)))} \tag{23}$$

and where $s_j(x)$ are the proper-time null roots. The potential A_{Coul} determines the Coulomb field and the potential A_{Rad} determines the radiation field. In order to calculate the radiation field, we must perform the following steps following our procedure $W^{\mu\nu}(x) = \partial^{[\mu}A^{\nu]} + i(*\partial^{[\mu}A^{\nu]})$, where the real part of W gives the physical Faraday tensor. Note that $*\partial^{[\mu}A^{\nu]} = i\partial^{[\mu}A^{\nu]} \Rightarrow W^{\mu\nu}(x) = 0$. We have for large r, $\partial^{[\mu}A^{\nu]} = \partial^{[\mu}A^{\nu]}_{Rad} + o(\frac{1}{r})$. If the A_{Rad} happens to be self-dual, then there will be no radiation. But the $w_j(x)$ are now assumed to be arbitrary functions, and we can choose them such that this is the case. The following potential functions are self dual in this sense $A_{SD}(x) = f_\pm(x^3 \mp x^0, x^1 \pm ix^2)a_\pm(0)$ where $f_\pm$ are arbitrary holomorphic functions of their arguments. So to cloak radiation for + or - chirality, we must choose the weighting functions to satisfy

$$\sum_{j=1}^{N} w_j(x) \frac{q i \tau_0 e^{-i s_j(x)/\tau_0}}{v(s_j(x)) \cdot (x - z(s_j(x)))} = f_\pm(x^3 \mp x^0, x^1 \pm i x^2); \quad \sum_{j=1}^{N} w_j(x) = 1 \tag{24}$$

$f_\pm = 0$ is an allowed case. Since two equations must be satisfied, at least two roots must be included in the solution. Denote two such roots by index 1 and 2. The equations can be solved simply with the result

$$\begin{pmatrix} w_1 \\ w_2 \end{pmatrix} = \frac{1}{C_1 - C_2} \begin{pmatrix} f_\pm - C_2 \\ -f_\pm + C_1 \end{pmatrix} \tag{25}$$

where the two C values in this formula are the functions multiplying the respective weight functions w in (24). To this solution might be added by superposition any combination of the weighting functions which satisfy the associated homogeneous equations. Thus, there are a large number of non-radiating solutions. When the cloaking conditions are satisfied, we can write $W^{\mu\nu} = \partial^{[\mu} A^{\nu]}_{Coul} + i(*\partial^{[\mu} A^{\nu]}_{Coul})$ since the radiation term vanishes. The asymptotic electric field will still vary as $q\hat{\mathbf{r}}/\mathbf{r}^2$, so the total charge is unchanged, but the charge near to the particle's origin will appear to be non zero and smeared out as viewed by a remote observer. One question that naturally arises is what is the state of minimum electrostatic energy? This will depend on the number of roots included in the sum, assuming that there are some solutions for which the energy is not infinite. Relevant to the question of emergence, the quantum mechanical wave equations for free particles are closely related to non-radiating electromagnetic sources [40]. The assumption of x dependent weighting factors thus seems to be a requirement if this theory is to be taken seriously. Without it, we could not avoid radiation.

9. Motion with weak external forces applied

In an external electromagnetic field, we take the equation of motion to be

$$m \frac{d^2 z^\mu}{ds^2} = q F^{\mu\nu}_{ext} \frac{dz_v}{ds} + i m \tau_0 \left(\frac{d^3 z^\mu}{ds^3} - \left(\frac{d^2 z}{ds^2} \right)^2 \frac{dz^\mu}{ds} \right) \tag{26}$$

The external Faraday tensor F_{ext} here is real on the real hyperspace, but is generally complex in the rest of the complex space-time by analytic continuation. We assume that the fields are weak and slowly varying on the time scale of τ_0. This reduces to simply the Lorentz equation if we ignore the Schott and radiation terms, which now have a factor of i in front of them. Consider the simplified case where we ignore the radiation term. The equation then becomes

$$m \frac{d^2 z^\mu}{ds^2} = q F^\mu_\nu \frac{dz^\nu}{ds} + i m \tau_0 \frac{d^3 z^\mu}{ds^3} \tag{27}$$

In this formula, the mass m is the observed mass divided by γ, so $m = m_e/\gamma = m_e/\sqrt{\frac{3}{4\alpha}}$.

10. The guiding center approximation

Assume that the particle moves in a highly localized oscillation at very high frequency, and the external electromagnetic fields are only very slowly varying in time and space on the scale of this oscillation. Then the weak field acts to weakly perturb the oscillating frequency and to guide the overall motion [15]. We start with a solution to the ordinary Lorentz equation for a trajectory of a time-like charged particle (the 4-position L is real)

$$m \frac{d^2 L^\mu}{ds^2} = q F^\mu_\nu \frac{dL^\nu}{ds}, \quad \left(\frac{dL}{ds} \right)^2 = \gamma^2 \tag{28}$$

and then we add a complex correction term δ to it which represents the high frequency oscillation

$$m\frac{d^2(L+\delta)^\mu}{ds^2} = qF^\mu_\nu\frac{d(L+\delta)^\nu}{ds} + im\tau_0\frac{d^3(L+\delta)^\mu}{ds^3} \tag{29}$$

Note that s is the proper time scaled by γ for the guiding center, and that the observed mass is γm in order to compensate for this. We make the following approximations: $F(z) \approx F(L)$, $d^3(L+\delta)^\mu/ds^3 \approx d^3\delta^\mu/ds^3$, and $|qF^{\mu v}/m| \ll 1/\tau_0$ to obtain $m\frac{d^2\delta^\mu}{ds^2} = im\tau_0\frac{d^3\delta^\mu}{ds^3}$. In general, these solutions will radiate energy. However, this equation will have multiple roots as did the Zitterbewegung solution, and so by using again the cloaking mechanism of position dependent weighting factors, the radiation rate should have a minimum and possibly be zero, and one would expect this to be a natural asymptotic state for the system after transients have died away.

10.1. Motion in constant fields

Look for solutions of the form $\delta = \delta(0)e^{i\omega s}$, so that $-m\omega^2\delta^\mu(0) - i\omega qF^\mu_\nu\delta^\nu(0) - m\tau_0\omega^3\delta^\mu(0) = 0$. There are eight roots for ω [15]. The most general solution can be written $\delta^\mu(s) = \sum_{k=1}^8 \delta^\mu_k(0)e^{i\omega_k s}$. The δ^μ_k must be constructed to solve the four homogeneous equations.

10.2. Motion in a 3D harmonic oscillator central force

$$m\frac{d^2z^j}{ds^2} = -kz^j + im\tau_0\left(\frac{d^3z^j}{ds^3} - \left(\frac{d^2z}{ds^2}\right)^2\frac{dz^j}{ds}\right),\ j = 1,2,3 \tag{30}$$

There will be guided-center approximate solutions to this equation, and also exact solutions with $a \cdot a = 0$ as follows. If the radiation term vanishes, then the equation has solutions of the form $z^j(s) = z^j(0)e^{i\omega s}$ with $\mathbf{z(0)} \cdot \mathbf{z(0)} = 0$. We can take $\mathbf{z(0)}$ to be proportional to either of the null vectors $\mathbf{a}_\pm\mathbf{(0)}$ 3-vectors for example for some arbitrary spin axis. Then the frequency equation becomes $\tau_0\omega^3 + \omega^2 - k/m = 0$. There are three roots to this which are approximately $\omega = \pm\sqrt{k/m}, -1/\tau_0$. A class of exact solutions are given by a linear superposition of these $\mathbf{z(s)} = \mathbf{a}_+\mathbf{(0)}\sum_{j=1}^3 c_j e^{i\omega_j s}$ and $z^0(s) = \gamma s$ for arbitrary complex constants c_j. These solutions are more complicated than the single-frequency Zitterbewegung solutions. Nevertheless, we expect that they will certainly have multiple null roots for the radiation calculation. Therefore, the same cloaking technique that was proposed to eliminate radiation in the Zitterbewegung case can be applied in the present circumstance to reduce or eliminate the radiation from these harmonic oscillator solutions as well.

10.3. Motion in a Coulomb field

Consider motion in a spherically symmetric Coulomb field of charge Q. On the real hyperspace we have $\mathbf{E}_{\text{ext}} = \frac{Q}{r^3}\mathbf{r}$, $\mathbf{B}_{\text{ext}} = 0$

$$m\frac{d^2z^\mu}{ds^2} = qF^{\mu 0}_{ext}\frac{dz_0}{ds} + im\tau_0\left(\frac{d^3z^\mu}{ds^3} - \left(\frac{d^2z}{ds^2}\right)^2\frac{dz^\mu}{ds}\right) \tag{31}$$

There will be guiding-center approximations to this equation, based on solutions to the Kepler problem. The simplest would be a circle. An exact analytic solution has not yet been found. There will undoubtedly be an infinite number of Riemann sheets for the retarded time function, and the radiation from these can be cloaked to the level of accuracy of the guiding-center approximation. Whether it is possible, using the cloaking method of x dependent weighting functions, to suppress all radiation in some cases, as would be required for the ground state of the Hydrogen atom for example, is not clear at the present time, but it is a good possibility.

11. Interpretation and Conclusion

The picture that emerges from this model is of a free particle which has a peculiar internal oscillation, and whose electromagnetic fields are not unique, being dependent on a set of complex weighting functions. It eliminates runaway solutions and replaces them with oscillating solutions that look similar to Zitterbewegung and give a model for the de Broglie clock . They allow a cloaking mechanism for radiationless accelerated motion, both in the self-oscillation case, the central harmonic force case, and possibly the Coulomb force case. There might be solutions which do not have any singularities on the real physical hyperspace (ie. the physical Riemann sheet), and some of these might have finite electromagnetic energy. The non-uniqueness of the fields for a particle have a vague similarity to the wave-particle duality of quantum mechanics. The particle would appear to an observer using an electromagnetic probe as possibly made up of a number of point-like constituents. In order to inhibit radiation, a new cloaking mechanism was invoked that relied on position dependent weighting functions for the different retarded times corresponding to different Riemann sheets superimposed. Some other possibilities and questions then naturally arise. For example, can the weighting functions change their values suddenly as the result of an observation being made? Thus giving a substantive explanation for wave-function collapse. Can different observers "see" different weighting functions for the same particle? Can we understand complementarity from this? What is a photon in this framework? How does the Riemann structure change when more than one particle is considered? What are the lowest energy states possible, assuming that there are solutions with finite electrostatic energy? What are the implications of this theory for non-abelian gauge theories? Could the multi-root solutions be related to the quark model for hadrons? Emergent quantum mechanics from Planck scale physics is a bottom-up approach whereas the model discussed here is more of a top-down one. It might thus be a useful stepping stone in the theoretical voyage from the Planck scale to the quantum atomic realm.

References

[1] Burinskii A 2007 *arxiv.org 0712.0577*
[2] Burinskii A 2008 *Gravitation and Cosmology* **14** 109–122
[3] Carter B 1968 *Phys. Rev.* **174** 1559
[4] Lynden-Bell D 2003 *Stellar Astrophysical Fluid Dynamics* p 369–375
[5] Newman E T 2002 *Phys. Rev. D* **65** 104005
[6] Pekeris C L and Frankowski K 1987 *Phys. Rev. A* **36** 5118
[7] Adamo T M, Kozameh C and Newman E T 2009 *Living Reviews in Relativity (www.livingreviews.org/lrr-2009-6)* **12**
[8] Newman E T 1973 *J. Math. Phys.* **14** 774
[9] Newman E T 1976 *Gen. Relativ. Gravit.* **7** 107–111
[10] Adler S L 2004 *Quantum theory as an emergent phenomenon* (Cambridge University Press)
[11] Carroll R 2004 *gr-qc/0406004*
[12] Markopoulou F and Smolin L 2004 *Phys. Rev. D* **70** 124029
[13] 't Hooft G 2000 *arxiv.org hep-th/0003005*
[14] 't Hooft G 2001 *arxiv.org hep-th/0104219*
[15] Davidson M 2011 *arXiv:1109.4923*
[16] Davidson M 2008 *arxiv.org 0807.1990*
[17] Davidson M 2010 *Physica E* **42** 317–322
[18] Schroedinger E 1930 *Sitzungsberichte der Preussischen Akademie der Wissenschaften* **24** 418–428
[19] Hestenes D 1990 *Found. Phys.* **20** 1213–1232
[20] Huang K 1952 *Am. J. Phys.* **20** 479–484
[21] Newman E T, Couch E, Chinnapared K, Exton A, Prakash A and Torrence R 1965 *J. Math. Phys.* **6** 918
[22] Newman E T 1973 *J. Math. Phys.* **14** 102
[23] Lynden-Bell D 2004 *Phys. Rev. D* **70** 105017
[24] Dirac P A M 1938 *P. Roy. Soc. Lond. A Mat.* **167** 148–169
[25] Jackson J D 1999 *Classical electrodynamics, 3rd Ed* (Wiley)
[26] Plass G N 1961 *Rev. Mod. Phys.* **33** 37

[27] Rohrlich F 2007 *Classical charged particles* (World Scientific)
[28] Teitelboim C, Villarroel D and Weert C G 1980 *La Rivista del Nuovo Cimento* **3** 1–64
[29] Landau L D and Lifshitz E M 1951 *The classical theory of fields* (Cambridge, MA: Addison-Wesley)
[30] Jackson J D 2007 *Am. J. Phys.* **75** 844
[31] Teitelboim C 1970 *Phys. Rev. D* **1** 1572
[32] Dirac P A M 1978 *The principles of quantum mechanics* (Clarendon Press)
[33] Barut A O and Bracken A J 1981 *Phys. Rev. D* **23** 2454
[34] de Broglie L 1924 *Researches sur la théorie des quanta, P.H. D. Thesis* Ph.D. thesis
[35] Bjorken J D and Drell S D 1998 *Relativistic Quantum Mechanics* 1st ed (McGraw-Hill)
[36] Guoanere M, Spighel M, Cue N, Gaillard M J, Genre R, Kirsch R, Poizat J C, Remillieux J, Catillon P and Roussel L 2008 *Annales de la Fondation Louis de Broglie* **33** 85–91
[37] Wheeler J A and Feynman R P 1945 *Rev. Mod. Phys.* **17** 157
[38] Scott T C, Mann R and Martinez II R E 2006 *Appl. Algebr. Eng. Comm.* **17** 41–47
[39] Corless R M, Gonnet G H, Hare D E G, Jerey D J and Knuth D E 1996 *Adv. Comput. Math.* **5** 329—359
[40] Davidson M P 2007 *Ann. Phys.* **322** 2195–2210

EmerQuM 11: Emergent Quantum Mechanics 2011 IOP Publishing
Journal of Physics: Conference Series **361** (2012) 012006 doi:10.1088/1742-6596/361/1/012006

Experimental tests of Quantum Mechanics: from Pauli Exclusion Principle Violation to spontaneous collapse models

C. Curceanu (Petrascu)[1,4], S. Bartalucci[1], A. Bassi[2], S. Bertolucci[3], M. Bragadireanu[1,4], M. Cargnelli[5], A. Clozza[1], S. Di Matteo[6], S. Donadi[2], J-P. Egger[7], C. Guaraldo[1], M. Iliescu[1], T. Ishiwatari[5], M. Laubenstein[8], J. Marton[5], E. Milotti[2], D. Pietreanu[1,4], M. Poli Lener[1,4],T. Ponta[4], A. Rizzo[1], A. Romero Vidal[1], A. Scordo[1], D.L. Sirghi[1,4], F. Sirghi[1,4], L. Sperandio[1], O. Vazquez Doce[1], E. Widmann[5] and J. Zmeskal[5]

[1] INFN, Laboratori Nazionali di Frascati, CP 13, Via E. Fermi 40, I-00044, Frascati (Roma), Italy
[2] Dipartimento di Fisica, Università di Trieste and INFN– Sezione di Trieste, Via Valerio, 2, I-34127 Trieste, Italy
[3] CERN, CH-1211, Geneva 23, Switzerland
[4] "Horia Hulubei" National Institute of Physics and Nuclear Engineering, Str. Atomistilor no. 407, P.O. Box MG-6, Bucharest - Magurele, Romania
[5] The Stefan Meyer Institute for Subatomic Physics, Boltzmanngasse 3, A-1090 Vienna, Austria
[6] Institut de Physique UMR CNRS-UR1 6251, Université de Rennes1, F-35042 Rennes, France
[7] Institut de Physique, Université de Neuchâtel, 1 rue A.-L. Breguet, CH-2000 Neuchâtel, Switzerland
[8] INFN, Laboratori Nazionali del Gran Sasso, S.S. 17/bis, I-67010 Assergi (AQ), Italy

E-mail: `catalina.curceanu@lnf.infn.it`

Abstract.
The Pauli exclusion principle (PEP) and, more generally, the spin-statistics connection, is at the very basis of our understanding of matter. The PEP spurs, presently, a lively debate on its possible limits, deeply rooted in the very foundations of Quantum Field Theory. Therefore, it is extremely important to test the limits of its validity. Quon theory provides a suitable mathematical framework of possible violation of PEP, where the q violation parameter translates into a probability of violating PEP. Experimentally, setting a bound on PEP violation means confining the q-parameter to a value very close to either 1 (for bosons) or -1 (for fermions). The VIP (Violation of the Pauli exclusion principle) experiment established a limit on the probability that PEP is violated by electrons, using the method of searching for PEP forbidden atomic transitions in copper. We describe the experimental method, the obtained results, both in terms of the q-parameter and as probability of PEP violation, we briefly discuss the results and present future plans to go beyond the actual limit by upgrading the experimental technique using vetoed new spectroscopic fast Silicon Drift Detectors. We mention as well the possibility of using a similar experimental technique to search for eventual X-rays generated as a signature of the spontaneous collapse of the wave function, predicted by continuous spontaneous localization type theories.

1. Introduction

The Pauli Exclusion Principle (PEP) plays a fundamental role in our understanding of many physical and chemical phenomena, from the periodic table of elements, to the electric conductivity in metals and to the degeneracy pressure (which makes white dwarfs and neutron stars stable). It is a consequence of the spin-statistics connection [1], and, as such, it is intimately connected to the basic axioms of quantum field theory [2]. Although the principle has been spectacularly confirmed by the number and accuracy of its predictions, its foundation lies deep in the structure of quantum theory and has defied all attempts to produce a simple proof, as nicely stressed by Feynman [3]. Given its basic standing in quantum theory, it is appropriate to carry out precise tests of the PEP validity and, indeed, mainly in the last 20 years, several experiments have been performed to search for possible small violations [4, 5, 6, 7, 8, 9, 10, 11, 12]. Often, these experiments were born as by-products of measurements with a different objective (e.g., dark matter searches, proton decay, etc.), and most of the recent limits on the validity of PEP have been obtained for nuclei or nucleons. Many (if not all) of these experiments are using methods which are not obeying the so-called Messiah-Greenberg superselection rule [13].

In 1988 Ramberg and Snow [14] performed a dedicated experiment, pioneering a technique avoiding the Messiah-Greenberg superselection rule by introducing new electrons in the system. They searched for anomalous X-ray transitions, that would point to a small violation of PEP in a copper conductor, circulating current into it. The result of the experiment was a probability [15] $\frac{\beta^2}{2}<1.7 \times 10^{-26}$ that the PEP is violated by electrons. The VIP Collaboration set up a much improved version of the Ramberg and Snow experiment, with a higher sensitivity apparatus [16]. Our final aim is to improve the PEP violation limit for electrons by 3-4 orders of magnitude, by using high resolution Charge-Coupled Devices (CCDs), as soft X-rays detectors [17, 18, 19, 20, 21], and decreasing the effect of background by a careful choice of the materials and sheltering the apparatus in the LNGS underground laboratory of the Italian Institute for Nuclear Physics (INFN).

In the next sections we describe the experimental setup and method, the results of a first measurement performed in the Frascati National Laboratories (LNF) of INFN, along with a the results obtained by VIP running at the underground Gran Sasso National Laboratory (LNGS) of INFN. We briefly discuss the results and present future plans to go beyond the present limit by using fast Silicon Drift Detectors (SDD) and a veto system.

We will then conclude the paper by presenting some ideas to use a similar experimental technique to perform measurements of X rays emitted by free electrons, predicted by the spontaneous collapse models.

2. The VIP experiment

VIP is a dedicated experiment for the measurement of the probability of the Pauli Exclusion Principle violation for electrons. The experiment uses the same methods of the Ramberg and Snow experiment (see below), with a much better X-ray detector and in a low-background experimental area - the INFN Gran Sasso underground laboratory. The detector is an array of Charge-Coupled Devices (CCDs), characterized by the excellent background rejection capability, based on pattern recognition and good energy resolution (320 eV FWHM at 8 keV in the present measurement).

2.1. *The VIP Experimental Method*

The experimental method, originally described in Ref. [14], consists in the introduction of "fresh" electrons into a copper strip, by circulating a current, and in the search for the X-rays resulting from the PEP forbidden radiative transitions that occur if one of these electrons is captured by a copper atom and cascades to a 1S state which is already filled by two electrons. In particular, we are looking for the $2P \to 1S$ (K_α)transition. The energy of this non-Paulian transition would

Figure 1. Two CCD detectors used by VIP together with the readout electronics

differ from the normal K_α transition energy by about 300 eV (7.729 keV instead of 8.040 keV) [22], providing an unambiguous signal of PEP violation. The measurement alternates periods without current in the copper strip, in order to evaluate the X-ray background in conditions where no PEP violating transitions are expected to occur, with periods in which current flows in the conductor, when we expect that the "fresh" electrons may lead to Pauli-forbidden transitions.

2.2. The VIP setup

The VIP setup consists of a copper cylinder, 45 mm radius, 50 μm thickness, and 88 mm height surrounded by 16 equally spaced "type 55" CCDs made by EEV [23]. The CCDs are at a distance of 23 mm from the copper cylinder, and paired one above the other (see fig. 1). The setup is enclosed in a vacuum chamber, and the CCDs are cooled to about 168 K by a cryogenic system. The current, provided by an external current supply, flows in the thin cylinder made of ultrapure (99.995%) copper foil from the bottom of the vacuum chamber. The CCDs surround the cylinder and are supported by cooling fingers which protrude from the cooling heads in the upper part of the chamber. The readout electronics is just behind the cooling fingers; the signals are sent to amplifiers on top of the chamber and the amplified signals are read out by ADC boards in the data acquisition computer. More details on CCD-55 performance, as well on the analysis method used to reject background events, can be found in reference [24, 25]. An overall schematic view of the setup is shown in fig. 2.

VIP improved orders of magnitude on the Ramberg and Snow measurement, thanks to the following features:

- use of CCD detectors instead of gaseous detectors, having much better energy resolution (4-5 times better) and higher stability;
- experimental setup located in the clean, low-background, environment of the underground LNGS Laboratory;
- collection of much higher statistics (longer DAQ periods).

We make full use of these features to obtain an improvement of several orders of magnitude on previous limits.

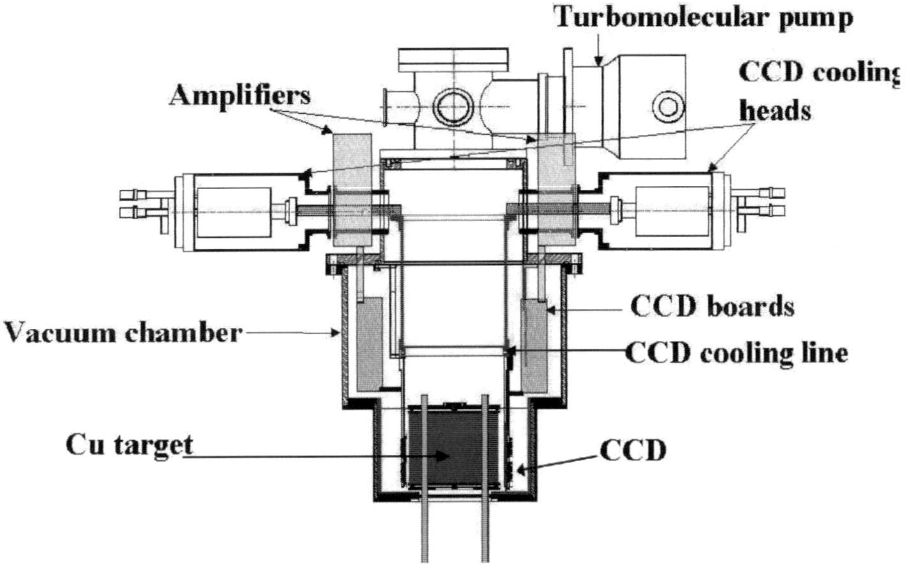

Figure 2. The VIP setup - schematic view

3. The VIP experimental results

The VIP setup was taking data in the low-background Gran Sasso underground laboratory of INFN from 2006 until 2010; presently data analysis is ongoing.

3.1. Results obtained at LNF-INFN

Before installation in the Gran Sasso laboratory, VIP was first prepared and tested at the LNF-INFN laboratory, where measurements were performed in the period 21 November - 13 December 2005. Two types of measurements were performed:

- 14510 minutes (about 10 days) of measurements with a 40 A current circulating in the copper target;
- 14510 minutes of measurements without current.

CCDs were read-out every 10 minutes. The resulting energy calibrated X-ray spectra are shown in figure 3.

These spectra include data from 14 CCD's out of 16, because of noise problems in the remaining 2. Both spectra, apart from the continuous background component, display clear Cu K_α and K_β lines due to X-ray fluorescence caused by the cosmic ray background and natural radioactivity. No other lines are present and this reflects the careful choice of the materials used in the setup, as for example the high purity copper and aluminium, the last one with K-complex transition energies below 2 keV. The subtracted spectrum is shown in Figure 4 a) (whole energy scale) and b) (a zoom on the region of interest). Notice that the subtracted spectrum fluctuates around zero within the statistical error, and is structureless. This not only yields an upper bound for a violation of the Pauli Exclusion Principle for electrons, but also confirms the correctness of the energy calibration procedure and points to the absence of systematic effects.

To extract the experimental limit on the probability that PEP is violated for electrons, $\beta^2/2$, from our data, we used the same arguments of Ramberg and Snow: see references [14] and [26] for details of the analysis. The obtained value is:

$$\frac{\beta^2}{2} < 4.5 \times 10^{-28} \tag{1}$$

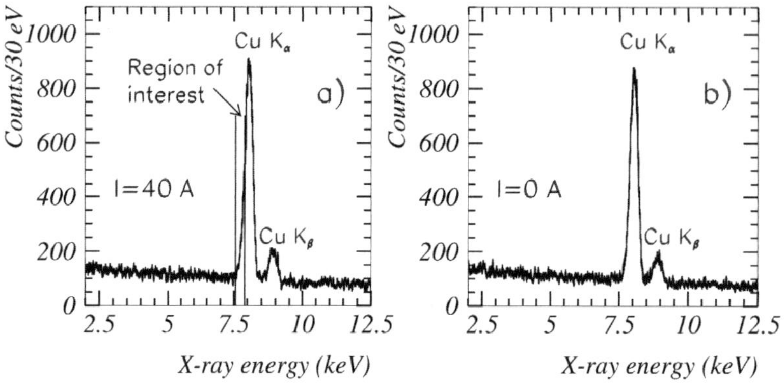

Figure 3. Energy spectra with the VIP setup at LNF-INFN: (a) with current (I = 40 A); (b) without current (I = 0).

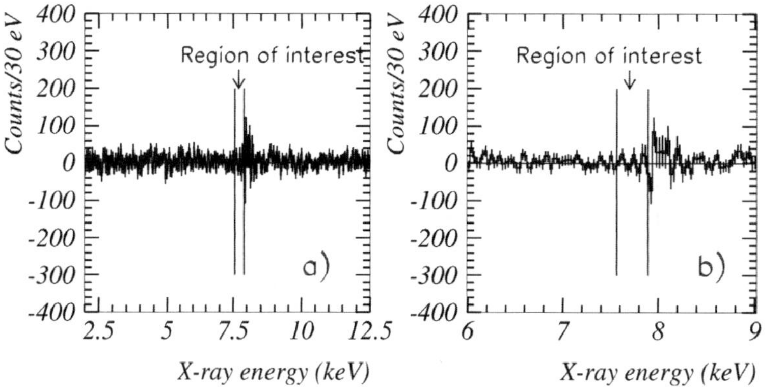

Figure 4. Subtracted energy spectra in the Frascati measurement, *current-on* minus *current-off*, giving the limit on PEP violation for electrons: a) whole energy range; b) expanded view in the region of interest (7.564 - 7.894 keV). No evidence for a peak in the region of interest is found.

Thus with this first measurement in an unshielded environment, we have improved the limit obtained by Ramberg and Snow by a factor ~ 40.

3.2. Experimental results from LNGS

The experiment was installed at LNGS-INFN in Spring 2006 - see figure 5, and was taking data until 2010, alternating period with current on (signal) to periods with current off (background).

A preliminary analysis of the LNGS data gives:

$$\frac{\beta^2}{2} < 4.7 \times 10^{-29} \tag{2}$$

Figure 5. The VIP setup at the LNGS laboratory during installation.

4. Discussion of the results

After the introduction of the straightforward quantum model of Ignatiev and Kuzmin in 1987 [15], Govorkov [27] showed that the model could not work in the wider framework of quantum field theory, because it led to many-particle states with negative norm. The situation changed with the introduction of quon theory [28], which turned out to be a consistent theory of *small* violations of PEP. The basic idea of quon theory is that (anti)commutators are replaced by weighted sums

$$\frac{1-q}{2}\left[a_i, a_j^+\right]_+ + \frac{1+q}{2}\left[a_i, a_j^+\right]_- = a_i a_j^+ - q a_j^+ a_i = \delta_{i,j} \tag{3}$$

where $q = -1$ ($q = 1$) gives back the usual fermion (boson) commutators. The statistical mixture in equation (3) also shows that the PEP violation probability is just $(1+q)/2$ and thus our best experimental bound on q is

$$\frac{1+q}{2} < 4.7 \times 10^{-29} \tag{4}$$

A consistent interpretation of the VIP results can thus be based on quon theory; however here we note that is not easy to devise tests of PEP, because of many conceptual difficulties (see, e.g. [29]), and VIP (and shares some problems of its own with its precursor, the Ramberg and Snow experiment). The main problem lies in the definition of "fresh" electrons: in fact it is unclear how an electron originally injected by the current source into the copper strip can be set apart from the other electrons already present in the strip. One possibility is that some of these "new" electrons have a "wrong wavefunction" in the sense suggested by Rahal and Campa [30]. Yet another possibility is that some localization effect really allows at least a partial identification of electrons: this localization was intuitively rather obvious in the experiment of Goldhaber and Scharff-Goldhaber [31], originally devised to test the identity of β-rays and electrons, later reinterpreted by Reines and Sobel as a test of PEP [32]. In that experiment electrons were injected by a radioactive source, rather than a power supply, and thus the "fresh" electrons were simply those electrons impinging on the target from the radioactive source. This concept of

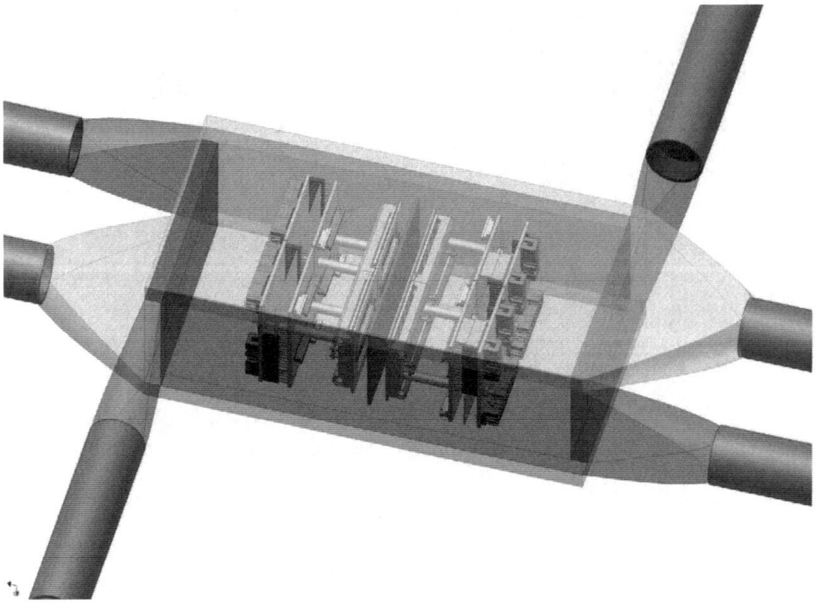

Figure 6. The possible implementation of the upgrade of the VIP experiment using SDD detectors and an external veto-system made of scintillators

novelty is related to electron localizability outside the target, and if an analogous process could be pinpointed for the power supply – which is required to achieve a large statistics, much larger than it is possible with a laboratory radioactive source – then this conceptual problem of VIP (and of the Ramberg and Snow experiment) would fade away. The required localization might be provided by some form of quantum decoherence: Yu and Eberly [33] have shown that in an idealized situation quantum coherence dies off in a finite time just because of quantum noise. Similarly we can conjecture that entanglement of the electron wavefunction in the copper strip could be limited in space and this could let us set apart "old" and "fresh" electrons.

We do not yet have a final answer to these conceptual problems, however we do strongly feel that the test is meaningful and we are now planning an improved version.

5. Upgrade of the VIP setup

The VIP setup used CCD detectors, which are excellent X-ray detectors (good energy resolution, background rejection based on pixel-size) but they are having no timing capability. We plan to switch to a new type of detectors for precision X-rays measurements, the triggerable Silicon Drift Detectors (SSD) which have a fast readout time ($\simeq 1\mu$s) and large collection area (100 mm^2). These detectors were successfully used in the SIDDHARTA experiment [34] for measurements of the kaonic atoms transitions at the DAΦNE accelerator of LNF-INFN; using a proper trigger system a background rejection factor of the order of 10^{-4} was achieved in SIDDHARTA.

With SDD detectors it might then be possible to realize a more compact system and to further reduce the background by using an external veto-system which would eliminate a large part of the background produced by charged particles coming from the outside the setup. A schematic layout of the new setup is shown in fig. 6. Presently, the construction of the setup is under way.

Table 1. Upper bounds on the λ parameter from [35]

Laboratory experiments	Distance (decades) from the enhanced CSL value	Cosmological data	Distance (decades) from the enhanced CSL value
Fullerene diffraction experiments	3-4	Dissociation of cosmic hydrogen	9
Decay of supercurrents (SQUIDS)	6	Heating of Intergalactic medium (IGM)	0
Spontaneous X-ray emission from Ge	-2	Heating of protons in the universe	4
Proton decay	10	Heating of Interstellar dust grains	7

6. Future perspectives: tests of spontaneous collapse models

Having such excellent X-ray detectors as CCDs and SDDs, we are presently considering the possibility to perform in the future measurements of X rays generated as spontaneous radiation predicted by (some) collapse models. The collapse models deal with the "measurement problem" in quantum mechanics, by introducing a new physical dynamics that naturally collapses the state vector. Collapse models make predictions which differ from those of standard quantum mechanics [35]. One of the most exciting task is to perform cutting-edge experiments, in order to asses whether quantum mechanics is exact, or an approximation of a deeper level theory.

In the nonrelativistic collapse model developed by Ghirardi, Rimini, Weber [36] and Pearle [37] (see also ref. [38] for a review), namely the continuous spontaneous localization (CSL) model, the state vector undergoes a nonunitary evolution in which particles interact with a fluctuating scalar field. This interaction has not only the effect of collapsing the state vector towards the particle number density eigenstates in position space, but it increases the expectation value of particle's energy as well. This means, for a free charged particle (as the electron) emission of electromagnetic radiation. This type of phenomenon is predicted by the CSL and is totally absent in the standard quantum mechanics.

Moreover, as shown in [38], the measurement of X rays spontaneously emitted in the collapse models is the most sensitive method to the λ parameter of the collapse; see Table 1 and [35] for related explanations.

In the paper [39] a pioneering work on this spontaneous emission of radiation was performed - the author analysed X-ray data measured in an underground experiment and interpreted them as a limit for the combination of the CSL parameters λ/a^2. It was shown that the highest sensitivity is at few keV X-rays, exactly in the range where our detectors are ideal. We plan to perform a feasibility study to arrive to define a dedicated experiment to measure the X rays predicted by spontaneous collapse models.

6.1. Acknowledgments

The VIP Collaboration wishes to thank all the LNGS laboratory staff for the precious help and assistance during all phases of preparation, installation and data taking. The support from the HadronPhysics FP6 (506078), HadronPhysics2 FP7 (227431), COST Action MP1006 and of the MIUR PRIN 2008 2008LH2X28 projects is acknowledged.

References

[1] W. Pauli, *Phys. Rev.* **58**, (1940), 716.
[2] G. Lüders and B. Zumino, *Phys. Rev.* **110** (1958) 1450.
[3] R. P. Feynman, R. B. Leighton and M. Sands *The Feynman Lectures on Physics*, Addison-Wesley, Reading, MA (1963).
87 (2000) 510.
[4] R. Arnold, *et al.*, *Eur. Phys. J. A***6** (1999) 361.
[5] H.O. Back, *et al.*, *Eur. Phys. J. C***37** (2004) 421.
[6] A. Barabash, *Found. of Phys.* **40** (2010) 703.
[7] A.S. Barabash,*et al.*, *JETP Lett.* **68** (1998) 112.
[8] P. Belli, *et al.*, **460** (1999) 236.
[9] G. Bellini, *et al.*, *Phys. Rev. C* **81** (2010) 034.
[10] R. Bernabei, *et al.*, *Phys. Lett. B* **408** (1997) 439.
[11] R. Bernabei, *et al.*, *Eur. Phys. J. C* **62** (2009) 327.
[12] Y. Suzuki, *et al.*, *Phys. Lett. B* **311** (1993) 357.
[13] A.M.L. Messiah, O.W. Greenberg, *Phys. Rev.* **136** (1964) B248.
[14] E. Ramberg and G. A. Snow, *Phys. Lett. B* **238** (1990) 438.
[15] A. Yu. Ignatiev and V. A. Kuzmin *Yad. Fiz.* **46** 786 (1987 *Sov. J. Nucl. Phys.* (1987) **47** 6); A. Yu. Ignatiev *Rad. Phys. Chem.* **75** (2006) 2090.
[16] The VIP proposal, LNF-LNGS Proposal, September, 2004, http://www.lnf.infn.it/esperimenti/vip.
[17] J. L. Culhane, *Nucl. Instrum. Methods A* **310** (1990) 1.
[18] J.-P. Egger, D. Chatellard and E. Jeannet, *Particle World* **3**, (1993) 139.
[19] G. Fiorucci, *et al. Nucl. Instrum. Methods A* **292** (1990) 141.
[20] D. Varidel, *et al.Nucl. Instrum. Methods A* **292** (1990) 147
[21] R. P. Kraft, *et al.Nucl. Instrum. Methods A* **372** (1995) 372.
[22] S. Di Matteo, L. Sperandio, VIP Note, IR-04, 26 April 2006; The energy shift has been computed by P. Indelicato, private communication.
[23] CCD-55 from EEV (English Electric Valve), Waterhouse Lane, Chelmsford, Essex CM1 2QU, UK.
[24] T. Ishiwatari , *et al. Phys. Lett. B* **593** (2004) 48; G. Beer, *et al. Phys. Rev. Lett.* **94** (2005) 212302.
[25] T. Ishiwatari, *et al. Nucl. Instrum. Methods Phys. Res. A* **556** (2006) 509.
[26] S. Bartalucci *et al.* (VIP Collaboration) *Phys. Lett. B* **641** (2006) 18.
[27] A. B. Govorkov *Phys. Lett. A* **137** (1989) 7.
[28] O. W. Greenberg *Phys. Rev. Lett.* **64** (1990) 705.
[29] R. D. Amado and H. Primakoff *Phys. Rev.* **C 64** (1980) 1338.
[30] V. Rahal and A. Campa *Phys. Rev.* **A 38** (1988) 3728.
[31] M. Goldhaber and G. Scharff-Goldhaber *Phys. Rev.* **73** (1948) 1472.
[32] F. Reines and H. W. Sobel *Phys. Rev. Lett.* **32** (1974) 954.
[33] Ting Yu and J. H. Eberly *Science* **323** (2009) 598.
[34] C. Curceanu *et al. Eur Phys* **J A31** (2007) 537-539; M. Bazzi *et al. Phys. Lett.* **B681** (2009) 310; M. Bazzi *et al. Phys. Lett.* **B704** (2011) 113.
[35] S.L. Adler, A. Bassi, *Science* **325** (2009) 275.
[36] G.C. Ghirardi, A. Rimini and T. Weber *Phys. Rev.* **D34** (1986) 470: ibid. (1987) 3287: *Found. Phys.* **18** (1988) 1.
[37] P. Pearle *Phys. Rev* **A39** (1989) 2277: G.C. Ghirardi, P. Pearle and A. Rimini *Phys. Rev.* **A42** (1990) 78.
[38] A. Bassi *Journal of Physics* **67** (2007) 012013.
[39] Q. Fu *Phys. Rev* **A56** (1997) 1806.

Quantum Phase from the Twin Paradox

G. N. Ord
Department of Mathematics, Ryerson University, Toronto, Canada M5B 2K3
E-mail: gord@ryerson.ca

Abstract. The modern concept of spacetime usually emerges from the consideration of moving clocks on the assumption that world-lines are continuous. In this paper we start with the assumption that natural clocks are digital and that events are discrete. By taking different continuum limits we show that the phase of non-relativistic quantum mechanics and the odd metric of spacetime both emerge from the consideration of discrete clocks in relative motion. From this perspective, the continuum limit that manifests itself in 'spacetime' is an infinite mass limit. The continuum limit that gives rise to the Schrödinger equation retains a finite mass as a beat frequency superimposed on the 'Zitterbewegung' at the Compton frequency. We illustrate this in a simple model in which a Poisson process drives a relativistic clock that gives rise to a Feynman path integral, where the phase is a manifestation of the twin paradox. The example shows that the non-Euclidean character of spacetime and the wave-particle duality of quantum mechanics share a common origin. They both emerge from the necessity that clocks age at rates that are path dependent.

1. Introduction

Historically and pedagogically, quantum mechanics and special relativity appear as two separate revolutions in physics. They were effectively merged by Dirac in his famous equation and it is customary to think of the merger as producing a relativistic form of Schrödinger's equation Fig.[1(A)]

Table 1. The statistical mechanics underneath the Classical phenomonological equations is well accepted. The formal similarity with the Quantum equations is rendered mysterious by the presence of i.

	Classical	**Quantum**
'Path' picture →	Wiener/Kac	Feynman
Diffusion / Schrödinger	$\frac{\partial U}{\partial t} = D\,\frac{\partial^2 U}{\partial x^2}$	$\frac{\partial \psi}{\partial t} = i\,D\,\frac{\partial^2 \psi}{\partial x^2}$
Telegraph / K. G.	$\frac{\partial^2 U}{\partial t^2} = c^2\frac{\partial^2 U}{\partial z^2} + a^2 U$	$\frac{\partial^2 \psi}{\partial t^2} = c^2\frac{\partial^2 \psi}{\partial z^2} + (i\,m)^2\psi$
Telegraph / Dirac	$\frac{\partial \mathbf{U}}{\partial t} = c\,\boldsymbol{\sigma}_z\frac{\partial U}{\partial z} + a\,\boldsymbol{\sigma}_x U$	$\frac{\partial \Psi}{\partial t} = c\,\boldsymbol{\sigma}_z\frac{\partial \Psi}{\partial z} + i\,m\,\boldsymbol{\sigma}_x\Psi$
Decay / Oscillate	$e^{-a\,t}$	$e^{\pm i\,m\,t}$
Clock	Thermodynamic	Deterministic

In this talk we reconsider the relationship between quantum mechanics and relativity. Specifically, we question whether the two theories 'live' separately in the continuum and have

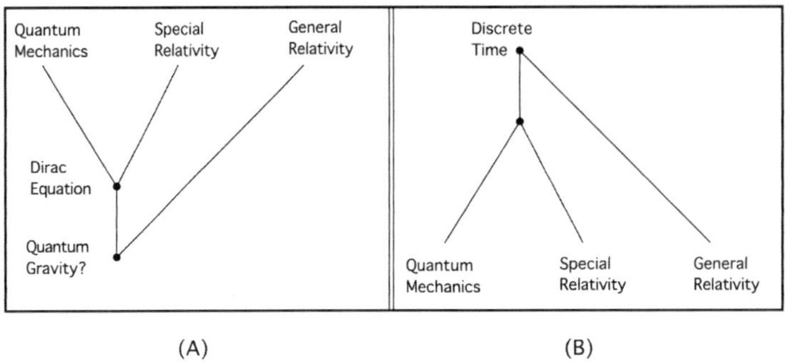

Figure 1. (A) Historically, quantum mechanics and special relativity developed independently and were successfully merged in the Dirac equation. (B) This talk takes an 'emergent' point of view, exploring the possibility that the two pictures are different continuum limits applied to discrete time.

ultimately to be merged into a comprehensive theory, as is commonly supposed, or whether they both emerge as separate continuum limits of a discrete theory.

To motivate an emergent picture, there are interesting similarities between the two theories in their relationships with their predecessors. Both theories mathematically employ a formal analytic continuation to convert mathematical expressions from within classical mechanics to expressions in the modified theory.

In Table (1) we see two sets of partial differential equations. The left-most set are phenomenological equations describing diffusive processes. All of these equations have well known statistical bases. The right hand column contains equations thought to be fundamental in the quantum context. They are obtained from the phenomenological equation by a formal analytic continuation that can be accomplished by the replacement of t by it. This replacement, while giving the quantum equations, does not directly yield a physically sensible statistical mechanics from which the equations emerge.

By comparison, the replacement of t by it takes a four dimensional Euclidean metric to Minkowski's spacetime, giving us the transition from classical mechanics to special relativity.

We shall explore the idea that classical special relativity and quantum propagation are not independent physical constraints to be married, but rather that they emerge as different continuum limits of simple clocks Fig.[1(B)].

We shall follow a hint from the last row of Table 1. Diffusive systems are examples of thermodynamic clocks. They measure time in terms of an approach to an equilibrium state. Quantum systems on the other hand are deterministic clocks that measure time via a periodic process. The clock frequency is ultimately determined by a 'mass' parameter. We shall exploit this feature by building a set of minimalist classical discrete clocks as models of 'point' particles. We shall see that by starting with discrete clocks, the odd metric of spacetime and quantum propagation are manifestations of the continuum limit. This is missed in conventional formulations that start in the continuum.

When we speak of classical digital clocks we mean that we can interpret everything about the clocks in a classical way and classical mechanics can be seen as a particular continuum limit.

The first clock we shall consider is a simple digital, deterministic, two-photon clock. The second clock,'Minkowski', is a continuum limit of the two-photon clock. The concept of 'Spacetime' arises from the consideration of this clock. The third clock,'Poisson', is a Minkowski clock driven by a Poisson process in a non-relativistic limit. It gives us Feynman's path integral *directly* in a context where the origin of phase is transparent. The phase is a remnant of the relativistic feature that objects, moving on different paths in spacetime, age at different rates.

Journal of Physics: Conference Series **361** (2012) 012007 doi:10.1088/1742-6596/361/1/012007

Feynman's phase is a remnant of the twin paradox in special relativity.

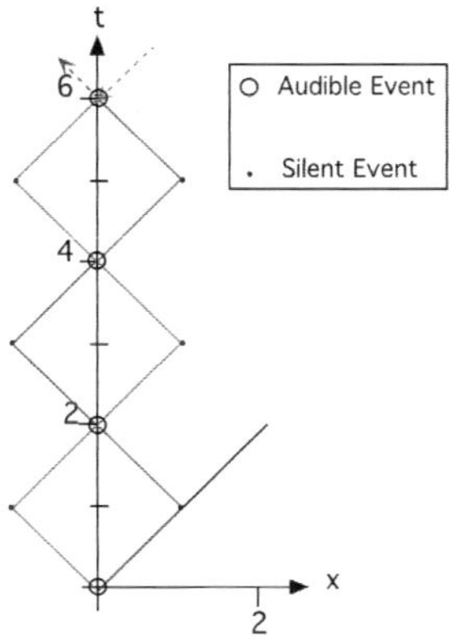

Figure 2. Two featureless photons confined to a box. They cross at $x = 0$ every 2 units of time giving an audible event. They move with speed $c = 1$. Points where paths change direction are considered silent events. Silent events are part of the spacetime geometry of the paths but are not part of the audible world line event sequence.

2. The Two-Photon Clock

Consider two featureless photons confined to a box Fig.[2]. We consider events of two types, audible and silent. Audible events line up along the t axis and may be regarded as ticks of a clock. The set of these audible ticks is a digital analog of the clock's worldline. Direction changes in the paths of the two photons are called silent events because they are important to make the clock tick, but are not observed directly.

To use the two-photon clock as a clock we have to count events. To handle boosts, we have to keep track of at least two components, Fig.[3]. This is done through the use of a two-component state variable that alternates sign and switches on and off with direction. The state variable s_k

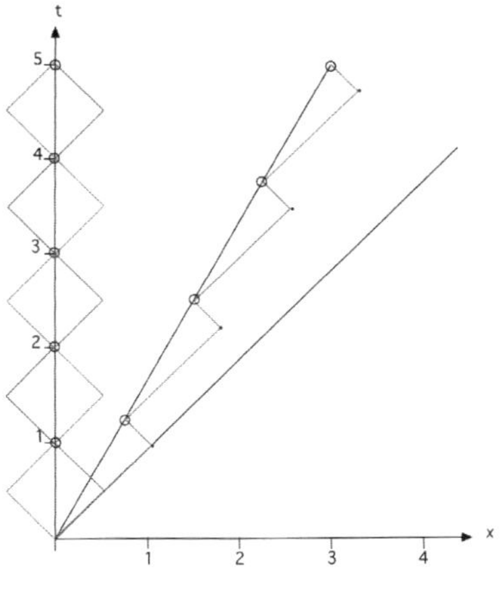

Figure 3. A two-photon clock from two different frames. The area between the paths between audible events is fixed, as is the speed c. This determines the shape of the rectangular areas and the position of the ticks on the space-time diagram. The change in the lengths of the sides of the rectangles with changing relative velocities requires a two-component characteristic function.

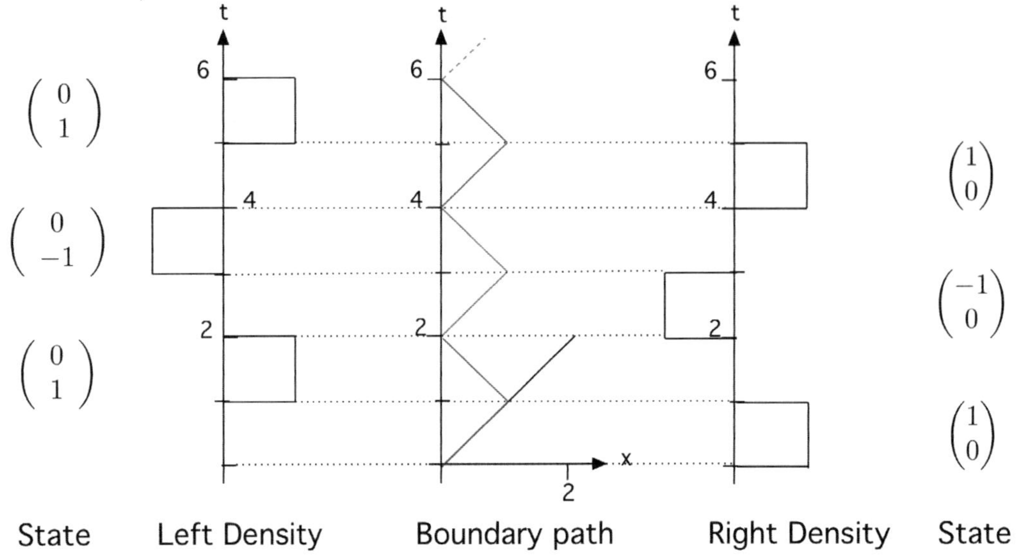

Figure 4. We use the right-hand boundary of the chain of rectangles to count events. A 2-component column vector cycles between the 4 possible states of the boundary. Each state gives rise to a density that changes sign with every two events.

is a characteristic function for orientation and it takes on one of four values:

$$s_k \in S = \left\{ \begin{pmatrix} 1 \\ 0 \end{pmatrix}, \begin{pmatrix} 0 \\ 1 \end{pmatrix}, \begin{pmatrix} -1 \\ 0 \end{pmatrix}, \begin{pmatrix} 0 \\ -1 \end{pmatrix} \right\} \tag{1}$$

If the two-photon clock is in state s_k at time k then the subsequent state, after the next event is

$$s_{k+1} = T s_k \tag{2}$$

where T is the transfer matrix:

$$T = \begin{pmatrix} 0 & -1 \\ 1 & 0 \end{pmatrix}. \tag{3}$$

Since $s_k = T^k s_0$ the power of the transfer matrix corresponds to the displacement in time! Fig.[4] shows the relation between s and the path boundary. The two components of s, made necessary by the need to be able to handle boosts, suggests that we can visualize the cycling of s by adding an extra dimension. Fig.[5] shows a visualization of the state variable with the extra dimension. The graph of the state function forms a four position spiral.

The digital clock in its current state would function well as a classical relativistic clock for any process that does not need time resolution on scales comparable with, or below, the digital period of the two-component clock. However special relativity needs analog reference frame clocks that are precise on all scales and we seek to find a relation between the two-photon clock and conventional analog time.

3. Minkowski's Clock

Periodic digital clocks measure time by counting the number of clock events between external events that happen on longer time scales. Although we cannot in practice construct clocks with arbitrarily high frequency, we can imagine sequences of clocks with increasing frequency and extract from this an approximation of an analog clock, Fig.[6].

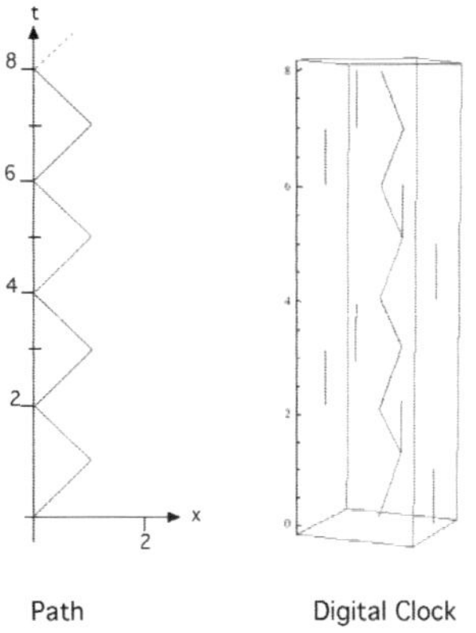

Figure 5. The two-photon clock digitizes a path boundary. Counting events with a two-component characteristic function suggests we include an extra dimension to visualize the event sequence. The path used by the clock is planar but the state function that is used for counting may be visualized with an extra dimension where it forms a four state spiral.

To this end, if T_ν is the transfer matrix for a high-frequency N-clock, synchronized with T, we must have an equivalence between the N-th power of T_ν and T itself. Thus

$$T_\nu^N = T \tag{4}$$

giving T_ν as an Nth root of T or

$$T_\nu = \begin{pmatrix} \cos\left(\frac{\pi}{2N}\right) & -\sin\left(\frac{\pi}{2N}\right) \\ \sin\left(\frac{\pi}{2N}\right) & \cos\left(\frac{\pi}{2N}\right) \end{pmatrix} = \cos\left(\frac{\pi}{2N}\right) I_2 + \sin\left(\frac{\pi}{2N}\right) T, \tag{5}$$

Fig.[7].

Taking a continuum limit gives a transfer matrix representing a reference clock with arbitrarily high precision. That is we define T_R as

$$\begin{aligned} T_R(t) = \lim_{N\to\infty} T_\nu^{Nt} &= \begin{pmatrix} \cos\left(\frac{\pi t}{2}\right) & -\sin\left(\frac{\pi t}{2}\right) \\ \sin\left(\frac{\pi t}{2}\right) & \cos\left(\frac{\pi t}{2}\right) \end{pmatrix} \\ &= \cos\left(\frac{\pi t}{2}\right) \mathrm{I}_2 + \mathrm{T} \sin\left(\frac{\pi \mathrm{t}}{2}\right) \end{aligned} \tag{6}$$

where I_2 is the 2×2 identity matrix. This clock agrees with the two-photon clock T at the integers but uses a rotation matrix to interpolate between events. That is $T_R(n) = T^n$ for integer n.

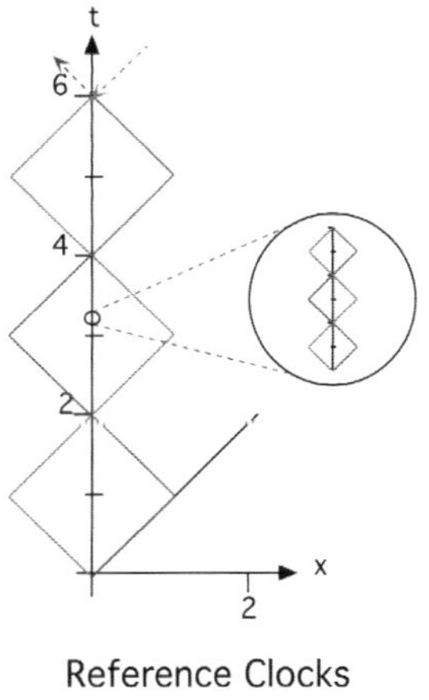

Figure 6. We can imagine a sequence of two-photon clocks of ever higher frequency, synchronized with the initial clock in such a way that all the events of the initial low-frequency clock coincide with a subset of the events of the high-frequency clocks.

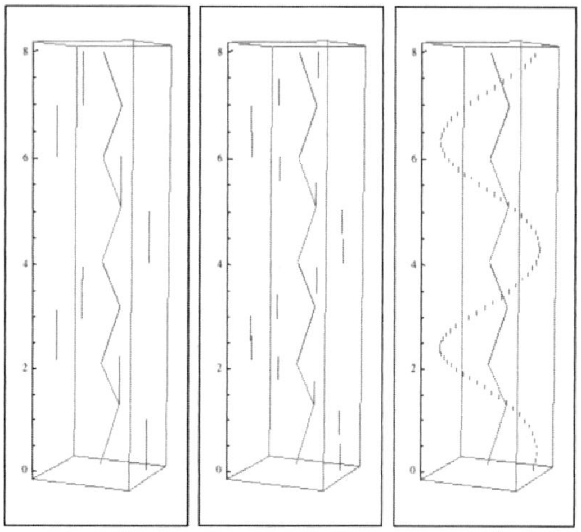

Figure 7. In the first frame the two-photon clocks 'counts' with the four-state indicator function. In the second frame the frequency is doubled. The eight states are just a 'rotation' of the original four. In the third frame the sampling frequency is 10 times the original. The approach to a spiral is evident.

The two-photon clock viewed from another inertial frame *preserves the areas* in the chain of areas and preserves the speed c. For the transfer matrix to accommodate the change in location of the audible events it has to take into account the relative velocity of the frame and the change in residence times in the two states Fig.[8,9]. For small time steps ϵ the modification of the Transfer matrix is:

$$T_M(\epsilon) = \begin{pmatrix} 1+\frac{\pi}{2}v\epsilon & -\frac{\pi}{2}\epsilon \\ \frac{\pi}{2}\epsilon & 1-\frac{\pi}{2}v\epsilon \end{pmatrix} = \mathrm{I}_2 - \frac{\pi}{2}\epsilon\,(\mathrm{i}\sigma_\mathrm{y} + \mathrm{v}\sigma_\mathrm{z}) \tag{7}$$

where we have used the Pauli matrices to show the structure of the matrix. Comparing this to a small t expansion of (5) we see that the only difference is the σ_z term.

If we take the limit of the transfer matrix raised to the power t/ϵ as ϵ goes to zero we get

$$\begin{aligned} T_M(t) &= \begin{pmatrix} \cos\left(\frac{\pi t}{2\gamma}\right) - v\gamma\sin\left(\frac{\pi t}{2\gamma}\right) & -\gamma\sin\left(\frac{\pi t}{2\gamma}\right) \\ \gamma\sin\left(\frac{\pi t}{2\gamma}\right) & \cos\left(\frac{\pi t}{2\gamma}\right) + v\gamma\sin\left(\frac{\pi t}{2\gamma}\right) \end{pmatrix} \\ &= \cos\left(\frac{\pi t}{2\gamma}\right) I_2 - \sin\left(\frac{\pi t}{2\gamma}\right)\gamma(v\sigma_z + i\sigma_y) \end{aligned} \tag{8}$$

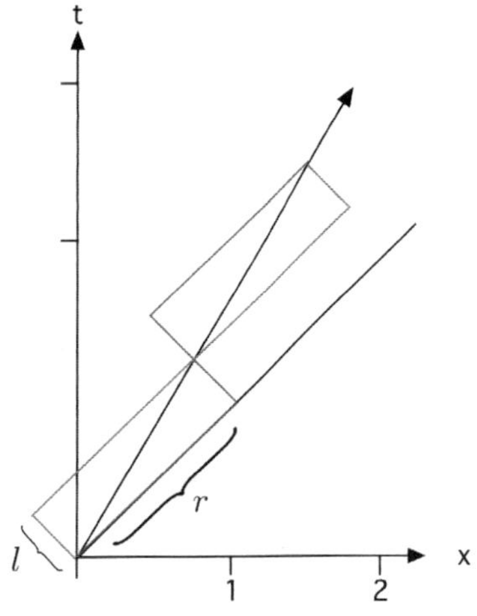

Figure 8. From a moving frame the two-photon clock is a chain of oriented areas. The areas and orientations of the areas do not change between inertial frames. However the boundaries of the areas change to accommodate the boost. It is for this reason that the two-photon clock requires at least two components. In the figure, the area boundaries spend a greater proportion of time in the $(1,1)$ direction than the $(-1,1)$ direction.

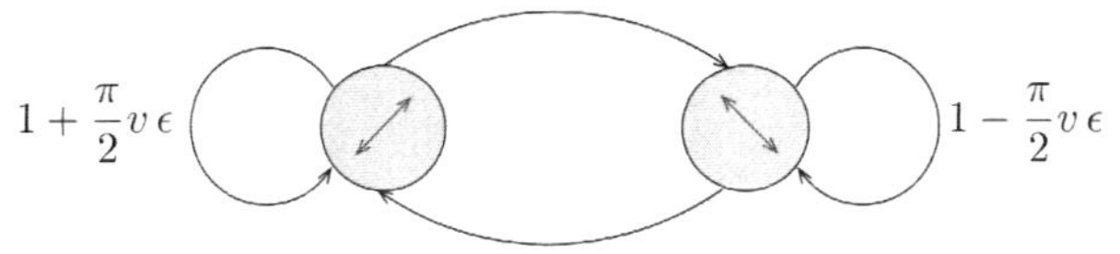

Figure 9. The boost alters the percentage of time the boundary paths spend in the different states. If $-v$ is the velocity of the frame with respect to the clock then the transfer matrix readjusts the residency times.

Algebraically, the transfer matrix consists of two parts. The even part is a 'scalar', the 2×2 identity matrix I_2. The odd part is a vector quantity $M = \gamma(v\sigma_z + i\sigma_y)$ that includes the time dilation factor $\gamma = \frac{1}{\sqrt{1-v^2}}$ as normalization. Notice that $M \cdot M = -I_2$ while $\sigma_z \cdot \sigma_z = I_2$ and $(i\sigma_y) \cdot (i\sigma_y) = -I_2$ The clock 'recognizes' the odd signature of spacetime through the transfer matrix. The vector M 'points' in the direction of the proper time axis of the moving frame. If we regard σ_z as the unit vector in space and $T = i\sigma_y$ as the unit vector in time, Minkowski's clock stores both the tick rate information needed to be a clock and *direction* information needed by special relativity. By comparison, a Newtonian clock that measures absolute time could simply be represented by a complex number.

T_M agrees with the two-photon clock at the *'dilated'* integers, $\{0, \gamma, 2\gamma, \ldots\}$ as required by special relativity.

4. Minkowski's Clock in 3-D Space

The 3-D transfer matrix functions in exactly the same way as in the previous version, except it codes for a reference frame that moves at constant velocity in a three dimensional space. The $(1,1)$ and $(2,2)$ blocks code for the Doppler shifts of the clock away from and towards the reference frame, as did the $(1,1)$ and $(2,2)$ elements in the two dimensional version. The result is that the vector in the odd part of (8) is embedded in four dimensions by $\sigma_z v \to \sigma_z \bigotimes(v_x\sigma_x + v_y\sigma_y + v_z\sigma_z)$ and $i\sigma_y \to i\sigma_y \bigotimes I_2$. The result is

$$T_M = \cos\left(\frac{\pi t}{2\gamma}\right) \mathrm{I}_4 - \sin\left(\frac{\pi \mathrm{t}}{2\gamma}\right) \gamma(\mathrm{v_x}\alpha_\mathrm{x} + \mathrm{v_y}\alpha_\mathrm{y} + \mathrm{v_z}\alpha_\mathrm{z} + \mathrm{i}\beta) \tag{9}$$

where the α 's and β anticommute and have unit norm. Notice again the implication of the odd spacetime signature and the natural setting of 'spacetime algebra'. We started with the simple digital two-photon clock in a two-dimensional Euclidean space. The demands of special relativity for the constancy of the speed of light have forced the clock, in a continuum limit, to recognize Minkowski space. As we shall shortly see, the continuum limit is in fact an infinite mass limit and by implication conventional 'spacetime' also emerges as an infinite mass limit!

5. The Poisson Clock

In the previous sections there was no mention of quantum mechanics. We simply built a classical relativistic digital clock and found that the odd signature of spacetime emerged in a continuum limit. Interpretation was not an issue. We could in principle build a two-photon clock with audible tick rates limited only by technology. Provided the audible ticks were indeed audible, we would not find evidence of quantum behaviour in the (discrete) worldline of the clock.

However, recognizing that conventional spacetime arrives as a continuum limit of a discrete clock, the natural question is "What happens between the ticks of a clock?". In the previous cases, we had a prescription for the photon paths between ticks and we used the same deterministic prescription down to arbitrarily small scales. Here we remove the determinism, not only between successive audible events but between a first and last observation many scales above the tick frequency of the clock. The result is a familiar one, but in a context that demonstrates the intimate link between special relativity and quantum mechanics.

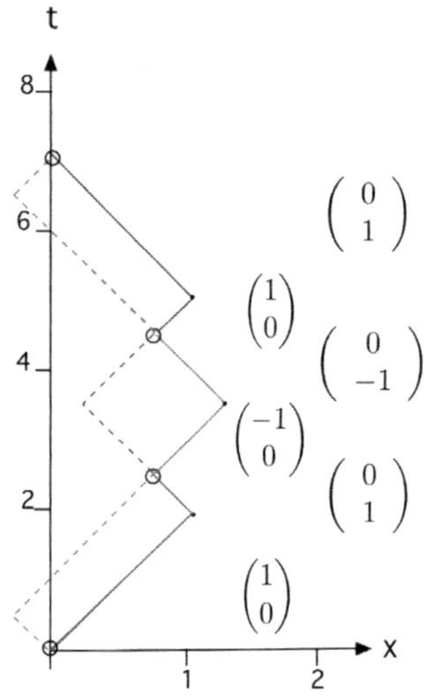

Figure 10. The Poisson clock is a stochastic version of the two-photon clock. The event sequence is determined by a Poisson process of fixed rate. The counting of events is implemented in the same way as for the two-photon clock. If we ultimately hear ticks only on scales very much larger than the natural 'wavelength', there are many possible discrete worldlines between these ticks.

Referring to Fig. [10] let $\mathbf{w}$ be the random variable generated by applying T to the clock state at every Poisson event. Thus if $\mathbf{n}(t)$ is the Poisson random variable, and the process starts out in s_1 we have

$$\mathbf{w}(t) = T^{n(t)} \begin{pmatrix} 1 \\ 0 \end{pmatrix}. \tag{10}$$

The expected value of $\mathbf{w}(t)$ is then

$$\begin{aligned} <\mathbf{w}(t)> &= e^{-\alpha t} \sum_{k=0}^{\infty} \frac{(\alpha t)^k}{k!} T^k \begin{pmatrix} 1 \\ 0 \end{pmatrix} \\ &= e^{-\alpha t} (I_2 \cos \alpha t + T \sin \alpha t) \begin{pmatrix} 1 \\ 0 \end{pmatrix} \\ &= e^{-\alpha t} e^{T\alpha t} \begin{pmatrix} 1 \\ 0 \end{pmatrix} \end{aligned} \tag{11}$$

The prefactor $e^{-\alpha t}$ signals an exponential decay superimposed on the Poisson clock. We thus define the random variable

$$\mathbf{P}(t) = e^{\alpha t} T^{n(t)} \begin{pmatrix} 1 \\ 0 \end{pmatrix}. \tag{12}$$

giving us a normalized 'Poisson clock'

$$\begin{aligned} <\mathbf{P}(t)> &= e^{T\alpha t} \begin{pmatrix} 1 \\ 0 \end{pmatrix} \\ &= (I_2 \cos \alpha t + T \sin \alpha t) \begin{pmatrix} 1 \\ 0 \end{pmatrix} \end{aligned} \tag{13}$$

whose expected length does not change with time.

From an inertial frame moving at speed $-v$ with respect to the frame of the clock, the matrix $e^{T\alpha t}$ becomes

$$\begin{aligned} e^{T\alpha t} &\rightarrow \begin{pmatrix} \cos\frac{\alpha t}{\gamma} + v\gamma \sin\frac{\alpha t}{\gamma} & -\gamma\sin\frac{\alpha t}{\gamma} \\ \gamma\sin\frac{\alpha t}{\gamma} & \cos\frac{\alpha t}{\gamma} - v\gamma\sin\frac{\alpha t}{\gamma} \end{pmatrix} \\ &= I_2 \cos\left(\frac{\alpha t}{\gamma}\right) + [\gamma(v\sigma_z + T)] \sin\left(\frac{\alpha t}{\gamma}\right) \\ &= e^{T_M \alpha t} \end{aligned} \tag{14}$$

where $T_M = [\gamma(v\sigma_z + T)]$ is the transfer matrix in the moving frame.

Notice that γ has two roles. It is the time dilation factor $\gamma = \frac{1}{\sqrt{1-v^2}}$ adjusting the frequency of the moving clock and it is the normalization factor that makes the vector T_M a unit vector with negative norm. T_M provides the directional information for the proper time axis and is responsible for the odd metric signature of Minkowski space for the clock.

In the non-relativistic limit, the unit vector T_M may be approximated by T and to first order in v^2 we have

$$\begin{aligned} < \hat{\mathbf{w}}(t) > &\approx e^{T\frac{\alpha t}{\gamma}} \begin{pmatrix} 1 \\ 0 \end{pmatrix} \\ &\approx e^{\alpha t T} e^{\alpha t T \frac{v^2}{2}} \begin{pmatrix} 1 \\ 0 \end{pmatrix}. \end{aligned}$$

The term $e^{\alpha t T}$ is a high frequency signal that is *independent of the speed and direction* of the reference frame. The remaining term is the relative phase of the moving clock.

Starting at the origin, if the clock is moved at constant speed v along the x-axis then at time t in the rest frame of the origin, the expected relative phase of the clock is

$$\exp[\alpha t T \frac{v^2}{2}] = \exp[\alpha T \frac{x^2}{2t}] \tag{15}$$

The only algebraic property of T we need in the relative phase $\exp[\alpha T \frac{x^2}{2t}]$ is that $T^2 = -1$. We can then replace the column vectors by complex numbers and T by the unit imaginary i. If we replace α by the real constant $\alpha = \frac{m}{\hbar}$ the expected phase is

$$\exp[i\frac{mx^2}{2\hbar t}] \tag{16}$$

As a unimodular complex number, this clock 'runs' at a relative rate appropriate to a clock moving at speed v with respect to the origin, provided $v << 1$.

To obtain a spacetime density of phase we average $\exp[i\frac{mx^2}{2\hbar t}]$ assuming all speeds between ± 1 are equally likely, the result in the non-relativistic limit is the density

$$K(0,0;x,t) = \left(\frac{2\pi i\hbar t}{m}\right)^{-\frac{1}{2}} \exp\left(i\frac{mx^2}{2\hbar t}\right). \tag{17}$$

This is just Feynman's free particle kernel. It appears here as the relative phase density due to an ensemble of relativistic clock paths! The path integral follows in the usual way by constructing polygonal paths using the above density as the phase of the individual segments.[1][2]

We pause here to notice a few things about the argument at this point. First of all, the replacement of α by $m/\hbar$ prior to equation (16) tells us that in the clock picture, what we

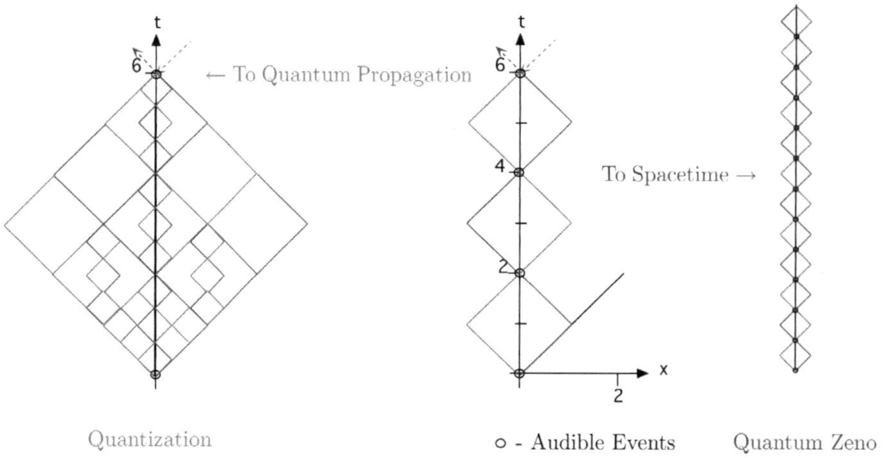

Figure 11. Spacetime and Quantum propagation emerge from a discrete clock via two *different* continuum limits. The classical limit produces a smooth world-line of audible events, but loses the geometric feature by which a particle can tell the time (Zitterbewegung). This corresponds to scaling up from deterministic clock behaviour. The quantum limit does not assume intermediate events are audible so world-lines between audible events are non-unique. This implicates the path-integral and leaves mass as a fundamental frequency that is observable through phase.

call mass is just the fundamental frequency of the clock. Referring to Fig. [6], this means that our continuum limit for Minkowski's clock was really an infinite mass limit. *Thus the usual formulations of Relativity employing spacetime frames assume an infinite mass limit for the frames themselves!* Association of these frames with particles whose mass is provided by dynamics decouples the direct association between mass and clock frequency, losing the implication of wave propagation that Poisson's clock demonstrates.

Minkowski's clock was classical by construction. Poisson's clock just adds a stochastic element (essentially Wiener paths). We might wonder, 'Is this the source of the appearance of wave propagation?' (cf. Stochastic mechanics and diffusion based models.) It is true that the Schrödinger form results from an ensemble average of stochastic paths and the non-relativistic approximation. On the other hand, it is ultimately the existence of phase that gives us wave propagation. This is 'hidden' in Minkowski's clock and originates in the determination of spacetime *events* through spacetime *areas*! Poisson's clock scales this feature *up* by *not fixing audible events between observations* (cf. superfluids.) By comparison, Minkowski's clock scales this down, ultimately hiding any stochastic element underneath a continuum limit.

Figure[11] illustrates the different limits involved in the emergence of special relativity and quantum propagation from two-photon clocks. Minkowski space involves the scaling down of the wavelength of the two-photon clock in order to reach a continuum limit. This is an infinite frequency limit which, from the Poisson clock perspective, is an infinite mass limit. Quantum Mechanics is not apparent in this limit in agreement with the Quantum Zeno effect. Adding mass as a background feature through dynamics, as is customary in special relativity, completes the classical picture of particles 'living' classically in a spacetime 'container'.

The quantum limit does not force the scale of observation to go to zero with the continuum limit. Instead, the audible events of Minkowski's clock are assumed unobserved and are left variable except at an initial and final event. In this picture, the fundamental frequency (Compton) survives through the phase of the beats of the expected characteristic function of oriented area (Feynman's kernel, equation (17)).

6. Discussion

With respect to the subject of this conference, this work has a number of points to offer. Ultimately, if we are looking for emergent quantum mechanics, we have to address the question 'Emergent from what?' The beautiful experiments of Couder's group[3] and Bush[4] show an intriguing physical picture of wave-particle duality. The theoretical framework of Grössing, Fussy, Mesa Pascasio and Schwabl (see articles in this volume and[5]) takes this picture and replaces the underlying wave by a structured vacuum. In both cases the quantum behaviour arises from the phase of waves that are literally and metaphorically 'underneath' the particle.

This work looks down in scale from the de Broglie wavelength to the Compton wavelength to 'find' the waves, or structured vacuum, that underlies the wave-particle duality we need for quantum mechanics to emerge.

The first result, Minkowski's clock, gives us spacetime as it emerges from the consideration of an extremely simple two-photon clock. By seeing how spacetime results from a continuum limit, we get one extra feature that is not apparent in the usual invocation of spacetime. We see that from the perspective of actual clocks, spacetime is an infinite frequency, infinite mass object. While this is obvious looking at special relativity through the perspective of the uncertainty principle ($\Delta E \Delta t \geq \hbar$), the uncertainty principle is usually assumed (incorrectly) to be outside the domain of classical physics, and as a result the usual pedagogy for relativity misses the connection between mass and frequency. With that connection gone, quantum propagation has to be invoked in a separate theory.

The second result, Poisson's clock, shows how the mechanism that enforces special relativity in a simple stochastic clock also gives rise to Feynman's phase in the path integral. The phase itself is a manifestation of the twin paradox familiar from special relativity. Two different spacetime paths between two events give rise to two different ages for particles that traverse the paths. The ages are measured in terms of the fundamental frequency (Compton) of the particles and, on observation, the synchronicity of the various paths affect the likelihood of an observation taking place. The sum over paths still motivates paradoxical questions like 'Did the clock move over one or all of the paths?'. However these questions are put into context by the realization that, from the perspective of actually *building a clock*, such questions are unanswerable even in the world of classical special relativity. Minkowski's clock, which appears to bypass wave propagation, illustrates this. It is classical because the continuum limit confines spacetime paths to scales below a fundamental period which itself is going to zero in the continuum limit.

7. Conclusions

The view that spacetime and quantum propagation both emerge from the consideration of simple clocks is a picture that intersects many attempts to understand why quantum mechanics is both precise and inscrutable.

One bridge that connects clock arguments to other pictures is the feature of dimensionality. Classical physics, including relativity, interpolates a sequence of events of particle observation by *smooth world-lines*. However, relativistic physics *implicitly* connects the events by invariant spacetime *areas*. The control of *events* by *areas* is a progenitor of 'wave-particle duality' in quantum mechanics and Minkowski space in relativity. The important role of dimension in quantum mechanics is emphasized in Fractal Spacetime approaches [6]-[8] and is apparent from the path integral perspective [9, 10]. The connection to special relativity arises from Feynman's chessboard model[11]-[12].

The light clock aspect of the model is similar to the Zig-Zag model of Penrose[13] with an emphasis on Clifford algebras as a geometric framework for the Dirac equation (see the papers of Hiley and de Gosson in this volume and[14, 15]). The clock model shares with deBroglie-Bohm pictures [16] the feature that it is 'particle *and* wave' as opposed to the 'particle or wave' of Copenhagen. Just as deBroglie-Bohm considers the particle and quantum potential as real aspects of an object, so the clock model considers the events and the inter-event regions as part of the model.

Ultimately, the clock model suggests that at least part of the problem regarding the interpretation of quantum mechanics resides in the premature invocation of real numbers. Real numbers are a mathematical convenience that are *not observable*. Quantum mechanics tells us that the stability of Nature reflects a predisposition for discrete energy levels. Mathematical models that start in a spacetime continuum presuppose that a coherent discrete picture should emerge from a continuum description. The clock model (see also Kauffman's paper in this volume) suggests that this may be putting the cart before the horse.

Thus it is from the clock model that we can see classical special relativity and quantum propagation as two different continuum limits of the same discrete model. After the limit is taken, the connection between the two pictures is severed, leaving only i as the ghost of discrete physics that haunts both theories.

References

[1] G.N. Ord. Quantum mechanics in a two-dimensional spacetime: What is a wavefunction? *Annals of Physics, Issue 6, June 2009*, 324(6):1211–1218, June 2009.

[2] G.N. Ord and R. B. Mann. How does an electron tell the time? *International Journal of Theoretical Physics*, pages 1–15, September 2011.

[3] Couder, Y., Boudaoud, A., Protière, S., and Fort, E. Walking droplets, a form of wave-particle duality at macroscopic scale? *Europhysics News*, 41(1):14–18, 2010.

[4] John W. M. Bush. Quantum mechanics writ large. *Proceedings of the National Academy of Sciences*, 107(41):17455–17456, 2010.

[5] G. Grössing, S. Fussy, J. Mesa Pascasio, and H. Schwabl. An explanation of interference effects in the double slit experiment: Classical trajectories plus ballistic diffusion caused by zero-point fluctuations. *Annals of Physics*, 327(2):421–437, February 2012.

[6] G.N. Ord. Fractal space-time a geometric analog of relativistic quantum mechanics. *J.Phys.A*, 16:1869–1884, 1983.

[7] El Naschie, M.S.: *The theory of Cantorian spacetime and high energy particle physics (an informal review)*. Chaos, Solitons & Fractals, 41(5):2635 – 2646, 2009.

[8] Laurent Nottale. *Scale Relativity and Fractal Spacetime: A New approach to unifying Relativity and Quantum Mechanics*. Imperial College Press, 2011.

[9] Richard P. Feynman. Nobel lecture.

[10] S.S. Schweber. Feynman and the visualization of space-time processes. *Review of Modern Physics*, 58:449, 1986.

[11] H. A. Gersch. Feynman's relativistic chessboard as an ising model. *Int. J. Theor. Physics*, 20:491, 1981.

[12] G.N. Ord. Quantum mechanics in a two-dimensional spacetime: What is a wavefunction? *Annals of Physics*, 324(6):1211–1218, June 2009.

[13] Roger Penrose. *The Road to Reality*. Jonathan Cape, 2004.

[14] David Hestenes. Mysteries and insights of dirac theory. *Annales de la Fondation Louis de Broglie*, 28:390–408, 2003.

[15] Chris Doran and Anthony Lasenby. *Geometric Algebra for Physicists*. Cambridge University Press, 2003.

[16] D. Bohm and B.J. Hiley. *The Undivided Universe*. Routledge, 1993.

The Quantum as an Emergent System

G Grössing[1], S Fussy[1], J Mesa Pascasio[1,2], H Schwabl[1],

[1] Austrian Institute for Nonlinear Studies, Akademiehof, Friedrichstr. 10, 1010 Vienna, Austria
[2] Institute for Atomic and Subatomic Physics, Vienna University of Technology, Operng. 9, 1040 Vienna, Austria

E-mail: ains@chello.at

Abstract. Double slit interference is explained with the aid of what we call "21^{st}century classical physics". We model a particle as an oscillator ("bouncer") in a thermal context, which is given by some assumed "zero-point" field of the vacuum. In this way, the quantum is understood as an *emergent system*, i.e., a steady-state system maintained by a constant throughput of (vacuum) energy. To account for the particle's thermal environment, we introduce a "path excitation field", which derives from the thermodynamics of the zero-point vacuum and which represents all possible paths a particle can take via thermal path fluctuations. The intensity distribution on a screen behind a double slit is calculated, as well as the corresponding trajectories and the probability density current. Further, particular features of the relative phase are shown to be responsible for nonlocal effects not only in ordinary quantum theory, but also in our classical approach.

1. Introduction: "Natural explanations" and emergence

In his book review of Steve Adler's *Quantum Theory as an Emergent Phenomenon* [1], Philip Pearle illustrated the concept of an "enlightenment task" by "...trying to explain the unnatural by the natural – in this case, the 'unnatural' being quantum physics and the 'natural' being classical physics. . . " [2]. Such tasks were already undertaken in the formative years of quantum theory, and Pearle characterized them very aptly in the following way:

> "Classical particles and their dynamics are re-introduced, but a strong element of the unnatural remains. In the deBroglie-Bohm and Madelung models, it is the mysterious quantum force. In the Nelson model, it is the mysterious backward diffusion process (which, together with the usual classical forward diffusion process, forces a particle's drift – its mean position – to be a dynamically determined quantity instead of, as classically, an independent variable set by external influences)."

Now, the "mysterious quantum force" ultimately derives from the "mysterious" wave function Ψ, a solution of the Schrödinger equation, which is an initial ingredient for any Bohmian-type model [3, 4, 5]. Whereas in the latter the existence of Ψ is usually not questioned and thus remains "unnatural" in Pearle's terms, the Nelsonian model actually is an attempt to rely on Newtonian physics only [6, 7, 8], or on what one may thus also call "natural explanations". Pearle's comment apparently leaves two possibilities to overcome the "mysterious" element of the Nelson model, i.e., of backward diffusion. One would be to substitute the combination of forward and backward diffusion processes by some other process where a particle's drift would turn out as a simple classical variable determined by external influences. As this is not viable since, upon the introduction of any usual forward diffusion, one does need a balancing process in opposition to it, the second possibility remains that a particle's drift actually *is* a dynamically determined quantity, albeit in a new framework. This framework, we propose, would have to be some kind of steady-state maintained by a throughput of energy from the ("contextual") environment.

In our "sub-quantum thermodynamics" approach to model quantum systems [9, 10, 11, 12, 13, 14, 15], we pursue the last option as a concrete possibility: not only do we consider *quantum theory* as emergent, but also, more specifically, the *quantum systems* themselves. This means that we refrain from any attempt to model the quantum on some singular, "basic" level only, but rather consider it as a "self-organizing system" whose description requires a more encompassing combination of levels. One of the first definitions in science of such a system was given by Heinz von Foerster, i.e., as "a system which maps order of its environment onto its own organization."[16] From the 1960ies onward, an ever increasing number of studies has been published, in a variety of different disciplines, on self-organization, and later on emergence, respectively. This tendency finally also entered the discussions on the foundations of physics, notably in the works by Ilya Prigogine [17], or in famous essays by P. W. Anderson ("More is Different" [18]), and later by Sam Schweber [19], as well as recently in popularized form by Robert B. Laughlin [20]. In a slight shift of emphasis from self-organization to emergence, Sam Schweber characterized the latter in the following way:

> "It is not enough to know the 'fundamental' laws at a given level. It is the solutions to equations, not the equations themselves, that provide a mathematical description of the physical phenomena. 'Emergence' refers to properties of the solutions – in particular, the properties that are not readily apparent from the equations." [19]

Schweber thus refers to the possibility that the "fundamental laws" on some "basic level" may by themselves not always be sufficient to grasp the essence of some physical phenomenon. This is reflected also in our group's approach where the quantum is modeled as a self-organizing, *dynamical* entity whose complete description needs more than just one (presumably "basic") level. As mentioned, we thus propose *that a quantum system be considered as a well-coordinated emergent system.* In doing so, we consider particle-like and wave-like phenomena as the result of both stochastic and regular dynamical processes. A prototype of such a system is well known from classical physics, viz., the "bouncing droplets" of Couder's group [21, 22, 23, 24, 25], which in fact exhibit a whole series of phenomena reminiscent of quantum ones. Analogously, our group has in recent years attempted to model a quantum as a non-equilibrium steady-state maintained by a permanent throughput of energy. Specifically, we consider a "particle" as a "bouncer" whose oscillations are phase-locked with those of the energy-momentum reservoir of the surrounding "vacuum", i.e., the zero-point field. (Note that the possible existence of a corresponding, underlying "medium" is *a priori* independent of quantum theory. For a similar view, compare, for example, the approach of "stochastic electrodynamics" by Cetto and de la Peña [26]; see also their contribution in this volume. For similar approaches in terms of assuming some sub-quantum, thermodynamic or hybrid-type variant of emergent quantum physics, see [1, 27, 28, 29, 30, 31, 32], and the papers by Adler, Carroll, Elze, Faber, Garbaczewski, Hofer, 't Hooft, Isidro, Khrennikov, Nelson, Ord, Scardigli, Schuch, and Wetterich in this volume.)

In other words, we attempt to model the quantum in a "classical" framework. Note, however, that we consider this framework more specifically to comprise what one may call "21st century classical physics", i.e., including all the recent developments in the fields of nonequilibrium thermodynamics, ballistic diffusion, diffusion wave fields, and the like. More concretely, in assuming that (part of) the "classical" zero-point fluctuations undergo regular oscillatory motion, where the latter is partly caused by and dynamically coupled to the oscillator's frequency ω, we are able to derive fundamental elements of quantum theory from a purely classical approach.

In Ref. [14], as well as in Schwabl et al. (this volume), we have shown how Planck's relation between the energy E and the frequency ω, $E = \hbar\omega$, can be derived from a sub-quantum physics, with Planck's (reduced) constant $\hbar$ indicating a universal angular momentum, and we have also shown that with this relation alone one can derive the exact Schrödinger equation from (modern) classical physics [9, 10]. Moreover, also the stochastic element of the zero-point fluctuations enters decisively into our model, such that in effect we obtain an exact description of free quantum motion via a combination of the propagation of classical Huygens-type waves with diffusion due to stochastic sub-quantum mechanics [11]. We particularly stress that the "particle" is considered as an off-equilibrium steady state oscillator maintained by a constant throughput of energy provided by the zero-point field. Thus, it is exactly this *energy throughput* which is responsible for a particle's *natural drift* in Pearle's sense: it is the permanent absorption and re-emission of kinetic energy which will be explicated below to provide a "natural" diffusion model, albeit one of a specific kind, which is called "ballistic diffusion". So, a quantum in our model emerges from the synchronized dynamical coupling between an oscillator ("bouncer") and its wave-like environment. In sum, with this ansatz, we have been able to derive from "21st century classical physics" the following quantum mechanical features:

- Planck's relation for the energy of a particle,
- the exact (one- and n-particle) Schrödinger equation for conservative and non-conservative systems,
- the Heisenberg uncertainty relations,
- the quantum mechanical superposition principle and Born's rule,
- the quantum mechanical decay of a Gaussian wave packet,
- quantum mechanical interference at the double slit.

Although in our model Huygens' principle applies, it does so only for an idealized combination of a "walking" motion of some velocity $\mathbf{v}$ with the centrally-symmetric diffusion waves' motions of velocity $\mathbf{u}$ orthogonal to it. Effectively, it only holds approximately, disturbed by that part of the accompanying diffusive process which is to be described by the corresponding velocity fluctuations $\delta\mathbf{u}$. In fact, as we have shown for a Gaussian slit, the exact quantum mechanical result can be described as a combination of classical wave mechanics with the addition of a corresponding stochastic diffusion process [11]. We have also shown [15] that the same modelling procedure also perfectly applies to a double slit system. In particular, to make our point as clear as possible, we provide here a more extensive discussion of a simple calculational tool related to what we call the "path excitation field". With it, one can easily derive results for quantum mechanics without ever using complex-valued functions such as wave functions, for example.

2. The path excitation field: A classical explanatory framework for Gaussian dispersion and double-slit interference

To begin with, we recall some of the basic results of our earlier work, including that on diffraction at a single Gaussian slit [11]. We claim for a particle of frequency ω embedded in a stochastic ("zero-point") environment that its *average total energy* is given by the average "total" energy $\overline{\hbar\omega}$ of the particle itself plus a kinetic energy term due to momentum changes $\mathbf{p}_u =: m\mathbf{u}$ which it receives from or gives off to the environment:

$$\overline{E_{\text{tot}}} = \overline{\hbar\omega} + \frac{\overline{p_u^2}}{2m} = \text{const}, \tag{1}$$

where the averaging (as denoted by the bars) is defined in n-dimensional configuration space as

$$\overline{p_u^2} := \int P p_u^2 \, \mathrm{d}^n x. \tag{2}$$

$P = P(\mathbf{x},t)$ refers to the probability density of some relevant distribution. For our model system, the latter is given as a solution of a generalized ("anomalous") diffusion equation, i.e., with a time-dependent diffusion coefficient. As can be seen from (1), the momentum changes can be either positive or negative, and actually will *on the average* balance each other, since they are *a priori* unbiased. The deeper reason for this balancing in our model is due to the fact that we consider the quantum to be a steady-state system in the sense that its "total" energy $\hbar\omega$ is maintained over times $t \gg 1/\omega$ by the permanent throughput of kinetic energy $p_u^2/2m$. In other words, to maintain the steady-state, during the intervals of the average order of $t \simeq 1/\omega$, there will both be an absorption of a momentum $p_u = mu$ and a release of the same amount, $p_u = -mu$, thus providing a "natural" explanation of the involved diffusion processes as envisaged in the introduction.

In our earlier papers [9, 10, 11], we have shown that, apart from the ordinary particle current $\mathbf{J}(\mathbf{x},t) = P(\mathbf{x},t)\mathbf{v}$, we are thus dealing with two additional, yet opposing, currents $\mathbf{J}_u = P(\mathbf{x},t)\mathbf{u}$, which are on average orthogonal to $\mathbf{J}$ [9, 10, 11, 14], and which are the emergent outcome from the presence of numerous corresponding velocities

$$\mathbf{u}_{\pm} = \mp\frac{\hbar}{2m}\frac{\nabla P}{P}. \tag{3}$$

We denote with $\mathbf{u}_+$ and $\mathbf{u}_-$, respectively, the two opposing tendencies of the diffusion process. In the reference frame of a single free particle, and starting at $t = 0$ at the center of the distribution P, the averages obey

$$\overline{\mathbf{u}}_-(\mathbf{x},t) = -\overline{\mathbf{u}}_+(-\mathbf{x},t). \tag{4}$$

Now let us consider an experimental setup with a particle source. To describe the velocity distribution, we introduce a velocity field with average velocity $\overline{\mathbf{v}}$, and amplitudes $R(\mathbf{x},t)$. As mentioned, we refer to their intensities $P = R^2$ as the solutions of a diffusion equation. These typically appear in the form of a Gaussian distribution $P(\mathbf{x},t)$ of possible particle locations $\mathbf{x}$, even if there is only one particle at a time emerging from the corresponding "Gaussian slit", in one dimension for simplicity,

$$P(x,t) = \frac{1}{\sqrt{2\pi}\,\sigma}\mathrm{e}^{-\frac{(x-x_0)^2}{2\sigma^2}}, \tag{5}$$

with the usual variance $\sigma^2 = \overline{(\Delta x)^2} = \overline{(x-x_0)^2}$, and where we choose $x_0 = 0$ furtheron. Regarding $\mathbf{u}$, even in this scenario of one-particle-at-a-time, we deal with an *ensemble* of velocity vectors $\mathbf{u}_\alpha(t)$ representing hypothetical motions on the sub-quantum level in a small volume around $\mathbf{x}$, whose mean value will be given by

$$\mathbf{u}(\mathbf{x},t) = \frac{1}{N(\mathbf{x},t)}\sum_{\alpha=1}^{N(\mathbf{x},t)}\mathbf{u}_\alpha(t). \tag{6}$$

Here, the (typically very large) number N refers to the number of possible path directions of the bouncer due to the existence of the wave-like excitations of the zero-point field. This may be reminiscent of Feynman's picture of photons *virtually* probing every possible path in an experimental setup, but in our case it is a configuration of *real* wave-like excitations, which in a resulting Brownian-type motion guide the bouncer along its path of average velocity $\mathbf{u}$. Again, we note that here we discuss not only a passive guidance of the "particle" by the surrounding wave configurations, but point out also the very active role of the "particle" in (partly) *creating* said wave configurations due to the effects of its bouncing. To account for (3), we split up $\mathbf{u}(\mathbf{x},t)$ according to

$$\mathbf{u}(\mathbf{x},t) = \frac{1}{2N}\left[\sum_{\alpha=1}^{N}\mathbf{u}_{\alpha,+} + \sum_{\alpha=1}^{N}\mathbf{u}_{\alpha,-}\right] = \frac{1}{2}\left[\mathbf{u}_+ + \mathbf{u}_-\right], \tag{7}$$

thus reflecting the isotropy of the diffusion process. Still, the uncontrollable and possibly unknowable velocity field $\mathbf{u}$ representing the Brownian motion of the bouncer may not be operational, but when we take the average according to the rule (2), we obtain a "smoothed-out" *average velocity field*

$$\overline{\mathbf{u}(\mathbf{x},t)} = \int P\mathbf{u}(\mathbf{x},t)\,\mathrm{d}^n x, \tag{8}$$

which is all that we need for our further considerations. Similarly, based on the fact that we have an initial Gaussian distribution of velocity vectors $\mathbf{v}(\mathbf{x},t)$, we define an average velocity field $\overline{\mathbf{v}}$ of the wave propagation as

$$\overline{\mathbf{v}(\mathbf{x},t)} = \int P\mathbf{v}(\mathbf{x},t)\,\mathrm{d}^n x, \tag{9}$$

and make use of an average orthogonality between the two velocity fields, $\mathbf{u}$ and $\mathbf{v}$, [9, 10, 11, 13],

$$\overline{\mathbf{v}\cdot\mathbf{u}} = \int P\mathbf{v}\cdot\mathbf{u}\,\mathrm{d}^n x = 0. \tag{10}$$

In effect, then, the combined presence of both velocity fields $\mathbf{u}$ and $\mathbf{v}$ can be denoted as a *path excitation field*: via diffusion, the bouncer in its interaction with already existing wave-like excitations of the environment creates an "agitated", or "heated-up", thermal "landscape", which can also be pictured by interacting wave configurations all along between source and detector of an experimental setup. Recall that our prototype of a "walking bouncer", i.e., from the experiments of Couder's group, is always driven by its interactions with a superposition of waves emitted at the points it visited in the past. Couder et al. denote this superposition of in-phase waves the "path memory" of the bouncer [25]. This implies, however, that the bouncers at the points visited in "the present" necessarily create new wave configurations which will form the basis of a "path memory" in the future. In other words, the wave configurations of the past determine the bouncer's path in the present, whereas its bounces in the present co-determine the wave configurations at any of the possible locations it will visit in the future. Therefore, we call the latter configurations the *path excitation field*, which may also be described as "heated-up" thermal field. As in the coupling of an oscillator with classical diffusion, diffusion wave fields arise with instantaneous field propagation [10, 33], one has elements of the whole setup which may be nonlocally oscillating ("breathing") in phase. This means that the Gaussian of (5) does represent a nonlocal path excitation field in that it is a physically existing and effective entity responsible for where the bouncing "particle" can possibly go. As we have shown [11], one can in this classical framework, along with the time-dependence of the diffusivity, effectively and easily describe the (sub-)quantum physics of diffraction at a Gaussian slit, which we now briefly recapitulate.

At first we note that Eq. (1) is an *average* energy conservation law only. This means that apart from the momentum changes $\mathbf{p}_u = \pm m\mathbf{u}$ discussed so far, also variations in $\mathbf{p}_u$ will have to be taken into account, and thus also variations in the "particle energy" $\hbar\omega$. If for the latter one just considers its kinetic energy term, $mv^2/2$, then said variations will lead to exchanges of velocity/momentum terms providing the net balance

$$m\delta\mathbf{v} = m\delta\mathbf{u}. \tag{11}$$

Using the expression (3) for $\mathbf{u}$, one obtains with the Gaussian (5), $\overline{\mathbf{p}_u^2}\,|_{t=0} =: m^2 u_0^2$, and with $\overline{(\nabla \ln P)^2} = -\overline{\nabla^2 \ln P}$, that

$$u_0^2 = \frac{D^2}{\sigma_0^2} = \overline{u^2} + \overline{(\delta u)^2} = \frac{D^2}{\sigma^2} + \overline{(\delta u)^2}, \tag{12}$$

where as usual $\sigma = \sigma(t) = \sqrt{\overline{x^2}}$ for $x_0(t=0) = 0$, and $\sigma_0 = \sigma(t=0)$. One can view the Gaussian distribution P of kinetic energy also as a sort of "heat accumulation", which has its maximum at the center. Considering now the application of momentum fluctuations (up to second order) to a particle with initial ($t=0$) distance $x(0)$ from said center, with the fluctuation term for $t>0$ defined as $\mathbf{p}_u \pm \delta\mathbf{p}_u = \pm m(\mathbf{u} \pm \delta\mathbf{u})$, one obtains at time t the envisaged "natural" drift as

$$\mathbf{x}(t) = \mathbf{x}(0) \pm (\mathbf{u} \pm \delta\mathbf{u})\, t. \tag{13}$$

Squaring Eq. (13) and forming the r.m.s. can then easily be shown to provide [11],

$$\overline{x^2}(t) = \overline{x^2}(0) + \left[\overline{u^2} + \overline{(\delta u)^2}\right] t^2 = \overline{x^2}(0) + u_0^2 t^2. \tag{14}$$

Comparing with Eq. (12) also provides the time evolution of the wave packet's variance as

$$\sigma^2 = \sigma_0^2 \left(1 + \frac{D^2 t^2}{\sigma_0^4}\right), \tag{15}$$

and finally the average velocity field of a Gaussian wave packet as

$$v_{\mathrm{tot}}(t) = v(t) + [x(t)]\, \frac{u_0^2 t}{\sigma^2}. \tag{16}$$

Note that Eqs. (15) and (16) are derived solely from classical physics. Still, they are in full accordance with quantum theory, and in particular with Bohmian trajectories [5]. Note also that one can rewrite Eq. (14) such that it appears like a linear-in-time formula for Brownian motion,

$$\overline{x^2} = \overline{x^2(0)} + D(t)\, t, \tag{17}$$

where a time dependent diffusivity

$$D(t) = u_0^2\, t = \frac{\hbar^2}{4m^2\sigma_0^2}\, t \tag{18}$$

characterizes Eq. (17) as *ballistic diffusion*. The appearance of a time-dependent $D(t)$ is essential, but also straightforward. The diffusivity is changed over time, because the "particle's" thermal environment changes: With the "heat" initially concentrated within the narrow spatial constraints determined by σ_0 of the source ("Gaussian slit"), $D(t)$ must become larger with time because of the gradually lower heat concentration due to dissipation into the unconstrained environment. (Note that a similar scenario was suggested by Garbaczewski [29]; others are presently intensively discussed in the context of the so-called "superstatistics" [34].) This makes it possible to simulate the dispersion of a Gaussian wave packet on a computer by simply employing coupled map lattices for classical diffusion, with the diffusivity given by Eq. (18). (For detailed discussions, see refs. [11] and [13].)

With the essentials of Gaussian dispersion at our disposal, it is very simple to now also describe and explain quantum interference with our approach. [15] We have chosen a textbook scenario in the form of the calculation of the intensity distribution and the particle trajectories in an electron interferometer. As we are also interested in the trajectories, we refer to, and compare our results with, the well-known work by Phillipidis et al. [35], albeit in the form as presented by Holland [5].

We choose similar initial situations as in [5], i.e., electrons (represented by plane waves in the forward y-direction) from a source passing through "soft-edged" slits 1 and 2 in a barrier (located along the x-axis) and recorded at a screen. In our model, we therefore note two

Gaussians representing the totality of the effectively "heated-up" path excitation field, one for slit 1 and one for slit 2, whose centers have the distances $+X$ and $-X$ from the plane spanned by the source and the center of the barrier along the y-axis, respectively.

As it is well known from classical wave mechanics, the total amplitude R of two coherent waves with (suitably normalized) amplitudes $R_i = \sqrt{P_i}$, $i = 1$ or 2, is given by

$$\begin{aligned} R(\mathbf{k_1}, \mathbf{k_2}, \mathbf{r}) &= R_1 \cos(\omega t - \mathbf{k}_1 \cdot \mathbf{r} + \varphi_0) + R_2 \cos(\omega t - \mathbf{k}_2 \cdot \mathbf{r} + \varphi_0) \\ &= R_1 \cos(\omega t - \varphi_1) + R_2 \cos(\omega t - \varphi_2), \end{aligned} \tag{19}$$

where φ_0 is some initial phase at $t = 0$, and $\varphi_i = \mathbf{k}_i \cdot \mathbf{r} - \varphi_0$. Considering the average of this expression over a multitude of similar wave superpositions, one obtains as usual the *averaged total intensity*

$$P_{\text{tot}} := R^2 = R_1^2 + R_2^2 + 2R_1R_2 \cos\varphi = P_1 + P_2 + 2\sqrt{P_1P_2}\cos\varphi, \tag{20}$$

where $\varphi = \varphi_1 - \varphi_2 = (\mathbf{k}_1 - \mathbf{k}_2) \cdot \mathbf{r}$. Note that the relative phase difference φ enters Eq. (20) only via the cosine function, such that, e.g., even if the total wave numbers (and thus also the total momenta) $\mathbf{k}_i$ were of vastly different size, the cosine effectively makes Eq. (20) independent of said sizes, but dependent only on an angle modulo 2π. This will turn out as essential for our discussion further below.

Now, the x-components of the centroids' motions from the two alternative slits 1 and 2, respectively, are given by the "particle" velocity components

$$v_x = \pm \frac{\hbar}{m} k_x, \tag{21}$$

respectively, such that the relative group velocity of the Gaussians spreading into each other is given by $\Delta v_x = 2v_x$. However, in order to calculate the phase difference φ descriptive of the interference term of the intensity distribution (20), one must take into account the total momenta involved, i.e, one must also include the wave packet dispersion as described in the previous Chapter. Thus, one obtains with the displacement $\pm x(t) = \mp(X + v_x t)$ in Eq. (16) the total relative velocity of the two Gaussians as

$$\Delta v_{\text{tot},x} = 2\left[v_x - (X + v_x t)\frac{u_0^2 t}{\sigma^2}\right]. \tag{22}$$

Therefore, the total phase difference between the two possible paths (i.e., through either slit) becomes

$$\varphi = \frac{1}{\hbar}(m\Delta v_{\text{tot},x}\, x) = 2mv_x \frac{x}{\hbar} - (X + v_x t)x \frac{1}{D}\frac{u_0^2 t}{\sigma^2}. \tag{23}$$

The Gaussians P_1 and P_2 for the corresponding slits are given as

$$P_1(x,t) = \frac{1}{\sqrt{2\pi\sigma^2}} \mathrm{e}^{-[x-(X+v_x t)]^2/2\sigma^2}, \tag{24}$$

and

$$P_2(x,t) = \frac{1}{\sqrt{2\pi\sigma^2}} \mathrm{e}^{-[x+(X+v_x t)]^2/2\sigma^2}. \tag{25}$$

With equal amplitudes $R = R_i = \sqrt{P_i}$, for $i = 1, 2$, of the Gaussians, and with normalization constant N, we thus obtain the usual interference pattern in the form of the intensity distribution:

$$P_{\text{tot}}(x,t) = R^2 N^2 \frac{1}{\sqrt{2\pi\sigma^2}} \mathrm{e}^{-[x^2+(X+v_x t)^2]/2\sigma^2} \left\{ e^{x(X+v_x t)/\sigma^2} + e^{-x(X+v_x t)/\sigma^2} + 2\cos\varphi \right\}, \tag{26}$$

with the relative phase φ given by Eq. (23). Whereas the first exponential describes the enveloping Gaussian, the term inside the curly brackets of Eq. (26) describes the interference fringes whose "dark" nodes are at the locations well-known from textbooks,

$$x = (n + \frac{1}{2})\pi / k_x, \tag{27}$$

if $X = -v_x t$, as the wave packets must approach each other ($v_x < 0$) for $t > 0$.

Fig. 1 depicts the interference of two beams emerging from Gaussian slits established with the aid of a purely classical simulation. As in the case of a Gaussian slit [11], we again simulate diffusion with a time-dependent diffusivity $D(t)$. To account for interference, we simply follow the classical rule for the intensities (20), with φ from Eq. (23). The averaged trajectories are the flux lines obtained by choosing a set of equidistant initial points at $y = 0$. Two adjacent flux lines thereby define regions of constant flux, i.e., $\int_A P \, \mathrm{d}A = \text{const.}$, with A being the cross section of a flux tube.

Just as with the dispersion of a single Gaussian, we want to stress that also the quantum interference pattern (26) has been derived here without the use of quantum mechanics, but solely on the basis of classical physics. Exploiting our concept of the "path excitation field", we have for this derivation implicitly used the assumption of velocity fields $\mathbf{u}(\mathbf{x}, t)$ and $\mathbf{v}(\mathbf{x}, t)$, respectively, which have entered the expression of a Gaussian's average velocity field, Eq. (16), and which in turn were shown to essentially contribute to the interference pattern (26). In what follows, we now want to make the use of these velocity fields more explicit, i.e., we shall now concentrate on understanding the emerging particle trajectories during quantum interference on the basis of our classical velocity fields. As it turns out, a thorough consideration of the nature of the relative phase φ will make it possible to obtain a deeper understanding of quantum interference in general.

3. Geometric meaning of the path excitation field

Let us consider a single, classical "particle" ("bouncer") following the propagation of a set of waves of equal amplitude R_i, each representing one of i possible alternatives according to our principle of path excitation. We first note that for the superposition of two weighted wave vectors, with resultant vector $\mathbf{k}$ and total amplitude R_{tot},

$$R_{\text{tot}} \mathbf{k} = R_1 \mathbf{k}_1 + R_2 \mathbf{k}_2, \tag{28}$$

where the *averaged scalar product* of the two associated unit vectors is given by $\hat{\mathbf{k}}_1 \cdot \hat{\mathbf{k}}_2 = \cos\varphi$ in accordance with (20).

We now focus on the specific role of the velocity fields $\mathbf{u}$, which were present in Section 2 only implicitly in the expressions (20) and (23). To describe the required details, each path i be occupied by a Gaussian wave packet with a "forward" momentum $\mathbf{p}_i = \hbar \mathbf{k}_i = m \mathbf{v}_i$. Moreover, due to the stochastic process of path excitation, the latter has to be represented also by a large number N of consecutive Brownian shifts, $\mathbf{p}_{u,\alpha} = m \mathbf{u}_\alpha$. Recalling (7), one obtains for the case of interference at a double slit the total averaged velocity field (with indices $i = 1$ or 2 referring to the two slits)

$$\overline{\mathbf{v}}_{\text{tot}} = \overline{\mathbf{v}}_{\text{tot},1} + \overline{\mathbf{v}}_{\text{tot},2} := \overline{\mathbf{v}}_1 + \frac{\overline{\mathbf{u}}_{1+}}{2} + \frac{\overline{\mathbf{u}}_{1-}}{2} + \overline{\mathbf{v}}_2 + \frac{\overline{\mathbf{u}}_{2+}}{2} + \frac{\overline{\mathbf{u}}_{2-}}{2}. \tag{29}$$

With two Gaussian distributions $P_1 = R_1^2$ and $P_2 = R_2^2$ as given in the previous Chapter, one has the corollary of (28), i.e.,

$$R_{\text{tot}} \overline{\mathbf{v}}_{\text{tot}} = R_1 \overline{\mathbf{v}}_{\text{tot},1} + R_2 \overline{\mathbf{v}}_{\text{tot},2} \, , \tag{30}$$

Figure 1. Classical computer simulation of the interference pattern: intensity distribution with increasing intensity from white through yellow and orange, with averaged trajectories (red) for two Gaussian slits, and with large dispersion (evolution from bottom to top; $\overline{v_{x,1}} = \overline{v_{x,2}} = 0$). The interference hyperbolas for the maxima characterize the regions where the phase difference $\varphi = 2n\pi$, and those with the minima lie at $\varphi = (2n+1)\pi$, $n = 0, 1, 2, \ldots$ Note in particular the "kinks" of trajectories moving from the center-oriented side of one relative maximum to cross over to join more central (relative) maxima. In our classical explanation of interference, a detailed micro-causal account of the corresponding kinematics can be given. The averaged trajectories follow a Bohm-type "no crossing" rule: particles from the left slit stay on the left side and *vice versa* for the right slit. This feature is explained here by a sub-quantum build-up of kinetic (heat) energy acting as an emergent repellor along the symmetry line.

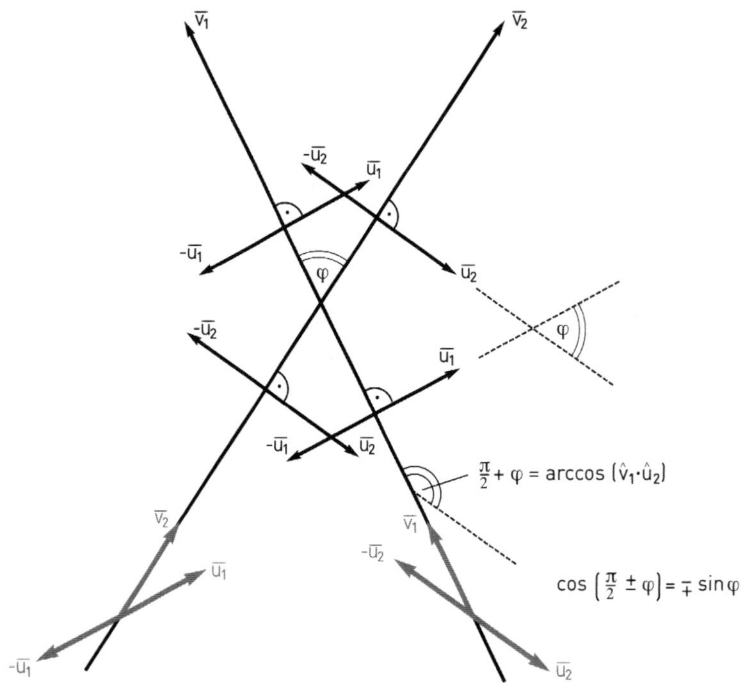

Figure 2. Schematic of the phase angles between various components of the path excitation field behind a double slit. $\overline{u_i}$ and $\overline{v_i}$, $i = 1$ or 2, denote the average velocity fields involved, with $\overline{v_i}$ referring to single free "particle" velocities (i.e., either through slit 1 or slit 2), and the $\overline{u_i}$ referring to additional diffusion velocities. The thick arrows (red online) indicate how the mixing of velocity fields $\overline{v_i}$ and $\overline{u_j}$, with $i \neq j$, produces the terms proportional to $\sin\varphi$ in Eq. (34). Note that angles such as φ in the schematic symbolize the actual phase angles between any two vectors, and must therefore be understood as emerging out of the interaction of all the velocity/momentum vectors involved in a particular point in space. So, the schematic depicts the totality of all the "excited" velocity/momentum fields involved, labeled separately for each channel. Upon actual superposition of the two channels in $x-$ space, i.e., in the course of the interactions among all these field excitations, one then obtains the final intensity distributions and, according to Eq. (35), the particle trajectories whose emergent behaviour turns out to be characterized by a "no crossing" rule (see Fig. 1). In other words, there is no simple linear superposition between the corresponding processes for each separate channel 1 and 2, respectively, but a complex evolution of all the involved sub-quantum processes that leads to the emergent interference pattern and the associated trajectories.

and thus

$$R_{\rm tot} = \left[R_1\left(\overline{\mathbf{v}}_1 + \frac{\overline{\mathbf{u}}_{1+}}{2} + \frac{\overline{\mathbf{u}}_{1-}}{2}\right) + R_2\left(\overline{\mathbf{v}}_2 + \frac{\overline{\mathbf{u}}_{2+}}{2} + \frac{\overline{\mathbf{u}}_{2-}}{2}\right)\right]\frac{\hat{\mathbf{v}}_{\rm tot}}{|\overline{\mathbf{v}}_{\rm tot}|}. \tag{31}$$

To help with the bookkeeping, the schematic of Fig. 2 displays all the relevant vectors and some of the corresponding phase angles. Taking into account the conservation of the "particle momentum" in both channels, we have $|\overline{\mathbf{v}}_{\rm tot}| = |\overline{\mathbf{v}}_{{\rm tot},1}| = |\overline{\mathbf{v}}_{{\rm tot},2}|$, leading to

$$R_{\rm tot} = \left[R_1\left(\hat{\mathbf{v}}_1 + \frac{\hat{\mathbf{u}}_{1+}}{2} + \frac{\hat{\mathbf{u}}_{1-}}{2}\right) + R_2\left(\hat{\mathbf{v}}_2 + \frac{\hat{\mathbf{u}}_{2+}}{2} + \frac{\hat{\mathbf{u}}_{2-}}{2}\right)\right]\hat{\mathbf{v}}_{\rm tot}. \tag{32}$$

With Eq. (4), and setting $\mathbf{u}_{1+} \to \mathbf{u}_1$ and $\mathbf{u}_{1-} \to -\mathbf{u}_1$, one obtains the total average current

$$\begin{aligned}
\overline{\mathbf{J}}_{\rm tot} &= R_{\rm tot}^2 \overline{\mathbf{v}}_{\rm tot} = R_1^2 \overline{\mathbf{v}}_1 + R_2^2 \overline{\mathbf{v}}_2 \\
&+R_1R_2 \Big\{ (\overline{\mathbf{v}}_1 + \overline{\mathbf{v}}_2)\cos(\hat{\mathbf{v}}_1, \hat{\mathbf{v}}_2) + \left(\overline{\mathbf{v}}_1 + \frac{\overline{\mathbf{u}}_2}{2}\right)\cos(\hat{\mathbf{v}}_1, \hat{\mathbf{u}}_2) - \left(\overline{\mathbf{v}}_1 - \frac{\overline{\mathbf{u}}_2}{2}\right)\cos(\hat{\mathbf{v}}_1, \hat{\mathbf{u}}_2) \\
&+ \left(\frac{\overline{\mathbf{u}}_1}{2} + \overline{\mathbf{v}}_2\right)\cos(\hat{\mathbf{u}}_1, \hat{\mathbf{v}}_2) - \left(-\frac{\overline{\mathbf{u}}_1}{2} + \overline{\mathbf{v}}_2\right)\cos(\hat{\mathbf{u}}_1, \hat{\mathbf{v}}_2) + \left(\frac{\overline{\mathbf{u}}_1}{2} + \frac{\overline{\mathbf{u}}_2}{2}\right)\cos(\hat{\mathbf{u}}_1, \hat{\mathbf{u}}_2) \\
&- \left(\frac{\overline{\mathbf{u}}_1}{2} - \frac{\overline{\mathbf{u}}_2}{2}\right)\cos(\hat{\mathbf{u}}_1, \hat{\mathbf{u}}_2) - \left(-\frac{\overline{\mathbf{u}}_1}{2} + \frac{\overline{\mathbf{u}}_2}{2}\right)\cos(\hat{\mathbf{u}}_1, \hat{\mathbf{u}}_2) + \left(-\frac{\overline{\mathbf{u}}_1}{2} - \frac{\overline{\mathbf{u}}_2}{2}\right)\cos(\hat{\mathbf{u}}_1, \hat{\mathbf{u}}_2) \Big\},
\end{aligned} \tag{33}$$

and thus finally

$$\overline{\mathbf{J}}_{\rm tot} = P_1 \overline{\mathbf{v}}_1 + P_2 \overline{\mathbf{v}}_2 + \sqrt{P_1 P_2}\,(\overline{\mathbf{v}}_1 + \overline{\mathbf{v}}_2)\cos\varphi + \sqrt{P_1 P_2}\,(\overline{\mathbf{u}}_1 - \overline{\mathbf{u}}_2)\sin\varphi. \tag{34}$$

Note that Eq. (34), upon the identification of $\overline{\mathbf{u}}_i = -\frac{\hbar}{m}\frac{\nabla R_i}{R_i}$ from Eq. (3) and with $P_i = R_i^2$, turns out to be in perfect agreement with a comparable "Bohmian" derivation [5, 36]. The formula for the averaged particle trajectories, then, simply results from Eq. (33), i.e.,

$$\overline{\mathbf{v}}_{\rm tot} = \frac{\overline{\mathbf{J}}_{\rm tot}}{P_{\rm tot}} = \frac{P_1 \overline{\mathbf{v}}_1 + P_2 \overline{\mathbf{v}}_2 + \sqrt{P_1 P_2}\,(\overline{\mathbf{v}}_1 + \overline{\mathbf{v}}_2)\cos\varphi + \sqrt{P_1 P_2}\,(\overline{\mathbf{u}}_1 - \overline{\mathbf{u}}_2)\sin\varphi}{P_1 + P_2 + 2\sqrt{P_1 P_2}\cos\varphi}. \tag{35}$$

In Fig. 1 one can observe a basic characteristic of the (averaged) particle trajectories, which, just because of the averaging, are identical with the Bohmian trajectories. In particular, due to the "no crossing" rule for Bohmian trajectories, the particles coming from, say, the right slit (and expected at the left part of the screen if a presumed "classical" momentum conservation should hold) actually arrive at the right part of the screen (and *vice versa* for the other slit). In our sub-quantum approach an explanation of the "no crossing" rule is actually a consequence of a detailed microscopic momentum conservation. As can be seen in Fig. 1, the (Bohmian) trajectories are repelled from the central symmetry line. However, in our case this is only implicitly due to a "quantum potential", but actually due to the identification of the latter with a kinetic (rather than a potential) energy: As has already been stressed in [10], it is the "heat of the compressed vacuum" that accumulates along said symmetry line (i.e., as reservoir of "outward" oriented kinetic energy) and therefore repels the trajectories.

The trajectories in Fig. 1 exactly obey Eq. (35) for the two Gaussian slits shown. The interference hyperbolas for the maxima characterize the regions where the phase difference $\varphi = 2n\pi$, and those with the minima lie at $\varphi = (2n+1)\pi$, $n = 0, 1, 2, \ldots$ Note in particular the "kinks" of trajectories moving from the center-oriented side of one relative maximum to cross over to join more central (relative) maxima. In addition to the full accordance with the trajectories obtained from the Bohmian approach (see [4], [5], [36] and [37], for example), in our classical explanation of interference a detailed "micro-causal" account of the corresponding kinematics can be given: Firstly, we note that the last term in Eq. (34), which is responsible for the genuinely "quantum" behaviour, determines the movement towards the symmetry line. This term is characterized by the product of the vector $\overline{\mathbf{u}}_1 - \overline{\mathbf{u}}_2$ and $\sin\varphi$, the combined effect of which results in the kinks typical for Bohmian trajectories (Fig. 1).

Thus, in the cases where the trajectories come from a relative maximum (bright fringe), the particles loose velocity/momentum in the direction towards the symmetry line and cross over into the area of the adjacent relative minimum (dark fringe). From there, they gain velocity/momentum in the direction towards the symmetry line and thus align with the other trajectories of the next bright fringe. In other words, one obtains areas where part of the current (along a relative maximum) is being removed ("depletion"), or where parts of currents flow together to produce a newly formed bright fringe ("accumulation"), respectively. This is in accordance with our earlier description of quantum interference, where the effects of diffusion wave fields were explicitly described by alternating zones of heat accumulation or depletion, respectively [10]. Towards the central symmetry line, then, one observes heat accumulation from both sides, and due to big momentum kicks from the central accumulation of heat energy, the forward particle velocities' directions align parallel to the symmetry axis. With the crossing-over of particle trajectories being governed by the last, diffusion-related, term on the right hand side of Eq. (34), one finds that for $\varphi = 0$ the resulting diffusive current is zero and thus, as total result of the overall kinematics, no crossing is possible. Further, we note that our results are also in agreement with the recently published experimental results by Kocsis et al. [38]. Here we just comment that, as opposed to the Bohmian interpretation, we give a micro-causal explanation of these results solely on classical grounds.

4. The meaning of the relative phase: Quantum superposition, modular momentum and the nonlocality of the path excitation field

Although we have obtained the usual quantum mechanical results, we have so far not used the quantum mechanical formalism in any way. However, upon employment of the Madelung transformation for each path j ($j = 1$ or 2),

$$\Psi_j = R e^{iS_j/\hbar}, \tag{36}$$

and thus $P_j = R_j^2 = |\Psi_j|^2 = \Psi_j^* \Psi_j$, with the definitions (3) and $v_j := \nabla S_j/m$, $\varphi = (S_1 - S_2)/\hbar$, and recalling the usual trigonometric identities such as $\cos\varphi = \frac{1}{2}\left(e^{i\varphi} + e^{-i\varphi}\right)$, etc., one can rewrite the total average current (34) immediately as

$$\begin{aligned} \overline{\mathbf{J}}_{\mathrm{tot}} &= P_{\mathrm{tot}} \overline{\mathbf{v}}_{\mathrm{tot}} \\ &= (\Psi_1 + \Psi_2)^*(\Psi_1 + \Psi_2)\frac{1}{2}\left[\frac{1}{m}\left(-i\hbar\frac{\nabla(\Psi_1 + \Psi_2)}{(\Psi_1 + \Psi_2)}\right) + \frac{1}{m}\left(i\hbar\frac{\nabla(\Psi_1 + \Psi_2)^*}{(\Psi_1 + \Psi_2)^*}\right)\right] \\ &= -\frac{i\hbar}{2m}\left[\Psi^*\nabla\Psi - \Psi\nabla\Psi^*\right] = \frac{1}{m}\mathrm{Re}\left\{\Psi^*(-i\hbar\nabla)\Psi\right\}, \end{aligned} \tag{37}$$

where $P_{\mathrm{tot}} = |\Psi_1 + \Psi_2|^2 =: |\Psi|^2$. The last two expressions of (37) are the exact well-known formulations of the quantum mechanical probability current, here obtained without any quantum mechanics, but just by a re-formulation of (34). In fact, it is a simple exercise to insert the wave functions (36) into (37) to re-obtain (34).

It is important to note that while the total wave-function $\Psi = \Psi_1 + \Psi_2$ obviously obeys the quantum mechanical superposition principle, no linear superposition principle holds for our total current $\overline{\mathbf{J}}_{\mathrm{tot}}$, as can easily be seen from Eq. (34). In accordance with 't Hooft's arguments at this conference ['t Hooft, this volume], we have thus demonstrated with an explicit model how the quantum mechanical superposition principle is only a calculatory means to describe the effects of sub-quantum processes, which are actually to be understood as complex behaviours of ontological microscopic states. This scenario is reassuring, as the constructed linearity of quantum mechanics appears to be "sandwiched" between the sub-quantum and the classical macro-levels, respectively, i.e., levels where the superposition principle does not hold. So, applying linear superposition to states other than those constructed for practical purposes only would lead to wrong predictions about the behaviours of ontological states.[1]

[1] We thus completely agree with 't Hooft's very basic statement: "Quantum wave functions were introduced for the convenience of the computations; linearity came as a handy tool for making calculations, but it so happens that quantum superpositions of ontological states themselves do not describe any real world, and this, as it turns out now, explains why we do not see quantum mechanical superpositions occuring in the macro world." ['t Hooft, this volume]

Now that the identity between the versions of our classically derived expression for the average current and the quantum one is established, we have to confront the claim that it was impossible to reproduce quantum results with a classical wave theory, because the meaning of a quantum phase apparently was very different from the meaning of a classical phase. This claim is detailed particularly clearly in [39], where the authors look for a mechanism to explain how the particle, say, at the right, "knows" what is happening at the left slit (i.e., whether the latter is closed or open, for example). Their explanation of interference from the single-particle perspective is based on non-local Heisenberg equations of motion for "modular variables" like the modular momentum, for example. With p denoting the usual momentum, the modular momentum is defined as $p_{mod} := p \bmod \frac{h}{d} = p - n\frac{h}{d}$, where d is the distance between the slits. As $p \bmod \frac{\mathrm{h}}{\mathrm{d}}$ has the topology of a circle, nothing changes if in the equations one replaces p with $p - n\frac{h}{d}$. The main argument now is that whereas the ordinary momentum (as well as all its higher moments) is independent of the relative phase between Ψ_1 and Ψ_2, the modular momentum is not. In other words, the authors claim that the sensitivity of the modular momentum towards changes in the relative phase makes it impossible to apply classical intuitions to double slit interference: As the relative phase is a truly non-local feature of quantum mechanics, they claim, classical notions would have to fail to describe the essence of interference. Physically, therefore, it is a non-local effect of having an open or closed slit to produce a shift in the particle's modular momentum while the expectation values of its ordinary momentum remain unaffected.

However, as we have shown, we have an identity in the outcomes of our classical calculations with the quantum ones. How can this come about? The answer is clearly given by the fact that the path excitation field is, by its very definition, a non-local field. Now, one may argue that non-local fields should have nothing to do with "classical" physics, but as we have repeatedly stressed, nothing speaks *a priori* against the assumption of some non-locally distributed zero-point oscillations in a purely classical context. Once this is accepted, the rest follows straightforwardly: instead of using the language of a "rotation in the space of a modular variable", like the "non-local exchange of modular momentum", for example, we have discussed and described interference with the non-local path excitation field, where the angle between two unit vectors representing the respective velocity fields originating from the two slits is given by the relative phase. As in [39] the parameter relevant to describe the effects of an open or a closed slit, respectively, is given by the distance between the slits appearing in the modular momentum approach, so does this same distance also appear in our description, i.e., it is explicitly given in the formula for the relative phase φ in Eq. (23). Thus, the appearance of the distances X in our expression for φ essentially demonstrates the latter to be a non-local one. For example, the closing of one slit at either $-X$ or $+X$ has an immediate effect in that for $v_x = 0$ the last term of the relative phase (23) is changed by a factor of $\frac{1}{2}$.

Moreover, if in Eq. (23) one discards the diffusion-related term, then φ becomes "classical" in the usual sense of the word, and it is only then that the quantum results could not be reproduced any more. Similarly, in Eq. (34) it is the last term proportional to $\sin\varphi$ which determines the genuinely quantum nature of the whole expression, and it is there where via the nonlocality of the "diffusive" velocities $\mathbf{u}_i$ the nonlocality of quantum mechanics becomes manifest. We have thus shown why our classical approach can produce the results in full accordance with quantum mechanics. A more detailed discussion of nonlocality and entanglement within our scheme will be the subject of a paper in preparation.

5. Conclusions

We have introduced a quantum as an emergent system by considering "particles" as oscillators ("bouncers") coupling to regular oscillations of the "vacuum's" zero-point field, which they also generate. Among other features, the dynamics between the oscillator and the "bath" of its thermal environment can be made responsible not only for Gaussian diffraction at a single slit [11], but also for the well-known interference effects at a double slit [15]. We have also shown how the model entails the existence of a path excitation field, i.e., a field spanned by the average velocity fields $\overline{\mathbf{v}}(\mathbf{x},t)$ and $\overline{\mathbf{u}}(\mathbf{x},t)$, respectively, where the latter refer to diffusion processes reflecting also the stochastic parts of the zero-point field. We have derived, on the basis of classical physics only, the exact intensity distribution at a screen behind a double slit, as well as the details of the more complicated particle current, or of the Bohmian particle trajectories, respectively. In a simple computer simulation, we have modeled quantum interference with simple classical rules employing well-known techniques to model diffusion processes.

Moreover, we have refuted claims about the impossibility to model quantum interference with any classical (and thus also our) model. The decisive feature of said claims, apparently without a classical, or "natural", explanation, is the non-local effect of opening or closing one slit on a particle going through the other slit, an effect which manifests itself in a changed expression for the relative phase. However, we were able to show explicitly within our classical approach that the path excitation field in this case must change to produce exactly the same effect on the relative phase. This both qualifies as a truly non-local effect within our approach and provides the identity with the usual quantum mechanical predictions.

Finally, upon comparison with the usual quantum mechanical formalism, we have demonstrated with our explicit model how the quantum mechanical superposition principle is only a calculatory means using non-ontological wave-functions to describe the effects of sub-quantum processes, which are actually to be understood as complex, nonlinear behaviours of ontological microscopic states.

References

[1] S. L. Adler, Quantum Theory as an Emergent Phenomenon: The Statistical Mechanics of Matrix Models as the Precursor of Quantum Field Theory, Cambridge University Press, Cambridge, 2004.

[2] P. Pearle, Stephen L. Adler, Quantum theory as an emergent phenomenon, Cambridge University Press, Cambridge, Isbn 0521831946, 2004, 238pp., Stud. Hist. Philos. Mod. Phys. 36 (2005) 716–723. `arXiv:quant-ph/0602078v1`.

[3] D. Bohm, A Suggested Interpretation of the Quantum Theory in Terms of "Hidden" Variables. I, Phys. Rev. 85 (1952) 166–179.

[4] D. Bohm, B. J. Hiley, The undivided universe: An ontological interpretation of quantum theory, repr. Edition, Routledge, London, 1993.

[5] P. R. Holland, The Quantum Theory of Motion: An account of the de Broglie-Bohm causal interpretation of quantum mechanics, Cambridge University Press, Cambridge, 1993.

[6] E. Nelson, Derivation of the Schrödinger Equation from Newtonian Mechanics, Phys. Rev. 150 (1966) 1079–1085.

[7] E. Nelson, Quantum fluctuations, Princeton series in physics, Princeton Univ. Press, Princeton, NJ, 1985.

[8] L. Fritsche, M. Haugk, A new look at the derivation of the Schrödinger equation from Newtonian mechanics, Ann. Phys. (Leipzig) 12 (2003) 371–403.

[9] G. Grössing, The Vacuum Fluctuation Theorem: Exact Schrödinger Equation via Nonequilibrium Thermodynamics, Phys. Lett. A 372 (2008) 4556–4563. `arXiv:0711.4954v2`.

[10] G. Grössing, On the thermodynamic origin of the quantum potential, Physica A 388 (2009) 811–823. `arXiv:0808.3539v1`.

[11] G. Grössing, S. Fussy, J. Mesa Pascasio, H. Schwabl, Emergence and collapse of quantum mechanical superposition: Orthogonality of reversible dynamics and irreversible diffusion, Physica A 389 (2010) 4473–4484. `arXiv:1004.4596v1`.

[12] G. Grössing, Sub-Quantum Thermodynamics as a Basis of Emergent Quantum Mechanics, Entropy 12 (2010) 1975–2044. `doi:10.3390/e12091975`.

[13] G. Grössing, S. Fussy, J. Mesa Pascasio, H. Schwabl, Elements of sub-quantum thermodynamics: quantum motion as ballistic diffusion, J. Phys.: Conf. Ser. 306 (2011) 012046. `doi:10.1088/1742-6596/306/1/012046`.

[14] G. Grössing, J. Mesa Pascasio, H. Schwabl, A Classical Explanation of Quantization, Found. Phys. 41 (2011) 1437–1453. `arXiv:0812.3561v4`.

[15] G. Grössing, S. Fussy, J. Mesa Pascasio, H. Schwabl, An explanation of interference effects in the double slit

experiment: Classical trajectories plus ballistic diffusion caused by zero-point fluctuations, Ann. Phys. 327 (2012) 421–437. arXiv:1106.5994v3.

[16] H. von Foerster, On Self-Organizing Systems and Their Environments, in: M. C. Yovits, S. Cameron (Eds.), Self-Organizing Systems, Pergamon Press, London, 1960, pp. 31–50.

[17] I. Prigogine, Non-equilibrium statistical mechanics, John Wiley & Sons Inc, 1962.

[18] P. W. Anderson, More Is Different, Science 177 (1972) 393–396.

[19] S. S. Schweber, Physics, Community and the Crisis in Physical Theory, Phys. Today 46 (1993) 34–40.

[20] R. B. Laughlin, A Different Universe: Reinventing Physics From The Bottom Down, Basic Books, 2006.

[21] Y. Couder, S. Protière, E. Fort, A. Boudaoud, Dynamical phenomena: Walking and orbiting droplets, Nature 437 (2005) 208–208.

[22] Y. Couder, E. Fort, Single-particle Diffraction and Interference at a macroscopic scale, Phys. Rev. Lett. 97 (2006) 154101.

[23] S. Protière, A. Boudaoud, Y. Couder, Particle-wave association on a fluid interface, J. Fluid Mech. 554 (2006) 85–108.

[24] A. Eddi, E. Fort, F. Moisy, Y. Couder, Unpredictable Tunneling of a Classical Wave-Particle Association, Phys. Rev. Lett. 102 (2009) 240401.

[25] E. Fort, A. Eddi, A. Boudaoud, J. Moukhtar, Y. Couder, Path-memory induced quantization of classical orbits, PNAS 107 (2010) 17515 –17520.

[26] L. de la Peña, A. M. Cetto, The quantum dice: an introduction to stochastic electrodynamics, Vol. 75 of Fundamental theories of physics, Kluwer Academic, Dordrecht, 1996.

[27] D. Acosta, P. F. de Cordoba, J. M. Isidro, J. L. G. Santander, An entropic picture of emergent quantum mechanics, pre-print. arXiv:1107.1898.

[28] H. T. Elze, G. Gambarotta, F. Vallone, General linear dynamics – quantum, classical or hybrid, J. Phys.: Conf. Ser. 306 (2011) 012010.

[29] P. Garbaczewski, Derivation of the quantum potential from realistic Brownian particle motions, Phys. Lett. A 162 (1992) 129–136.

[30] A. Khrennikov, Prequantum classical statistical field theory: background field as a source of everything?, J. Phys.: Conf. Ser. 306 (2011) 012021. doi:10.1088/1742-6596/306/1/012021.

[31] G. 't Hooft, Emergent Quantum Mechanics and Emergent Symmetries, AIP Conf. Proc. 957 (2007) 154–163. arXiv:0707.4568.

[32] C. Wetterich, Zwitters: particles between quantum and classical, pre-print. arXiv:0911.1261.

[33] A. Mandelis, Structure and the Reflectionless/Refractionless Nature of Parabolic Diffusion-Wave Fields, Phys. Rev. Lett. 87 (2001) 020801.

[34] C. Beck, Superstatistics: Theoretical Concepts and Physical Applications, in: R. Klages, G. Radons, I. M. Sokolov (Eds.), Anomalous Transport: Foundations and Applications, Wiley-VCH, 2008, pp. 433–457.

[35] C. Philippidis, C. Dewdney, B. J. Hiley, Quantum interference and the quantum potential, Nuov. Cim. B 52 (1979) 15–28.

[36] A. S. Sanz, S. Miret-Artés, A trajectory-based understanding of quantum interference, J. Phys. A: Math. Gen. 41 (2008) 435303. arXiv:0806.2105.

[37] A. S. Sanz, F. Borondo, Contextuality, decoherence and quantum trajectories, Chem. Phys. Lett. 478 (2009) 301–306. arXiv:0803.2581.

[38] S. Kocsis, B. Braverman, S. Ravets, M. J. Stevens, R. P. Mirin, L. K. Shalm, A. M. Steinberg, Observing the Average Trajectories of Single Photons in a Two-Slit Interferometer, Science 332 (2011) 1170 –1173.

[39] J. Tollaksen, Y. Aharonov, A. Casher, T. Kaufherr, S. Nussinov, Quantum interference experiments, modular variables and weak measurements, pre-print. arXiv:0910.4227.

EmerQuM 11: Emergent Quantum Mechanics 2011 IOP Publishing
Journal of Physics: Conference Series **361** (2012) 012009 doi:10.1088/1742-6596/361/1/012009

A causal net approach to relativistic quantum mechanics

R D Bateson [1]

London Centre of Nanotechnology, University College London, 17-19 Gordon St., London, WC1H 0AH, UK.

E-mail: richarddbateson@gmail.com

Abstract.
In this paper we discuss a causal network approach to describing relativistic quantum mechanics. Each vertex on the causal net represents a possible point event or particle observation. By constructing the simplest causal net based on Reichenbach-like conjunctive forks in proper time we can exactly derive the 1+1 dimension Dirac equation for a relativistic fermion and correctly model quantum mechanical statistics. Symmetries of the net provide various quantum mechanical effects such as quantum uncertainty and wavefunction, phase, spin, negative energy states and the effect of a potential. The causal net can be embedded in 3+1 dimensions and is consistent with the conventional Dirac equation. In the low velocity limit the causal net approximates to the Schrödinger equation and Pauli equation for an electromagnetic field. Extending to different momentum states the net is compatible with the Feynman path integral approach to quantum mechanics that allows calculation of well known quantum phenomena such as diffraction.

1. Introduction

Causality, or the concept of "cause and effect", has long been viewed by philosophers as a fundamental and often an *a priori* principle in our understanding and interpretation of nature. Although Newton's laws were clearly written in causal terms, the indeterminacy of quantum mechanics lead to confusion of the role of causality in describing quantum systems. Several attempts have been made to fuse relativity and quantum indeterminism, for example, where branching space-time can be built on the primitives of a set of "possible point events" and causal relations [1] and the recent causal set approach to quantum gravity [2]. The fundamental equation of relativistic quantum mechanics is the Dirac equation for a fermion [3] and is based on the concept of continuous space and time. Richard Feynman [4] presented a discrete space-time derivation of the 1+1 dimension Dirac equation for a free particle – the "Feynman chequerboard" – since a luminal velocity massive particle is viewed in the calculation as "zig-zagging" diagonally forwards through space-time in a similar manner to a bishop in chess. Numerous attempts have been made to achieve a discrete quantum mechanics [5,6,7,8,9] but an exact lattice based formulation has never been achieved. In this paper we discuss a causal network discretisation approach which exactly derives the full 4-vector Dirac equation and provides all the common fermion features, such as spin, negative energy states, action of a potential and summation of

[1] Present address: Oxford Man Institute, University of Oxford, Eagle House, Walton Well Road, OX2 6ED, UK.

paths. The most basic causal net describes a plane wave solution with the space axis aligned along the direction of momentum. This 1+1 dimension net can be embedded in 3+1 dimension space-time using the Pauli matrices and is consistent with the full Dirac equation and quantum mechanical statistics.

2. Reichenbach's principle of common cause and causal networks

The application of probability theory to causality and its relation to the direction of time was developed by Hans Reichenbach. His *principle of common cause* (PCC) [10] was summarized as follows: "If coincidences of the two events A and B occur more frequently than would correspond to their independent occurrence, that is, if these events satisfy $P(A.B) > P(A)P(B)$ then there exists a common cause C for these events that the fork ACB is conjunctive." That is the probability of A and B occurring together is greater than the product of the individual probabilities of A and B. A conjunctive fork ACB (see Fig. 1) between events is open on one side where C is earlier in time than A or B. This asymmetry Reichenbach argued provides a definition of the flow of the direction of time in terms of microstatistics. Essentially a common cause is expected when coincidences or correlations between events occur repeatedly with greater frequency than complete statistical independence $P(A.B) = P(A)P(B)$. The principle of common cause provides a definition of simultaneity, since if A and B are simultaneous there cannot be a causal linkage between them except through the earlier event C. Reichenbach's principle of common cause can be readily extended into a relativistic framework [11] and a standard simultaneity condition developed [12] where simultaneous events lie on a hyperplane orthogonal to the particle world-line. It is widely accepted that quantum mechanics cannot be developed from a basic application of Reichenbach's principle of common cause with a single conjunctive fork since general quantum mechanical statistics violate the principle. The principle of common cause involving a single conjunctive fork can even actually be used to formulate the well known Bell's theorem [13,14], which motivated the famous experiments by Alain Aspect [15] to exclude the possibilities of certain types of hidden variables. Here we shall consider a modified framework where a complete causal network of possible events is comprised of conjunctive forks such that each possible event has two effective local common causes or screening factors. It appears that, at least in the simple case we consider, this allows our common cause principle - based on the simultaneity of neighbouring possible events - to be applied consistently with quantum statistics.

We shall adopt a relational view of time as an ordered series of closely spaced "events". Now if we consider time as a series of closely spaced events then from this perspective a classical particle trajectory could appear as a statistically correlated series of events in space-time (for example, a series of actual observations). If the correlation is perfect then one may loosely say that an event at one point in space "causes" the event at the next point, providing a Newtonian trajectory. However, if the correlation of events is imperfect, but greater than that resulting from statistical independence, then adjacent events in space are implied to have a common cause originating at a previous time. A trajectory becomes probabilistic in nature and we would have to involve a statistical interpretation. A network of Reichenbach's conjunctive forks [10] constitutes a causal net in which time is ordered and events may be considered simultaneous only when they share a common cause. To construct the causal net for a particle motion in space-time, we consider a 1 dimensional space aligned with the direction of particle motion, and embedded in 3 dimensional space. In this 1 dimensional space the simplest causal net that satisfies our definition of simultaneity is a 1+1 dimensional "diamond" lattice with causal links connecting the lattice points as in Figure 2. Each causal connection is defined by a connecting arrow giving a definite lineal order and an associated probability. Each vertex on the causal net represents a possible event - meaning a possible observation of the particle - and has two incoming and two outgoing causal connections so that each event has two effective possible common causes.

Starting at a vertex and following an outgoing arrow at random at each subsequent vertex describes a "causal chain" as a series of possible events. Measurement or observation at a vertex or a region of the net provides, through Bayesian statistics, a re-evaluation of these probabilities after a measurement. This is illustrated in Figure 2 where an event at A is more likely to have been caused by an event at B than C and D is an impossibility due to zero connectivity between the paths. Bayes theorem provides a way of translating this common sense concept into a formal probabilistic context since $P(A \mid B) > P(A \mid C) > P(A \mid D)$.

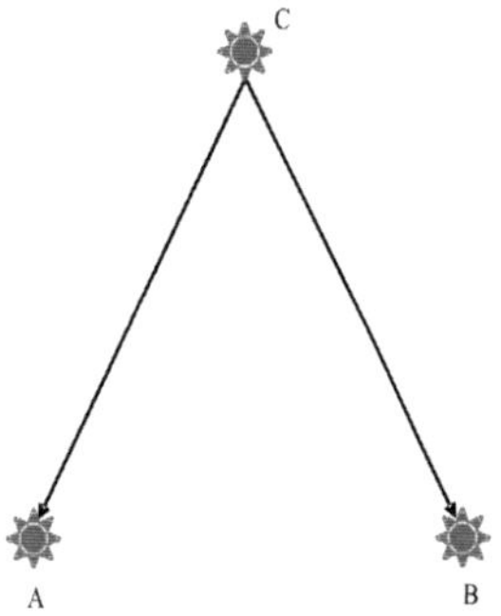

Figure 1. Reichenbach conjunctive fork linking events A and B with common cause C. C is earlier in time than simultaneous events A and B.

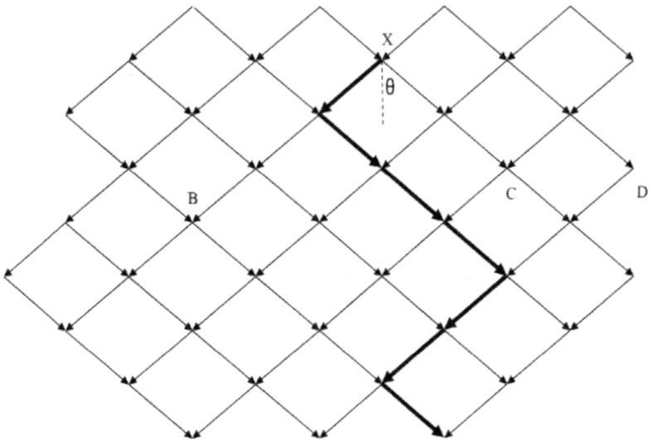

Figure 2. Causal net showing causal chain from X to Y.

3. Relativistic causal nets for a free particle

First, we will consider the simple case of a particle randomly diffusing on the causal net shown in Figure 2. In this model time and space can be discretely "counted" by attaching an integer to each of the vertex points but there is no underlying continuous space-time. To relate to conventional mechanics we interpolate this set of integers by a set of real number coordinates. Expecting that space and time have different dimensions we need to introduce a constant c with dimensions [space/time]. The net is then made up of elementary triangles labelled with $(\Delta x, c\Delta t, c\Delta\tau)$ as shown in Figure 3. We have not yet added any specific interpretation to these quantities. However, to guarantee invariance of causality on the net we impose c as the speed of light [16]. Since, from geometry, $\Delta x/c\Delta t = \sin\theta \leq 1$, we then identify Δx and Δt as relativistic space-time intervals in an observer frame S' and Δt as the particle proper time interval in its rest frame S. The net geometry guarantees the invariant space-time interval

$$(c\Delta\tau)^2 = (c\Delta t)^2 - (\Delta x)^2. \tag{1}$$

Having abandoned the concept of absolute and continuous space-time we need to define the observed velocity in terms of finite differences. The definition we shall adopt is $v = \Delta x/\Delta t$ which we equate to the expectation of the velocity on the causal net. The two time intervals are then related by $\Delta\tau = \Delta t/\gamma$ where $\gamma = 1/\sqrt{1 - v^2/c^2}$ is the Lorentz factor specifying the net angles $\cos\theta = 1/\gamma$ and $\sin\theta = v/c$.

We now specialise to the case of the motion of a free particle. Clearly Eq. (1) and thus the net can be scaled by a factor. If we identify this with the particle rest mass m then from Eq. (1) we then have the relativistic dispersion relation $E^2 = p^2c^2 + m^2c^4$ where E is the particle energy $E = \gamma mc^2$ and $p = \gamma mv$ the momentum. We can further rearrange to derive a second useful invariant relation $-mc^2\Delta\tau = p\Delta x - E\Delta t$ and a third, the Lagrangian for a free particle $L = -mc^2/\gamma = pv - E = pv - H$ where H is the Hamiltonian.

From our definition of simultaneity and the geometry of the net we can see that the invariant relations provide an action $\sum p\Delta x$ which is the same on the lattice for all paths between two events. This is a restatement of Maupertuis principle, which is a weak form of the well known principle of least action, that is the integral $\int p\, dx$ is stationary. On our net if the action $\sum p\Delta x$ differed for different trajectories then this would rule some trajectories as physically inadmissible. Therefore we conclude that $\sum p\Delta x$ is the same on the lattice for all paths between two events which implies that $p\Delta x$ is a constant η for a valid causal net.

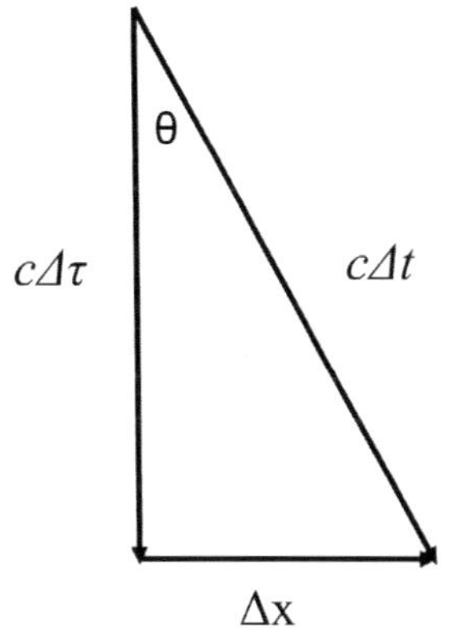

Figure 3. The elementary space-time triangle for the causal net.

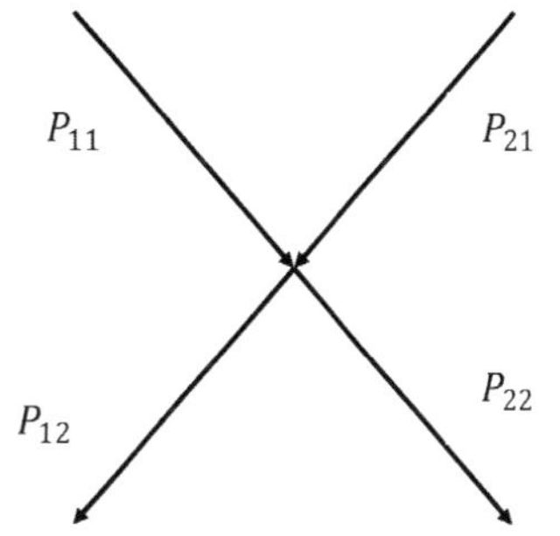

Figure 4. A vertex (1,2) on the causal net with associated probabilities.

To impose our imperfect correlation of events we shall assume that there is an indeterminism or randomness to the particle motion at each net vertex. We shall make the assumption that this indeterminism is governed by Eq. (1) on the causal net. Thus a particle in its own rest frame S over interval $\Delta\tau$ moving at a speed $|v|$ in inertial frame S' can move to a position $\pm\Delta x$ in time Δt. This produces a random trajectory in space-time (Fig. 2).

Initially we shall consider "classical" or non–quantum probabilities in construction of the causal net. Consider an individual vertex on the net and label the incoming probabilities on row 1 P_{11} and P_{21} and outgoing probabilities on row 2 P_{12} and P_{22} (Fig. 4). Probability is conserved at the vertex and the total probability at a vertex is given by $\Omega_{12} = P_{11} + P_{21}$. If the average velocity measured on the lattice is uniform then $P_{11} = P_{22}$ and $P_{12} = P_{21}$. This implies that the probabilities "cross" at each vertex without actually interfering although the probabilities are coupled. We shall see that this non-interacting case corresponds to the equilibrium case of a free particle. If we consider normalised branching probabilities at the vertex defined as $\hat{P}_{11} + \hat{P}_{21} = 1$ then since expected velocity at the vertex is defined to be v we have

$$E[v] = \gamma\frac{\Delta x}{\Delta t}[\hat{P}_{11} - \hat{P}_{21}] = v. \tag{2}$$

The branching probabilities are then given by

$$\hat{P}_{11} = \frac{E + mc^2}{2E} \qquad \hat{P}_{21} = \frac{E - mc^2}{2E}. \tag{3}$$

From this we can see that in the low velocity limit $|v| \to 0$ then $\hat{P}_{11} \to 1$ and $\hat{P}_{21} \to 0$ and in the high velocity limit $|v| \to c$ then $\hat{P}_{11} \to \hat{P}_{21} \to 1/2$. The branching ratio Γ can be written as a function of γ or the net angle θ.

$$\Gamma = \frac{\hat{P}_{11}}{\hat{P}_{21}} = \frac{E + mc^2}{E - mc^2} = \frac{\gamma + 1}{\gamma - 1} = \frac{1 + \cos\theta}{1 - \cos\theta}. \tag{4}$$

Using the branching probabilities (Eq. 3) we can write a non trivial matrix equation linking the probabilities

$$\vec{P} = \begin{pmatrix} \hat{P}_{22} \\ \hat{P}_{12} \end{pmatrix} = \begin{pmatrix} \hat{P}_{11} \\ \hat{P}_{21} \end{pmatrix} = \begin{pmatrix} 0 & \Gamma \\ 1/\Gamma & 0 \end{pmatrix} \begin{pmatrix} \hat{P}_{11} \\ \hat{P}_{21} \end{pmatrix}. \tag{5}$$

4. Relativistic quantum mechanics on the causal net

We shall now see how the causal net is compatible with the quantum mechanics of the Dirac equation for a free particle. We notice that identifying the net constant η with Planck's constant h provides the de Broglie relation $\lambda p = h$ [17] with $\Delta x = \lambda/2$, and a Heisenberg like relation $\Delta p \Delta x \sim h/2$ [18,19]. The discrete nature of the net automatically entails a de Broglie relation and an uncertainty principle. The probability is invariant in the rest frame S of the particle and we can write each probability by combining complex probability amplitudes $P_{ij} = \phi_{ij}.\phi_{ij}^*$ with

$$\phi_{ij} = \sqrt{P_{ij}} e^{-imc^2\tau/\hbar} = \sqrt{P_{ij}} e^{(px - Et)/\hbar}, \tag{6}$$

which depends on the proper time τ at the net vertices and $x = x_{ij}$ and $t = t_{ij}$ are defined at the discrete net vertices. The phase is independent of position x for a particular τ with equivalent phase thus defining simultaneity on the net and is equivalent to a $U(1)$ global gauge relationship. We can rewrite Eq. (5) as

$$\Phi = \begin{pmatrix} \phi_{22} \\ \phi_{12} \end{pmatrix} = \begin{pmatrix} \phi_{11} \\ \phi_{21} \end{pmatrix} = \begin{pmatrix} 0 & \sqrt{\Gamma} \\ 1/\sqrt{\Gamma} & 0 \end{pmatrix} \begin{pmatrix} \phi_{11} \\ \phi_{21} \end{pmatrix}, \tag{7}$$

which can be alternatively expressed for Eq. (7) in terms of a unique transfer matrix M

$$\Phi = \begin{pmatrix} \phi_{22} \\ \phi_{12} \end{pmatrix} = \begin{pmatrix} \phi_{11} \\ \phi_{21} \end{pmatrix} = M \begin{pmatrix} \phi_{11} \\ \phi_{21} \end{pmatrix}, \tag{8}$$

defined as

$$M = \begin{pmatrix} \cos\theta & \sin\theta \\ \sin\theta & -\cos\theta \end{pmatrix} = \begin{pmatrix} 1/\gamma & v/c \\ v/c & -1/\gamma \end{pmatrix} = \frac{1}{E} \begin{pmatrix} mc^2 & pc \\ pc & -mc^2 \end{pmatrix} = \frac{H_D}{E}. \tag{9}$$

Here we recognise H_D as the Dirac Hamiltonian for a free particle [3, 20] with defined momentum p. To connect with the complete quantum mechanics we note that Eq. (9) can be put in the conventional form [21] by assuming that space-time is locally differentiable at the vertex, allowing us to use the usual momentum operator $\hat{p}$ to replace the momentum eigenvalues p. This assumption of differentiability is satisfied if we consider space-time to be a continuum of causal nets, all infinitesimally displaced in space and proper time. We can write

$$\begin{pmatrix} mc^2 & c\hat{p} \\ c\hat{p} & -mc^2 \end{pmatrix} \Psi = E\Psi = i\hbar \frac{\partial \Psi}{\partial t}, \tag{10}$$

where we have replaced the probability amplitudes Φ with the familiar 2 component Dirac spinor Ψ for the free particle [21]

$$\Psi(x,t) = \begin{pmatrix} \psi_1 \\ \psi_2 \end{pmatrix} = \Phi(x,t) = A \begin{pmatrix} 1 \\ \frac{pc}{E + mc^2} \end{pmatrix} e^{(px - Et)/\hbar}, \tag{11}$$

and A is an appropriate normalisation constant.

5. Causal net quantum symmetries, spin and Lorentz invariance

Note that the unique matrix M above is a unitary, orthogonal matrix which provides an SU(2) group transformation corresponding to an improper rotation – that is a rotation $R(\theta)$ followed by an inversion β so $M = \beta R(\theta)$. The matrix provides the transformations for the probabilities $\vec{P} = M^2\vec{P} = I\vec{P}$ and probability amplitudes $\Psi = M\Psi$. Importantly because $M^3 = M$ there exists only two levels of symmetry at the vertex and the causal net provides simultaneously both the probabilities and the underlying probability amplitudes. Since it is an improper rotation the symmetry determines a preferred axis which provides helicity along the axis of movement. If we revisit Eq. (7) and consider both possible positive and negative roots we can see that even and odd solutions with helicity $\lambda = \pm 1$ and positive energy $\epsilon = +1$ are given by

$$\Phi_{\substack{\epsilon=+1\\ \lambda=+1}} = \begin{pmatrix} \sqrt{P_{11}} \\ \sqrt{P_{21}} \end{pmatrix} \qquad \Phi_{\substack{\epsilon=+1\\ \lambda=-1}} = \begin{pmatrix} \sqrt{P_{11}} \\ -\sqrt{P_{21}} \end{pmatrix}, \tag{12}$$

corresponding to transfer matrices $M_{\epsilon=+1,\lambda=\pm 1} = \beta R(\pm\theta)$. Note that in the above and following discussion we have omitted the normalisation constant and phase factor $e^{-imc^2\tau/\hbar}$ since this cancels in both sides of Eq. (8). Until now we have considered only the positive energy states, but negative energy solutions arise from the negative solution of the relativistic dispersion relation $E = \epsilon\sqrt{p^2c^2 + m^2c^4} = \epsilon|E|$ with $\epsilon = \pm 1$. This results in a reversal of the branching probabilities in Eq. (3) and two additional possible even and odd spinor solutions

$$\Phi_{\substack{\epsilon=-1\\ \lambda=+1}} = \begin{pmatrix} -\sqrt{P_{21}} \\ \sqrt{P_{11}} \end{pmatrix} \qquad \Phi_{\substack{\epsilon=-1\\ \lambda=-1}} = \begin{pmatrix} \sqrt{P_{21}} \\ \sqrt{P_{11}} \end{pmatrix}, \tag{13}$$

for transfer matrices $M_{\epsilon=-1,\lambda=\pm 1} = -\beta R(\pm\theta)$. If we include the negative energy states then, by combining all 4 net solutions above (Eq. 13 and Eq. 14), we can write 4 orthogonal 4-vectors which for helicity $\lambda = \pm 1$ and energy $E = \epsilon|E|$ with $\epsilon = \pm 1$ are

$$\Psi_{p,\epsilon,\lambda=+1} = A \begin{pmatrix} 1 \\ 0 \\ \frac{cp}{E+mc^2} \\ 0 \end{pmatrix} e^{(px-Et)/\hbar} \qquad \Psi_{p,\epsilon,\lambda=-1} = A \begin{pmatrix} 0 \\ 1 \\ 0 \\ \frac{-cp}{E+mc^2} \end{pmatrix} e^{(px-Et)/\hbar}. \tag{14}$$

Using the 4 possible transfer matrices $M_{\epsilon=\pm 1,\lambda=\pm 1}$ then the 1+1 dimension 4-matrix Dirac equation is

$$\begin{pmatrix} mc^2 & c\beta\hat{p} \\ c\beta\hat{p} & -mc^2 \end{pmatrix} \Psi = E\Psi = i\hbar\frac{\partial\Psi}{\partial t}. \tag{15}$$

Now this Dirac equation and the spinor wavefunction Eq. (15) correspond to exactly the conventional 3+1 dimension Dirac spinor for the special case of the particle moving along the x-axis and with a well defined spin aligned parallel and antiparallel with the x-axis [21].

The causal net is also consistent with the extraordinarily simple Foldy-Wouthuysen representation [22] of the Dirac equation where the positive and negative energy states are decoupled through a rotation of θ (the net angle) of the Dirac Hamiltonian. For example, one Foldy-Wouthuysen 1+1 dimension state is given by the rotation through $\theta/2$ of the Dirac state Eq. (12) since $cos(\theta/2) = \sqrt{P_{11}}$ and $sin(\theta/2) = \sqrt{P_{21}}$

$$\Phi_{\substack{FW\\ \epsilon=+1}} = R(\theta/2)\Phi_{\substack{Dirac\\ \epsilon=+1\\ \lambda=+1}} = \begin{pmatrix} \sqrt{P_{11}} & \sqrt{P_{21}} \\ -\sqrt{P_{21}} & \sqrt{P_{11}} \end{pmatrix} \begin{pmatrix} \sqrt{P_{11}} \\ \sqrt{P_{21}} \end{pmatrix} = \begin{pmatrix} 1 \\ 0 \end{pmatrix}. \tag{16}$$

Importantly, in this representation, establishing an exact particle position is impossible (there is only a mean position operator) and a particle is viewed as spread out over a finite region of about a wavelength which is consistent with our causal net picture.

The causal net can also be constructed by a Lorentz boost of the Foldy-Wouthuysen states. Setting $tanh\omega = v/c = sin\theta$ then the Dirac spinor Lorentz operator $\hat{S}_L(\omega)$ acts for example as

$$\Phi_{\substack{Dirac\\ \epsilon=+1\\ \lambda=+1}} = \hat{S}_L \Phi_{\substack{FW\\ \epsilon=+1}} = \begin{pmatrix} \cosh\omega/2 & -\sinh\omega/2 \\ -\sinh\omega/2 & \cosh\omega/2 \end{pmatrix} \Phi_{\substack{FW\\ \epsilon=+1}} = \gamma^{1/2} \begin{pmatrix} \sqrt{P_{11}} & \sqrt{P_{21}} \\ \sqrt{P_{21}} & \sqrt{P_{11}} \end{pmatrix} \begin{pmatrix} 1 \\ 0 \end{pmatrix}. \quad (17)$$

Applying successive Lorentz boosts reconstructs a symmetric causal network of connected events in any reference frame so, for example $\Phi'(x', t') = \hat{S}_L(\hat{\omega})\Phi(x, t)$, for boost $\hat{\omega}$ with $\omega' = \omega + \hat{\omega}$. The Lorentz boost to any frame preserves the conservation and composition of probabilities $P_{11} + P_{21} = \gamma(P_{11} - P_{21}) = 1$ as Eq. (2) to satisfy simultaneity on the net.

6. The 3+1 dimension Dirac equation

To extend to the general 3+1 dimension case we must consider transformations of the causal net that leave it invariant under spatial direction of velocity $\vec{v}$. Using polar coordinates then for momentum $\vec{p} = |\vec{p}|(sin\vartheta cos\varphi, sin\vartheta sin\varphi, cos\vartheta)$ we can expect that the wavefunction components become dependent on the coordinates (ϑ, φ) so $\sqrt{P_{ij}}$ becomes $\sqrt{P_{ij}}\chi(\vartheta, \varphi)$. Following Dirac's convention [3] we can replace the 1 dimension momentum operator $\hat{p}$ with the 3 dimensional momentum operator $(\vec{\sigma}.\vec{p})$, formed from Pauli matrices σ_k $(k = 1, 2, 3)$. By definition this momentum operator provides the relation $(\vec{\sigma}.\vec{p})\chi_{\pm} = |\vec{p}|\chi_{\pm}$ with two eigenvectors $\chi_+ = (\cos\vartheta/2, e^{i\varphi}\sin\vartheta/2)$, $\chi_- = (-e^{-i\varphi}\sin\vartheta/2, \cos\vartheta/2)$. The general solutions for the wavefunction then become from (Eq. 12 and Eq. 13) four 4-component orthogonal vectors corresponding to up and down spin $S = \pm 1/2$ with positive and negative energies $\epsilon = \pm 1$. Omitting the phase factors and normalisation constant these are

$$\Psi_{\substack{\epsilon=+1\\ S=+1/2}} = \begin{pmatrix} \sqrt{P_{11}}\chi_+ \\ \sqrt{P_{21}}\chi_+ \end{pmatrix} \qquad \Psi_{\substack{\epsilon=+1\\ S=-1/2}} = \begin{pmatrix} \sqrt{P_{11}}\chi_- \\ -\sqrt{P_{21}}\chi_- \end{pmatrix}, \quad (18)$$

$$\Psi_{\substack{\epsilon=-1\\ S=+1/2}} = \begin{pmatrix} -\sqrt{P_{21}}\chi_+ \\ \sqrt{P_{11}}\chi_+ \end{pmatrix} \qquad \Psi_{\substack{\epsilon=-1\\ S=-1/2}} = \begin{pmatrix} \sqrt{P_{21}}\chi_- \\ \sqrt{P_{11}}\chi_- \end{pmatrix}. \quad (19)$$

These are the general solutions to the conventional 3+1 dimension Dirac equation

$$\begin{pmatrix} mc^2 & c(\vec{\sigma}.\vec{p}) \\ c(\vec{\sigma}.\vec{p}) & -mc^2 \end{pmatrix} \Psi = E\Psi = i\hbar\frac{\partial\Psi}{\partial t}. \quad (20)$$

Thus for a 3 dimensional space we require all 3 Pauli matrices to construct the vector $\vec{\sigma}$ and the dimensionality of space defines the Pauli matrices. A plane wave solution to the Dirac equation has a unique velocity or space direction and the causal net is constructed along this direction in 3 dimensional space.

7. The effect of a potential on a causal net and the Pauli equation

The case we have examined is that of a free particle but we could include a potential V on the causal net since E can be replaced by $E - V$ in the construction of the lattice and the branching ratios. Returning to the 1+1 dimension case, between two media with different scalar potentials the net is compressed or stretched in space in the potential region with a form similar to Snell's law $cos\theta_2/cos\theta_1 = E/(E - V)$. We can write Eq. (10) conveniently as

$$M = \begin{pmatrix} \cos\theta_2 & \sin\theta_2 \\ \sin\theta_2 & -\cos\theta_2 \end{pmatrix} = \frac{1}{E - V} \begin{pmatrix} mc^2 & pc \\ pc & -mc^2 \end{pmatrix} = \frac{H_D}{E - V}. \quad (21)$$

If we consider invariance under a local gauge transformation U then in general $M(U\Phi) \neq (U\Phi)$ so to retain invariance we must add an additional term to the causal net Dirac equation or

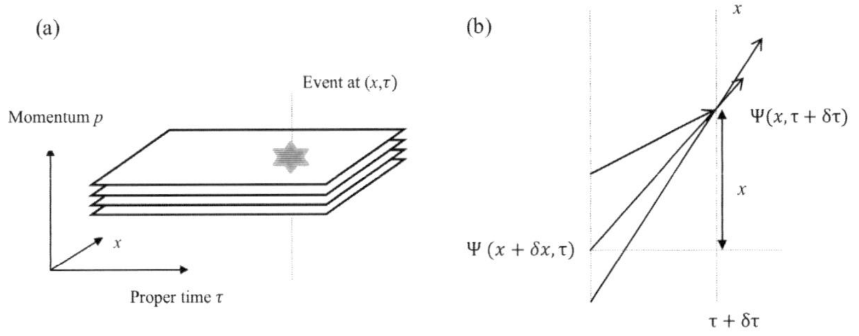

Figure 5. (a) A stacked "deck" of causal nets for different momentum states and (b) causally connected paths traversing a single space-time event.

Lagrangian which corresponds to a gauge potential term. The simple conjunctive fork (Fig. 1) is effectively replaced with an "interactive" fork [23], representing the interaction between different causal nets. To illustrate, consider the special case of a transformation where the proper time interval $\Delta\tau$ is unchanged by a potential. The triangle in Figure 3 is deformed by an amount $\delta\tau$ in time and δx in space $(c\Delta\tau)^2 = (c(\Delta t - \delta t))^2 - (\Delta x - \delta x)^2$. If we write $eA_0 = \gamma mc\delta t/\Delta t$ and $eA_1 = \gamma mc\delta x/\Delta t$ then we have the dispersion relation for an electron of charge e in an electromagnetic field (A_0, A_1) as $(E - eA_0)^2 = (pc - eA_1)^2 + m^2c^4$ and the corresponding transfer matrix M is given by

$$M = \begin{pmatrix} \cos\theta_2 & \sin\theta_2 \\ \sin\theta_2 & -\cos\theta_2 \end{pmatrix} = \frac{1}{E - eA_0}\begin{pmatrix} mc^2 & pc - eA_1 \\ pc - eA_1 & -mc^2 \end{pmatrix}. \tag{22}$$

If as Eq. (20) we embed the causal net in a continuous 3 dimensional space we can replace p with the 3 dimensional momentum operator and can consider the non-relativistic case of motion in a weak field. If we neglect the smaller component of the spinor and $E' = E - mc^2$ we have, following [20], the Pauli equation for a non-relativistic spin-1/2 particle.

$$\left(\frac{(\vec{p} - e\vec{A})^2}{2m} + eA_0 - \frac{e\hbar}{2mc}(\vec{\sigma}.\vec{H})\right)\psi = E'\psi. \tag{23}$$

8. Momentum states and the Feynman path integral

It is interesting to consider the more general case of a range of momentum states with each momentum state occupying an individual causal net. This can be visualised in Figure 5(a) as a stacked "deck" of infinitely extended causal nets. Consider an event in at $(x, \tau + \delta\tau)$ and the prior events that are causally connected from a earlier slice of proper time at τ, which are given by different space points x from each momentum net. If we sum the different spinor components contributing to the overall probability amplitude at $(x, \tau + \delta\tau)$ and include the change in phase over interval $\delta\tau$ from Eq. (6) we have

$$\Psi(x, \tau + \delta\tau) = \sum_{\substack{causally \\ connected \\ points}} \Psi(x + X, \tau)e^{-imc^2\delta\tau/\hbar}, \tag{24}$$

where X is the relative space coordinate (Fig. 5b). For one casual net with a free particle with a single momentum state Eq. (24) is trivial since velocity and probability are uniform across the net with only the phase varying with τ so $\Psi(x,\tau+\delta\tau) = \Psi(x+X,\tau)e^{-imc^2\delta\tau/\hbar}$ providing a simple delta function propagator for proper time interval $\delta\tau$, $K(x, x+X; \tau+\delta\tau) = \delta(X)e^{-imc^2\delta\tau/\hbar}$ However, if there is a continuum of momentum nets by geometry the sum in Eq. (24) selects a single probability amplitude contribution from each net with momentum $p = mX/\delta t$ for a given relative position X. Writing the relativistic infinitesimal action $S_{rel}(\delta\tau) = -mc^2\delta\tau$ we can write Eq. (24) as an integration

$$\Psi(x,\tau+\delta\tau) = \int_{-\infty}^{\infty} \Psi_p(x+X,\tau)e^{iS_{rel}(\delta\tau)/\hbar}\mathrm{d}X, \tag{25}$$

where Ψ_p denotes the spinor with momentum $p = mX/\delta\tau$. If we consider the non-relativistic limit where one spinor component ψ dominates and $\tau \to t$ we can use the semi-classical action $S_{cl} = \int(m/2)(\mathrm{d}x/\mathrm{d}t)^2\mathrm{d}t$ and write this infinitesimal path integral in the limit of large time interval T to give the conventional Feynman path integral [4]

$$\psi(x,t+T) = \int_{-\infty}^{\infty} K_0(x,X;T)\psi(X,t)\mathrm{d}X, \tag{26}$$

where $K_0(x,X;T) = \sqrt{(m/2\pi i\hbar T)}e^{imX^2/2\hbar T}$ is the free particle propagator for the Schrödinger equation. The causal net model is thus consistent with the quantum mechanical summation of paths and solutions to various problems such as slit diffraction using Feynman integrals [4].

9. Non-Euclidean space-time, general relativity and mass

In non-Euclidean curved space-time our elementary triangles (Fig. 3) comprising our causal net will become distorted and we can no longer apply Pythagoras' theorem to evaluate the space-time interval. In the language of general relativity the space-time interval is given by the metric $g_{\mu\nu}$ so $ds^2 = g_{\mu\nu}dx^\mu dx^\nu$. Previously, we have considered the special case of the Minkowski metric $\eta_{\mu\nu}$ for flat space-time but general relativity considers Riemann spaces that have quadratic metric equations and are characterised as locally flat. Considering a small displacement in space Δx from a point x using Taylor expansion we have the metric $g_{\mu\nu}(x+\Delta x) = \eta_{\mu\nu} + \frac{1}{2}g_{\mu\nu,\rho\sigma}\Delta x^\rho\Delta x^\sigma$. We might assume that variation to the scalar Minkowski action S_{rel} would produce a correction S_g which, to be a scalar under general coordinate transformations, can only include second order derivatives of the metric. Mathematically, the simplest curvature scalar is the Ricci scalar $R = g^{\mu\nu}R_{\mu\nu}$ formed from the Ricci curvature tensor $R_{\mu\nu}$. We can postulate that the simplest additional action might be of the form $S_g = B\int R\mathrm{d}^4x$. If we arbitrarily set the constant $B = -\varepsilon_0/16\pi Gc^4$ where ε_0 is a "density" and G Newton's gravitational constant then we have Einstein's *unimodular* gravity. If we further impose the density as $\varepsilon_0 = -\sqrt{-g}$ where $g = det(g_{\mu\nu})$ then we recover the Einstein–Hilbert action

$$S_g = -\frac{1}{16\pi Gc^4}\int R\sqrt{-g}\mathrm{d}^4x. \tag{27}$$

Thus the causal net model, although a microscopic theory, would appear to be consistent with the macroscopic theory of general relativity if the elementary triangles of our causal net model are distorted to "tile" curved space–time between causally connected possible events to preserve our definition of simultaneity.

Finally, in our model the particle mass m is an arbitrary scaling constant but if we assume a 1+1 dimension Laplace equation describes the diffusive causal paths on the net, then by imposing a Dirichlet boundary condition, the uncharged mass term of the kth fermion follows an exponential spectrum with $m_k = exp(ak+b)$ where a and b are constants.

10. Discussion

By considering simple casual connections between elementary events, based on Reichenbach's principle of common cause, we have constructed a causal model where the Dirac equation and the fermion particles it describes are seemingly "emergent" properties. The simplest causal network describes exactly the Dirac equation and provides quantum mechanical statistics and major quantum phenomena (diffraction, wave-particle duality, uncertainty ...). In an inertial frame an observer will view an ordered series of events in space–time as an entity behaving as either a wave or a particle, depending on how the experimental measuring setup is conceived, and thus exhibits wave-particle duality exactly as proposed by de Broglie. Geometrical quantities of the causal net correspond to measurable physical qualities: mass (scaling factor), momentum and energy (net angle and geometry) and potentials and forces (change of net angle). The global gauge symmetry of the net provides quantum phase and the other degenerate solutions arising from the symmetries of the net are equivalent to the Dirac spin and negative energy states. The discretisation of the net infers similarities to the Heisenberg uncertainty principle and a "stack" of nets provides, consistent with the Feynman path integral approach, quantum phenomenon such as diffraction. Distorting the causal net in non-Euclidean space-time suggests analogies with general relativity. Also imposing extra internal Lie symmetries corresponds to the case of "interacting" causal forks and leads to the introduction of gauge forces such as electromagnetism. Conventional continuous physics, with its well known differential equations, can perhaps be viewed as being "emergent" from a discrete underlying causal net. None of these emergent aspects of our causal net would have been apparent from our simple starting point of an equilibrium distribution of ordered events. Fermion particles can be considered as quasiparticles of the causal network composed of possible and actual events analogous to holes, phonons and recent magnetic monopoles in spin-ice. The quantum measurement "problem" is reduced to simple Bayesian statistics and there is no wavefunction "collapse". The quantum to classical transition becomes a straightforward feature of the resolution of the net and conventional quantum mechanics can be tentatively viewed as nature's causal "equilibrium" condition. Lastly, the model is compatible with both subjective (Kant, Hume) and objective (Reichenbach) philosophical models of causality.

References

[1] Belnap N 1992 *Synthese* **92** 385
[2] Sorkin R D 1997 *Int. J. Theor. Phys.* **36** 2759
[3] Dirac P A M. 1928 *Proc. Royal Soc.* A 117
[4] Feynman R P and Hibbs A R 1965 *Quantum Mechanics and Path Integrals* (McGraw-Hill)
[5] Kauffman L H and Noyes H P 1996 Discrete Physics and the Dirac Equation *Preprint SLAC PUB 7115*
[6] Gaveau B, Jacobson T, Kac M and Schulman L S 1984 *Phys. Rev. Lett.* **53** 419
[7] Coddens G 1995 *Found. Physics Lett.* **8** 3
[8] Bialynicki Birula I 1994 *Phys. Rev.* D **49** 12
[9] Schotte K D, Iwabuchi S and Truong T T 1985 *Z. Phys.* B **60** 255
[10] Reichenbach H 1956 *The Direction of Time* (University California Press)
[11] Penrose O and Percival I C 1962 *Proc. Phys. Soc.* **79** 605
[12] Malament D 1977 *Nous* **11** 293
[13] Bell J 1966 *Rev. Mod. Phys.* **38** 447
[14] Grasshof G, Portmann S and Wuthrich A 2005 *Brit. J. Phil. Sci.* **56** 663
[15] Aspect A, Dalibard J, Roger G 1982 *Phys. Rev. Lett.* **49** 1804
[16] Rindler W 1982 *Introduction to Special Relativity* (Oxford)
[17] De Broglie L 1925 *Ann. Phys.* **3** 22
[18] Heisenberg W 1927 *Z. Phys.* **43** 172
[19] Schurmann T and Hoffmann I 2009 *Found. Phys.* **39** 958
[20] Davydov A S 1965 *Quantum Mechanics* (Pergamon Press)
[21] Coulter B and Adler C 1971 *AJP* **39** 305
[22] Foldy L and Wouthuysen S 1950 *Phys. Rev.* **78** 1
[23] Salmon W C 1998 *Causality and Explanation* (Oxford)

EmerQuM 11: Emergent Quantum Mechanics 2011 IOP Publishing
Journal of Physics: Conference Series **361** (2012) 012010 doi:10.1088/1742-6596/361/1/012010

Remarks on Osmosis, Quantum Mechanics, and Gravity

Robert Carroll
University of Illinois, Urbana, IL 61801

Abstract. Some relations of the quantum potential to Weyl geometry are indicated with applications to the Friedmann equations for a toy quantum cosmology. Osmotic velocity and pressure are briefly discussed in terms of quantum mechanics and superfluids with connections to gravity.

1. Remarks on Weyl Geometry

We begin with some features of Weyl geometry and the Dirac-Weyl theory following [54, 55, 56, 84] (cf. also [1, 2, 3, 4, 8, 12, 15, 16, 17, 19, 20, 23, 24, 25, 29, 77, 81, 85, 86, 87, 95]). Note that the paper is in part an expansion of some ideas in [24] and we correct some notational confusion from [17, 24]. There is a curious issue here of too much or too little, which is compounded by various, sometimes conflicting, notations and we feel that it is well advised to simply follow the extensive development of Israelit-Rosen which will keep the notation under control and benefit from insights therein. The background involves looking at a metric tensor $g_{ab} = g_{ba}$ and a length connection vector w_μ. If a vector is displaced by dx^ν then

$$dB^\mu = -B^\sigma \hat{\Gamma}^\mu_{\sigma\nu} dx^\nu;\ dB = Bw_\nu dx^\nu \tag{1.1}$$

One can also write (**1A**) $d(B^2) = 2B^2 w_\nu dx^\nu$ and this all requires that (**1B**) $g_{\mu\sigma}\hat{\Gamma}^\sigma_{\nu\sigma} + g_{\nu\sigma}\hat{\Gamma}^\sigma_{\mu\sigma} = \partial_\sigma g_{\mu\nu} - 2g_{\mu\nu}w_\sigma$. Then assuming that $\hat{\Gamma}^\sigma_{\mu\nu} = \Gamma^\sigma_{\mu\nu}$ there results

$$\hat{\Gamma}^\sigma_{\mu\nu} = \Gamma^\sigma_{\mu\nu} + g_{\mu\nu}w^\sigma - \delta^\sigma_\nu w_\mu - \delta^\sigma_\mu w_\nu \tag{1.2}$$

where $\Gamma^\sigma_{\mu\nu}$ denotes the standard Riemannian Christoffel symbol. The covariant derivative for g_{ab} is written ∇_α or e.g. $B^\mu_{;\nu} = \nabla_\nu B^\mu$ (and involves $\Gamma^\sigma_{\mu\nu}$) whereas with the Weyl connection (1.2) one forms a co-covariant derivative of a vector

$$B^\mu_{:\nu} = \partial_\nu B^\mu + B^\sigma \hat{\Gamma}^\mu_{\sigma\nu};\ B_{\mu,:\nu} = \partial_\nu B_\mu - B_\sigma \hat{\Gamma}^\sigma_{\mu\nu} \tag{1.3}$$

so that

$$B^\mu_{:\nu} = B^\mu_{;\nu} + B^\mu(g_{\sigma\nu}w^\mu - \delta^\mu_\nu w_\sigma - \delta^\mu_\sigma w_\nu) \tag{1.4}$$

Note also

$$g_{\mu\nu:\sigma} = 2g_{\mu\nu}w_\sigma;\ g^{\mu\nu}_{:\sigma} = -2g^{\mu\nu}w_\sigma \tag{1.5}$$

Now under Weyl gauge transformations one has a length change (**1C**) $B \to \hat{B} \sim \tilde{B} = e^{\lambda(x)}B$ with $g_{\mu\nu} \to \tilde{g}_{\mu\nu} = e^{2\lambda}g_{ab}$ and $g^{ab} \to e^{-2\lambda}g^{ab}$. Writing $e^{2\lambda(x)} = f(x)$ there is a calculation from [28] for lengths $\ell \to \hat{\ell} = f^2\ell$

$$2\hat{\ell}d\hat{\ell} \sim f_\alpha dx^\alpha \ell^2 + 2f\ell d\ell \Rightarrow \frac{d\hat{\ell}}{\hat{\ell}} \sim \frac{1}{2}\partial_\alpha log(f)dx^\alpha + w_\alpha dx^\alpha \Rightarrow \hat{w}_\alpha = \tag{1.6}$$

$$= w_\alpha + \frac{1}{2}\partial_\alpha log(f) = w_\alpha + \partial_\alpha \lambda(x)$$

REMARK 1.1. In Weyl geometry consider now a scalar S of Weyl weight n defined via $\tilde{S} = exp(n\lambda)S$ (e.g. $B = (g_{ab}B^aB^b)^{1/2} = \ell \to \hat{\ell} = e^\lambda \ell$) and $\partial_\mu \tilde{S} = exp(n\lambda)(\partial_\mu S + nS\partial_\mu \lambda)$. This is not covariant with respect to Weyl gauge transformations (WGT) unless $n = 0$ but one can define a gauge covariant derivative (**1D**) $S_{||\mu} = \partial_\mu S - nSw_\mu$ with $\tilde{S}_{||\mu} = exp(n\lambda)S_{||\mu}$ (using (1.6)). For a gauge covariant derivative of a vector S^μ of Weyl weight n, where $\tilde{S}^\alpha = exp(n\lambda)S^\alpha$, there results (**1E**) $S^\mu_{||\nu} = S^\mu_{;\mu} - nS^\mu w_\nu = exp(n\lambda)S^\mu_{||\nu}$ (note $\hat{\Gamma}^\mu_{\sigma\nu} - n\delta^\mu_\sigma w_\nu$ is actually a connection). ■

Dirac [28] (1973) introduced a β field of weight -1 (as a Lagrange type multiplier) in order to have a satisfactory scalar tensor action of the form (cf. [54], Chap. 4)

$$I = \int \sqrt{-g}d^4x[-\beta^2 R + \sigma\beta^2 w^c w_c + (\sigma+6))\partial_c\beta\partial^c\beta + 2\sigma\beta\partial_c\beta w^c + 2\Lambda\beta^4] \tag{1.7}$$

Note here that $\sqrt{-g}$ is of weight 4 and R of weight -2; electromagnetic terms have been omitted (also $\hbar = c = 1$). It is this factor β which was identified as a quantum mass $\mathfrak{M} = mexp(\mathfrak{Q})$ in [87] (and subsequently utilized in [15, 16, 17, 19, 20, 23, 24]). It's quantum significance was also noticed in [8] (cf. also [81]) in connection with conformal GR (GR stands for general relativity). Unfortunately a different notation was used in [8] and in [81] (0009169), namely (**1F**) $\hat{g}_{ab} = \Omega^2(x)g_{ab}$ with $\Omega^2 = exp(-\psi)$ in [8] and $\Omega^2 = exp(\psi)$ in [81] (0009109) - a fact which I occasionally overlooked in subsequent work (cf. [24] for some errata). The latter paper [81] (0009169) contains some important material relating various actions in Einstein frame (EF) and Jordan frame (BD type equations). It was shown that an action

$$S_4 = \int d^4x\sqrt{-\hat{g}}e^{-\psi}\left[\hat{R} - \left(\alpha - \frac{3}{2}\right)|\hat{\nabla}\psi|^2 + 16\pi e^{-\psi}L_M\right] \tag{1.8}$$

was the only one (of 4 considered) which was invariant under transformations of units (length, time, and mass) and this was then dubbed a string frame action (under the assumption that strings are ultimate and inevitable - cf. [44]). Note that $\Omega^2 = exp(\psi) = \phi$ implies

$$\hat{g}_{ab;c} = \partial_c\psi\hat{g}_{ab} \Rightarrow w_c = \frac{1}{2}\partial_c\psi \tag{1.9}$$

In particular conformal Riemannian geometry as in S_4 is a WIST (Weyl integrable space time) as noted in [81] (0009169) and [73, 74].

Now S_4 arises in our work via a conformal map $g_{ab} \to \hat{g}_{ab} = \Omega^2 g_{ab}$ from an EF action (cf. [68]) (**1G**) $S = \int \sqrt{-g}d^4x[R - \alpha|\nabla\psi|^2 + 16\pi L_M]$ in the form (**1H**) $\hat{S} = \int \sqrt{-\hat{g}}d^4x[\hat{\phi}\hat{R} - (\omega/\hat{\phi})|\hat{\nabla}\hat{\phi}|^2 + 16\pi\hat{\phi}^2 L_M]$ where we will write out the theory here in terms of $\Omega^2 = \exp(\psi)$ following [19] (cf. also [15, 16] and note a few sign adjustments are needed in [17]). Recall first that $\phi \sim \beta^2/m^2 = \mathfrak{M}^2/m^2$ and $\hat{\phi} = \phi^{-1} = exp(-\psi)$ and note that $\sqrt{-g} = \hat{\phi}^2\sqrt{-\hat{g}}$ (cf. [15, 16]) and

$$\sqrt{-\hat{g}}\hat{\phi}\hat{R} = \hat{\phi}^{-1}\sqrt{-\hat{g}}\hat{\phi}^2\hat{R} = \hat{\phi}^{-1}\sqrt{-g}\hat{R} = \frac{\beta^2}{m^2}\sqrt{-g}\hat{R} \tag{1.10}$$

But via [28, 54] we can write (cf. also [17] for the pattern here)

$$\beta^2\hat{R} = \beta^2 R - 6\beta^2\nabla_\kappa w^\kappa + 6\beta^2 w^\kappa w_\kappa \tag{1.11}$$

and via $-\beta^2\nabla_{gk}w^\kappa = -\nabla_\kappa(\beta^2 w^\kappa) + 2\beta\partial_\kappa\beta w^\kappa$, with a vanishing divergence term upon integration, the first integral in $\hat{S}$ becomes

$$I_1 = \int\sqrt{-g}d^4x\left[\frac{\beta^2}{m^2}R + 12\beta\partial_\kappa\beta w^\kappa + 6\beta^2 w^\kappa w_\kappa\right] \tag{1.12}$$

The second integral is ($\gamma = \alpha - (3/2)$)

$$I_2 = -\gamma\int\sqrt{-\hat{g}}d^4x\hat{\phi}\frac{|\hat{\nabla}\hat{\phi}|^2}{|\hat{\phi}|^2} = -\frac{4\gamma}{m^2}\int\sqrt{-g}d^4x|\hat{\nabla}\beta|^2 \tag{1.13}$$

since $\hat{\phi}_c = -\psi_c e^{-\psi}$, $\hat{\nabla}\psi = -\hat{\nabla}\hat{\phi}/\hat{\phi}$, $\hat{\phi} = \phi^{-1} = m^2\beta^{-2}$, and $\hat{\phi}/\hat{\phi} = -2\hat{\nabla}\beta/\beta$ (along with (1.10)). Via (**1D**) $\hat{\nabla}_c\beta = \partial_c\beta + w_c\beta$ for $\Pi(\beta) = -1$ (Π denotes the Weyl weight) and from $\beta^2 = m^2 exp(\psi)$ one has $\psi_c exp(\psi) = (\beta^2/m^2)\psi_c \Rightarrow \psi_c = 2\beta_c/\beta$. Hence via (1.9) $w_c = \beta_c/\beta$ (correcting [17]) and consequently $\hat{\nabla}_c\beta = 2\beta_c$. In any event

$$\hat{S}_{GR} = \frac{1}{m^2}\int\sqrt{-g}d^4x[\beta^2 R + 12\beta\partial_c\beta w^c + 6\beta^2 w^c w_c - 4\gamma|\hat{\nabla}\beta|^2 + 16\pi m^2 L_M] \tag{1.14}$$

(as in [17], p. 59 (6.6) and p. 237 (3.32)).

REMARK 1.2. There was some sign confusion in [17] which we have adjusted above. Thus equations (6.7)-(6.8) on p. 59 and (3.33))-(3.34) on p. 237 should be altered as indicated below. ■

First we write $|\hat{\nabla}\beta|^2 = \beta^c\beta_c$ and from (1.9) and (1.14)

$$I_1 = \frac{1}{m^2}\int\sqrt{-g}d^4x[\beta^2 R + 18\partial_c\beta\partial^c\beta];\ I_2 = -16\gamma\int\sqrt{-g}d^4x\beta^c\beta_c \tag{1.15}$$

Consequently (recall $\gamma = \alpha - (3/2)$)

$$\hat{S} = \frac{1}{m^2}\int\sqrt{-g}d^4x[\beta^2 R + 16\pi m^2 L_M + 2(21 - 8\alpha)\beta^c\beta_c] \tag{1.16}$$

Then we rewrite (1.7) in the form

$$I = \int\sqrt{-g}d^4x[-\beta^2 R + 3(\sigma + 2)\beta_c\beta^c + 2\Lambda\beta^4] \tag{1.17}$$

Ignoring L_M and Λ for the moment we see that $I = m^2\hat{S}$ provided that (**1I**) $3\sigma = 8(3 - \alpha)$ and, a priori, α or σ can be arbitrary (modulo e.g. possible cosmological restrictions on α).

REMARK 1.3 The Brans-Dicke (BD) type theory of S_4 (written also as $\hat{S}_4$) was used for the Friedman equations in [24] for example where a relation $\sigma \sim -4\alpha$ was used (based on some sign confusion). Hence one should convert the Weyl-Dirac action (1.1) of [24] to (1.7) and use the (σ, α) relation (**1I**) in order to obtain the correct Friedmann equations for the toy quantum cosmology considered there (cf. Section 2). ■

2. Remarks on Cosmology

In [24] the Friedmann equations were based on an S_4 type action

$$\tilde{S} = \frac{1}{16\pi}\int \sqrt{-g}d^4x\left[R\Phi - \omega\frac{|\nabla\Phi|^2}{\Phi} + L_M\right] \tag{2.1}$$

where (**2A**) $L_M = -V(\Phi) + 16\pi\mathfrak{L}$ with Λ assumed to be suitably inserted in $\mathfrak{L}_M$ (cf. [33, 38]). An FRW metric ($c = 1$)

$$ds^2 = dt^2 - a^2(t)\left[\frac{dr^2}{1-kr^2} + r^2d\Sigma^2\right] \tag{2.2}$$

is used and one has field equations as in [24]

$$G_{ab} = \frac{8\pi}{\Phi}T^M_{ab} + \frac{\omega}{\Phi^2}\left[\nabla_a\Phi\nabla_b\Phi - \frac{1}{2}g_{ab}\nabla^c\Phi\nabla_c\Phi +\right] \tag{2.3}$$

$$+\frac{1}{\Phi}(\nabla_a\nabla_b\phi - g_{ab}\Box\Phi) - \frac{V}{2\Phi}g_{ab}$$

where (**2B**) $T^M_{ab} = -(2/\sqrt{-g})(\delta/\delta g^{ab})(\sqrt{-g}\mathfrak{L}_M)$ (we ignore a possible shift $\sqrt{-g} \to \sqrt{g}$, etc. due to the signature in (2.2)). Variation of the action with respect to Φ gives

$$\frac{2\omega}{\Phi}\Box\Phi + R - \frac{\omega}{\Phi^2}\nabla^c\Phi\nabla_c\Phi - \frac{dV}{d\Phi} = 0 \tag{2.4}$$

with trace

$$R = -\frac{8\pi T_M}{\Phi} + \frac{\omega}{\Phi^2}\nabla^c\Phi\nabla_c\Phi + \frac{3\Box\Phi}{\Phi} + \frac{2V}{\Phi} \tag{2.5}$$

and using (2.5) to eliminate R from (2.4) yields

$$\sqcup\Phi = \frac{1}{2\omega+3}\left[8\pi T^M + \Phi\frac{dV}{d\Phi} - 2V\right] \tag{2.6}$$

Then using (2.2) one obtains (**2C**) $\nabla^c\Phi\nabla_c\Phi = -(\dot{\Phi})^2$ and $\Box\Phi = -(\ddot{\Phi} + 3H\dot{\Phi}) = -(1/a^3)(d/dt)(a^3\dot{\Phi})$. Assume now that (**2D**) $T^M_{ab} = (P^M + \rho^M)u_au_b + P^Mg_{ab}$ and then the time dependent component of the BD field equations gives a constraint equation

$$H^2 = \frac{8\pi}{3\Phi} + \frac{\omega}{6}\left(\frac{\dot{\Phi}}{\Phi}\right)^2 - H\frac{\dot{\Phi}}{\Phi} - \frac{k}{a^2} + \frac{V}{6\Phi} \tag{2.7}$$

Then, using $R = 6[\dot{H} + 2H^2 + (k/a^2)]$, there results

$$\dot{H} + 2H^2 + \frac{k}{a^2} = -\frac{4\pi T^M}{3\Phi} - \frac{\omega}{6}\left(\frac{\dot{\Phi}}{\Phi}\right)^2 + \frac{1}{2}\frac{\Box\Phi}{\Phi} + \frac{V}{3\Phi} \tag{2.8}$$

From (2.6), (2.7), and the trace equation (**2E**) $T^M = 3P^M - \rho^M$ one has then

$$\dot{H} = \frac{-8\pi}{(2\omega+3)\Phi}\left[(\omega+2)\rho^M + \omega P^M\right] - \frac{\omega}{2}\left(\frac{\dot{\Phi}}{\Phi}\right)^2 + \tag{2.9}$$

$$+2H\frac{\dot{\Phi}}{\Phi} + \frac{k}{a^2} + \frac{2}{2(2\omega+3)\Phi}\left(\Phi\frac{dV}{d\Phi} - 2V\right)$$

and (2.6) reduces to

$$\ddot{\Phi} + 3H\dot{\Phi} = \frac{1}{2\omega + 3}\left[8\pi(\rho^M - 3P^M) - \Phi\frac{dV}{d\Phi} + 2V\right] \tag{2.10}$$

In order to apply this to our model (1.7) we look at these Friedmann equations for $\Phi = \hat{\phi}$ (with $\mathfrak{Q} \sim \psi$) where $(\beta^2/m^2) = \hat{\phi}^{-1} = exp(\psi) = exp(\mathfrak{Q})$. Then $\hat{\phi} = exp(-\mathfrak{Q}) = \Phi$ and equations (2.3)-(2.10) can be written in terms of the quantum potential $\mathfrak{Q}$ from **(2F)** $\mathfrak{Q} = (\hbar^2/m^2)(\Box|\Psi|/|\Psi|)$ with Ψ arising in the quantum process underlying (1.2) (cf. [2, 3, 4, 15, 16, 17, 20, 23, 24, 25, 85, 87, 95] and see also [6, 32]). We assume first that $\mathfrak{L}_M = 0$ and then one can write **(2G)** $\dot{\Phi} = -\dot{\mathfrak{Q}}\Phi$, $\ddot{\Phi} = (\dot{\mathfrak{Q}}^2 - \ddot{\mathfrak{Q}})\Phi$ and thence e.g.

$$(\dot{\mathfrak{Q}}^2 - \ddot{\mathfrak{Q}})\Phi - 3H\dot{\mathfrak{Q}}\Phi = \frac{1}{2\omega + 3}\left[8\pi(\rho^M - 3P^M) - \Phi\frac{dV}{d\Phi} + 2V\right] \tag{2.11}$$

Note that **(2H)** $\omega = \alpha - (3/2)$ and $3\sigma = 8(3-\alpha) \Rightarrow \sigma = 12 - (8/3)\omega$. We see that the dynamics of Φ is determined in part by $V(dV/d\Phi) - 2V$ (which vanishes for $V = c\Phi^2 = cexp(-2\mathfrak{Q})$) and by ρ^M, P^M. Thus, in particular, it is possible to envision some cosmological behaviour provided by a quantum background as in (1.7). We note also that Mannheim refers to intrinsically quantum mechanical gravity associated to general Weyl geometry with the Weyl tensor, etc. (cf. [5, 69]). We especially recommend Padmanabhan's paper [76] (1012.4476) and references there for a discussion of quantum mechanics and spacetime. Even more striking is the discussion in Grössing's book [46] (p. 123) which indicates (assuming an aether) how Einsteinian gravity can be considered a pure quantum phenomenon.

REMARK 2.1. It is interesting to note that via $\hat{\phi} = exp(-\mathfrak{Q})$ one has $\mathfrak{Q} \sim \psi$ and hence by (1.9) the Weyl vector is $w_c = (1/2)\partial_c\mathfrak{Q}$ implying a geometric role for $\mathfrak{Q}$. In particular this seems to show a direct connection of quantum mechanics with gravity. Of course this is not surprising since the geometry of (1.7) was generated via quantum mechanics but it suggests that Weyl geometry itself may have an implicit quantum connection via $w_a = (1/2)\partial_a\mathfrak{Q}$ with $\mathfrak{Q}$ a quantum potential. This seems to be a new observation (although connections of QM to the Weyl- Ricci curvature arise (cf. [2, 3, 15, 16, 20, 25, 85, 95]). ■

REMARK 2.2. In [87] there are also joint field equations involving the conformal factor Ω^2 (expressed via ϕ or ϕ^{-1}) and for the quantum potential $\mathfrak{Q}$ (expressed separately via a Lagrangian) with the goal of thereby deriving connections between $\mathfrak{Q}$ and ϕ. Also Bohmian aspects of the theory are emphasized. Some of this is also reviewed in [15, 16]. A main conclusion of this is that the quantum potential is a dynamical field and interactions between Λ and $\mathcal{Q}$ represent a connection between large and small scale structures. We note also from [73] that Λ plays an important role in generating mass for scalar fields and the understanding of Λ is considerably deepened and enhanced via work of Klinkhamer and Volovik (cf. [61, 62, 90, 91]) some of which is described briefly below. ■

REMARK 2.3. Relations between thermodynamics and gravity have been extensively studied following early work of Bekenstein and Hawking plus more recently Jacobson [57], and Padmanabhan [75, 76]. One can derive general Einstein field equations via thermodynamic principles and gravity itself seems to be characterized via thermodynamics (see [17] for a brief sketch of some of this and there is much more information in works of Padmanabhan et al. The work of Verlinde [89] has triggered another explosion of interest in entropy and gravity and we mention here only a few articles, e.g. [9, 10, 13, 93] (related to the Friedmann equations) and [67, 70, 72, 88] (relations to quantum mechanics). ■

We would like to mention here that entropy in quantum mechanics is normally connected to momentum fluctuations and the quantum potential gives rise to an entropy functional $\int P\mathfrak{Q}\,dx$ corresponding to Fisher information (here $P = |\psi|^2$ where ψ is a wave function and $\mathfrak{Q}$ represents a 3-dimensional quantum potential). Generally $\mathfrak{Q}$ can be described e.g. via an osmotic velocity or a thermalization of this (cf. [15, 16, 17, 27, 34, 39, 45, 49, 50, 51] and Section 3). There is also a gravitational version of this related to the Wheeler-deWitt (WDW) framework in the form

$$\int \mathcal{D}h\, P\mathfrak{Q} = \int \mathcal{D}h \frac{\delta P^{1/2}}{\delta h_{ij}} G_{ijk\ell} \frac{\delta P^{1/2}}{\delta h_{k\ell}} \tag{2.12}$$

(cf. [16, 17, 21, 34, 51]) and see also [40, 92] for WDW). We note also the Perelman entropy functional

$$\mathfrak{F} = \int_M (R + |\nabla f|^2) e^{-f} dV \tag{2.13}$$

and corresponding Ricci flows are related to a so called Nash entropy (**2I**) $S = \int u\, log(u)\, dV$ where $u = exp(-f)$ and various aspects of quantum mechanics related to the Schrödinger equation and Weyl geometry arise (cf. [17, 18, 19, 43, 53, 59]).

3. Some Background

First it is interesting and perhaps enlightening to have some 3-D background here on the quantum potential. Thus, referring to [15, 16, 17, 22, 23, 26, 34, 35, 36, 37, 39, 45, 47, 48, 49, 50, 51, 71, 72, 82, 83, 88] we recall some results involving the QP relative to a 3-D SE. Various notations are used and we deal with a Schrödinger equation (SE) (**3A**) $i\hbar\psi_t = \Delta\psi + V\psi$ with $\psi = Rexp(iS/\hbar)$ and

$$\partial_t S + \frac{1}{2m}|\nabla S|^2 + V + \mathfrak{Q} = 0;\ \partial_t(R^2) + \frac{1}{m}\nabla(R^2\nabla S) = 0 \tag{3.1}$$

where the quantum potential (QP) is defined via

$$\mathfrak{Q} = -\frac{\hbar^2}{2m}\frac{\Delta R}{R} = -\frac{\hbar^2}{4m}\left[\frac{1}{2}\left(\frac{\nabla P}{P}\right)^2 - \frac{\Delta P}{P}\right] \tag{3.2}$$

(note this is different from $\mathfrak{Q}$ in e.g. (**2F**)). Here $|\psi|^2 = R^2 = P$ and $\rho = mP$. Setting $D = (\hbar/2m)$ one can write (**3B**) $\mathbf{u} = -D(\nabla P/P)$ (osmotic velocity) which is expressed in terms of a "canonical" momentum fluctuation $\delta p/m = \mathbf{u}$ (cf. [15, 16, 17] for references and more details). Now setting $\mathbf{k}_u = -(1/2)\mathbf{u}$ one can write (**3C**) $\mathfrak{Q} = (\hbar^2/2m)(|\mathbf{k}_u|^2 - \nabla\cdot\mathbf{k}_u) = (m/2)\mathbf{u}^2 - (\hbar/2)(\nabla\cdot\mathbf{u})$. There is also a thermalization of the QP based on [45] (cf. also [17, 22]) whereby one refers to a basic particle energy $E = \hbar\omega$ (where $kT = \hbar\omega = 1/\alpha$ with $k \sim k_B$ the Boltzman constant). One then defines a thermal function $\mathcal{Q}$ where $\Delta\mathcal{Q} = \mathcal{Q}(t) - \mathcal{Q}(0) < 0$ determines a heat dissipation with an osmotic velocity (**3D**) $\mathbf{u} = -D(\nabla P/P) = (1/2\omega m)\nabla\mathcal{Q}$. Hence and from (3.2), (**3B**), and (**3D**) there results

$$\int \mathfrak{Q}Pd^3x = \frac{\hbar^2}{8m}\int\left(\frac{\nabla P}{P}\right)^2 Pd^3x = \frac{m}{2}\int|\mathbf{u}|^2 Pd^3x = \tag{3.3}$$

$$= \frac{1}{8\omega^2 m}\int|\nabla\mathcal{Q}|^2 Pd^3x = \tilde{c}(t)\int(\nabla\mathcal{Q})^2 exp^{-\alpha\mathcal{Q}}d^3x$$

(the last equation via $log(P) = -\alpha\mathcal{Q} + c(t) \Rightarrow P = \hat{c}(t)exp(-\alpha\mathcal{Q})$ where $\alpha = 1/\hbar\omega$ - cf. [17, 22, 45, 47, 48] for more on this). One should also note that (**3E**) $\int P\mathfrak{Q}d^3x = (\hbar^2/8m)\int[(|\nabla P|^2/P]d^3x = (\hbar^2/8m)FI$ where FI denotes Fisher information.

Now we note that an idea of osmotic pressure arises in the Klinkhamer-Volovik theory of superfluids (cf. [61, 62, 90, 91]) from which we extract here (see especially gr-qc 0711.3170, 0806.2805, 0907.4887, 0811.4347, and 1102.3152; hep-th 0907.4887 and 1101.1281; cond-mat 1004.0597; physics 0909.1044 and cf. also [30, 52, 79, 80]). This is a relativistic theory and the relativistic quantum vacuum is considered as a self sustained medium described by a variable $\mathfrak{q}$ which is a conserved quantity analogous to a particle density n in condensed matter theory but $\mathfrak{q}$ is the relativistic invariant quantity (see also Section 4). There is a Gibbs-Duhem relation $\epsilon_{vac}(\mathfrak{q}) - \mu\mathfrak{q} = -P_{vac}$ where ϵ_{vac} is the vacuum energy density and μ is the vacuum chemical potential (thermodynamically conjugate to $\mathfrak{q}$). Dynamical equations for $\mathfrak{q}$ show that $\Lambda = \epsilon_{vac} - \mu\mathfrak{q} = -P_{vac}$ so the vacuum variable is automatically self-tuned to nullify in equilibrium any contribution from different quantum fields. In [91] (0909.1044) one considers a dilute solution of ^3He quasiparticles in a superfluid vacuum ^4He at $T = 0$. The negative contribution of the vacuum to the osmotic pressure of ^3He is given by the same equation

$$P_{os} = P_{mat} + P_{vac} = P_{mat} - \frac{1}{2}\chi_{vac}\left[\mathfrak{q}\frac{\partial\epsilon_{mat}(n,\mathfrak{q})}{\partial\mathfrak{q}}\right]^2 \tag{3.4}$$

where χ_{vac} is the vacuum compressibility

$$\chi_{vac}^{-1} = \left[\mathfrak{q}^2\frac{d^2\epsilon_{vac}(\mathfrak{q})}{d\mathfrak{q}^2}\right]_{\mathfrak{q}=\mathfrak{q}_0} \geq 0 \tag{3.5}$$

(where now χ_{vac} corresponds to the compressibility of liquid ^4He. This negative contribution to the osmotic pressure has been experimentally verified (according to 0909.1044).

REMARK 3.1 . Note that the formula in (3.3) involving the 3-D integral $\int|\nabla\mathcal{Q}|^2 exp(-\alpha\mathcal{Q})d^3x$ reminds one of the I_2 formula (1.13) from S_4 where there is a 4-D integral involving $exp(-\psi)|\hat{\nabla}\psi|^2$; also the S_4 or I_2 represent "action" terms in the Weyl theory as does $\int P\mathfrak{Q}d^3x$ in the 3-D theory (cf. [15, 16, 17]). However there is no immediate extrapolation here from 3-D to 4-D or vice-versa. In fact we know from remarks before (2.11) that $\psi \sim \mathfrak{Q}$ in the 4-D theory where $\mathfrak{Q}$ is given by (**2F**) and in the 3-D theory there is a formula (cf. [17, 45])

$$\mathfrak{Q} = -\frac{\hbar^2}{4k_BTm}\left[\nabla^2\mathcal{Q} - \frac{1}{D}\partial_t\mathcal{Q}\right] \tag{3.6}$$

We refer to Section 4 for some remarks on the KG equation and recall that there is related information via the Wheeler-deWitt (WDW) or Klein-Gordon (KG) equations in e.g. [6, 17, 21, 32, 34, 46, 51]. ∎

REMARK 3.2. The role of information theory and entropy in QM has been clarified by e.g. Caticha [26] (see also in this connection [7, 71, 72, 88]). Osmotic velocity plays an important role in the diffusion theory of course and we mention also some recent work of Munkhammar [70] related to [89]. One can argue that sometimes QM is signalled simply by the entrance of $\hbar$ into the theory. For example in the holographic force scenario of Verlinde [89] it seems to arise via the Unruh temperature relation to acceleration (**3F**) $k_BT = (1/2\pi)(\hbar a/c)$. Padmanabhan et al have written deeply and extensively about matters of entropy, entanglement, equipartition, horizons, and gravity as an emergent phenomenon (cf. [75, 76]). Entanglement entropy is also studied by J.W. Lee et al in [60, 63, 64] and one theme involves entropy and information erasure. An interesting Machian point of view is developed in [41, 42] and in [97] there is a microscopic approach to entropic gravity via the idea of coherence length. In [45] one defines a thermal term s via $\nabla\mathcal{Q} = 2\omega\nabla(\delta s) = 2\omega m\mathbf{u} = -\omega\hbar(\nabla P/P)$, providing the osmotic velocity, and in [26] this is combined with a relative entropy term $\mathfrak{S}$, based on "hidden" variables, for the "drift" velocity and a current velocity determined via ϕ where $\phi = \mathfrak{S} - log(\rho^{1/2})$ with velocity connection $v^a = b^a + u^a$ (see [26] for details). In [58] it is also shown that in entropic quantum dynamics gravitational potential and accelerating frames are informationally equivalent. ■

4. Superfluids

The superfluid universe of Klinkhamer and Volovik (cf. especially [91] (cond-mat 1004.0597) and [62] (0811.4347) is a fascinating theme and we mention a few mathematical points here which were designed to give some structure to the superfluid world. We have a very imperfect and fragmentary idea of the physics here and only look at certain mathematical features which seem related to the material sketched above (and in [15, 16, 17] etc.). As an example of the arena for $\mathfrak{q}$ theory one looks for $\mathfrak{q}^{\mu\nu}$ which in a homogeneous vacuum has the form (**4A**) $\mathfrak{q}^{\mu\nu} = \mathfrak{q}g^{\mu\nu}$ or (**4B**) $\mathfrak{q}_{\mu\nu\alpha\beta} = \mathfrak{q}(g_{\alpha\mu}g_{\beta\nu} - g_{\alpha\nu}g_{\beta\mu})$ (generally here $\hbar = c = k_B = 1$). Another example is (**4C**) $\mathfrak{q}^{\mu\nu\alpha\beta} = \mathfrak{q}\epsilon^{\mu\nu\alpha\beta}$ where $\epsilon^{\mu\nu\alpha\beta}$ is the fully antisymmetric Levi-Civita tensor (cf. [14]). More specifically one can consider a chiral condensate of gauge fields (e.g. a gluonic condensate in QCD). Assume e.g. that $F_{\alpha\beta}$ represents a color magnetic field with (**4D**) $< F_{\alpha\beta} >= 0$ and $< F_{\alpha\beta}F_{\mu\nu} >= (\mathfrak{q}/24)\sqrt{-g}\epsilon_{\alpha\beta\mu\nu}$ where $\mathfrak{q}$ is the anomaly driven topological condensate

$$\mathfrak{q} =< \tilde{F}^{\mu\nu}F_{\mu\nu} >= \frac{1}{\sqrt{-g}}\epsilon^{\alpha\beta\mu\nu} < F_{\alpha\beta}F_{\mu\nu} > \tag{4.1}$$

Then one chooses a vacuum action (**4E**) $S_{\mathfrak{q}} = \int d^4x\sqrt{-g}\epsilon(\mathfrak{q})$ leading to a stress-energy tensor

$$T^{\mathfrak{q}}_{\mu\nu} = -\frac{2}{\sqrt{-g}}\frac{\delta S_{\mathfrak{q}}}{\delta g^{\mu\nu}} = \epsilon(\mathfrak{q})g_{\mu\nu} - 2\frac{\partial\epsilon}{\partial\mathfrak{q}}\frac{\partial\mathfrak{q}}{\partial g^{\mu\nu}} \Rightarrow \frac{\partial q}{\partial g^{\mu\nu}} = \frac{1}{2}qg_{\mu\nu} \Rightarrow \tag{4.2}$$

$$\Rightarrow T^{\mathfrak{q}}_{\mu\nu} = g_{\mu\nu}\rho_{vac}(\mathfrak{q});\ \rho_{vac}(\mathfrak{q}) = \epsilon(\mathfrak{q}) - \mathfrak{q}\frac{\partial\epsilon}{\partial\mathfrak{q}}$$

In terms of the Einstein equations one has then (**4F**) $T^{\mathfrak{q}}_{\mu\nu} = \Lambda g_{\mu\nu};\ \Lambda = \rho_{vac}(\mathfrak{q}) = \epsilon(\mathfrak{q}) - \mathfrak{q}\frac{\partial\epsilon}{\partial\mathfrak{q}}$. Other possibilities are indicated in [62, 91] but fulfilment of (4.2) and (**4F**) is essential. Recall also from Section 3 that $\epsilon_{vac} - \mu\mathfrak{q} = -P_{vac}$ where μ is the chemical potential and one can argue that $-P_{vac} \sim \Lambda \sim \rho_{vac}$.

We recall now, from the discussion in Section 3, that given an osmotic velocity $\mathbf{u}$ in some region $\Omega \in \mathbf{R}^3$ there will be a local osmotic pressure

$$P_{loc} = \frac{1}{2}\rho\mathbf{u}^2 = \frac{m}{2}PD^2\frac{|\nabla P|^2}{P^2} = \frac{\hbar^2}{8m}\frac{|\nabla P|^2}{P} \tag{4.3}$$

so that (**4G**) $P_{osm} = \int_\Omega P_u d^3x = \int \mathfrak{Q}Pd^3x = (\hbar^2/8m)FI$. In the superfluid universe mentioned above (with vacuum ^4He) one might to suggest that this could correspond to a vacuum osmotic pressure induced by a quantum generated "vacuum" based on Sections 1-2. In order to develop this one should examine the osmotic pressure idea in a 4-D context, in view of the expression (**2F**) for $\mathfrak{Q}$.

We sketch now from [46] (quant-ph 0201035) where the Nelsonian perspective is extended to a deBroglie-Bohm (BB) in the context of describing a KG framework (cf. also [65]). Perhaps somewhat prophetically it is mentioned in [46] (quant-ph 0201035) that the quantum vacuum has to play a major role in the evolution of the universe and one must take Λ seriously (which directive is being followed today, especially in the theory of superfluids). The introduction of an aether is appropriate in a relativistic (BB) theory and enters into the KG construction. We extract now from [46] (quant-ph 0201035) a few equations regarding osmotic velocity in hopes of implementing in some manner the superfluid theory. Thus, following [46], we write (**4H**) $\psi(x,t) = R(x,t)exp(-iS/\hbar)$; $[\Box + (m^2c^2/\hbar^2)]\psi = 0$ leading to ($\Box = (1/c^2)\partial_t^2 + \partial_x^2 + \partial_y^2 + \partial_z^2$)

$$\frac{\Box R}{R} + \frac{1}{\hbar}\frac{\partial_\mu R}{R}\partial^\mu S + \frac{i}{\hbar}\frac{\partial^\mu R}{R}\partial_\mu S - \frac{1}{\hbar^2}\partial_\mu S\partial^\mu S + \frac{i}{\hbar}\Box S + \frac{m^2c^2}{\hbar^2} = 0 \tag{4.4}$$

from which follows (for $P = R^2$ and $J_\mu = P\partial_\mu S$)

$$2\frac{\partial^\mu R}{R}\partial_\mu S + \Box S = 0;\ \partial^\mu J_\mu = \partial^\mu P\partial_\mu S + P\Box S = 0 \tag{4.5}$$

along with (**4I**) $\partial_\mu S\partial^\mu S = M^2c^2 = m^2c^2 + \hbar^2(\Box R/R)$. Now to include vacuum energy one writes (**4J**) $S^0 = E_{vac}t$ along with (**4K**) $\Psi(x,t) = R(x,t)exp[-i(S+S^0)/\hbar]$ for a solution of the KG equation. Then instead of (4.5) one involves a "hidden" probability current $J_\mu^0 = P\partial_\mu S^0$ to get a conservation law for the total probability current (**4L**) $\partial^\mu(J_\mu + J_\mu^0) = 0$ where (**4M**) $\partial^\mu J_\mu = \partial^\mu P\partial_\mu S + P\Box S = -\partial^\mu P\partial_\mu S^0 - P\Box S^0 = -\partial^\mu J_\mu^0$. Further S^0 also modifies (**4I**) leading to a variable total mass

$$M^2c^2 \to \left[mc + \frac{E_{vac}}{c}\right]^2 + \hbar^2\frac{\Box R}{R} \tag{4.6}$$

Thus a ("hidden") diffusion current J_μ^0 is introduced, due to the assumed stochastic aether dynamics and a corresponding all-pervading vacuum energy E_{vac}. One expects that timelike trajectories will ensue, i.e. the zero component $J_{vac}^0 = P\partial_t(S+S^0) \geq 0$.

Now in order to relate this causal KG theory with Nelson's stochastic theory one defines (**4N**) $S = Et - \mathbf{px} = \hbar p_\mu x^\mu$ such that $\mathbf{v}^\mu = \partial^\mu S/M$. Also an external potential V is assumed (which could include the vacuum energy E_{vac}) with M given by (**4I**) as ($\bullet$) $M = \{[m + (V/c^2)]^2 + (\hbar^2/c^2)(\Box R/R)]\}^{1/2}$. Thus the two real valued equations corresponding to the KG equation involves (**4O**) $\partial^\mu J_\mu = \partial^\mu P\partial_\mu S + P\Box S = 0$ and a Hamilton- Jacobi-Bohm equation

$$\partial_\mu S\partial^\mu S = M^2\mathbf{v}_\mu\mathbf{v}^\mu = M^2c^2 = \left[mc + \frac{V}{c}\right]^2 + \mathfrak{Q} \tag{4.7}$$

where (**4P**) $\mathfrak{Q} = \hbar^2(\Box R/R)$. This leads to an "osmotic velocity" (**4Q**) $\mathbf{u}^\mu = D[\partial^\mu P/P]$ with $D = \hbar/2m$ (cf. also [46] - book). One can then rewrite (**4O**) and (4.7) respectively as

$$\partial_\mu\mathbf{v}^\mu = -\frac{1}{D}\mathbf{u}_\mu\mathbf{v}^\mu - \frac{\partial_\mu M}{M}\mathbf{v}^\mu;\ M^2c^2 = \left[mc + \frac{V}{c}\right]^2 + \mathfrak{Q} \tag{4.8}$$

where (**4R**) $\mathfrak{Q} = m^2c^2 + m\hbar\partial_\mu\mathbf{u}^\mu$ showing that the dynamics is essentially governed by the 4-gradient of the osmotic velocity. Also there is then a local osmotic pressure based on $(1/2)M|\mathbf{u}|^2$ with M as in ($\bullet$). In the sense of [65] (but now in explicit dependence on the osmotic velocity) equations (4.8) are the two basic equations of relativistic stochastic mechanics (cf. [46] for more on this). Note that one can also rewrite (4.8-A) as

$$\partial_\mu\mathbf{v}^\mu = -\frac{1}{D}(\mathbf{u}_\mu + \mathbf{u}_\mu^M)\mathbf{v}^\mu;\ \mathbf{u}_\mu^M = D\frac{\partial_\mu M}{M} \tag{4.9}$$

We refer to [46] for more equations and discussion and anticipate possible interesting connections of the analysis here to superfluid theory.

The obvious question now is to determine a possible relation between M and $\mathfrak{M}$ where $\mathfrak{M}$ is the quantum mass of Section 1 and M is the relativistic mass of [46]. We recall that the Weyl geometry involved $\hat{g}_{ab} = \Omega^2(x)g_{ab}$ with(**4S**) $\mathfrak{M}^2 = m^2 exp(\mathfrak{Q}$ and (**4T**) $\mathfrak{Q} = (\hbar^2/m^2)(\Box|\Psi|/|\Psi|)$. On the other hand there is no Weyl geometry explicit in Section 4 and one has from (•), (4.7), (4.8)

$$M^2c^2 = \left[mc + \frac{V}{c}\right]^2 + \mathfrak{Q};\ \mathfrak{Q} = \hbar^2\frac{\Box R}{R} \tag{4.10}$$

Now we recall that when working directly with a KG equation it is natural to use an approximation (**4U**) $exp(\mathfrak{Q}) \sim 1 + \mathfrak{Q}$ (cf. [15, 16, 87]) and (**4S**) becomes then (**4V**) $\mathfrak{M}^2 \sim m^2(1+\mathfrak{Q})$. Setting $c = 1$ in (4.10) then one can compare

$$M^2 = (m+V)^2 + \hbar^2\frac{\Box R}{R}\ \ and\ \ \mathfrak{M}^2 = m^2\left[1 + \frac{\hbar^2}{m^2}\frac{\Box R}{R}\right] \tag{4.11}$$

Evidently $\mathfrak{M}^2 \sim M^2$ with $m \to m + V$ where V contains E_{vac}. If we take $V = E_{vac}$ then M only differs from $\mathfrak{M}$ by an E_{vac} term, i.e. for $\tilde{\mathfrak{Q}} = \hbar^2\Box R/R$

$$\sqrt{M^2 - \tilde{\mathfrak{Q}}} - \sqrt{\mathfrak{M}^2 - \tilde{\mathfrak{Q}}} = E_{vac} \tag{4.12}$$

In a certain sense then the effect of relativity here is to display the influence of a quantum vacuum.

References

[1] R. Adler, M. Bazin, and M. Schiffer, Introduction to general relativity, McGraw Hill, 1965
[2] J. Audretsch, Phys. Rev. D, 24 (1981), 1470-1477; 27 (1983), 2872-2884
[3] J. Audretsch, F. Gäehler, and N. Straumann, Comm. Math. Phys., 95 (1984), 41-51
[4] J. Audretsch and C. Lämmerzahl, Class. Quant. Gravity, 5 (1988), 1285-1295
[5] C. Bender and P. Mannheim, Phys. Rev. D, 78 (2008), 025022; Phys. Rev. Lett., 100 (2008), 110402; hep-th 0807.2607
[6] G. Bertoldi, A. Faraggi, and M. Matone, Class. Quant. Grav., 17 (2000), 3965
[7] D. Bohm and B. Hiley, Phys. Repts., 172 (1989), 93-122
[8] R. Bonal, I. Quiros, and R. Cardenas, gr-qc 0010010
[9] R. Cai and S. Kim, hep-th 0501055
[10] R. Cai, L. Cao, and N. Ohta, hep-th 1001.3470 and 1002.1136
[11] S. deCampo, R. Herrera, and P. Lebrana, gr-qc 0711.1559
[12] V. Canuto, P. Adams, S. Hsieh, and E. Tsiang, Phys. Rev. D, 16 (1977), 1643-1663
[13] Q. Cao, Y. Chen, and K. Shao, hep-th 1001.2597
[14] M. Carmelli, Classical fields, World Scientific, 2001
[15] R. Carroll, Fluctuations, information, gravity, and the quantum potential, Springer, 2006
[16] R. Carroll, On the quantum potential, Arima Publ., 2007
[17] R. Carroll, On the emergence theme of physics, World Scientific, 2010
[18] R. Carroll, Progress in Physics, 4 (2007), 22-24; 1 (2008), 21-24 and 2 (2008), 89-90
[19] R. Carroll, math-ph 0701077. 0703065, 0705.3921, 0712.3251; 0507027
[20] R. Carroll, Foundations of Physics, 35 (2005), 131-154
[21] R. Carroll, Teor. i Mat. Fiz., 152 (2007), 904-914
[22] R. Carroll, math-ph 0807.1320 and 0807.4158
[23] R. Carroll, Quantum Potential as Information: A mathematical survey, in: New trends in quantum information, Eds. Felloni, Singh, Licata, and Sakaji, Aracne Editrice, 2010, pp. 155-189
[24] R. Carroll, math-ph 1007.4744; gr-qc 1010.1732
[25] C. Castro, Found. Phys., 22 (1992), 569-615 and 37 (2007), 366-409; Found. Phys. Lett., 4 (1991), 81-99; Jour. Math. Phys., 31 (1990), 2626-2633
[26] A. Caticha, quant-ph 0907.4335, 1005.2357, and 1011.0746
[27] L. Crowell, Quantum fluctuations of spacetime, World Scientific, 2005
[28] P. Dirac, Proc. Royal Soc. London, A 333 (1973), 403 and 338 (1974), 439
[29] W. Drechsler, gr-qc 9901030

[30] Z. Ezawa, Quantum Hall effects, World Scientific, 2008
[31] F. Falciano, M. Novello, and J. Salim, Found. Phys., 40 (2010), 1885-1901
[32] A. Faraggi and M. Matone, Inter. Jour. Mod. Phys. A, 15 (2000), 1869-2017
[33] V. Faraoni, Cosmology in scalar-tensor theory, Kluwer, 2004; gr-qc 0906.1901 and 1005.2327
[34] B. Frieden, Science from Fisher information, Springer, 2004; Physics from Fisher information, Cambridge Univ. Press, 1998
[35] B. Frieden, A. Plastino, A. R. Plastino, and B. Soffer, Phys. Rev. E, 60 (1999), 48-53 and 66 (2002), 046128; Jour. Phys. A, 304 (2002), 73-78
[36] B. Frieden and R. Gatenby, Exploratory data analysis using Fisher information, Springer, 2007
[37] B. Frieden and B. Soffer, Physica A, 388 (2009), 1315-1330
[38] Y. Fuji and K. Maeda, The scalar-tensor theory of gravitation, Cambridge Univ. Press, 2003
[39] P. Garbaczewski, Entropy, 7 (2005), 253-299; Jour. Stat. Phys. 123 (2007), 315-355; quant-ph 0408192, 0612151, 0706.2481, and 0805.1536; cond-mat 0211362, 0604538, 0703147, 0703204, and 0811.3856
[40] L. Glinka, gr-qc 0612079, 0707.3341, 0711.1380, 0712.1624, 0712.2769, 0801.4157, 0803.0533, and 0808.1533; physics 1102.5002 (book)
[41] M. Gogberashvili, physics 0910.0169; gr-qc 0807.2439 and1008.2544
[42] M. Gogberashvili and I. Kanatchikov, gr-qc 0807.2439 and 1009.2266; physics 1012.5914
[43] W. Graf, gr-qc 0602054
[44] B. Greene, The hidden reality, Knopf, 2011
[45] G. Grössing, Entropy, 12 (2010), 1975-2044; quant-ph 0201035, 0205048, 0311109, 0404030, 0410236, 0508079, 0711.4954, 0806.4462, and 0808.3539; Phys. Lett. A, 372 (2008), 4556-4562; Found. Phys. Lett., 17 (2004), 343-362
[46] G. Grössing, Quantum cybernetics, Springer, 2000; quant-ph 0201035 and 0205047
[47] G. Grössing, S. Fussy, J. Mesa Pascasio, and H. Schwabl, quant-ph 1004.4596 and 1005.1058
[48] G. Groṡsing and J. Mesa Pascasio, quant-ph 0812.3561
[49] M. Hall, quant-ph 9802052, 9806013, 9903045, 9912055, 0103072, 0107149, 0302007; gr-qc 0408098; Phys. Rev. A, 55 (1997), 100-113 and 62 (2000), 012107, Jour. Phys. A, 37 (2004), 7799; Gen. Relativ. Grav., 37 (2005), 1505-1515
[50] M. Hall and M. Reginatto, Jour. Phys. A. 35 (2002), 3289-3303; Fortschrift Phys., 50 (2002), 646-651
[51] M. Hall, K. Kumar, and M. Reginatto, Jour. Phys. A, 36 (2003), 9779-9794 (hep-th 0206235 and 0307259)
[52] T. Harko, gr-qc 1101.3655
[53] J. Isidro, J. Santander, and P. Fernandez de Cordoba, hep-th 0902.0143 and 0912.1535
[54] M. Israelit, The Weyl-Dirac theory and our universe, Nova Sci. Publ., 1999
[55] M. Israelit, Found. Phys., 29 (1999), 1303; 32 (2002), 295 and 945; 28 (1998), 205; 35 (2005), 1725 and 1769; 37 (2007), 1628; gr-qc 9609035, 0710.3709, 0710.3910, 07103913, 07103923, and 0905.2482; astro-ph 1008.0767; Inter. Jour. Mod. Phys. A, 17 (2002), 4229
[56] M. Israelit and N. Rosen, Found. Phys., 13 (1983), 1023
[57] T. Jacobson, Phys. Rev. Lett., 75 (1995), 1260; hep-th 1001.4853; gr-qc 0711.3852 and 0801.1547
[58] D. Johnson and A. Caticha, quant-ph 1010.1467
[59] A. Kholodenko, gr-qc 0110064; hep-th 0303334 and 0701084
[60] H. Kim, J. W. Lee, and J. Lee, 0709.3573
[61] F. Klinkhamer, hep-th 1001.1939, 1006.2094, 1101.1281, and 1101.5370; gr-qc 0703009
[62] F. Klinkhamer and G. Volovik, gr-qc 0711.3170, 0807.3896, 0811.4347, and 1102.3152; hep-th 0905.1919 and 0907.4887
[63] J.W. Lee, hep-th 1003.1878, 1003.4464,1005.2739, and 1011.1657
[64] J.W. Lee, H. Kim and J. Lee, hep-th 0701199, 1001.5445 and 1002.4568
[65] W. Lehr and J. Park, Jour. Math. Phys., 18 (1977), 1235-1240
[66] B. Liu, Y. Dai, X. Hu, and J. Deng, gr-qc 1012.2560
[67] J. Makela, gr-qc 0506087, 0605098, 061078, 0805.3952, 0805.3955, 0810.4901, 1001.3808, and 1011.2052
[68] G. Mangano and L. Sokolowski, gr-qc 9312008
[69] P. Mannheim, hep-th 0707.2283, 0809.1200, 0909.0212, 0912.2635, 1005.5108, and 1101.2186; astro-ph 0505266 gr-qc 0001011 and 0703037
[70] J. Munkhammar, hep-th 1003.1262 and 1006.0723; physics 1003.4981 and 1101.1417
[71] M. Nagasawa, Schroedinger equations and diffusion theory, Birkhäuser, 1993; Stochastic processes in quantum physics, Birkhäuser, 2000
[72] E. Nelson, Quantum fluctuations, Princeton Univ. Press, 1985; Dynamical theory of Brownian motion, PUP, 1967; Phys. Rev., 130 (1966), 1079-1085
[73] M. Novello, astro-ph.CO, 1003.5126; physics 1004.3195 and 1008.2371
[74] M. Novello, J. Salim, and F. Falciano, gr-qc 0901.3741; Found. Phys., 40 (2010), 1885-1901

[75] T. Padmanabhan, Gravitation, Cambridge Univ. Press, 2010
[76] T. Padmanabhan, gr-qc 0202078, 0209088, 0308070, 0311036, 0408051, 0412068, 0606061, 0910.0839, 0911.1403, 0911.5004, 0912.3165, 1001.3380, 1003.5662, 1007.5066, 1012.0119, and 1012.4476; hep-th 0205278
[77] G. Papini, gr-qc 0111005 and 0304082
[78] G. Papini and W. Wood, Phys. Lett. A, 202 (1995), 45-49
[79] C. Pethick and H. Smith, Bose Einstein condensation in dilute gases, Cambridge Univ. Press, 2008
[80] L. Pitaevskij and S. Stringari, Bose-Einstein condensates, Oxford Univ. Press, 2003
[81] I. Quiros, gr-qc 9904004, 9905071, and 0004014; hep-th 0009169
[82] M. Reginatto, Phys. Rev. A, 58 (1998), 1775; quant-ph 9909065; gr-qc 0501030
[83] M. Reginatto and F. Lengyel, cond-mat 9910039
[84] N. Rosen, Found. Phys., 12 (1982), 213
[85] E. Santamato, Phys. Rev. D, 29 (1984), 216-222 and 32 (1985), 2615-2621; Jour. Math. Phys., 25 (1984), 2472-2480; Phys. Lett. A, 130 (1988), 199-202
[86] E. Scholz, astro-ph 0403446 and 0805.2557; hep-th 1102.3478; gr-qc 0703102 and 0710.0269
[87] F. and A. Shojai, gr-qc 0306099, 0404102, 0409053 and 001.3496; Inter. Jour. Mod. Phys. A, 15 (2000), 1859; Class. ,Quant. Grav., 21 (2004), 1-9
[88] L. Smolin, hep-th 0605052; quant-ph 0609109; gr-qc 1001.3668; Class. Quant. Grav., 3 (1986), 347-359; Phys. Lett., 113A (1986), 408-412
[89] E. Verlinde, hep-th 1001.0785
[90] G. Volovik, The universe in a helium droplet, Oxford Univ. Press, 2003
[91] G. Volovik, gr-qc 0004049, 0405012, 0406005, 0505104, and 0612134; cond-mat 0507454 and 1004.0597; physics 0909.1044; Found. Phys., 41 (2011), 516-528
[92] C. Wang, gr-qc 0603062 and 0605124
[93] S. Wei, Y. Liu, and Y. Wang, hep-th 1001.5238
[94] S. Weinberg, Cosmology, Oxford Univ. Press, 2008
[95] J. Wheeler, Phys. Rev. D, 41 (1990), 431-441 and 44 (1991), 1769-1773
[96] W. Wood and G. Papini, Phys. Rev. D, 45 (1992), 3617-3627; Found. Phys. Lett., 6 (1993), 207-223; Phys. Lett. A, 202 (1995), 45-49; gr-qc 9612042
[97] H. Xiong, hep-th 1012.5858, 1101.0525, and 1101.1270

Journal of Physics: Conference Series **361** (2012) 012011 doi:10.1088/1742-6596/361/1/012011

Review of stochastic mechanics

Edward Nelson

Department of Mathematics, Princeton University, Princeton, NJ 08544, USA

E-mail: nelson@math.princeton.edu

Abstract. Stochastic mechanics is an interpretation of nonrelativistic quantum mechanics in which the trajectories of the configuration, described as a Markov stochastic process, are regarded as physically real. The natural stochastic generalization of classical variational principles leads to a derivation of the Schrödinger equation. A brief review of the successes and failures of the theory is given, with references.

Stochastic mechanics has been the work of many people. It is described in the books [1] and [2], the second of which contains many references. It is an attempt to derive and explain nonrelativistic quantum mechanics as an emergent theory in which particle trajectories are physically real and governed by stochastic laws of motion.

1. Classical mechanics

Consider n particles in d space dimensions, with configuration space R^{dn}. Let m_k be the mass of the kth particle and let m_{ij} be the diagonal matrix with entries m_k on the diagonal when $(k-1)d < i = j < kd$. Then m_{ij} is a Riemann metric, and even though it is flat it is convenient to use tensor notation with the summation convention.

Then $T = \frac{1}{2}m_{ij}\dot{x}^i\dot{x}^j$ (where $\dot{x}$ is the velocity) is the kinetic energy. Let V be the potential energy, so $L = T - V$ is the Lagrangian, and

$$F_i = -\frac{\partial v}{\partial x^i} + \frac{d}{dt}\frac{\partial V}{\partial v^i}$$

is the force (where $v^i = \dot{x}^i$).

The *state* of the system is (x, v) and a *dynamical variable* is a function of the state. We assume that $V = \varphi - A_i\dot{x}^i$, where φ is the *scalar potential* and A is the *covector potential.* This form of the potential energy is to ensure that the force will be a dynamical variable. From a variational equation one derives the Hamilton-Jacobi equation

$$\frac{\partial S}{\partial t} + \frac{1}{2}(\nabla^i S - A^i)(\nabla_i S - A_i) + \varphi = 0$$

with the Newton equation $F_i = m_{ij}\ddot{x}^j$.

2. Kinematics of stochastic mechanics

A stochastic process is a *Markov process* in case the past and future are conditionally independent given the present. That is, knowledge of the past gives no more information about the future than does knowledge of the present, and vice versa. The notion is time symmetric.

In stochastic mechanics, the trajectory $x(t)$ is a Markov process governed by a stochastic differential equation of the form

$$dx(t) = dw(t) + b(x(t), t)dt$$

Here w is a Wiener process (Brownian motion process) with

$$\mathrm{E}_t dw^i(t) dw^j(t) = \hbar m^{ij} dt$$

where E_t is the conditional expectation with respect to the present at time t. Then $dw(t)$ is singular, of order $dt^{1/2}$. Such a process has continuous but nowhere differentiable trajectories.

The assumption that physical particles behave in this way is just on the borderline of being falsifiable. We can measure the position at two times t_1 and t_2 with an error given by the uncertainty principle. With a constant bigger than $\hbar$ in the diffusion tensor $\hbar m^{ij}$ we could determine that particles do not move in the way described by stochastic mechanics.

The *mean forward derivative* $\mathrm{D}F(t)$ of a stochastic process F is defined by

$$\mathrm{D}F(t) = \lim_{dt \to 0+} \mathrm{E}_t \frac{F(t+dt) - F(t)}{dt}$$

and the *mean backward derivative* is

$$\mathrm{D}_* F(t) = \lim_{dt \to 0+} \mathrm{E}_t \frac{F(t) - F(t-dt)}{dt}$$

Then $b(x(t), t)$ is $\mathrm{D}x(t)$, the *mean forward velocity* of the process. It is one substitute for the derivative, which does not exist. By time reversal symmetry, there is also a *mean backward velocity* $b_*(x(t), t)$, and we also form the *current velocity* $v = \frac{1}{2}(b + b_*)$ and the *osmotic velocity* $u = \frac{1}{2}(b - b_*)$.

Let $\rho(x, t)$ be the probability density of the configuration at time t. Then we have the *osmotic equation*

$$u^i = \frac{1}{2} \frac{\nabla^i \rho}{\rho}$$

and the *current equation*

$$\frac{\partial \rho}{\partial t} = -\nabla_i (v^i \rho)$$

3. Dynamics of stochastic motion

Classical dynamics comes from a variational principle applied to action integrals. How can we formulate the action in the absence of derivatives? The contribution $\int \varphi(x(t), t)dt$ from the scalar potential is an ordinary Riemann integral, and the contribution from the covector potential can be expressed as a Fisk-Stratonovich time-symmetric stochastic integral $\int A_j(x(t), t)dx^j(t)$. The kinetic action is more subtle and was formulated by Francesco Guerra and Laura Morato [3]. There is an account of this is §9 of [2].

Briefly, let dx^i/dt be a difference quotient with $dt > 0$, not a derivative. We need to calculate

$$\mathrm{E}_t \frac{1}{2} \frac{dx^i}{dt} \frac{dx_i}{dt}$$

to o(1). Let

$$W^k = \int_t^{t+dt} [w^k(r) - w^k(t)]dr$$

We find

$$dx^i dx_i = b^i b_i dt^2 + 2b^i dw_i dt + 2\nabla_k b^i W^k dw_i + o(dt^2)$$

First miracle: the term $2b^i dw_i dt$ is singular, of order $dt^{3/2}$, but $\mathrm{E}_t 2b^i dw_i dt = 0$ since $2b^i$ lives in the present and dw_i is in the future with mean 0.

Now we use the fact that w has orthogonal increments and calculate further, finding

$$\mathrm{E}_t \frac{1}{2}\frac{dx^i}{dt}\frac{dx_i}{dt} = \frac{1}{2}b^i b_i + \frac{1}{2}\nabla_i b^i + \frac{nd}{2dt} + o(1)$$

Second miracle: the singular term $nd/2dt$ is a constant, independent of the trajectory, so it drops out when a variational principle is applied to the action.

Let $R = \frac{\hbar}{2}\log\rho$, so $\nabla^i\rho$ is the osmotic velocity u^i. Apply the variational principle to the expected action. We find the stochastic Hamilton-Jacobi equation

$$\frac{\partial S}{\partial t} + \frac{1}{2}(\nabla^i S - A_i)(\nabla_i S - A_i) + \varphi - \frac{1}{2}\nabla^i R \nabla_i R - \frac{\hbar}{2}\nabla^i \nabla_i R = 0$$

which without the terms containing R is the classical Hamilton-Jacobi equation. Write the current equation in terms of R and S and find

$$\frac{\partial R}{dt} + \nabla_i R(\nabla^i S - A^i) + \frac{\hbar}{2}\nabla^i \nabla_i S - \frac{\hbar}{2}\nabla_i A^i = 0$$

We have a pair of coupled nonlinear partial differential equations.

Third miracle: Let $\psi = e^{\frac{1}{\hbar}(R+iS)}$. Then these equations are equivalent to the Schrödinger equation

$$i\hbar\frac{\partial\psi}{\partial dt} = \left[\frac{1}{2}\left(\frac{\hbar}{i}\nabla^j - A^j\right)\left(\frac{\hbar}{i}\nabla_j - A_j\right) + \varphi\right]\psi$$

We also have the *stochastic Newton equation* $F_i = m_{ij}a^j$ where a is the *mean acceleration* given by $\frac{1}{2}(\mathrm{DD}_* x + \mathrm{D}_*\mathrm{D}x)$.

We can apply the same procedure to any Riemannian manifold. We find the Schrödinger equation with an additional term $(\hbar^2/12)\bar{R}$ where $\bar{R}$ is the scalar curvature. This is the Bryce DeWitt term [4].

4. Successes of stochastic mechanics

Here is a list of the main successes of stochastic mechanics.

- A classical derivation of the Schrödinger equation, by Guerra and Morato [3].
- The probability density ρ of the Markov process agrees with $|\psi|^2$ at all times.
- A stochastic explanation of the relation between momentum and the Fourier transform of the wave function, by David Shucker [5].
- A proof of the existence of the Markov process under the physically natural assumption of finite action, by Eric Carlen [6]. This is perhaps the most technically demanding work in the entire subject.
- A stochastic explanation of why identical particles satisfy either Bose-Einstein or Fermi-Dirac statistics if $d \geq 3$, with parastatistics possible if $d = 2$. This is not contained in §20 of [2], but it follows from the discussion there.
- A stochastic explanation of spin and why it is integral or half-integral, work of Thaddeus Dankel [7], Timothy Wallstrom [8], and of Daniela Dohrn and Francesco Guerra jointly [9].
- If the force is time-independent, the expected stochastic energy $\mathrm{E}_t(\frac{1}{2}u^i u_i + \frac{1}{2}v^i v_i + \varphi)$ is conserved; see §14 of [2].
- A stochastic picture of the two-slit experiment, explaining how particles have trajectories going through just one slit or the other, but nevertheless produce a probability density as for interfering waves; see §17 of [2].

5. Failures of stochastic mechanics

In quantum mechanics, if there are two dynamically uncoupled systems, an alteration of the second system in no way affects the first, even if the two systems are entangled. This is not so in stochastic mechanics. With two dynamically uncoupled particles, a force applied to one can immediately affect the motion of the other, in a way independent of their spatial separation; see §23 of [2]. This makes it unrealistic to regard the trajectories as physically real.

There is a more serious problem. Since $\rho = |\psi|^2$ at all times, stochastic mechanics gives the same prediction as quantum mechanics for a measurement performed at a single time. But it can give wrong predictions for measurements performed at two different times; see Chapter 10 of [10]. Consider two entangled but dynamically uncoupled harmonic oscillators. Let $X_i(t)$ be the Heisenberg position operator of oscillator i at time t. Each is periodic in t, so the correlation of $(X_1(t_1), X(t_2))$ does not decay as $t_2 \to \infty$. Let $x_i(t)$ be the position of oscillator i at time t according to stochastic mechanics. Then $x_i(t)$ has the same probability distribution as $X_i(t)$ for each i and each t, but $(x_1(t_1), x_2(t_2))$ does not have the same probability distribution as $(X_1(t_1), X_2(t_2))$. In fact, the correlation of $(x_1(t_1), x_2(t_2))$ decays to 0 as $t_2 \to \infty$. The oscillators are uncoupled, so $X_1(t_1)$ and $X_2(t_2)$ commute, and according to quantum mechanics, the probability distribution is that of $(X_1(t_1), X_2(t_2))$. If $(x_1(t_1), x_2(t_2))$ represented the real physical situation, theirs would be the probability distribution. Thus stochastic mechanics and quantum mechanics give different predictions for the result. Why do I not suggest that the experiment be done? Because if a record of the observation of the first oscillator at time t_1 is made by some physical means, and similarly for the second oscillator, and the two records are compared at a common later time t_3, this is an observation at a single time, for which quantum mechanics and stochastic mechanics agree. The nonlocality of stochastic mechanics conspires to bring the records into agreement.

How can a theory to be so right and yet so wrong? The most natural explanation is that stochastic mechanics is an approximation to a correct theory of quantum mechanics as emergent. But what is the correct theory?

References

[1] Nelson E 1967 *Dynamical Theories of Brownian Motion* (Princeton, NJ: Princeton University Press) http://www.math.princeton.edu/~nelson/books/bmotion.tex
[2] Nelson E 1985 *Quantum Fluctuations* (Princeton, NJ: Princeton University Press) http://www.math.princeton.edu/~nelson/books/qf.pdf
[3] Guerra F and Morato L 1983 Quantization of dynamical systems and stochastic control theory *Phys. Rev.* D **27** pp 1774-86
[4] DeWitt B 1957 *Rev. Mod. Phys.* **29** 377
[5] Shucker D 1980 *J. Functional Analysis* **38** pp 146-55
[6] Carlen E 1986 Existence and sample path properties of the diffusion in Nelson's stochastic mechanics *Stochastic Processes in Classical and Quantum Mechanics* Albeverio S *et al* eds. Lecture Notes in Physics **262** (New York: Springer-Verlag) pp 139-47
[7] Dankel T 1977 *J. Math. Phys.* **18** 255
[8] Wallstrom T 1988 *On the Derivation of the Schrödinger Equation from Stochastic Mechanics* (Princeton, NJ: Princeton University Press)
[9] Dohrn D and Guerra F 1978 *Lettere al Nuovo Cimento* **22** pp 271-77
[10] Nelson E 2006 *Diffusion, Quantum Theory, and Radically Elementary Mathematics* Faris W ed. (Princeton, NJ: Princeton University Press) pp 229-32

Probabilistic whereabouts of the "quantum potential"

Piotr Garbaczewski
Institute of Physics, University of Opole, 45-052 Opole, Poland
E-mail: pgar@uni.opole.pl

Abstract. We review major appearances of the functional expression $\pm\Delta\rho^{1/2}/\rho^{1/2}$ in the theory of diffusion-type processes and in quantum mechanically supported dynamical scenarios. Attention is paid to various manifestations of "pressure" terms and their meaning.

1. From ρ to $\pm\frac{\Delta\rho^{1/2}}{\rho^{1/2}}$

For clarity of presentation, we restrict considerations to probability density functions (pdfs) in one spatial dimension, with the time label suppressed for a while (albeit we have in mind not only random variables, but stochastic processes as well). Given a continuous, at least twice differentiable pdf $\rho(x)$, $\int_R \rho(x)\,dx = 1$, we infer a hierarchy of functions $-\ln\rho(x)$, $-\nabla\ln\rho$ and $-\Delta\ln\rho$ whose meaning will soon become more transparent.

An information theoretic notion of the Shannon entropy of ρ (with varied interpretations, among which the notions of disorder and/or uncertainty seem to prevail) reads

$$\mathcal{S}(\rho) = -\langle\ln\rho\rangle = -\int \rho(x)\ln\rho(x)dx \tag{1}$$

thus giving $\ln\rho$ the meaning of (Shannon) entropy density. An affiliated measure of disorder/uncertainty named the Fisher information of ρ has the form

$$\mathcal{F}(\rho) \doteq \langle(\nabla\ln\rho)^2\rangle = \int \frac{(\nabla\rho)^2}{\rho}dx \tag{2}$$

and in view of $\langle\nabla\ln\rho\rangle = 0$ stands for a mean square deviation of a function $\nabla\ln\rho(x)$ of the random variable X, with values $x \in R$.

These two information theoretic measures are interrelated, as for example can be seen through so-called isoperimetric inequality: $\mathcal{F} \geq (2\pi e)\exp(2\mathcal{S})$. More than that, it is the Fisher information which is directly involved in the primordial form of the indeterminacy (uncertainty) principle. Namely, $\langle\nabla\ln x\rangle = 0$ and $Var(x) = \sigma^2 = \langle(x - \langle x\rangle)^2\rangle$ imply an indeterminacy relationship (no quantum context as yet):

$$Var(\nabla\ln x) = \mathcal{F}(\rho) \geq 1/\sigma^2 > 0\,. \tag{3}$$

We can get a deeper insight into the indeterminacy rule, by noting that actually $\rho^{1/2}$ is a square integrable function. Therefore, standard $L^2(R)$ Fourier transform techniques $\psi \to \tilde{\psi}$ can be here adopted. In a slightly more general notation, which encompasses the previous case, $\rho = |\psi|^2, \psi \in L^2 \Longrightarrow \tilde{\psi}$, $\tilde{\rho} = |\tilde{\psi}|^2$, one infers (in self-explanatory notation)

$$(1/\sigma^2) \leq \mathcal{F} \leq 16\pi^2\tilde{\sigma}^2 \tag{4}$$

and

$$(4\pi/\tilde{\sigma}) \leq (1/\sqrt{2\pi e})\exp[\mathcal{S}] \leq \sigma. \tag{5}$$

We can continue our discussion of $-\ln \rho(x)$, $-\nabla \ln \rho$ and $-\Delta \ln \rho$ by simply playing with them, to reveal a number of emergent quantities, like e.g.

$$-\Delta \ln \rho = -\frac{\Delta \rho}{\rho} + \frac{(\nabla \rho)^2}{\rho^2} \Longrightarrow -\langle \Delta \ln \rho \rangle = \langle \frac{(\nabla \rho)^2}{\rho^2} \rangle = \mathcal{F}(\rho). \tag{6}$$

At this point we introduce a potential $\frac{\Delta \rho^{1/2}}{\rho^{1/2}}$ of a Newton-type "force field" with vanishing mean value $\langle \nabla (\frac{\Delta \rho^{1/2}}{\rho^{1/2}}) \rangle = 0$

$$-\frac{\Delta \rho^{1/2}}{\rho^{1/2}} = \frac{1}{2}[-\frac{\Delta \rho}{\rho} + \frac{1}{2}\frac{(\nabla \rho)^2}{\rho^2}] \implies +\nabla (\frac{\Delta \rho^{1/2}}{\rho^{1/2}}) = \frac{1}{2\rho}\nabla(\rho \Delta \ln \rho) \tag{7}$$

and make explicit its links with the Fisher information of the pdf ρ, through

$$-\langle \frac{\Delta \rho^{1/2}}{\rho^{1/2}} \rangle = -\frac{1}{4}\langle \Delta \ln \rho \rangle = \frac{1}{4}\mathcal{F}(\rho) \geq \frac{1}{4Var(x)} > 0\,, \tag{8}$$

We emphasize a conspicuous absence of any specific physical context. Nonetheless, while accounting for a temporal evolution of $\rho = \rho(x,t)$, a number of physically interesting quantities can be easily identified. They are omnipresent in local conservation laws for diffusion-type stochastic processes, as well as in the hydrodynamical formulation of the Schrödinger picture quantum dynamics.

2. Emergence of $\pm\frac{\Delta \rho^{1/2}}{\rho^{1/2}}$ in hydrodynamical (local) conservation laws

2.1. Quantum hydrodynamics

Taking as obvious the standard wisdom about a hydrodynamical representation of the Schrödinger picture quantum dynamics, we merely recall that the Schrödinger equation

$$i\hbar\partial_t\psi = \left[-\frac{\hbar^2}{2m}\Delta + V\right]\psi \tag{9}$$

involves a Hamiltonian $\hat{H}$ that is a self-adjoint operator in a suitable Hilbert space domain. Since we shall be dealing with bounded from below operators, for later convenience we impose an additive renormalization of the Hamiltonian so that $\hat{H} \geq 0$. Further notations are reproduced for the record: the pdf is $\rho(x,t) = |\psi|^2(x,t)$, $v = (\hbar/2mi)[(\nabla\psi/\psi) - (\nabla\psi^*/\psi^*)]$ stands for the current velocity field. With the polar (Madelung) decomposition of ψ being implicit, we get:

$$\partial_t\rho = -\nabla(\rho v); \quad \partial_t s + \frac{1}{2m}(\nabla s)^2 + (V + Q) = 0 \Longrightarrow \tag{10}$$

$$\partial_t v + (v\nabla v) = -\frac{1}{m}\nabla(V + Q)$$

where $v = \frac{1}{m}\nabla s$ and $Q = Q[\rho] = -\frac{\hbar^2}{2m}\frac{\Delta\rho^{1/2}}{\rho^{1/2}}$ has a folk name of a "quantum potential".
We note that the ground state condition for $\hat{H}$, with bottom eigenvalue 0, given $|\psi| = \rho_*^{1/2}$, directly involves the "quantum potential", presently evaluated with respect to the ground state pdf ρ_*:

$$V = +\frac{\hbar^2}{2m}\frac{\Delta\rho_*^{1/2}}{\rho_*^{1/2}} = -Q[\rho_*]. \tag{11}$$

Denote $u(x,t) \doteq (\hbar/2m)\nabla\ln\rho$. It is well known that the dynamics arises via the $\{\rho, s\}$ extremum principle for

$$I(\rho, s) = \int_{t_1}^{t_2} \langle\Big[\partial_t s + \frac{m}{2}(u^2+v^2) + V\Big]\rangle(t)dt. \tag{12}$$

In terms of valid solutions $\rho(x,t)$, $s(x,t)$, we arrive at a strictly positive constant of motion: $-\langle\partial_t s\rangle = H = \langle\big[\frac{m}{2}(u^2+v^2) + V\big]\rangle > 0$ (a finite energy condition).

2.2. Brownian hydrodynamics
The semigroup dynamics and the emergent generalized diffusion equation (note that by setting $V = 0$ we pass to the standard heat equation)

$$\exp(-t\hat{H}/2mD)\Psi_0 = \Psi_t \implies \partial_t\Psi = \left[D\Delta - \frac{V}{2mD}\right]\Psi \tag{13}$$

is a self-adjoint relative of the more familiar Fokker-Planck equation $\partial_t\rho = D\Delta\rho - \nabla(b\rho)$ and likewise, although indirectly, determines the evolution of $\rho(x,t)$. Here $\hat{H}$ is self-adjoint, $\hat{H} \geq 0$, $t \geq 0$. (We keep in mind a re-definition $\hbar \equiv 2mD$.)

Let $\Psi(x,t) \to \rho_*^{1/2}$ as $t \to \infty$. Define $\rho(x,t) = \Psi(x,t)\rho_*^{1/2}(x)$ with $b = D\nabla\ln\rho_*$, $u = D\nabla\ln\rho$ and $v = b - u = (1/m)\nabla s$. The connection between the Fokker-Planck and semigroup dynamics is being established, provided a compatibility condition

$$\frac{V(x)}{2mD} = +D\frac{\Delta\rho_*^{1/2}}{\rho_*^{1/2}} \doteq mD\left[\frac{b^2}{2D} + \nabla b\right] \tag{14}$$

holds true. The (rescaled) "quantum potential" appears in the above (c.f. also (11)), as well as in Hamilton-Jacobi type equations of motion:

$$\partial_t\rho = D\Delta\rho - \nabla(b\rho) \Longleftrightarrow \partial_t\rho = -\nabla(v\rho) \tag{15}$$

$$\partial_t s + (1/2m)(\nabla s)^2 - (V + Q) = 0 \implies \partial_t v + (v\nabla v) = +\frac{1}{m}\nabla(V+Q)$$

The $\{\rho, s\}$ extremum principle for

$$I(\rho, s) = \int_{t_1}^{t_2} \langle\big[\partial_t s + (m/2)(v^2 - u^2) - V\big]\rangle \tag{16}$$

yields the previous Hamilton-Jacobi type dynamics. In terms of dynamically admitted fields $\rho(x,t)$ and $s(x,t)$, we have $-\langle\partial_t s\rangle = H = \langle\big[\frac{m}{2}(v^2-u^2) - V\big]\rangle \equiv 0$.

2.3. Time dependence of Shannon and Fisher functionals

The dynamics of $\rho(x,t)$ is dictated by the continuity equation and this equation alone sets the evolution rule for the Shannon entropy $\mathcal{S}[\rho](t)$. Indeed, there holds (provided the fall-off of ρ at spatial infinities ensures the vanishing of $v\rho$):

$$\frac{d\mathcal{S}}{dt} = \langle \nabla v \rangle = -\frac{1}{D} \langle v\, u \rangle \tag{17}$$

where $D \equiv \hbar/2m$ corresponds to the quantum case. Obviously, there is an evolution of the velocity field $v(x,t)$ to be accounted for, c.f. the corresponding Hamilton-Jacobi type equations and their gradient versions.

By exploiting the Hamilton-Jacobi type equations, in case of time independent external potentials, we easily demonstrate that

$$\frac{d\mathcal{F}}{dt} = \mp 2 \langle v \nabla P \rangle \tag{18}$$

where $P = \rho \Delta \ln \rho$. The minus sign in the above refers to diffusion processes, while the plus sign to the quantum motion.

A more detailed discussion of the pressure-type term P and its functioning will be given later. We note a clear parallel with a classical power release expression $dE/dt = v\,F$, where $F = -\nabla V$ is a standard Newtonian force.

3. Dynamical duality - illusion of "Euclidean time"

In the light of our previous discussion it appears quite persuasive to execute (at least formally) the Wick rotation in the complex time plane $it \to t \geq 0;\ \ \hbar \to 2mD$

$$\exp(-i\hat{H}t/\hbar)\psi_0 = \psi_t \Longrightarrow \exp(-t\hat{H}/2mD)\Psi_0 = \Psi_t \tag{19}$$

that maps between diffusion-type and quantum mechanical patterns of dynamical behaviour. Given the spectral solution for $\hat{H} = -\Delta + V$, the integral kernel of $\exp(-t\hat{H})$ reads

$$k(y,x,t) = \sum_j \exp(-\epsilon_j t)\, \Phi_j(y)\Phi_j^*(x). \tag{20}$$

Remember that $\epsilon_0 = 0$ and the sum may be replaced by an integral in case of a continuous spectrum, (with complex-valued generalized eigenfunctions). Set $V(x) = 0$ identically. Then we end up with a familiar heat kernel:

$$k(y,x,t) = [\exp(t\Delta)](y,x) = (2\pi)^{-1/2} \int \exp(-p^2 t)\, \exp(ip(y-x)\, dp = \tag{21}$$

$$(4\pi t)^{-1/2}\, \exp[-(y-x)^2/4t].$$

Consider $\hat{H} = (1/2)(-\Delta + x^2 - 1)$ (e.g. the rescaled harmonic oscillator Hamiltonian). The integral kernel of $\exp(-t\hat{H})$ is given by the classic Mehler formula:

$$k(y,x,t) = k(x,y,t) = [\exp(-t\hat{H})(y,x) = \tag{22}$$

$$[\pi(1-\exp(-2t))^{-1/2} \exp[-(1/2)(x^2-y^2) - (1-\exp(-2t))^{-1}\, (x\exp(-t) - y)^2].$$

The normalization condition $\int k(y,x,t) \exp[(y^2-x^2)/2]\, dy = 1$ actually defines the transition probability density of the Ornstein-Uhlenbeck process

$$p(y,x,t) = k(y,x,t)\, \rho_*^{1/2}(x)/\rho_*^{1/2}(y) \tag{23}$$

with $\rho_*(x) = \pi^{-1/2}\exp(-x^2)$. A more familiar form of the kernel reads (note the presence of $\exp(t/2)$ factor)

$$k(y,x,t) = \frac{\exp(t/2)}{(2\pi\sinh t)^{1/2}}\exp\left[-\frac{(x^2+y^2)\cosh t - 2xy}{2\sinh t}\right]. \tag{24}$$

To conform with the statistical physics lore of the 50-ies and 60-ties, we can easily pass to an integral kernel of the density operator, labeled by equilibrium values of the temperature. To this end one should set e.g. $t \equiv \hbar\omega/k_BT$ for a harmonic oscillator with a proper frequency ω and remember about evaluating the normalization factor $1/Z_T$ where Z_T stands for a partition function of the system.

Concerning the "Euclidean issue", we note that by formally executing $t \to it$ one arrives at the free Schrödinger propagator

$$K(y,x,t) = [\exp(it\Delta)](y,x) = (2\pi)^{-1/2}\int \exp(-ip^2t)\exp(ip(y-x)\,dp = \tag{25}$$

$$(4\pi it)^{-1/2}\exp[+i(y-x)^2/4t]$$

and likewise, at that of (here -1 renormalized) harmonic oscillator propagator

$$K(y,x,t) = \frac{\exp(it/2)}{(2\pi i\sin t)^{1/2}}\exp\left[+i\frac{(x^2+y^2)\cos t - 2xy}{2\sin t}\right] \tag{26}$$

Learn a standard Euclidean (field) theory lesson concerning multi-time correlation functions. In the exemplary harmonic oscillator case, $t > t' > 0$; $t \to it$ results in:

$$E[X(t')X(t)] = \int \rho_*(x')\,x'\,p(x',t',x,t)\,x\,dzdx' = (1/2)\exp[-(t-t')] \Longrightarrow \tag{27}$$

$$W(t',t) = \langle\psi_0, \hat{q}_H(t)\hat{q}_H(t')\psi_0\rangle = (1/2)\exp[-i(t-t')]$$

where $\hat{q}_H(t)$ stands for the position operator in the Heisenberg picture. In passing, we note that this appealing correspondence breaks down in R^n, $n > 1$, in the presence of electromagnetic fields.

The major message of our discussion is that we encounter two distinct dynamical patterns of behaviour that follow equally real (realistic) clocks. The Euclidean mapping (Wick rotation) is merely a mathematical artifice connecting the pertinent dynamical models, or rather transforming one model into another.

4. Comments on variational extremum principles

4.1. (Shannon) Entropy extremum principle

Given $V = V(x)$, fix a priori $\langle V\rangle = \zeta$. Extremize $\mathcal{S} = -\langle\ln\rho\rangle$ under this constraint, i.e. seek an extremum of

$$\mathcal{S}(\rho) + \alpha\langle V\rangle = \langle -\ln\rho + \alpha V\rangle$$

where α is a Lagrange multiplier. As an outcome we get the α-family of pdfs $\rho_\alpha = A_\alpha\exp[\alpha V(x)]$, provided $(A_\alpha)^{-1} = \int\exp[\alpha V(x)]\,dx$ exists. The Lagrange multiplier α-value must be inferred from the constraint $\langle V\rangle_\alpha = \zeta$.

4.2. Fisher information extremum principle

Fix a priori $\langle V \rangle = \zeta$. Extremize the Fisher information measure $\mathcal{F}(\rho)$ under that constraint:

$$\mathcal{F}(\rho) + \lambda \langle V \rangle = \langle (\nabla \ln \rho)^2 + \lambda V \rangle$$

Remember that $-\langle \frac{\Delta \rho^{1/2}}{\rho^{1/2}} \rangle = \frac{1}{4}\mathcal{F}(\rho)$. The extremizing pdf $\rho(x) \doteq \rho_*(x)$ comes out from

$$V(x) = \frac{2}{\lambda}[\frac{\Delta \rho}{\rho} - \frac{1}{2}\frac{(\nabla \rho)^2}{\rho^2}] = +\frac{4}{\lambda}\frac{\Delta \rho^{1/2}}{\rho^{1/2}}.$$

Outcome: λ-family of pdfs; λ gets fixed by $\langle V \rangle_\lambda = \zeta$. Setting $\lambda = 2/mD^2$, we recover the Brownian framework; $\lambda = 8m/\hbar^2$ is admitted as a special case.

4.3. Hamilton-Jacobi route

Think of a purely classical case $H = p^2/2m + V(x)$, $\{\dot{q} = p/m,\ \dot{p} = -\nabla V(q)\}$. Next, assign random initial data (here, in space) $\rho_0(x) \Rightarrow \mathcal{S}(\rho)$ and $\mathcal{F}(\rho)$. By extremizing the action functional we deduce the standard Hamilton-Jacobi description of an ensemble of classical systems:

$$I_0(\rho, s) = \int_{t_1}^{t_2} \langle \left[\partial_t s + \frac{1}{2m}(\nabla s)^2 + V \right] \rangle \, dt \Longrightarrow \partial_t s + \frac{1}{2m}(\nabla s)^2 + V = 0 \qquad (28)$$

plus the continuity equation $\partial_t \rho = -\nabla(v\rho)$. Here an assumption $v = (1/m)\nabla s$ implies $\partial_t v + (v\nabla v) = -\nabla V$.

4.4. Constrained Fisher information

Fix a priori $\int_{t_1}^{t_2} \mathcal{F}(\rho)(t)\, dt = \zeta$. Extremize

$$I_\gamma(\rho, s) = \int_{t_1}^{t_2} dt \, \langle \left[\partial_t s + \frac{(\nabla s)^2}{m} \pm V \right] + \gamma \frac{(\nabla \rho)^2}{\rho^2} \rangle \Longrightarrow$$

$$\partial_t \rho = -\nabla(v\rho) \qquad (29)$$

$$\partial_t s + \frac{(\nabla s)^2}{m} \pm V + 4\gamma \frac{\Delta \rho^{1/2}}{\rho^{1/2}} = 0$$

where by denoting $\pm V$ we intend to make a distinction between confining (generically bounded from below) and scattering potentials.

Outcomes (an admissible case of $\gamma = 0$ is left aside):

(i) $\gamma = -mD^2/2$, eventually followed by setting $D = \hbar/2m$, leads to the D-labelled quantum hydrodynamics (before, we have referred to $+V$ only)

$$\partial_t s + \frac{1}{2m}(\nabla s)^2 \pm V + Q = 0$$

(ii) $\gamma = +mD^2/2$, with the potential term $-V$ only, leads to the Brownian hydrodynamics

$$\partial_t s + (1/2m)(\nabla s)^2 - (V + Q) = 0$$

Note: $t \to it$ relationship can be secured for $+V$, where V is a confining potential.

$$\partial_t s + \frac{1}{2m}(\nabla s)^2 + (V + Q) = 0$$

c.f. $t \to it \Longrightarrow \exp(-t\hat{H}/2mD)\Psi_0 = \Psi_t \longrightarrow \exp(-it\hat{H}/2mD)\psi_0 = \psi_t$ issue.

We demand $\hat{H}$ to have a bottom eigenvalue equal zero (to yield a contractive semigroup). For a bounded from below Hamiltonian this can be always achieved, like e.g. in case of $\hat{H} = (1/2)(-\Delta + x^2 - 1)$.

4.5. Hamilton-Jacobi route - a catalogue of "standards"

In what follows we emphasize a relevance of the sign of the external potential (positive - confinement, negative-scattering) in Lagrangian densities.
(i) $\mathcal{L}^+ = -\rho\left[\partial_t s + (m/2)(v^2+u^2) + V\right] \Longrightarrow \partial_t s + (1/2m)(\nabla s)^2 + (V+Q) = 0$
(ii) $\mathcal{L}^\pm_{cl} = -\rho\left[\partial_t s + (m/2)v^2 \pm V\right] \Longrightarrow \partial_t s + (1/2m)(\nabla s)^2 \pm V = 0$
(iii) $\mathcal{L}^- = -\rho\left[\partial_t s + (m/2)(v^2-u^2) - V\right] \Longrightarrow \partial_t s + (1/2m)(\nabla s)^2 - (V+Q) = 0.$
A continuity equation $\partial_t\rho = -\nabla(v\,\rho)$ is shared by all listed cases, provided we set $v = (1/m)\nabla s$. On dynamically admitted fields $\rho(t)$ and $s(x,t)$, $L(t) = \int dx\,\mathcal{L} \equiv 0$, i.e. we have $\langle\partial_t s\rangle = -H$. The respective Hamiltonian functionals have the form:
(i) $H^+ \doteq \int dx\,\rho\left[(m/2)v^2 + V + (m/2)u^2\right] > 0$, is a (quantum) constant of motion
(ii) $H^\pm_{cl} \doteq \int dx\,\rho\left[(m/2)v^2 \pm V\right] = E,\;\; E = (p^2/2m) \pm V(x)$, constant on each path
(iii) $H^- \doteq \int dx\,\rho\left[(m/2)v^2 - V - (m/2)u^2\right] = 0$, identically in Brownian motion.
We emphasize that, from the start, $V(x)$ is chosen to be confining. A class of continuous and bounded from below functions allows to secure $\hat{H} \geq 0$. Eventually, after subtracting the lowest eigenvalue of the bounded from below energy operator.

5. Kinetic theory lore: Brownian analogies and hints

Consider free phase-space Brownian motion in the large friction regime. $W(x,u,t)$ stands for phase-space (velocity-position) probability distribution with suitable initial data at $t = 0$. Denote by $w(u,t)$ and $w(x,t)$, the marginal pdfs.
We set $D = k_BT/m\beta$ and observe that actually, in the large friction regime, $w(x,t)$ stays in the vicinity of (and ultimately converges to) the heat kernel solution $w(x,t) \sim (4\pi Dt)^{-1/2}\exp(-x^2/4Dt)$ of $\partial_t w = D\Delta w$.
We introduce moments and local moments of the pdf $w(x,t)$ in the large friction regime: $\langle u\rangle = \int du\,u\,W(x,u,t) \to \langle u\rangle = (x/2t)w(x,t)$, $\langle u\rangle_x = \langle u\rangle/w(x,t) = x/2t = -D(\nabla w)/w$, $\langle u^2\rangle_x = \langle u^2\rangle/w(x,t) = (D\beta - D/2t) + \langle u\rangle_x^2$.
The Kramers-Fokker-Planck equation

$$\partial_t W + u\nabla_x W = \beta\nabla_u(Wu) + q\Delta_u W \tag{30}$$

with $q = D\beta^2$, implies the local conservation laws

$$\partial_t w + \nabla(\langle u\rangle_x w) = 0 \tag{31}$$

$$\partial_t(\langle u\rangle_x w) + \nabla_x(\langle u^2\rangle_x w) = -\beta\langle u\rangle_x w.$$

Introducing the kinetic pressure notion $P_{kin}(x,t) = [\langle u^2\rangle_x - \langle u\rangle_x^2]w(x,t)$ we arrive at

$$\partial_t + \langle u\rangle_x\nabla)\langle u\rangle_x = -\beta\langle u\rangle_x - \nabla P_{kin}/w\,. \tag{32}$$

In the large friction regime we have

$$-\frac{\nabla P_{kin}}{w} = +\beta\langle u\rangle_x - \frac{\nabla P_{osm}}{w} \tag{33}$$

where $P_{osm} = D^2 w\Delta\ln w$ denotes an osmotic pressure of the Brownian motion.
We note that $\nabla P_{osm} = -w\,\nabla Q/m$ with $Q = -2mD^2\frac{\Delta w^{1/2}}{w^{1/2}}$.
Actually $-\nabla P_{osm} = (D/2t)\nabla w$. Thus, letting $\langle u\rangle_x = v(x,t)$ we arrive at:

$$(\partial_t + v\nabla)v = -\frac{\nabla P_{osm}}{w} = +\frac{1}{m}\nabla Q \tag{34}$$

to be compared with the general Brownian hydrodynamics result

$$\partial_t v + (v\nabla v) = +\frac{1}{m}\nabla(V + Q) \tag{35}$$

In the past (1992) I have named all that a "derivation of the quantum potential from realistic Brownian particle motions".

5.1. Functioning of pressure terms P_{kin} and P_{osm}

In view of $-\langle \Delta \ln \rho \rangle = \mathcal{F}(\rho) > 0$, the osmotic pressure P_{osm} is predominantly negative-definite. To the contrary, the kinetic pressure P_{kin} is positive definite. That imposes limitations on the validity of the large friction regime, to become operational after times $t > (2\beta)^{-1}$.
Let us introduce the notion of kinetic temperature:

$$0 \leq \Theta_{kin} = m\frac{P_{kin}}{w} \sim (k_B T - \frac{mD}{2t}) < k_B T \tag{36}$$

whose (large time limit) asymptotic value, $k_B T$ actually is. Since $P_{osm}/w = D^2 \Delta \ln w = -D/2t$, we learn that a (predominantly) positive-definite quantity

$$\Theta_{osm} = -m\frac{P_{osm}}{w} = -mD^2 \Delta \ln w \Longrightarrow \Theta_{kin} \sim (k_B T - \Theta_{osm}) \tag{37}$$

gives account of the deviation from thermal equilibrium, in terms of the local "thermal energy" (agitation) Θ_{osm}.
One more useful identity (not an independent equation) is here valid. It expresses the "thermal energy" conservation law (observe that no thermal currents are hereby induced):

$$(\partial_t + v\nabla)\Theta_{osm} = -2(\nabla v)\Theta_{osm} \Longrightarrow \partial_t \Theta_{osm} = -2(\nabla v)\Theta_{osm} \tag{38}$$

5.2. Meaning of the pressure term in Brownian hydrodynamics ($P_{osm} \doteq P$)

We come back to local conservation laws of the Brownian "fluid". This is an ensemble picture of the Brownian motion: imagine the Pablo Picasso art of placing one upon another hundreds of transparent foils, each carrying a drawing of one complete Brownian trajectory. All random paths are supposed to start from the same point and next allowed to run a pre-defined time period $[0, T]$, common for all repetitions. Brownian hydrodynamics is about statistical properties of such an ensemble:

$$\partial_t v + (v\nabla v) = +\frac{1}{m}\nabla(V + Q) = \frac{1}{m}F - \frac{\nabla P}{w} \tag{39}$$

$$-\frac{\nabla P}{w} = +\frac{1}{m}\nabla Q; \quad F \doteq -\nabla(-V)$$

In normal liquids the pressure is exerted upon any control volume (here, by an imagined small droplet), thus involving its compression. Just to the contrary, in case of Brownian motion we deal with a definite decompression.
Let us consider a reference volume (control interval, finite droplet) $[\alpha, \beta]$ in R^1 (or $\Lambda \subset R^1$) which at time $t > 0$ comprises a certain fraction of particles (it is a loose term designating, whatever they would be, the Brownian "fluid" constituents).
The time rate of particles loss or gain by the volume $[\alpha, \beta]$ at time t, is equal to the flow outgoing through the boundaries

$$-\partial_t \int_\alpha^\beta \rho(x,t)dx = \rho(\beta,t)v(\beta,t) - \rho(\alpha,t)v(\alpha,t)$$

To analyse the momentum balance, let us slightly deform the boundaries $[\alpha, \beta]$ to compensate the mass imbalance: $[\alpha, \beta] \to [\alpha + v(\alpha, t)\triangle t, \beta + v(\beta, t)\triangle t]$. Effectively, we pass to a locally co-moving (droplet) frame; that is the Lagrangian picture.
(i) The mass balance has been thus established in the moving droplet:

$$lim_{\triangle t\downarrow 0}\frac{1}{\triangle t}\Big[\int_{\alpha+v_\alpha\triangle t}^{\beta+v_\beta\triangle t}\rho(x,t+\triangle t)dx - \int_\alpha^\beta \rho(x,t)dx\Big] = 0$$

(ii) For local matter flows $(\rho v)(x,t)$, in view of $\partial_t(\rho v) = -\nabla(\rho v^2) + (1/m)\rho\nabla(V+Q)$, the time rate of change of momentum (per unit of mass) of the droplet, reads

$$lim_{\triangle t\downarrow 0}\frac{1}{\triangle t}\Big[\int_{\alpha+v_\alpha\triangle t}^{\beta+v_\beta\triangle t}(\rho v)(x,t+\triangle t) - \int_\alpha^\beta (\rho v)(x,t)\Big] = \int_\alpha^\beta \rho\frac{1}{m}\nabla(V+Q)dx$$

However, $\nabla Q/m = -\frac{\nabla P}{\rho}$ and $P = D^2\rho\triangle ln\rho$. Therefore:

$$\int_\alpha^\beta \rho\frac{1}{m}\nabla(V+Q)dx = \int_\alpha^\beta \rho\nabla\Omega dx - \int_\alpha^\beta \nabla P dx = \frac{1}{m}E[\nabla V]_\alpha^\beta + P(\alpha,t) - P(\beta,t)$$

(iii) The time rate of change of the kinetic energy of the droplet is:

$$lim_{\triangle t\downarrow 0}\frac{1}{\triangle t}\Big[\int_{\alpha+v_\alpha\triangle t}^{\beta+v_\beta\triangle t}\frac{1}{2}(\rho v^2)(x,t+\triangle t) - \int_\alpha^\beta \frac{1}{2}(\rho v^2)(x,t)\Big] = \int_\alpha^\beta \frac{1}{m}(\rho v)\nabla(V+Q)dx$$

Note that $\int_\alpha^\beta \rho v\nabla Q dx = -\int_\alpha^\beta v\nabla P dx$ (c.f. the standard notion of power release $\frac{dE}{dt} = F\cdot v$)

5.3. Meaning of the pressure term in quantum hydrodynamics ($-P_{osm} \doteq P$)
We do not care about a specific "trajectory" representation of the quantum motion and pass directly to local conservation laws:

$$\partial_t v + (v\nabla v) = -\frac{1}{m}\nabla(V+Q) = \frac{F}{m} - \frac{\nabla P}{\rho} \Longrightarrow \tag{40}$$

$$-\frac{1}{m}\nabla Q = +\frac{\nabla P_{osm}}{\rho} \doteq -\frac{\nabla P}{\rho}$$

which enforce $-P_{osm} = -D^2\rho\Delta\ln\rho \doteq P$, $D = \hbar/2m$, while $F = -\nabla V$. If compared to the Brownian hydrodynamics all $(V+Q)$ contributions come with an inverted sign. This carries over to the mass, momentum and kinetic energy rates.
Quite at variance with the Brownian $P = P_{osm}$, the quantum pressure term $P = -P_{osm}$ is predominantly positive. We recall that $-\langle\Delta\ln\rho\rangle = \langle\frac{(\nabla\rho)^2}{\rho^2}\rangle = \mathcal{F}(\rho) > 0$.
We note in passing that quantum mechanically derivable heat transfer equation

$$(\partial_t + v\nabla)\Theta_{osm} = -2\frac{\nabla q}{\rho} - 2(\nabla v)\Theta_{osm} \tag{41}$$

with $\Theta_{osm} = -m\frac{P_{osm}}{\rho} = -mD^2\Delta\ln\rho$ and $q = -2mD^2\rho\Delta v$, reproduces the Brownian form, at least for generic free Schrödinger wave packets with $\Delta v = 0$. We get $\partial_t\Theta_{osm} = -2(\nabla v)\Theta_{osm}$ as well. There is no heat current in such case.

6. Hamilton-Jacobi related hydrodynamics and (Bohmian) trajectory descriptions

At this point it seems instructive to make a comment on uses of the Eulerian picture (passive control) vs Lagrangian picture (active control in a co-moving frame) of hydrodynamical equations of motion. Let us simply give our droplet, previously considered as a co-moving control volume, an infinitesimal size. We readily identify the (fairly small sized) droplet dynamics: each droplet "looks" particle-like, while following Bohm-type trajectories.
Since, $f(x,t) \rightarrow f(x(t+\Delta t), t+\Delta t) \sim [\partial_t f + (v\nabla)f]\Delta t$; $\dot{x} = v = v(x,t)_{|x(t)=x}$, with $x(t+\Delta t) \sim v\Delta t$, $v = (1/m)\nabla s$ and $\partial_t s = \frac{ds}{dt} - mv^2$, we are in fact bound to work with:
(i) Classical hydrodynamics: (droplet) paths in the Lagrangian frame

$$\frac{d\rho}{dt} = -\rho\nabla v \longrightarrow \rho(x(t+\Delta t), t+dt) \sim \exp[-(\nabla v)\Delta t]\, \rho(x,t) \tag{42}$$

$$\frac{ds}{dt} = \frac{1}{2m}(\nabla s)^2 - (\pm V) \Longrightarrow m\frac{dv}{dt} = -\nabla(\pm V)$$

(ii) Brownian hydrodynamics: (droplet) paths in the Lagrangian frame

$$\frac{d\rho}{dt} = -\rho\nabla v \tag{43}$$

$$\frac{ds}{dt} = \frac{1}{2m}(\nabla s)^2 + (V+Q) \Longrightarrow m\frac{dv}{dt} = +\nabla(V+Q)$$

We need to recall a purely random (Wiener noise) background of the hydrodynamical formalism. We encounter here a primordial description in terms of random variables and paths, The latter may cross the droplet (e.g. enter from the outside, leave or simply stay within for a while): $dX(t) = b(X(t))dt + \sqrt{2D}dW(t) \Longrightarrow \partial_t\rho = D\Delta\rho - \nabla(b\rho)$; $\frac{V(x)}{2mD} = mD\left[\frac{b^2}{2D} + \nabla b\right]$
(iii) Quantum hydrodynamics: (droplet) paths in the Lagrangian frame $\Longrightarrow$ Bohmian trajectories

$$\frac{d\rho}{dt} = -\rho\nabla v \tag{44}$$

$$\frac{ds}{dt} = \frac{1}{2m}(\nabla s)^2 - (V+Q) \Longrightarrow m\frac{dv}{dt} = -\nabla(V+Q)$$

7. Acceleration concept in random motion: IInd Newton law

Consider a Markovian diffusion process on R, for times $t \in [0,T]$: $dX(t) = b(X(t),t)dt + \sqrt{2D}dW(t)$, where $W(t)$ stands for the Wiener noise and $X(t_0) = x_0$. Given $p(y,s,x,t), s \leq t$ and $\rho_0(x)$, we can infer a statistical future of the process:

$$\rho(x,t) = \int \rho(y,s)p(y,s,x,t)dy \Longrightarrow \partial_t\rho = D\Delta - (\nabla b\rho) \tag{45}$$

$$b(x,t) = \lim_{\Delta\to 0}\frac{1}{\Delta t}\int (y-x)p(x,t,y,t+\Delta t)dy = v(x,t) + (D\nabla\rho/\rho)(x,t)$$

We can as well reproduce a statistical past of the process, by means of

$$p_*(y,s,x,t) \doteq p(y,s,x,t)\frac{\rho(y,s)}{\rho(x,t)} \Longrightarrow \rho(y,s) = \int p_*(y,s,x,t)\rho(x,t)dx \tag{46}$$

$$b_*(y,s) = \lim_{\Delta s\to 0}\frac{1}{\Delta s}\int (y-s)p_*(x,s-\Delta s,y,s)dx = v(y,s) - (D\nabla\rho/\rho)(y,s)$$

Making notice of $v = (1/2)(b + b_*)$, we get:

$$\partial_t \rho = -\nabla(v\rho) = D\Delta\rho - (\nabla b\rho) = -D\Delta\rho - \nabla(b_*\rho). \tag{47}$$

Consider $b = DX$ and $b_* = D_*X$ as special cases of forward (predictive) and backward (retrodictive) time derivatives of functions of the random variable $X(t)$:

$$(Df)(X(t), t) = (\partial_t + b\nabla + D\Delta)f(X(t), t) \tag{48}$$
$$(D_*f)(X(t), t) = (\partial_t + b_*\nabla - D\Delta)f(X(t), t).$$

Analyse acceleration formulas for diffusion-type processes, in terms of the time rate of change (forward and backwards) drift functions and their local averages. Their response to small time increments reads:

(i) $b(x,t) \to b(X(t + \Delta t), t + \Delta t) \Longrightarrow \langle b\rangle(x, t + \Delta t)$
$\langle b\rangle(x, t + \Delta t) = \int p(x, t, z, t + \Delta t)b(z, t + \Delta t)dz \sim b(x, t) + (D^2X)(t)\Delta t.$
(ii) $b(X(t - \Delta t), t - \Delta t) \Longrightarrow \langle b\rangle(x, t - \Delta t) \to b(x, t)$
$\langle b\rangle(x, t - \Delta t) = \int b(y, t - \Delta t)p_*(y, t - \Delta t, x, t)dy \sim b(x, t) - (D_*DX)(t)\Delta t.$
(iii) $b_*(x, t) \to b_*(X(t + \Delta t), t + \Delta t) \Longrightarrow \langle b_*\rangle(x, t + \Delta t).$
$\langle b_*\rangle(x, t + \Delta t) \sim b_*(x, t) + (DD_*X)(t)\Delta t$
(iv) $b_*(X(t - \Delta t), t - \Delta t) \Longrightarrow \langle b_*\rangle(x, t - \Delta t) \to b_*(x, t)$ $\langle b_*\rangle(x, t - \Delta t) \sim b_*(x, t) - (D_*^2X)(t)\Delta t.$

Accordingly, we can associate various acceleration formulas with general diffusion-type processes. They may be interpreted as stochastic analogues of the IInd Newton law in the (local) mean: accelerations are related to conservative volume forces of external origin.
We indicate that by over-emphasizing the Newtonian viewpoint, we get somewhat blurred an important intrinsic acceleration input, due to the background random motion. It is encoded in the $\pm\frac{1}{m}\nabla Q$ contribution to the current velocity time rate of change.
Following the standard association of accelerations with external (volume) forces, one is inclined to attribute the second Newton law meaning to the formula

$$(D^2X)(t) = (\partial_t + v\nabla)v - \frac{1}{m}\nabla Q = (D_*^2X)(t) = +\frac{1}{m}\nabla V \tag{49}$$

in case of the Brownian motion, while

$$\frac{1}{2}[(DD_* + D_*D)X](t) = (\partial_t + v\nabla)v + \frac{1}{m}\nabla Q = -\frac{1}{m}\nabla V \tag{50}$$

in case of Nelson's stochastic mechanics, i.e. the diffusion-type probabilistic counterpart of the Schrödinger picture quantum motion.

It is the mid-term in the formulas (49) and (50), where the role of $\mp\nabla Q$ must be strongly emphasized. Set $\nabla V = 0$; the mid-term still accounts for definite acceleration phenomena, that are intrinsic to the random motion proper (c.f. a discussion of Section 5):

$$(i)\ \frac{dv}{dt} - \frac{1}{m}\nabla Q = 0, \quad (ii)\ \frac{dv}{dt} + \frac{1}{m}\nabla Q = 0 \tag{51}$$

Therefore, as appropriate candidates for the IInd Newton law, we promote not (47) or (48), but rather the (hydrodynamical) local conservation laws:

$$(iii)\ (\partial_t + v\nabla)v = \pm\frac{1}{m}\nabla(Q + V) \tag{52}$$

where the (seemingly minor) sign difference of the right-hand-side terms is hereby crucial and will set grounds below to the IIIrd Newton law in the local mean.

8. IIIrd Newton law

8.1. Impulse-momentum change law for for small times

Both the acceleration and impulse-momentum change concepts have been borrowed directly from classical mechanics. Nonetheless, an exploitation of properly tailored local mean values allows to extend the meaning of purely mechanical concepts to the theory of random motion (e.g. diffusion-type processes, with continuous, but generically non-differentiable sample paths). That was the case in connection with the IInd Newton law. It could have been explicitly rooted in the impulsive (short time increments) behaviour of drifts in the Brownian motion:

$$b_*(x,t) - \langle b_* \rangle (x, t-\Delta t) \sim \langle b \rangle (x, t+\Delta t) - b(x,t) \sim \frac{1}{m} \nabla V \, \Delta t \tag{53}$$

and/or an impulsive behaviour of drifts in stochastic mechanics (probabilistic counterpart of the Schrödinger picture evolution)

$$b_*(x,t) - \langle b_* \rangle (x, t-\Delta t) \sim \langle b \rangle (x, t+\Delta t) - b(x,t) \sim \frac{1}{m} \nabla (V + 2Q) \, \Delta t \, . \tag{54}$$

However, there is no other than aesthetic reason to associate the Brownian acceleration with the first and not the second (stochastic mechanics) formulas above. Indeed, more careful examination proves that the Brownian motion can be characterized on an equal footing by both acceleration definitions:

$$(D^2X)(t) = (D_*^2X)(t) = +\frac{1}{m}\nabla V \Longleftrightarrow \frac{1}{2}[(DD_* + D_*D)X](t) = \frac{1}{m}\nabla(V + 2Q) \tag{55}$$

and likewise, in the stochastic description of quantum motion (e.g. stochastic mechanics):

$$(D^2X)(t) = (D_*^2X)(t) = -\frac{1}{m}\nabla(V + 2Q) \Longleftrightarrow \frac{1}{2}[(DD_* + D_*D)X](t) = -\frac{1}{m}\nabla V \tag{56}$$

It is now clear that the Brownian motion and the stochastic transcription of quantum motion (stochastic mechanics) do differ fundamentally in their response to both random noise and external forces. We may stay happy with that observation, e.g. a clearly identified difference between two types of random motion. Accordingly, they can be interpreted as totally divorced from each other, except for incidental formal connections (c.f. the Euclidean time and the Wick transformation issue).

8.2. Action-reaction rule

There is still another possibility, that may be given a physical status of relevance, albeit on sufficiently small time-scales only. Considering the impulse-momentum change law as a valid property of random motion, we may as well interpret the two considered dynamical patterns of behaviour as being involved in a perpetual Brownian – anti-Brownian acceleration intertwine. That is, via the IIIrd Newton law in the mean. In the past (1992, 1999) we have heuristically formalized this idea in the concept of the *Brownian recoil principle.*

Indeed, all acceleration expressions in Eqs. (55) can be mapped into those of Eqs. (56), and in reverse, by addition/subtraction of the force term:

$$\Downarrow \pm \frac{2}{m}\nabla(V + Q) \quad \Uparrow \tag{57}$$

provided, at a given time instant we per force attribute the same values of $\rho(x,t)$ and $v(x,t)$ to both the Brownian and quantum hydrodynamical fields.

If the two motions are coupled by the action-reaction principle (IIIrd Newton law) at time t, then an impulse - momentum (here, velocity field) change law can be used to predict $+\Delta t$ updated values of ρ and v. Thus e.g. the Brownian impulse in a co-moving frame (given ρ and v) induces

$$\Delta\rho = -[(\nabla v)\,\Delta t]\,\rho \quad m\Delta v = +\nabla(V+Q)\,\Delta t \tag{58}$$

while an accompanying anti-Brownian impulse in a co-moving frame (given ρ and v) reads

$$\Delta\rho = -[(\nabla v)\,\Delta t]\,\rho \quad m\Delta v = -\nabla(V+Q)\,\Delta t\,. \tag{59}$$

We do not attempt to address the celebrated "egg-before-hen" dilemma. The formulas (58) and (59) are regarded as a statistically relevant record of the *recoil* effect where an "anti-Brownian" impulse associated with a quantum particle (we do not bother what is actually meant under this notion) induces, and in turn gets induced, by the Brownian motion pulse excited in a dissipative random medium (any conceivable notion of a surrounding medium, like e.g. the "vacuum", zero point radiation field etc.). The dissipation assumption destroys the symmetry of the action-reaction picture, since the Brownian pulse should quickly decay.

8.3. A concept of the Brownian recoil principle

Consider $\Delta t \ll 1$. Within $[t, t+\Delta t]$, let the action-reaction coupling between the "vacuum" (whatever that may be) and matter particles sets rules of the game $\Longrightarrow \langle\Delta p\rangle_{vacuum} + \langle\Delta p\rangle_{matter} = 0$.

The "vacuum turbulence" propels matter particles by transferring them an anti-Brownian recoil impulse (set $D = \hbar/2m$), whose "vacuum" trace (and reason) is the Brownian impulse (may quickly die out due to dissipation, we track the matter data).

Step I. Given the matter data $\rho(x,t)$ and $v(x,t)$. At $t+\Delta t$ we have $\rho+\Delta\rho = \exp[-(\nabla v)\,\Delta t]\,\rho$ and $v \to v + \Delta v$, where the action ("vacuum" impulse)

$$\Delta v = +\frac{1}{m}\nabla(V+Q)\,\Delta t \;\; (Brownian) \tag{60}$$

is paralleled by the reaction (matter impulse): ($\Downarrow$ - subtract; $\Uparrow$ - add: $\frac{2}{m}\nabla(V+Q)$!)

$$\Delta v = -\frac{1}{m}\nabla(V+Q)\,\Delta t \;\; (anti-Brownian, e.g.\ quantum) \tag{61}$$

Step II. Update the matter data to $\rho(x, t+\Delta t)$, $v(x, t+\Delta t)$, leave aside those referring to the "vacuum" and to the preceding Brownian impulse, turn to the next Δt episode when both impulses are excited anew. The new (updated) values of ρ and v at time $t+\Delta t$ are presumed to be determined by the anti-Brownian impulse again.

Any physical justification of the Brownian recoil principle needs a double-medium picture:
(i) an active "vacuum" (background random field, non-equilibrium reservoir, zero-point fluctuations) that is generating and supporting Brownian pulses. These may be interpreted in terms virtual particles
(ii) matter particles, whose dynamics is governed by the IIIrd Newton law and the resultant recoil effect.
A detailed theory of the "vacuum"-particle coupling is obviously necessary to go beyond heuristics.

With an inspiration coming from Yves Couder's lecture on particle-wave associations, and from deepened studies of the role of trajectory descriptions in diffraction/interference phenomena, as a residual subfield of quantum chemistry, we end up with a simple statement

that there is plenty of room down there. Indeed, before going into any sophisticated formal arguments, one should always keep in mind the quantum mechanical scales of interest: atomic nucleus size $\sim 10^{-15} - 10^{-14}m$, atom size $\sim 10^{-10} - 10^{-9}m$, electron size (whatever that means for inspired theoreticians) $\sim 10^{-15}m$, possibly down to $\sim 10^{-18}m$.
Presumably it is not devastatingly naive to address an issue of the (Schrödinger's wave function) ψ-ness of the electron "cloud" in the atom, while realizing that, with or without the second quantization and with or without quantum electro- or chromodynamics, the "vacuum" (not an empty void) functioning in quantum physics is still an open territory.

9. Bibliographic notes

There is no way to give justice to all contributors in the field of quantum hydrodynamics or stochastic mechanics. Our selection of references will be less then modest and in addition to Nelson's, Holland's and Wyatt's monographs, will mainly concentrate on a sample of my own research in this area. More references (with a bibliography of the subject) can be found and retrieved from my personal Web page: http://www.fiz.uni.opole.pl/pgar/.

My own hunch is that the Schrödinger picture quantum motion admits a consistent representation in terms of diffusion-type processes. That was the main idea of Nelson's stochastic mechanics. However, in search for "reasons of randomness" we have found unavoidable to admit a coupled double-medium picture.

Its qualitative features seem to be not distant from the particle-wave association picture (with all reservations raised in the original paper due to Y. Couder and E. Fort, (2006)), in which a physically "real" particle induces a physically relevant "realistic" wave, and that wave in turn is capable of affecting further motion of the particle. A conceptual input of diffusion waves (c.f. A. Mandelis, (2001)) might be useful at this point.

References

[1] Sonego S 1991 *Found. Phys.* **21** 1135
[2] Holland P 1993 *Quantum Theory of motion* (Cambridge University Press, Cambridge)
[3] Wyatt R E 2005 *Quantum Dynamics with Trajectories* (Springer-Verlag, Berlin)
[4] Coffey T M, Wyatt R E and Schieve W C 2011 *Phys. Rev. Lett.* **107** 240403
[5] Couder Y and Fort E 2006 *Phys. Rev. Lett.* **97** 154101
[6] Eddi A, Sultan E, Moukhtar J, Fort E, Rossi M and Couder Y 2011 *J.Fluid Mech.* **647** 433
[7] Genovese M 2005 *Physics Reports* **413** 319
[8] Nelson E 1985 *Quantum Fluctuations* (Princeton University Press, Princeton)
[9] Faris W G (ed.) 2006 *Diffusion, Quantum Theory and Radically Elementary Mathematics* (Princeton University Press, Princeton)
[10] Garbaczewski P and Vigier J P 1992 *Phys. Rev.* A **46** 4634
[11] Garbaczewski P 1999 *Phys. Rev.* E **59** 1498
[12] Garbaczewski P 2008 *Phys. Rev.* E **78** 031101
[13] Czopnik R and Garbaczewski P 2001 *Phys. Rev.* E **63** 021105
[14] Garbaczewski P 2006 *J. Stat. Phys.* **123** 315
[15] Fritsche L and Haugk M 2003 *Ann. Physik. (Leipzig)* **12** 371
[16] Grössing G, Mesa Pascasio J and Schwabl H 2011 *Found. Phys.* **41** 1437
[17] Mandelis A, 2001 *Diffusion-Wave Fields* (Springer-Verlag, Berlin)

Quantization as an emergent phenomenon due to matter–zeropoint field interaction

A M Cetto, L de la Peña and A Valdés-Hernández

Instituto de Física, Universidad Nacional Autónoma de México, Ap. postal 20-364, 01000 México DF, Mexico

E-mail: ana, luis, andreavh@fisica.unam.mx

Abstract. Quantization is derived as an emergent phenomenon, resulting from the permanent interaction between matter and radiation field. The starting point for the derivation is the existence of the (continuous) random zero-point electromagnetic radiation field (zpf) of mean energy $\hbar\omega/2$ per normal mode. A thermodynamic and statistical analysis leads unequivocally (and without quantum assumptions) to the Planck distribution law for the complete field in equilibrium. The problem of the quantization of matter is then approached from the same perspective: A detailed study of the dynamics of a particle embedded in the zpf shows that when the entire system eventually reaches a situation of energy balance thanks to the combined effect of diffusion and dissipation, the particle has acquired its characteristic quantum properties. To obtain the quantum-mechanical description it has been necessary to do a partial averaging and take the radiationless approximation. Consideration of the neglected radiative terms allows to establish contact with nonrelativistic quantum electrodynamics and derive the correct formulas for the first-order radiative corrections. Quantum mechanics emerges therefore as a partial, approximate and time-asymptotic description of a phenomenon that in its original (pre-quantum) description is entirely local and causal.

1. Introduction

Since the pioneering works in stochastic electrodynamics (SED) [1], [2], quantum mechanics has been conceived of as an emergent theory arising from a deeper-level dynamics. At the more fundamental level the world is assumed to be causal, deterministic and local: particles interact (locally) with (and through) the random, continuous electromagnetic zero-point radiation field (zpf), and follow well-defined (even if unknown stochastic) trajectories. At the quantum level the description becomes nonlocal, particle trajectories seem to have lost their sense (and for many even their existence), discreteness and indeterminacy enter the picture, and the classical algebra is replaced by a non-commuting one.

The program of SED has gone a long way in retrieving and explaining quantum mechanics, starting with the harmonic oscillator and including other characteristic 'quantum' phenomena such as diamagnetism and van der Waals forces, among others (see [2]). However, establishing the connection between these apparently incompatible worlds — the classical and the quantum — for the general case (and for the atomic case in particular), has not been an easy matter.

Several factors contribute to this difficulty. On the one hand the classical formalism used as a starting point, and the mathematics that come with it, do not offer a clear indication as to how to effect the transition to the quantum level from physical principles alone. On the other hand, while the emerging (quantum) theory contains characteristic novel phenomena, some central features of the underlying dynamics disappear altogether in the transition. This is the case, in particular, with the random zpf, which is considered in SED to be precisely the ultimate cause for the emergence of the quantum phenomenon.

To try to 'complete' the description and recover causality or locality (or both) by simply adding hidden elements or re-introducing the lost ones into the quantum description, seems an exercise as futile as trying to reconstruct the raw eggs out of a baked cake. But if the cake turns out to be delicious one accepts it. The great successes of quantum mechanics are normally considered a strong enough reason to be satisfied with the theory as it is, and to renounce any attempt at looking for explanations beyond (or underneath) it.

In this paper we go back to the origins of quantum mechanics, in line with the program of SED. This means taking the existence of the random electromagnetic zpf as the starting point and carefully analysing some important consequences of its existence. We will show that quantization emerges naturally, which is in itself a remarkable outcome. But equally interesting is what one learns along the way in the transition from the deeper, 'pre-quantum' level to the higher quantum one. Because, as could be expected, quantum mechanics turns out to represent only a partly averaged, asymptotic approximation (although a very accurate one). The assumptions and approximations made in this transition help us, thus, to better understand the origin of some basic quantum features and assign a physical meaning to them.

2. Planck's law as a consequence of the zero-point field

In this first part of the paper it is shown that Planck's distribution law for the equilibrium radiation field can be derived without introducing any discontinuity in the field or in its interaction with matter. The cornerstone of the derivation is the existence of the zpf as an integral part of the radiation field. The presentation draws in part from previous work [3], [4].

2.1. The equilibrium radiation field

The existence of a nonthermal field with energy different from zero implies an extension of the classical thermodynamic description. The starting point for this extension is Wien's law for the mean energy U of the oscillators of frequency ω in equilibrium at temperature T,

$$U(\omega, T) = \omega f(\omega/T), \tag{1}$$

which is a general result of thermodynamics. At $T = 0$ this equation takes the form

$$\mathcal{E}_0 \equiv U(\omega, 0) = A\omega, \tag{2}$$

with A constant. In classical thermodynamics the constant A is taken as zero, thereby excluding a priori the existence of a nonthermal energy. In contrast, the extended treatment admits a zero-point energy (zpe) for the oscillators that is different from zero. Clearly this simple introduction of the zpe, being contrary to classical energy equipartition, opens the door to interesting physical consequences.

The thermal energy distribution of the field oscillators in thermal equilibrium inside a cavity can be written according to thermodynamics in the general form

$$W_g(\mathcal{E})d\mathcal{E} = \frac{1}{Z_g(\beta)} g(\mathcal{E}) e^{-\beta\mathcal{E}} d\mathcal{E}, \tag{3}$$

where $Z_g(\beta) = \int g(\mathcal{E}) e^{-\beta\mathcal{E}} d\mathcal{E}$ is the partition function, $\beta = 1/(k_B T)$, and $g(\mathcal{E})$ is the structure function. The mean value of any function $f(\mathcal{E})$ is then given by

$$\langle f(\mathcal{E}) \rangle = \int_0^\infty W_g(\mathcal{E}) f(\mathcal{E}) d\mathcal{E} \tag{4}$$

which for $f(\mathcal{E}) = \mathcal{E}$ gives $\langle\mathcal{E}\rangle \equiv U$.

The (classical) Boltzmann distribution is obtained from (3) with $g(\mathcal{E})$ independent of $\mathcal{E}$, in which case

$$W_{\rm cl}(\mathcal{E}) = W_{g={\rm cl}}(\mathcal{E}) = \frac{e^{-\beta\mathcal{E}}}{\int_0^\infty e^{-\beta\mathcal{E}} d\mathcal{E}} \tag{5}$$

and $\langle\mathcal{E}\rangle_{\rm cl} = U_{\rm cl}(\beta) = \beta^{-1} = k_B T$. This means that to allow for a zpe, $g(\mathcal{E})$ must depend on the energy; this will lead to a different distribution $W_g(\mathcal{E})$ and hence a different function $U(\beta)$.

The general form for the distribution $W_g(\mathcal{E})$ implies the recurrence relation

$$\langle\mathcal{E}^{r+1}\rangle = U\langle\mathcal{E}^r\rangle - \langle\mathcal{E}^r\rangle', \tag{6}$$

where the prime denotes derivative with respect to β. For $r = 1$ this gives for the thermal energy variance

$$\sigma^2 \equiv \overline{\mathcal{E}^2} - U^2 = -U', \tag{7}$$

which can be rewritten as the well-known relation $\sigma^2 = kT^2 (\partial U/\partial T)_\omega = kT^2 C_\omega$. Because the heat capacity C_ω remains finite at $T = 0$, it follows that $\sigma^2(T = 0) = 0$, meaning that the dispersion of the energy is suppressed at $T = 0$, even when $\mathcal{E}_0 \neq 0$. In other words, a purely thermodynamic treatment is limited in that it does not allow for the inclusion of a *fluctuating* zpe.

To include the zero-point fluctuations it becomes necessary to look for a statistical distribution $W_s(\mathcal{E})$ that maximizes the entropy S_s for a given mean energy U, defined (up to an arbitrary additive constant) as $S_s = -k_B \int W_s(\mathcal{E}) \ln W_s(\mathcal{E}) d\mathcal{E}$. In addition to complying with the constraints of W_s being normalized to unity and $\langle\mathcal{E}\rangle_s = U(\beta)$, the results derived from $W_s(\mathcal{E})$ should be consistent with those derived from the thermal distribution W_g, In contrast to the thermodynamic entropy S which is defined in the phase space of the particles, the *statistical entropy* S_s is interpreted as a measure of the disorder present in the system [5]. The maximum-entropy formalism [6] leads to the following result for W_s:

$$W_s(\mathcal{E}) = \frac{1}{U} e^{-\mathcal{E}/U}. \tag{8}$$

This distribution gives for the energy moments $\langle\mathcal{E}^r\rangle_s = r!U^r$, whence the total energy variance is given by

$$\sigma_s^2 = \langle\mathcal{E}^2\rangle_s - U^2 = U^2. \tag{9}$$

At $T = 0$ this result reduces to $\sigma_s^2(T = 0) = \mathcal{E}_0^2$; these are the nonthermal energy fluctuations. The purely thermal energy variance is therefore given at any temperature by

$$\sigma^2 = \sigma_s^2 - \mathcal{E}_0^2 = U^2 - \mathcal{E}_0^2. \tag{10}$$

2.2. The Planck distribution

From Eq. (7) it follows that $U(\beta)$ can be obtained from the thermal energy variance σ^2 given in Eq. (10) as a function of U, by integrating

$$d\beta = -dU/\sigma^2(U) \tag{11}$$

and inverting $\beta = \beta(U)$. The result (subject to the limit condition $U \to \infty$ as $\beta \to 0$) is

$$U(\beta) = \mathcal{E}_0 \coth \mathcal{E}_0\beta. \tag{12}$$

This contains the classical case as a particular solution, corresponding to $A = 0$ in Eq. (2). With $A = \hbar/2$, corresponding to an energy $\mathcal{E}_0 = \hbar\omega/2$ for the zpf modes of frequency ω, (12) gives the well-known formula

$$U(\beta,\omega) = \tfrac{1}{2}\hbar\omega \coth \tfrac{1}{2}\hbar\omega\beta, \tag{13}$$

which shows that the (complete) Planck law is an *unavoidable* consequence of the existence of the nonthermal energy $\mathcal{E}_0$. This establishes Planck's law as a physical result whose ultimate meaning — or source — is the existence of this zpe.

The corresponding partition function, obtained from Eqs. (3) and (4), is

$$Z_g = \frac{1}{2\sinh \mathcal{E}_0\beta} = \sum_{n=0}^{\infty} e^{-\beta\mathcal{E}_n} = \int_0^{\infty} \sum_{n=0}^{\infty} \delta(\mathcal{E} - \mathcal{E}_n) e^{-\beta\mathcal{E}} d\mathcal{E}, \tag{14}$$

whence $g(\mathcal{E}) = \sum_{n=0}^{\infty} \delta(\mathcal{E} - \mathcal{E}_n)$, with the energy values $\mathcal{E}_n$ given by $\mathcal{E}_n = \mathcal{E}_0(2n+1) = \hbar\omega(n+\frac{1}{2})$. The distribution (3) turns out to be

$$W_g(\mathcal{E}) = \frac{1}{Z_g} \sum_{n=0}^{\infty} \delta(\mathcal{E} - \mathcal{E}_n) e^{-\beta\mathcal{E}}, \tag{15}$$

giving for the mean value of any function $f(\mathcal{E})$

$$\langle f(\mathcal{E})\rangle = \int_0^{\infty} W_g(\mathcal{E}) f(\mathcal{E}) d\mathcal{E} = \frac{1}{Z_g} \sum_{n=0}^{\infty} f(\mathcal{E}_n) e^{-\beta\mathcal{E}_n} = \sum_{n=0}^{\infty} w_n f(\mathcal{E}_n), \tag{16}$$

with canonical weights $w_n = e^{-\beta\mathcal{E}_n}/Z_g$. This shows that the mean value of a function of the *continuous* variable $\mathcal{E}$ (the energy) calculated with $W_g(\mathcal{E})$, can be obtained equivalently by averaging over the set of *discrete states* n, with respective weights w_n. Because we are dealing with a canonical ensemble, the structure of w_n suggests identifying $\mathcal{E}_n$ with discrete energy levels of the quantum oscillators (including the zpe). Both descriptions hold simultaneously; they are the two formal sides of the same coin.

It is important to stress that Planck's law has been obtained as a consequence of the fluctuating zero point energy without introducing any explicit quantum or discontinuity requirement.[1] The fact that Wien's law (with $A \neq 0$) is the one that opens the door to the zpe compels us to view this law as an extension of classical physics that enters into the quantum domain. Strictly speaking, as a precursor of Planck's distribution it should be considered to contain the seed of the first quantum law. Furthermore, since Wien's law is satisfied by field and material oscillators alike, Eq. (13) and the ensuing consequences are of general validity, regardless of the nature of the oscillators. The confirmation that the law which gave birth to quantum theory stems from the existence of a zpe brings thus to the fore the crucial importance of this energy for the understanding of quantum mechanics and more generally, of quantum theory.

Note that in terms of the thermal part U_T of the energy, given according to Eq. (12) by

$$U_T = U - \mathcal{E}_0 = \frac{2\mathcal{E}_0}{e^{2\mathcal{E}_0\beta} - 1}, \tag{17}$$

Eq. (10) reads

$$\sigma^2 = U_T^2 + 2\mathcal{E}_0 U_T, \tag{18}$$

[1] See [3], [4] for previous derivations. It is interesting to note that an independent derivation of Planck's law has been obtained by Dimitrov [7], using 'averaged' statistical thermodynamics and the assumption that (classical) particles possess a random (ultrarelativistic) unobservable, unfreezeable motion.

in contrast to the classical (purely wave) expression $\sigma_{\rm cl}^2 = U_T^2$. The identification of the term U_T^2 as a product of the interference of the modes (of frequency ω) of the thermal field suggests to interpret (the 'particle' term) $2\mathcal{E}_0 U_T$ as due to additional interferences, now between the thermal field and the zpf, whose mean energy is just $\mathcal{E}_0$.

3. Quantum mechanics as a consequence of the zero-point field

In this second part of the paper it will be shown that also the quantum laws governing the dynamics of a particle can be derived without introducing a priori any discreteness or discontinuity in the description. The cornerstone of the derivation is again the existence of the zpf in permanent interaction with an otherwise classical particle subject to an external force. The presentation made here draws in part from previous work [2],[8],[9].

3.1. The particle-zpf system in phase space

We consider a system composed of a charged particle immersed in the zpf. The motion of the mechanical subsystem (in one dimension, for simplicity) is governed in the nonrelativistic limit by

$$\dot{x} = p/m, \qquad \dot{p} = f(x) + m\tau\dddot{x} + eE(t), \tag{19}$$

where $f(x)$ is the external force, $E(t)$ is the electric component of the zpf in the long-wavelength approximation and $m\tau\dddot{x}$ is the radiation reaction force, with $\tau = 2e^2/(3mc^3)$ ($\approx 10^{-23}$ s for the electron).[2] The density of points in the particle's phase space is described by

$$\frac{\partial}{\partial t}R + \frac{\partial}{\partial x}(\dot{x}R) + \frac{\partial}{\partial p}\left(f(x) + m\tau\dddot{x}\right)R = -\frac{\partial}{\partial p}E(t)R. \tag{20}$$

We are ultimately interested in describing the particle motion, not for a specific realization of the field but for the ensemble of its realizations, or alternatively an ensemble of equivalent systems. We therefore average $R(x,p,t)$ over the realizations of E to obtain the averaged density in the phase space of the particle, $\overline{R(x,p,t)}^E \equiv Q(x,p,t)$. With the help of the projector technique (see e.g. [8], [10]), this leads after a somewhat lengthy calculation to the Fokker-Planck-type equation

$$\frac{\partial}{\partial t}Q + \hat{L}Q = e^2\frac{\partial}{\partial p}\hat{D}(t)Q, \tag{21}$$

where $\hat{L}$ is the Liouville operator

$$\hat{L} = \frac{1}{m}\frac{\partial}{\partial x}p + \frac{\partial}{\partial p}(f + m\tau\dddot{x}) \tag{22}$$

and $\hat{D}(t)$ is a complicated integro-differential diffusion operator. More precisely, the diffusion term can be written in the form [8]

$$e^2\frac{\partial}{\partial p}\hat{D}(t)Q = e\frac{\partial}{\partial p}\hat{P}E\sum_{k=0}^{\infty}\left[e\hat{G}\frac{\partial}{\partial p}(1-\hat{P})E\right]^{2k+1}Q, \tag{23}$$

where $\hat{P}$ is the smoothing operator that averages over the realizations of E ($\hat{P}A = \overline{A}^E$) and the operator $\hat{G}$ is defined by

$$\hat{G}A(x,p,t) = \left(\frac{\partial}{\partial t} + \hat{L}\right)^{-1}A(x,p.t) = \int_0^t e^{-\hat{L}(t-t')}A(x,p,t')dt'. \tag{24}$$

[2] Note that the correct (integro-differential) version of the radiation reaction term is fully causal. The noncausal behaviour that comes with the $m\tau\dddot{x}$ term is merely an artefact of the approximation made in neglecting all higher-order derivatives. For an extensive discussion of this point see [2].

Note that the diffusion term comprises at least a correlation of the form $\overline{E_i(t)E_k(t')}^E$ and hence contains a factor proportional to the spectral (energy) density of the zpf,

$$\rho_{\mathrm{zpf}}(\omega) = \frac{\hbar\omega^3}{2\pi^2 c^3} = \frac{\omega^2}{\pi^2 c^3} \cdot \frac{\hbar\omega}{2}. \tag{25}$$

The interaction between particle and field is therefore represented in Eq. (21) by both the radiation reaction term and the diffusion term, and it is through the latter that Planck's constant enters the picture.

3.2. Particle dynamics in configuration space

Equation (21) provides in principle a (partly averaged) statistical description of the dynamics of the particle for any time t. It is clear that initially, after particle and field have started to interact, there is an irreversible process of energy and momentum exchange during which the field has a randomizing and dissipative effect on the particle dynamics, due to the terms with E and $\dddot{x}$ in Eq. (21). However, to explore the connection with quantum mechanics we are specifically interested in the solution in the time-asymptotic limit, when *and if* a regime is attained in which this exchange is no longer irreversible. In such situation a balance should be reached between the mean power radiated and that absorbed by the particle from the field; this balance condition is discussed in detail in section 3.4. When this occurs, the radiative and dissipative terms mentioned have produced their main (irreversible) effect in taking the system to such regime, and one may expect any additional effect arising from these terms to represent just a radiative correction.

In order to establish the connection with quantum mechanics, let us first go from Eq. (21) to a description in configuration space. An orderly way to make this transition is by means of the characteristic (moment-generating) function $\widetilde{Q}$ associated with the density Q,

$$\widetilde{Q}(x, z, t) = \int Q(x, p, t) e^{ipz} dp. \tag{26}$$

The marginal probability density $\rho(x,t)$ is then given by

$$\rho(x,t) = \int Q(x,p,t)dp = \widetilde{Q}(x,0,t), \tag{27}$$

and the (partially averaged) local moments of p are

$$\langle p^n \rangle (x) = \langle p^n \rangle_x = \frac{1}{\rho} \int p^n Q dp = (-i)^n \left(\frac{1}{\widetilde{Q}} \frac{\partial^n \widetilde{Q}}{\partial z^n} \right) \Bigg|_{z=0} \tag{28}$$

for any integer n. The Fourier transform of Eq. (21) is (with all surface terms assumed to vanish at infinity)

$$\frac{\partial \widetilde{Q}}{\partial t} - i\frac{1}{m}\frac{\partial^2 \widetilde{Q}}{\partial x \partial z} - izf(x)\widetilde{Q} - \frac{\tau}{m} f' z \frac{\partial \widetilde{Q}}{\partial z} = -ie^2 z (\widetilde{\hat{D}Q}). \tag{29}$$

By expanding this equation into a power series around $z = 0$ and separating the coefficients of z^k $(k = 0, 1, 2, \ldots)$ one obtains an infinite hierarchy of equations containing momenta of p of increasing order. The first two equations (coefficients of z^0 and z) read

$$\frac{\partial \rho}{\partial t} + \frac{1}{m}\frac{\partial}{\partial x} (\langle p \rangle_x \rho) = 0, \tag{30}$$

$$\frac{\partial}{\partial t}\left(\langle p\rangle_x \rho\right)+\frac{1}{m}\frac{\partial}{\partial x}(\langle p^2\rangle_x \rho)-f\rho-\frac{\tau}{m}f'\langle p\rangle_x \rho=-e^2\left.(\widetilde{\hat{D}Q})\right|_{z=0}. \tag{31}$$

Equation (30) is the continuity equation, with the local (partially averaged) momentum given by $\langle p\rangle_x = -i(\partial \ln \widetilde{Q}/\partial z)_{z=0}$ (see Eq. (28) with $n=1$). Equation (31) describes the transfer of momentum and contains, in addition to ρ and $\langle p\rangle_x$, the second moment $\langle p^2\rangle_x$, which reappears in the third equation along with $\langle p^3\rangle_x$, and so on. This coupling between successive equations creates a highly difficult mathematical problem. However, in the case of interest here the first two equations decouple from the rest of the hierarchy, as shown below. Therefore only the first local moments of p intervene in the description of interest, which means that one may concentrate on the behaviour of $\widetilde{Q}(x,z,t)$ for small values of z.

From Eq. (28) one obtains

$$\langle p^2\rangle_x - \langle p\rangle_x^2 = -\left(\frac{\partial^2}{\partial z^2}\ln \widetilde{Q}\right)_{z=0}. \tag{32}$$

To find an expression for the right-hand-side term as a function of x we introduce the change of variables

$$z_+ = x+\eta z, \qquad z_- = x-\eta z, \tag{33}$$

where η is an as yet undetermined parameter having dimensions of action. We now write $\widetilde{Q}$ in the general form

$$\widetilde{Q}(x,z,t) = q_+(z_+,t)q_-(z_-,t)\chi(x,z,t), \tag{34}$$

where, according to Eq. (26),

$$q_+(z_-,t)=q_-^*(z_-,t), \quad q_-(z_+,t)=q_+^*(z_+,t), \quad \chi^*(x,z,t)=\chi(x,-z,t). \tag{35}$$

Equations (27) and (32) become then

$$\rho(x,t)=\widetilde{Q}(x,0,t)=q_+^*(x,t)q_+(x,t)\chi_0,$$

$$\langle p^2\rangle_x - \langle p\rangle_x^2 = -\eta^2\frac{\partial^2}{\partial x^2}\ln\rho+\Sigma, \tag{36}$$

with $\Sigma = 4\eta^2(\partial_{z+}\partial_{z-}\ln\chi)_{z=0}$, $\chi_0=\chi(x,0,t)$. Inserting these results into Eq. (31) one obtains

$$m\frac{\partial}{\partial t}(\langle p\rangle_x\rho)+m\frac{\partial}{\partial x}\left(\langle p\rangle_x^2\rho\right)-\frac{\eta^2}{m}\frac{\partial}{\partial x}\left(\rho\frac{\partial^2}{\partial x^2}\ln\rho\right)-f\rho=$$
$$=\frac{\tau\langle p\rangle}{m}\frac{\partial f}{\partial x}\rho-e^2\left.(\widetilde{\hat{D}Q})\right|_{z=0}-\frac{1}{m}\frac{\partial}{\partial x}\Sigma\rho. \tag{37}$$

3.3. Quantum mechanics in the radiationless approximation

Following the discussion at the beginning of the previous section, we consider that in the time-asymptotic limit the effect of the terms in Eq. (37) due to radiation reaction (proportional to $\tau\sim e^2$) and diffusion (also proportional to e^2) has become a mere radiative correction of order $\alpha\sim e^2$ (the fine structure constant). In addition, we introduce the assumption that the contribution of the nonfactorizable function $\chi(z_+,z_-,t)$, and hence of Σ in Eq. (37), has also become small, of order e^2. We shall come back to this assumption below (section 4.3). Under these conditions one may write $\chi\simeq\chi_0=1$ (by absorbing the dependence on x and t in q_- and q_+) and neglect the three terms on the right-hand side of Eq. (37); this is the meaning of the radiationless approximation.

In the radiationless approximation Eq. (36) becomes

$$\left\langle p^2\right\rangle_x = \langle p\rangle_x^2 - \eta^2\frac{\partial^2}{\partial x^2}\ln\rho, \tag{38}$$

so that $\left\langle p^2\right\rangle_x$ decouples from the rest of the hierarchy. Equation (29) reduces then to

$$\frac{1}{\widetilde{Q}}\partial_t\widetilde{Q} - \frac{i}{m\widetilde{Q}}\partial_x\partial_z\widetilde{Q} = izf(x). \tag{39}$$

Writing explicitly $\widetilde{Q}(x,z,t) = q_+(z_+,t)q_-(z_-,t)$, according to (34), this equation gives

$$\frac{1}{q_+q_-}\partial_t(q_+q_-) - \frac{i\eta}{m}\left[\frac{1}{q_+}\partial_+^2(q_+) - \frac{1}{q_-}\partial_-^2(q_-)\right] = \frac{i}{2\eta}(z_+ - z_-)f[\tfrac{1}{2}(z_+ + z_-)]. \tag{40}$$

Recall that we are interested in small values of z only. Assuming $f(z)$ to be well behaved in the interval $[z_-, z_+]$, the mean-value theorem applied to the term on the right-hand side gives to lowest order in z

$$(z_+ - z_-)f[\tfrac{1}{2}(z_+ + z_-)] = \int_{z_-}^{z_+} f(u)du = -\left[V(z_+) - V(z_-)\right]. \tag{41}$$

This result holds for any (well-behaved) potential $V(x)$. One thus gets from Eq. (40)

$$-\frac{i\eta}{m}\frac{1}{q_+}\partial_+^2 q_+ + \frac{i}{2\eta}V(z_+) + \frac{1}{q_+}\partial_t q_+ = -\frac{i\eta}{m}\frac{1}{q_-}\partial_-^2(q_-) + \frac{i}{2\eta}V(z_-) - \frac{1}{q_-}\partial_t q_-. \tag{42}$$

The two sides of this equation depend on different variables and hence separate; let K be the (real) separation constant. In the limit $z \to 0$ both z_+ and z_- reduce to x, whence putting $q_+(x,t) \equiv e^{iKt}\psi(x,t)$, $q_-(x,t) \equiv e^{-iKt}\psi^*(x,t)$ (see Eq.(35)) one gets in this limit

$$-2\frac{\eta^2}{m}\frac{\partial^2\psi}{\partial x^2} + V(x)\psi = 2i\eta\frac{\partial\psi}{\partial t} \tag{43}$$

and its complex conjugate, with $\rho(x,t) = \psi^*(x)\psi(x)$. Except for the (constant) factor η to be determined (see the discussion in the next section), we recognize in (43) the Schrödinger equation for $\psi(x,t)$. This means that, indeed, a regime of unitary (time-reversible) evolution has been attained.

3.4. Detailed energy balance

The full equivalence between Eq. (43) and the Schrödinger equation entails the relation $\eta = \hbar/2$. To prove that this value is correctly predicted by the theory we resort to the condition of energy balance, which has been assumed to hold in the time-asymptotic limit (see section 3.2).

From the Fokker-Planck-type equation (21) multiplied by p^2 and integrated over the entire phase space it follows (assuming again all surface terms to vanish at infinity) that

$$\frac{1}{2m}\frac{d}{dt}\left\langle p^2\right\rangle = \frac{1}{2m}\frac{d}{dt}\int p^2 Q dx dp = \frac{1}{m}\left\langle fp + m\tau p\dddot{x} - \frac{e^2}{2}p\hat{D}\right\rangle, \tag{44}$$

where it is understood that $\hat{D}$ acts on $Q(x,p,t)$. Since $d\langle V\rangle/dt = -\langle fp\rangle/m$, the total energy gained or lost in the mean by the particle through radiation exchange is

$$\frac{d}{dt}\langle H\rangle = \frac{d}{dt}\left\langle\frac{1}{2m}p^2 + V\right\rangle = m\tau\langle\dot{x}\dddot{x}\rangle - \frac{e^2}{2m}\left\langle p\hat{D}\right\rangle, \tag{45}$$

with H representing the mechanical Hamiltonian. The terms on the right-hand side represent the average power dissipated by the particle due to Larmor radiation and the average power absorbed by the particle from the field, respectively. When these terms compensate each other,

$$m\tau\left\langle \dot{x}\,\dddot{x}\right\rangle = \frac{e^2}{2m}\left\langle p\hat{D}\right\rangle, \tag{46}$$

$\langle H\rangle$ becomes constant and energy balance is reached.

In line with the radiationless approximation used above to arrive at Eq. (43), it suffices to calculate both sides of (46) to lowest order in $\tau \sim e^2$. The left-hand side is readily calculated using the solutions of Eq. (43) (in terms of η) and gives for the ground state

$$m\tau\left\langle \dot{x}\,\dddot{x}\right\rangle = -m\tau_k\omega_{0k}^4\left|x_{0k}\right|^2, \tag{47}$$

where $\omega_{0k} = (\mathcal{E}_0 - \mathcal{E}_k)/2\eta$, $x_{0k} = \int \psi_0^* x \psi_k dx$, $\mathcal{E}_k$ are the energy eigenvalues and ψ_k the corresponding eigenfunctions. To calculate the right-hand side of (46) one must resort to Eq. (23) to lowest order in e^2, introduce Eq. (25) for the spectral density of the zpf, and take into account that energy balance is established in the time-asymptotic limit, so that the integral over time extends to infinity. Using again the solutions of Eq. (43), one arrives finally at

$$\frac{e^2}{2m}\left\langle p\hat{D}\right\rangle = -\frac{\hbar m\tau}{2\eta}\sum_k \omega_{k0}^4\left|x_{0k}\right|^2. \tag{48}$$

Equating this result with Eq. (47) one obtains for the parameter η the value $\eta = \hbar/2$, thus recovering the Schrödinger equation

$$i\hbar\frac{\partial\psi}{\partial t} = -\frac{\hbar^2}{2m}\nabla^2\psi + V\psi \tag{49}$$

for any potential $V(x)$. Notice that the Planck constant in Schrödinger's equation comes directly from the zpf through Eq. (48). It is notable that the dipole approximation for the zpf has been sufficient to arrive at this result. Equations (47) and (48) have exactly the same structure, which means that balance is achieved term by term; one speaks therefore of *detailed* energy balance. In fact a spectrum proportional to ω^3, being responsible for the ω_{k0}^4 factor in Eq. (48), is the *only* one that guarantees detailed energy balance. This establishes a fundamental difference with classical systems, for which equilibrium is reached only for the Rayleigh-Jeans law, proportional to ω^2. Since a spectrum $\sim \omega^3$ is the single Lorentz-invariant one [1],[11], it is verified that Lorentz invariance and detailed equilibrium with matter are intimately related properties of the zpf. In particular, atomic stability is reached for those specific (stationary and discrete) states for which the Larmor radiation along the orbit is compensated in the mean by the energy absorbed from the field and turned into fluctuating motions.

3.5. Quantum nonlocality

With the vanishing of the radiative terms on the right-hand side, Eq. (37) reduces (now with $\eta = \hbar/2$) to

$$m\frac{\partial}{\partial t}(\langle p\rangle_x\rho) + m\frac{\partial}{\partial x}\left(\langle p\rangle_x^2\rho\right) - \frac{\hbar^2}{4m}\frac{\partial}{\partial x}\left(\rho\frac{\partial^2}{\partial x^2}\ln\rho\right) - f\rho = 0. \tag{50}$$

The behaviour of the quantum system is therefore fully described (in the radiationless approximation) by this equation along with the continuity equation (30) (plus the energy-balance condition), or alternatively by the Schrödinger equation and its complex conjugate, with $\rho(x,t) = \psi^*\psi$.

Note that the only term that distinguishes Eq. (50) from the corresponding classical equation is the one containing $\ln\rho$. From Eq. (38) it is clear that this term originates in the fluctuations of the momentum, transcribed to configuration space. The source of these fluctuations is the zpf, which also determines their size through Planck's constant. Hence the (momentum) quantum fluctuations are conventional fluctuations having a *causal* origin. Their mean value is given by

$$\left\langle \sigma_p^2(x) \right\rangle = \left\langle -\frac{\hbar^2}{4m}\frac{\partial^2}{\partial x^2}\ln\rho \right\rangle = \frac{\hbar^2}{4}\int \rho\left(\frac{1}{\rho}\frac{\partial\rho}{\partial x}\right)^2 dx. \tag{51}$$

The connection of this term with the quantum potential is readily established by applying the Madelung transformation $\psi(x,t) = \sqrt{\rho}\exp(iS/\hbar)$ to the Schrödinger equation, which leads to the Hamilton-Jacobi-type equation

$$\frac{\partial S}{\partial t} + \frac{1}{2m}\left(\frac{\partial S}{\partial x}\right)^2 - \frac{\hbar^2}{2m}\frac{1}{\sqrt{\rho}}\frac{\partial^2\sqrt{\rho}}{\partial x^2} + V = 0. \tag{52}$$

The term $Q_p = \left(-\hbar^2/2m\right)(1/\sqrt{\rho})(\partial^2\sqrt{\rho}/\partial x^2)$, the so-called *quantum potential*, ascribes to this (otherwise classical) equation a statistical and nonlocal meaning. This is precisely the term responsible for the nonlocality of Bohm's theory, where it is also known as 'Bohm potential' [12].[3] As is clear from the present derivation, however, it is of *kinematic* origin.[4] It is important to observe that this nonlocality is generated by the reduction of the statistical description to the configuration space, and appears even in the case of a single particle — with the exception of the free particle described by a plane wave, ρ = const. For a system composed of two or more particles, additional nonlocalities arise. For instance, in the two-particle case the phase-space density described by the respective Fokker-Planck-type equation — which is an extension of Eq. (21) — depends on the variables of both particles. This will be reflected in general in the configuration-space description by means of a nonfactorizable joint probability density $\rho(x_1,x_2) = \rho_1(x_1)\rho_2(x_2)\rho_{12}(x_1,x_2)$. The corresponding 'quantum potential' for particle 1, for instance, will be of the form

$$Q_{p1}(1,2) = -\frac{\hbar^2}{2m}\left[\frac{\partial_1^2\sqrt{\rho_1}}{\sqrt{\rho_1}} + \frac{\partial_1^2\sqrt{\rho_{12}}}{\sqrt{\rho_{12}}} + \frac{1}{2}\frac{\partial_1\rho_1}{\rho_1}\frac{\partial_1\rho_{12}}{\rho_{12}}\right], \tag{53}$$

containing two additional terms that depend on the presence of particle 2. These terms represent a nonlocal action of particle 2 on particle 1. Indeed, a more detailed analysis of the two-particle problem, following a procedure slightly different from the one exposed here, has led to a clarification of the mechanism leading to the entanglement of states which is normally associated with Bell's inequalities [14].

4. Radiative corrections

This third part of the paper is devoted to the analysis of further (first-order) effects deriving from the particle-field interaction in the quantum regime. Firstly the condition of energy balance is analysed for a particle in an excited state. More generally, absence of detailed balance is shown to give rise to radiative transitions, with the respective transition probabilities exactly as given by nonrelativistic quantum electrodynamics (QED). Finally, other radiative corrections to first order include the Lamb shift and a (nonrelativistic) mass correction.

[3] As mentioned in section 3.2, $\langle p\rangle_x/m$ is the local mean velocity of the subensemble of particles moving in the neighbourhood of x at time t; hence the instantaneous velocity of an individual particle differs from it by a stochastic component $\delta\dot{x}$. This is important with regard to Bohm's theory, which gives information only about the mean trajectories by neglecting $\delta\dot{x}$. Due to this stochastic component, quantum trajectories are entirely free to intersect.

[4] This result is in line with previous conclusions arrived at independently by other authors; see Ref. [13].

4.1. Energy balance for a particle in an excited state

In section 3.4, detailed energy balance was shown to hold with both particle and field in their ground state. Assume now that the particle has by some external mechanism been transferred to an excited state n, the background field still being in its ground state. Then both terms on the right-hand side of (45) must be recalculated. For the first term one obtains instead of (47)

$$m\tau \langle \dot{x}\,\dddot{x}\rangle_n = -m\tau \sum_k \omega_{nk}^4 \left|x_{nk}\right|^2 . \tag{54}$$

For the second term, on the other hand, one obtains instead of (48) (with $\eta = \hbar/2$)

$$\frac{e^2}{2m}\left\langle p\hat{D}\right\rangle_n = -m\tau \sum_k \omega_{kn}^4 \left|x_{nk}\right|^2 \mathrm{sign}\omega_{kn}. \tag{55}$$

Both expressions have again the same structure, except that the latter contains now a mixture of positive and negative terms, whilst in Eq. (54) all terms have the same sign. The upshot is that there cannot be detailed balance between the zpf and a particle in an excited state. This confirms that only the ground state of the particle is stable in the sole presence of the zpf.

Let us now investigate whether there is *any* background field of spectral density $\rho(\omega) = \rho_{\mathrm{zpf}}(\omega)\gamma(\omega)$ (with $\gamma(\omega)$ to be determined), with which a mechanical system in a particular excited state n can be in equilibrium. The radiation-reaction term is again given by Eq. (54). For the 'diffusion' term, which depends on the spectral density, we obtain instead of Eq. (55)

$$\frac{e^2}{2m}\left\langle p\hat{D}\right\rangle_n = -m\tau \sum_k \omega_{kn}^4 \left|x_{nk}\right|^2 \gamma(\left|\omega_{kn}\right|)\mathrm{sign}\omega_{kn}. \tag{56}$$

Both Eqs. (54) and (56) contain in general mixtures of terms of different frequencies (ω_{nk} for different values of k), but with different signs, so that detailed balance cannot be satisfied in general. Only for the particular case in which all values of $|\omega_{nk}|$ coincide, the possibility of detailed balance exists in principle. This possibility is realized for the harmonic oscillator, the single system for which all $|\omega_{nk}|$ are equal (and coincide with the oscillator frequency ω_0). Using again the solutions provided by the Schrödinger equation, Eqs. (54) and (56) give

$$m\tau \langle \dot{x}\,\dddot{x}\rangle_n = -\frac{1}{2}\hbar\tau\omega_0^3(2n+1), \quad \frac{e^2}{2m}\left\langle p\hat{D}\right\rangle_n = -\frac{1}{2}\hbar\tau\omega_0^3\gamma(\omega_0). \tag{57}$$

On equating these two results we conclude that indeed, detailed balance exists between a harmonic oscillator in its excited state n and a background field with spectral energy density

$$\rho_n(\omega) = \rho_{\mathrm{zpf}}(\omega)(2n+1) = \frac{\hbar\omega^3}{2\pi^2c^3}(2n+1) \tag{58}$$

This result should not come as a surprise, since it corresponds to a field with precisely an energy per normal mode $\frac{1}{2}\hbar\omega(2n+1)$ as derived in section 2, equal to the energy of the mechanical oscillators with which it is in equilibrium. This is the secret of the Planck distribution and field quantization: equilibrium with matter is reached only for those oscillators that have energy $\frac{1}{2}\hbar\omega(2n+1)$.

4.2. Radiative transitions: Einstein A and B coefficients

Now we investigate some important implications of the absence of detailed balance. This can be done by introducing Eqs. (54) and (56) into (45), to determine the average rate of energy loss or gain for a particle in state n; the result is

$$\frac{dH_n}{dt} = -m\tau \sum_k \omega_{nk}^4 \left|x_{nk}\right|^2 \left[1 - \gamma(\left|\omega_{nk}\right|)\mathrm{sign}\omega_{kn}\right]. \tag{59}$$

With $\gamma(\omega) = 1 + \gamma_a(\omega)$, where $\gamma_a(\omega)$ represents the contribution from the excited (or additional) background field to the spectral energy density, this equation takes the form

$$\frac{dH_n}{dt} = m\tau \sum_k \omega_{nk}^4 \left|x_{nk}\right|^2 \left[(\gamma_a)_{\omega_{kn}>0} - (2 + \gamma_a)_{\omega_{kn}<0}\right]. \tag{60}$$

The first term within brackets represents the absorptions and the second one, the emissions. It is clear from this result that there can be absorptions only when the background field is excited (or there is an external component), whilst the emissions can be either 'spontaneous' (in presence of just the zpf, the term with coefficient 2) or stimulated by the additional field. The coefficients appearing in the various terms determine the respective rates of energy gain and energy loss; therefore, they should be directly related with Einstein's A and B coefficients.

Indeed, for the coefficient A, which is defined as the time rate for 'spontaneous' emissions via $dH_n = -\hbar \sum_k \omega_{nk} A_{nk} dt$, Eq. (60) with $\gamma_a = 0$ gives

$$A_{nk} = \frac{4e^2 \left|\omega_{nk}\right|^3}{3\hbar c^3} \left|x_{nk}\right|^2 = \frac{2\tau m}{\hbar} \left|\omega_{nk}\right|^3 \left|x_{nk}\right|^2, \tag{61}$$

which coincides with the result given by nonrelativistic QED [15]. For the coefficients B, which give the rate of energy gain or loss due to transitions induced by the external field and are defined through $dH_n = \pm\hbar \sum_k \omega_{nk} B_{nk} \rho_a(\omega_{nk}) dt$, Eq. (60) gives

$$B_{nk} = B_{kn} = \frac{m\tau\omega_{nk}^4 \left|x_{nk}\right|^2 \gamma_a(\omega_{nk})}{\hbar \left|\omega_{nk}\right| \rho_a(\omega_{nk})} = \frac{4\pi^2 e^2}{3\hbar^2} \left|x_{nk}\right|^2, \tag{62}$$

which again coincides with the respective formula of QED. This confirms the key role played by *both* radiative terms in determining the rates of transition. According to traditional wisdom, the A coefficient is due to the commutator of the creation and annihilation field operators. As a mathematical rule this is of course a nice description. From a present point of view its meaning becomes transparent: the factor 2 comes directly from the combined effect of Larmor radiation and the zpf (in equal parts) in 'inducing' emissions from an excited state.

The expressions for the Einstein coefficients involve each a single frequency $\left|\omega_{nk}\right|$, which confirms that the system as a whole reaches a state of *detailed balance*, i.e., energy balance of matter with the field at every separate frequency. The theory has thus led us from global balance to detailed balance in the quantum regime. We recall that this demand was one of Einstein's major hypotheses in his pioneering work where he introduced the absorption and emission coefficients [16].

4.3. Lamb shift and mass correction

In the context of the present work, the radiative corrections can be traced to the terms that were discarded from the right-hand side of Eq. (37) when taking the radiationless approximation that led to Eq. (50). The first and the second terms are the ones that intervene in the energy balance equation (45), being responsible for the finite lifetimes of excited states and the transitions between states. The third term, the one containing Σ in Eq. (37), represents a contribution to the momentum fluctuations, whence to first order in perturbation theory it adds to the mean kinetic energy of a stationary state an amount given by $\langle\Sigma\rangle/2m$. This term accounts for another important radiative correction, namely the Lamb shift of the atomic levels (and an accompanying mass correction).

Because of the complexity of the calculations (which are the subject matter of a future paper) we report here the results for a simple system, namely the harmonic oscillator. An elementary procedure can be followed in this case to calculate the value of the energy shift, by going back to the Langevin equation (19) and carrying out a perturbative treatment [8] (see also Refs. [1], [2]). The correction to the mean kinetic energy can be shown to be given by the term $\delta E = -\frac{1}{2}\langle \mathbf{d}\cdot\mathbf{E}\rangle$, which originates in the coupling of the instantaneous fluctuations of the electric dipole **d** to the electric component of the zpf. Qualitatively this corresponds to one of the most popular interpretations of the Lamb effect in QED, and in particular, to the noted model of Welton [17]. However, by noting that for the harmonic oscillator under energy balance $\langle \mathbf{d}\cdot\mathbf{E}\rangle = (\tau m/\omega)\langle \ddot{\mathbf{x}}^2\rangle$, one can write this shift in the alternative form $\delta E = -(\tau m/2\omega)\langle \ddot{\mathbf{x}}^2\rangle$ and interpret it as due to Larmor radiation. This is the kind of ambiguity that appears in QED due to the freedom to reorder the operators.

The calculations of the above term are readily performed to lowest order of approximation and give two corrections, namely a mass renormalization of value $\delta m = 3m/(8\pi\alpha)$ and a level (Lamb) shift of value

$$\delta E = \frac{\alpha\hbar^2\omega^2}{\pi mc^2}\ln\frac{\omega_c}{\omega}, \tag{63}$$

where $\alpha = e^2/\hbar c$ and ω_c is some cutoff frequency. The reason for the large value of both results is the long tail of the spectral density $\rho(\omega) \sim \omega^3$ at high frequencies. Fixing the cutoff at $\omega_c = mc^2/\hbar$, reasonable for a nonrelativistic calculation, one obtains $\delta m/m \sim \alpha$ and for δE the same value as predicted by nonrelativistic QED [15],

$$\frac{\delta E}{\frac{1}{2}\hbar\omega} = \frac{2\alpha}{\pi}\frac{\hbar\omega}{mc^2}\ln\frac{mc^2}{\hbar\omega}. \tag{64}$$

This represents a very small correction,[5] but one that played an important role during the foundation of QED and is frequently considered one of the best successes of QED. With this, the approximation $\chi = 1$ made above becomes fully justified.

5. Concluding remarks

Several important conclusions and observations can be drawn from the results here presented. Of course the main one is that the quantum behaviour can be understood as emerging from the interaction matter-zpf. The ubiquitous presence of the zpf implies thus the omnipresence of the quantum phenomenon. The quantum properties are not intrinsic to matter or field, but are something more interesting and rich: they belong to a world of emerging properties.

In the present account the quantum fluctuations of matter lose their cryptic meaning to become a causal phenomenon, produced by the action of the random zpf. Analogously, quantum indeterminism arises as an anticipated property of the statistical description of the stochastic dynamics. The Heisenberg inequalities find thus a natural explanation: the minimal variances of the dynamical variables determined by the inequalities are those impressed by the zpf, which itself fluctuates with an intensity proportional to $\hbar$.

Nonlocality, that other quantum mystery, is explained in the present context as an outcome of the description at the quantum level. At the deeper level the theory is entirely local, yet an effective nonlocality arises as a result of the reduction of the statistical description from phase space to configuration space. This nonlocality is encapsulated in the quantum potential, which represents a kinetic contribution to the energy due to the momentum fluctuations. The correlations among dynamical variables that pertain to two or more entangled particles give rise to additional nonlocalities, e.g. of the EPR or Bell type.

The quantum behaviour of matter emerges in the time-asymptotic limit, when the combined diffusive and dissipative action of radiation has driven the system to a regime of unitary time

[5] Note that this result is independent of the state, hence it does not lead to any observable shift for the harmonic oscillator.

invariance. Herein lies the secret of the stability of atomic states, which quantum mechanics is unable to reveal due to its unawareness of the zpf: stationarity is reached for certain orbital motions only. Quantum mechanics is derived once the radiative terms can be neglected. Taking due account of these neglected terms to first order provides correct results for the rates of transition between stationary states and the corresponding radiative corrections of (nonrelativistic) QED.

In between the initial time (when particle and zpf start to interact) and the quantum regime there is a new physics, which is neither the purely classical one — since it contains Planck's constant and the zpf as important ingredients — nor the quantum one — since for short times even the Heisenberg inequalities can be violated. In particular the function $Q(x, p, t)$ should furnish a better phase-space description, an important matter that requires further consideration. Moreover, there are other properties of quantum systems that from the present perspective should emerge from a deeper level, such as the spin of particles and all that comes with it. In summary, there is a wide interesting field open to study.

One should of course bear in mind that ours is not the only possible deeper-level explanation of the quantum phenomenon. In recent times the notion that quantization is an emergent phenomenon, possibly rooted on a classical dynamics, has gained much ground and several interesting — alternative or complementary — theories have been put forward [18]. Doubtlessly a finer and richer picture of the quantum world, and hopefully a physically more satisfactory one, will emerge as a result of these developments.

References

[1] Marshall T W 1963 *Proc. Roy. Soc.* A **276** 475, 1965 *Proc. Camb. Phil. Soc.* **61** 537
[2] de la Peña L and Cetto A M 1996 *The Quantum Dice, an Introduction to Stochastic Electrodynamics* (Dordrecht, Holland: Kluwer)
[3] Boyer T H 2003 *Am. J. Phys.* **71** 866
[4] de la Peña L, Valdés-Hernández A and Cetto A M 2008 *Am. J. Phys.* **76** 947; Valdés-Hernández A, de la Peña L and Cetto A M 2010 *New trends in statistical physics,* eds A. Macías and L. Dagdug (Singapore: World Scientific) p 113
[5] Callen H 1985 *Thermodynamics and an Introduction to Thermostatistics* (New York: John Wiley and Sons)
[6] Montroll E W and Shlesinger M F 1983 *J Stat Phys* **32** 209; Pathria R K 1996 *Statistical Mechanics* (Oxford: Buttenworth-Heinemann)
[7] Dimitrov V I 1997 arXiv:9707007v1 [quant-ph]
[8] de la Peña L and Cetto A M 1977 *J. Math. Phys.* **18** 1612
[9] de la Peña L, Valdés-Hernández A and Cetto A M 2009 *Found. Phys.* **39** 1240; de la Peña L, Valdés-Hernández A, Cetto A M and França H M 2011 *Physics Letters* A **375** 1720; de la Peña L, Cetto A M, Valdés-Hernández A and França H M 2011 arXiv:1110-4641v1 [quant-ph]
[10] Frisch U 1968 *Probabilistic Methods in Applied Mathematics* Vol. I, ed. A. T. Barucha-Reid (New York: Academic Press)
[11] Boyer T H 1969 *Phys. Rev.* **182** 1374
[12] Holland P R 1993 *The Quantum Theory of Motion* (Cambridge: Cambridge U P)
[13] Garbaczewski P 1992 *Phys. Lett.* A **162** 129; Grössing G 2008 *Phys. Lett.* A **372** 4556.
[14] Valdés-Hernández A, de la Peña L and Cetto A M 2011 *Found. Phys.* **41** 843
[15] Milonni P W 1994 *The Quantum Vacuum. an Introduction to Quantum Electrodynamics* (San Diego: Academic Press)
[16] Einstein A 1917 *Phys. Zeitschr.* **18** 121
[17] Welton T A 1948 *Phys. Rev.* **74** 1157
[18] Adler S 2004 *Quantum Theory as an Emergent Phenomenon* (Cambridge: Cambridge UP); 't Hooft G 2005 *Quo Vadis Quantum Mechanics?* eds A. C. Elitzur et al (Berlin: Springer) p. 99; Nieuwenhuizen T M 2005 *Quantum Theory. Reconsideration of Foundations-3*, AIP Conf. Proc. 810, eds. G. Adenier et al (New York: AIP) p. 198; Carroll R 2010 *On the Emergence Theme in Physics* (Singapore: World Scientific); Khrennikov A Yu 2010 *J. Math. Phys.* 51, 082106

EmerQuM 11: Emergent Quantum Mechanics 2011 IOP Publishing
Journal of Physics: Conference Series **361** (2012) 012014 doi:10.1088/1742-6596/361/1/012014

Weak Values: Approach through the Clifford and Moyal Algebras

B. J. Hiley
TPRU, Birkbeck, University of London, Malet Street,
London WC1E 7HX.
E-mail: `b.hiley@bbk.ac.uk`

Abstract. In this paper we calculate various transition probability amplitudes, TPAs, known as 'weak values' for the Schrödinger and Pauli particles. It is shown that these values are related to the Bohm momentum, the Bohm energy and the quantum potential in each case. The results for the Schrödinger particle are obtained in three ways, the standard approach, the Clifford algebra approach of Hiley and Callaghan, and the Moyal approach. To obtain the results for the Pauli particle, we combine the Clifford and Moyal algebras into one structure. The consequences of these results are discussed.

1. Introduction

Let us begin by considering the mean value of some Hermitian operator $\hat{A}$ in the state $|\psi\rangle$. By introducing a complete orthonormal set $|\phi_i\rangle$, we can write

$$\langle\psi|\hat{A}|\psi\rangle = \sum_i \langle\psi|\phi_i\rangle\langle\phi_i|\hat{A}|\psi\rangle = \sum_i \rho_\psi(\phi_i)\frac{\langle\phi_i|\hat{A}|\psi\rangle}{\langle\phi_i|\psi\rangle} \tag{1}$$

In recent literature, the quantity $\langle\phi_i|\hat{A}|\psi\rangle/\langle\phi_i|\psi\rangle$ has been called a 'weak value' because its magnitude can be found in a so called 'weak measurement' if certain criterion are met [1] [11]. However in the context of this paper the concept of 'weakness' has no meaning. It is merely a transition probability amplitude, [TPA], which, in general, will be a complex number. Nonetheless we will, following convention, continue to call it a weak value.

The choice of $|\phi_i\rangle$ is known as 'post selection'. Suppose we post select $|a_i\rangle$, an eigenket of $\hat{A}$, we then find

$$\frac{\langle a_i|\hat{A}|\psi\rangle}{\langle a_i|\psi\rangle} = a_i, \tag{2}$$

so that equation (1) becomes simply

$$\langle\psi|\hat{A}|\psi\rangle = \sum_i \rho_\psi(a_i)a_i$$

This is an extremely well known result showing that in this particular case the TPA (2) is simply an eigenvalue a_i. Here it is meaningful to call (2) a 'value' as it has that meaning, but again there is nothing 'weak' about it.

Let us now consider a more interesting case where we post select $|b_i\rangle$, an eigenket of an operator $\hat{B}$ which does not commute with $\hat{A}$. Now form

$$\langle\psi|\hat{A}|\psi\rangle = \sum_i \rho_\psi(b_i)\frac{\langle b_i|\hat{A}|\psi\rangle}{\langle b_i|\psi\rangle}.$$

The last term in this equation can be written as

$$\frac{\langle b_i|\hat{A}|\psi\rangle}{\langle b_i|\psi\rangle} = \sum_j a_j \frac{\langle b_i|a_j\rangle\langle a_j|\psi\rangle}{\langle b_i|\psi\rangle}$$

Thus we see that in this case, the meaning of the TPA is not a simple 'value' of anything. Here it is about a relationship between values.

A more interesting example is when we replace $\hat{A}$ and $\hat{B}$ by a position operator $\hat{X}$ and the momentum operator $\hat{P}$. In this case we have

$$\langle\psi(t)|\hat{P}|\psi(t)\rangle = \int \rho(x,t)\frac{\langle x|\hat{P}|\psi(t)\rangle}{\langle x|\psi(t)\rangle}dx.$$

To take this further, we find

$$\langle x|\hat{P}|\psi(t)\rangle = \int \langle x|\hat{P}|x'\rangle\langle x'|\psi(t)\rangle dx'$$

But $\langle x|\hat{P}|x'\rangle = -i\delta(x-x')\nabla_{x'}$,[1] so that

$$\langle x|\hat{P}|\psi(t)\rangle = -i\nabla_x\psi(x,t).$$

Let us now write $\psi(x,t) = R(x,t)e^{iS(x,t)}$ and find

$$\frac{\langle x|\hat{P}|\psi(t)\rangle}{\langle x|\psi(t)\rangle} = \nabla_x S(x,t) - i\nabla_x\rho(x,t)/2\rho(x,t). \tag{3}$$

Here $\rho(x,t)$ is the usual expression for the probability, $\rho(x,t) = |\psi(x,t)|^2$. Thus we see this weak value is complex, the real part simply being the Bohm momentum.

This fact has already been pointed out by Leavens [23], but he makes no comment on the imaginary part. In fact the imaginary part is identical to the 'osmotic velocity' [2] introduced by Bohm and Hiley [3] in their exploration of the relation of the Bohm model to a diffusion model first introduced by Nelson [26].

Before going on to discuss the possible relationship between a diffusion model of quantum phenomena to the standard approach, I want to draw attention to a similar result obtained by Wiseman [29] in a slightly different manner.

He starts by recalling that, in the Heisenberg picture, the velocity can be obtained from the equation

$$i\frac{d\hat{X}}{dt} = [\hat{X}, \hat{H}].$$

[1] We have put $\hbar = 1$ through out this paper.

[2] The notion of an 'osmotic velocity' arises whenever there is a probability gradient no matter what is the cause of this gradient. Here we offer no underlying physical reason for the existence of this gradient in quantum processes.

By using the definition of a weak value, we find

$$v(x,t) = Re\frac{\langle x|i[\hat{H},\hat{X}]|\psi(t)\rangle}{\langle x|\psi(t)\rangle}. \tag{4}$$

Then if we put $\hat{H} = \hat{P}^2/2m + V(\hat{X})$ and $\psi(x,t) = R(x,t)e^{iS(x,t)}$, we again get an expression of the Bohm velocity $\nabla_x S(x,t)/m$. Wiseman goes on to discuss the physical and philosophical implications that stem from this result. Here we will only concentrate on the mathematical structure lying behind these weak values, extending them to the case of a non-relativistic particle with spin.

2. Relationship of these results to Nelson's discussion

2.1. The left-right derivatives, the Bohm Momentum and the Moyal Marginal Momentum

Let us explore the relation to Nelson's work [26]. He writes equation (3) as a velocity in the form

$$v(x,t) = \frac{1}{m}\nabla_x S(x,t) - iD\frac{\nabla_x \rho(x,t)}{\rho(x,t)}. \tag{5}$$

If we write $D = 1/2m$ in equation (5), we immediately see that the last term has exactly the same form as the osmotic velocity introduced by Nelson [27] and also later by Bohm and Hiley [3]. However we have derived the result directly from a standard weak value and have no need to speculate about underlying diffusive motion.

To continue let us write the Bohm velocity, $v_B(x,t) = \nabla_x S(x,t)/m$, in the form

$$v_B(x,t) = \frac{1}{2m}\left[\frac{\langle\psi(t)|\hat{P}|x\rangle}{\langle\psi(t)|x\rangle} + \frac{\langle x|\hat{P}|\psi(t)\rangle}{\langle x|\psi(t)\rangle}\right] \tag{6}$$

while the osmotic velocity, v_O, is

$$v_O(x,t) = \frac{1}{2m}\left[\frac{\langle\psi(t)|\hat{P}|x\rangle}{\langle\psi(t)|x\rangle} - \frac{\langle x|\hat{P}|\psi(t)\rangle}{\langle x|\psi(t)\rangle}\right] \tag{7}$$

Now let us compare these results with those obtained in Bohm and Hiley [3]. They write

$$v_B(x,t) = [b(x,t) + b_*(x,t)]/2 \quad \text{and} \quad v_O(x,t) = b(x,t) - b_*(x,t). \tag{8}$$

Here $b(x,t)$ is the mean 'forward' velocity, which is based on Nelson's mean forward derivative [26],

$$Dx(t) = \lim_{\Delta t \to 0^+} E_i \frac{x(t+\Delta t) - x(t)}{\Delta t}$$

and $b_*(x,t)$ is the mean 'backward' velocity, which in turn is based on Nelson's mean backward derivative

$$D_*x(t) = \lim_{\Delta t \to 0^+} E_i \frac{x(t) - x(t-\Delta t)}{\Delta t}$$

The need to introduce two derivatives in a diffusive or stochastic process arises because $x(t)$ is not differentiable. It is then necessary to distinguish between the probability that a particle initially at position $x - \Delta x$ will arrive at x at time t in the time interval Δt and the probability that a particle at $x + \Delta x$ would have been found at x at an earlier time $t - \Delta t$.

Comparing the above results, we find that, in the quantum formalism, we may identify these velocities with a weak value in the following way,

$$b(x,t) = \frac{1}{m}\frac{\langle\psi(t)|\hat{P}|x\rangle}{\langle\psi(t)|x\rangle} \quad \text{and} \quad b_*(x,t) = \frac{1}{m}\frac{\langle x|\hat{P}|\psi(t)\rangle}{\langle x|\psi(t)\rangle}$$

This difference, rather than arising from a diffusive motion, arises from formal noncommutativity. Symbolically we can say that right action $\overrightarrow{P}|\psi(t)\rangle$ is different from left action $\langle\psi(t)|\overleftarrow{P}$. However as can be seen from equations (6) and (7), the physically meaningful quantities arise from sums and differences of weak values, so we expect terms like

$$\frac{\langle\psi(t)|\hat{P}|x\rangle}{\langle\psi(t)|x\rangle} \pm \frac{\langle x|\hat{P}|\psi(t)\rangle}{\langle x|\psi(t)\rangle} = \frac{\psi^*(x)\overleftarrow{\nabla}_x\psi(x) \pm \psi^*(x)\overrightarrow{\nabla}_x\psi(x)}{\rho(x)} \tag{9}$$

to play a significant role.

2.2. *The Moyal Algebra*

Let us now see how these weak can be treated by using the Moyal algebra [28], [24] [30]. Recall this algebra does not use operators but instead uses ordinary functions of x and p. Here the distinction between left and right action arises as a consequence of the noncommutativity of the Moyal $\star$-product. We recall that the $\star$-product, [24], [30], is defined by

$$\star = \exp\left[\frac{i}{2}\left(\overleftarrow{\partial_x}\overrightarrow{\partial_p} - \overleftarrow{\partial_p}\overrightarrow{\partial_x}\right)\right]$$

Then we find[3]

$$p \star f(x,p) = \left(p - \frac{i}{2}\overrightarrow{\partial_x}\right) f(x,p)$$

and the dual

$$f(x,p) \star p = f(x,p)\left(p + \frac{i}{2}\overleftarrow{\partial_x}\right)$$

Here we see clearly the need for a left and right derivatives. The sums and differences of the type shown in equation (9) suggests we form the Baker[4] and the Moyal brackets which are defined as

$$[p, f(x,p)]_{BB} = \frac{f \star p + p \star f}{2} = pf(x,p)$$

and

$$[p, f(x,p)]_{MB} = \frac{f \star p - p \star f}{i} = \nabla_x f(x,p).$$

To this point we have assumed $f(x,p)$ is some general phase space function. Let us now choose $f(x,p)$ to be the Wigner-Moyal distribution, given by

$$f(x,p) = 2\pi \int \psi^*(x - y/2)\psi(x + y/2)e^{ip \cdot y}dy \tag{10}$$

[3] We will suppress the time dependence for succinctness

[4] This is just the Jordan product.

Alternatively, as Moyal shows in [28], we may also express this relationship in terms of a momentum representation. This is just the Fourier transform of equation (10). Thus

$$f(x,p)=\left(\frac{1}{2\pi}\right)\int\phi^*(p+\theta/2)\phi(p-\theta/2)e^{-ix.\theta}d\theta.$$

This equation can also be written as

$$f(x,p)=2\pi\int\int\phi^*(p_1)\phi(p_2)\delta\left(p-\frac{p_2+p_1}{2}\right)e^{ix.(p_2-p_1)}dp_1dp_2 \tag{11}$$

To make contact with the results of subsection (2.1), we simply project to marginals by either integrating over the momentum, p, or over the position x, to find

$$\int f(x,p)dp=|\psi(x)|^2=\rho(x)\quad\text{and}\quad\int f(x,p)dx=|\phi(p)|^2=\bar{\rho}(p),$$

which are just the standard probability densities. The marginal momentum, $\overline{\overline{p}}(x)$, introduced by Moyal [28], is given by

$$\rho(x)\overline{\overline{p}}(x)=\int pf(x,p)dp.$$

Using equation (11) we find,

$$\begin{aligned}\rho(x)\overline{\overline{p^n}}(x)&=2\pi\int\int p^n\phi^*(p_1)\phi(p_2)\delta\left(p-\frac{p_2+p_1}{2}\right)e^{ix.(p_2-p_1)}dp_1dp_2\\&=\left(\frac{1}{2i}\right)^n\left[\left(\frac{\partial}{\partial x_1}-\frac{\partial}{\partial x_2}\right)^n\psi(x_1)\psi^\star(x_2)\right]_{x_1=x_2=x}\end{aligned} \tag{12}$$

In the case when $n=1$ and $\psi=Re^{iS}$ it can easily be shown that

$$\overline{\overline{p}}(x)=\nabla_xS(x).$$

which is, of course, the Bohm momentum.

The corresponding integration of the Moyal bracket gives

$$\frac{\int[p,f(x,p)]_{MB}dp}{2\rho(x)}=\frac{\nabla_x\left(\int f(x,p)dp\right)}{2\rho(x)}=\frac{\nabla_x\rho(x)}{2\rho(x)}$$

showing clearly that what has been called the 'osmotic' term arises from the failure of the Moyal bracket to vanish, i.e., the noncommutativity of the Moyal product.

Thus we have derived the same results as obtained from the left-right derivatives $\overrightarrow{P}|\psi\rangle$ and $\langle\psi|\overleftarrow{P}$. In this particular case the Baker and Moyal brackets are simply a way of separating the real and the imaginary parts of weak values. It should also be noticed that the Bohm model and the Moyal approach produce identical results as has already been shown by Hiley [16].

2.3. Consequences for the Kinetic Energy

To complete this part of the discussion and show a further relation of weak values to the Bohm model, let us examine what happens to the kinetic energy. Consider the second order derivative p^2, and form

$$p^2\star f(x,p)=\left[p^2-ip\overrightarrow{\nabla_x}-\frac{1}{4}\overrightarrow{\nabla_x}^2\right]f(x,p).$$

together with

$$f(x,p)\star p^2 = f(x,p)\left[p^2 + ip\overleftarrow{\nabla}_x - \frac{1}{4}\overleftarrow{\nabla}_x^2\right]$$

Then forming the Baker bracket we find

$$[p^2, f(x,p)]_{BB} = p^2 f(x,p) - \frac{1}{4}\nabla_x^2 f(x,p).$$

If we again integrate over p to obtain the marginal kinetic energy[5],

$$\int [p^2, f(x,p)]_{BB}dp = \rho(x)\overline{\overline{p^2}}(x) - \frac{1}{4}\nabla_x^2\rho(x).$$

Using (12) for $n = 2$, we find

$$\overline{\overline{p^2}}(x) = (\nabla_x S(x))^2 + \frac{1}{4}\frac{\nabla_x^2\rho(x)}{\rho(x)} - \frac{\nabla_x^2 R(x)}{R(x)},$$

so that

$$\int [p^2, f(x,p)]_{BB}dp/\rho(x) = (\nabla_x S(x))^2 - \nabla_x^2 R(x)/R(x).$$

The Bohm kinetic energy for a particle of mass 1/2 is $KE_B = (\nabla_x S)^2$, while the second term is the quantum potential $Q = -\nabla_x^2 R/R$. Thus we see that the integral over p of the Baker bracket, produces both the Bohm kinetic energy and the quantum potential for the free particle again showing another way to establish a relation between the weak values and the Bohm variables.

This leaves us with the Moyal bracket which we write down for completeness is

$$\int [p^2, F(x,p)]_{MB}dp/\rho(x) = \nabla_x^2 S(x) + \left(\frac{\nabla_x\rho(x)}{\rho(x)}\right)\nabla_x S(x).$$

It should also be noted that these results are identical with those obtained from $\left(\langle\psi(t)|\hat{P}^2|x\rangle \pm \langle x|\hat{P}^2|\psi(t)\rangle\right)/2$, showing that the Baker and Moyal brackets give exactly the same results as sums and differences of weak values.

We end this section by making two remarks.

(1) Exactly the same results can be obtained using cross Wigner functions as has been recently shown by de Gosson and de Gosson [13].

(2) All these results show that the TPA or 'weak' value approach, the Bohm approach and the Moyal approach produce identical results.

3. Extension to Spin-half Particles

Let us now move on to consider weak values for a non-relativistic particle with spin-half, namely the Pauli particle. In this case there are two methods that we could use. The straight forward but somewhat inelegant conventional method using matrices or the approach using Clifford algebras, the greater details of which were presented in Hiley and Callaghan [19]. However this approach introduces the concept of a generalised Dirac operator [12] which was imported from outside the algebraic structure. Furthermore it does not show that these derivatives introduce a generalised symplectic structure.

[5] For simplicity we assume a mass of the particle to be 1/2.

What we will do now is to show how the generalised Dirac derivative can emerge from a larger algebra formed by combining the Clifford and Moyal algebras. Thus both the orthogonal and symplectic (Moyal) structures are united into a single algebraic structure [7].

In our previous work [17] [19], we used the elements $\Phi_L(x)$ and its conjugate, $\widetilde{\Phi}_L(x)$, formed by Clifford reversion, to construct an entity which we have called the Clifford density element $\rho_c(x) = \Phi_L(x)\widetilde{\Phi}_L(x)$. In this paper we will generalise this object to include wider class of entities by writing

$$\rho_{\nu,\mu}(x_1, x_2) = \Phi_{L_\nu}(x_1)\widetilde{\Phi}_{L\mu}(x_2)$$

so that $\rho_c(x) = \rho_{\nu,\nu}(x_1 = x_2 = x)$[6]. Then we construct a generalisation of equation (10)

$$F(x,p) = 2\pi \int Tr[\Phi_L(x_1)\widetilde{\Phi}_L(x_2)]e^{-ip.y}dy.$$

where $x = (x_2 + x_1)/2$ and $y = (x_2 - x_1)$ and Tr is a trace. It should be noted that we have chosen the notation $F(x,p)$ as it is a generalisation of the Moyal density $f(x,p)$ for spin-less particles as we will explain below.

In the previous papers [17], [19], we formed an element of a minimal left ideal in the Pauli Clifford algebra $\mathcal{C}_{3,0}$ by choosing a particular primitive idempotent $\epsilon = (1 + e_3)/2$ and then constructed elements of the form

$$\Phi_L(x) = \phi_L(x)\epsilon = [g_0(x) + g_1(x)e_{23} + g_2(x)e_{13} + g_3(x)e_{12}]\epsilon. \quad (13)$$

Here $e_{ij} = e_i \circ e_j$ where the e_k are the generating elements of $\mathcal{C}_{3,0}$, the Pauli Clifford algebra[7]. However, for our purposes here, there is a more convenient, but equivalent way of writing equation (13), namely,

$$\Phi_L(x) = [(g_0 - g_3e_{123}) + (g_2 - g_1e_{123})e_1]\epsilon,$$

so that if we use

$$\begin{aligned} 2g_0 &= \phi_1^* + \phi_1 \qquad 2g_2 = \phi_2^* + \phi_2 \\ 2e_{123}g_3 &= \phi_1^* - \phi_1 \quad 2e_{123}g_1 = \phi_2^* - \phi_2, \end{aligned} \quad (14)$$

Φ_L becomes simply

$$\Phi_L(x) = [\psi_1(x) + \psi_2(x)e_1]\epsilon.$$

Actually we can make things somewhat easier if we express the result in terms of the p-representation. Then

$$F(x,p) = \left(\frac{1}{2\pi}\right) \int Tr[\Xi_L(p_1)\widetilde{\Xi}_L(p_2)]e^{-ix.\pi}d\pi. \quad (15)$$

where $p = (p_2 + p_1)/2$ and $\pi = p_2 - p_1$. In this case we write

$$\Xi_L(p) = \xi(p)\epsilon = [\phi_1(p) + \phi_2(p)]\epsilon \quad (16)$$

[6] For the most general case we would write $\rho_{\Phi,\Xi} = \Phi_L(x_1)\Xi_R(x_2)$, but we will not pursue this generalisation here.

[7] The generators of the Clifford algebra, e_k, combine using the Clifford product,
$[e_i \circ e_j + e_j \circ e_i] = 2\delta_{ij}$

where, of course, the $\phi(p)$s are Fourier transforms of the $\psi(x)$s,

To find the Bohm momentum in the Moyal algebra, we use

$$F(x,p) = \left(\frac{1}{2\pi}\right)\int Tr[\xi_L(p_2)\epsilon\widetilde{\xi}_L(p_1)]e^{-ix.\pi}d\pi$$

with $\epsilon = (1+e_3)/2$. Then using equation (16), together with its conjugate

$$\widetilde{\Xi}_L(p) = \epsilon[\phi_1^*(p) + \phi_2^*(p)],$$

we find

$$F(x,p) = \int\int M(p_2,p_1)\delta\left(p - \frac{p_2-p_1}{2}\right)e^{ix.(p_2-p_1)}dp_1dp_2, \tag{17}$$

where

$$M(p_2,p_1) = [\phi_1(p_2)\phi_1^*(p_1) + \phi_2(p_2)\phi_2^*(p_1)]. \tag{18}$$

Having obtained an expression for $F(x,p)$, let us now consider

$$p\star F(x,p) = pF(x,p) - \frac{i}{2}\nabla_x F(x,p)$$

and

$$F(x,p)\star p = pF(x,p) + \frac{i}{2}\nabla_x F(x,p)$$

where we have written $p = \sum_{j=1}^{3} p_j e_j$. The Baker bracket and the Moyal bracket are then

$$[p,F]_{BB} = pF(x,p), \quad \text{and} \quad [p,F]_{MB} = \nabla_x F(x,p).$$

Finally to obtain the expression for the Bohm momentum, we take the marginal

$$\rho(x)P_B(x) = \int [p,F]_{BB}dp = \int pF(x,p)dp.$$

Now we need to evaluate the last integral in this equation. To do this we note

$$\int pF(x,p)dp = \int\int\int pM(p_2,p_1)\delta\left(p - \frac{p_2-p_1}{2}\right)e^{ix.(p_2-p_1)}dp_1dp_2dp.$$

The RHS of this equation is the same form as the second term in equation (12) except that it contains the sum of two pairs of $\phi(p_2)\phi^*(p_1)$. Each pair can be evaluated separately, so that finally we obtain

$$\rho(x)P_B(x) = \int [p,F]_{BB}dp = \rho_1(x)\partial_x S_1(x) + \rho_2(x)\partial_x S_2(x), \tag{19}$$

where we have, as usual, used $\psi = Re^{iS}$. This is exactly the expression for Bohm momentum presented in Bohm and Hiley [4]. As a final confirmation we need to show this corresponds to the expression obtained by Bohm, Schiller and Tiomno [2]. We do this by first writing the spinor wave function, $\Psi(x)$, in the form of the Caley-Klein parameters, viz,

$$\Psi(x) = R(x)e^{iS(x)/2}\begin{pmatrix}\cos\theta(x)/2e^{i\phi(x)/2}\\ i\sin\theta(x)/2e^{-i\phi(x)/2}\end{pmatrix} \tag{20}$$

then it is straight forward to show that the expression (19) reduces to

$$P_B = \nabla_x S/2 + \cos\theta\ \nabla_x\phi/2,$$

which is just the expression found in Bohm, Schiller and Tiomno [2]

3.1. The Bohm Kinetic Energy from the Moyal Algebra

Let us complete this section by using the Moyal algebra to find the Bohm kinetic energy for the Pauli particle and an expression for its quantum potential. The method is straight forward. Using $p^2 = (\sum p_j e_j)^2 = p_1^2 + p_2^2 + p_3^2$. we find

$$p^2 \star F(x,p) = \left[p^2 - ip\,\overrightarrow{\nabla_x} - \frac{1}{4}\,\overrightarrow{\nabla_x}^2\right] F(x,p).$$

While

$$F(x,p) \star p^2 = F(x,p)\left[p^2 + ip\,\overleftarrow{\nabla_x} - \frac{1}{4}\,\overleftarrow{\nabla_x}^2\right].$$

Then forming the Baker bracket we find

$$[p^2, F(x,p)]_{BB} = p^2 F(x,p) - \frac{1}{4}\nabla_x F(x,p).$$

As we have seen in the previous subsection, $F(x,p)$ behaves as if it a sum of two $f(x,t)$, one corresponding to component ψ_1 of the spinor and the other corresponding to ψ_2. Thus we find the marginal

$$\int [p^2, F(x,p)_{BB}]dp = \rho_1(x)\overline{\overline{p_1^2}} - \frac{1}{4}\nabla_x^2 \rho_1(x) + \rho_2(x)\overline{\overline{p_2^2}} - \frac{1}{4}\nabla_x^2 \rho_2(x)$$

which leads directly to the result

$$\begin{aligned} KE_B = \rho_1(x)(\nabla_x S_1(x))^2 + \rho_2(\nabla_x S_2(x))^2 \\ -R_1(x)\nabla_x^2 R_1(x) - R_2(x)\nabla_x^2 R. \end{aligned} \tag{21}$$

Unfortunately this equation does not fall easily into the form $P_B^2/2m + Q$ so that it can be compared directly with the expression obtained in Bohm, Schiller and Tiomno [2]. However using Cayley-Klein expression for the spinor $\Psi(x)$, the RHS of equation (21) takes the form

$$P_B^2 + Q = P_B^2 - \frac{\nabla_x^2 R}{R} + (\nabla_x \theta/2)^2 + \sin^2\theta(\nabla_x \phi/2)^2 \tag{22}$$

This is exactly the form given in Bohm, Schiller and Tiomno [2] and also in Dewdney, Holland, Kyprianidis and Vigier [8] where there is a simple discussion of the whole approach.

This completes the demonstration of how the Moyal approach coupled with the orthogonal Clifford can be used to obtain weak values.

4. Conclusions

In this paper we have shown how a set of weak values can be given a simple meaning in terms of the Bohm approach [4] to quantum mechanics. Not only have we shown that weak values can be given a meaning for the non-relativistic Schrödinger particle, but they are also given meaning for a non-relativistic Pauli particle with spin-half.

In the case of the Schrödinger particle, we have shown how these weak values can be obtained from the standard approach and through a approach using the Moyal algebra [28]. To extend the approach to include a spin-half particle, we have combined the Clifford algebra approach of Hiley and Callaghan [17] [19] [20] with the Moyal algebra. This gives a more general mathematical structure in which spin can naturally be included. Space has prevented this method being applied to the relativistic Dirac particle. These results, which are based on the work reported in [20] will be presented elsewhere.

A union of the Clifford and Moyal algebras does not merely present a unified mathematical structure, but it is the very nature of this algebra that has significance. The orthogonal Clifford was constructed by Clifford himself [6], not by considering quantum mechanics, but by arguing that it arises from movements or processes in Euclidean space-time. Clifford used terms like 'versors' (literally 'turners'), 'rotors' and 'motors' to try to emphasise his overarching philosophical idea; an idea that claims everything starts with activity, with movement, i.e. from process – and that the processes themselves condition the geometry of space-time [5] [18].

In this sense the algebra is a geometric algebra and quantum phenomena are conditioned by, and help condition the geometry as has been stressed by Hestenes [15] and by Doran and Lasenby [9]. These authors considered only the orthogonal Clifford, but in this paper we have shown how the symplectic Clifford algebra [7], namely, the Moyal algebra, must also be included. As remarked by Dubin, Hennings and Smith [10], this algebraic structure brings out the deeper connection between the Moyal algebra and Weyl quantisation.

The existence of this intimate relation between the extended algebra and quantum phenomena suggests that the latter are not something exotic, but deeply related to subtle topological structures within space-time itself [21]. In addition this more general mathematical structure has the advantage that it shows a much closer relationship between quantum and classical phenomena than normally assumed. This point is discussed in detail in a recent paper by de Gosson and Hiley [14]. A further merit of our approach is that the classical limit emerges naturally without any need to appeal to decoherence [16].

One of the key features of the approach using the orthogonal Clifford is the central role played by the Clifford density element $\rho_c = \Phi_L(x)\widetilde{\Phi}_L(x)$, where the tilde denotes Clifford reversion [19]. This object is the algebraic analogue to the standard density matrix. Its extension in order to evaluate weak values follows immediately by a generalisation which involves replacing $\widetilde{\Phi}_L(x)$ by an element of a different right ideal. The further extension is needed to include the Moyal algebra. This extension is

$$\rho_M(x,p) = \int \Phi_L(x_1)\Xi_R(x_2)e^{ipy}dy,$$

where $x = (x_2 + x_1)/2$ and $y = (x_2 - x_1)$. This is just the cross-Wigner function used in de Gosson and de Gosson [13]. We see that we are constructing a non-local description of quantum phenomena. The locality of the classical world then emerges as a limiting procedure which neglects the more subtle features of quantum processes [5].

One final remark. The recent weak measurement experiments of Kocsis, Braverman, Ravets, Stevens, Mirin, Shalm, and Steinberg [22] and of Lundeen, Sutherland, Patel, Stewart and Bamber [25] opens up the possibility of actually measuring the Bohm momentum, the Bohm energy and hence the quantum potential. This possibility moves the Bohm approach from a mere philosophical speculation into the realm of experimental physics. However welcome as these results are, our discussion as to how these weak values arise strongly suggests that the so-called 'Bohmian mechanics' is a pale reflection of the deeper algebraic structure underlying quantum processes.

Acknowledgements
I wish to thank Bob Callaghan, Maurice de Gosson, David Robson and Graham Yendell for many stimulating discussions.

References

[1] Aharonov, Y. and Vaidman, L., Properties of a quantum system during the time interval between two measurements, *Phys. Rev.* **41** (1990) 11-19.
[2] Bohm, D., Schiller, R. and Tiomno, J., A Causal Interpretation of the Pauli Equation (A) and (B), *Nuovo Cim.* Supp. **1**, (1955) 48-66 and 67-91.
[3] Bohm, D. and Hiley, B. J., Non-locality and Locality in the Stochastic Interpretation of Quantum Mechanics, *Phys. Reports*, **172**, (1989), 93-122.
[4] Bohm, D. and Hiley, B. J., *The Undivided Universe: an Ontological Interpretation of Quantum Theory*, Routledge, London (1993).
[5] Bohm, D. and Hiley, B. J., Ibid, section 15.10.
[6] Clifford W.K., *Mathematical Papers*, XLII, Further note on biquaternions, 385-94, Macmillan, London (1882)
[7] Crumeyrolle, A., *Orthogonal and Symplectic Clifford Algebras: Spinor Structures*, ch. 17, Kluwer, Dordrecht (1990) .
[8] Dewdney, C., Holland, P.R., Kyprianidis, A. and Vigier, J.P., Spin and non-locality in quantum mechanics, *Nature* **336** (1988) 536-544.
[9] Doran, C. and Lasenby, A., *Geometric algebra for physicists*, Cambridge University Press,Cambridge, 2003.
[10] Dubin, D., Hennings, M. A., and Smith, T. B., *Mathematical Aspects of Weyl Quantization and Phase*, World Scientific, 2000.
[11] Duck, I. M., Stevenson, P. M. and Sudarshan, E. C. G., The sense in which a "weak measurement"of a spin-1/2 particle's spin component yields a value 100, *Phys. Rev.*, **40** (1989) 2112-17.
[12] Gilbert, J. E., and Murray, M. A. M., *Clifford algebras and Dirac operators in harmonic analysis* Cambridge University Press, Cambridge (1991).
[13] de Gosson, M.A. and de Gosson, S. M., Weak values of a quantum observable and the cross-Wigner distribution, accepted for publication in Physics Letters, (2011), quant-ph:1109.3665.
[14] de Gosson, M. A., and Hiley, B. J., Imprints of the Quantum World in Classical Mechanics, *Found. Phys.* **41**, (2011), 1415-1436. DOI 10.1007/s 10701-011-9544-5.
[15] Hestenes, D., Space-time physics with geometric algebra, *Am. J. Phys.*, **71**, (2003), 691-714
[16] Hiley, B. J., On the Relationship between the Wigner-Moyal and Bohm Approaches to Quantum Mechanics: A step to a more General Theory? *Found. Phys.* **40** (2010) 356-67. DOI 10.1007/s10701-009-9320-y
[17] Hiley. B. J. and Callaghan, R. E., Clifford Algebras and the Dirac-Bohm Quantum Hamilton-Jacobi Equation, *Accepted by Found. Phys.* (2011) DOI: 10.1007/s10701-011-9558-z
[18] Hiley, B. J., Process, Distinction, Groupoids and Clifford Algebras: an Alternative View of the Quantum Formalism, in *New Structures for Physics*, ed Coecke, B., *Lecture Notes in Physics*, **813** pp. 705-750, Springer (2011).
[19] Hiley. B. J. and Callaghan, R. E., The Clifford Algebra approach to Quantum Mechanics A: The Schrödinger and Pauli Particles. *Maths-ph/1011.4031.*
[20] Hiley, B. J. and Callaghan, R. E., The Clifford Algebra Approach to Quantum Mechanics B: The Dirac Particle and its relation to the Bohm Approach. *Maths-ph/1011.4033.*
[21] Kauffman, L. H., New invariants in the theory of knots, *Amer. Math. Monthly,* **95**, (1988), 195-242.
[22] Kocsis, S., Braverman, B., Ravets, S., Stevens, M. J., Mirin, R. P., Shalm, L.K., Steinberg, M. A., Observing the Average Trajectories of Single Photons in a Two-Slit Interferometer, *Science.* **332** (2011) 1170-73.
[23] Leavens, C. R., Weak Measurements from the Point of View of Bohmian Mechanics, *Found. Phys.* **35** (2005) 469-491. DOI:10.1007/s10701-004-1984-8.
[24] Littlejohn, R. G., The semiclassical evolution of wave packets, *Phys. Rep.* **138** (1986) 193-291.
[25] Lundeen, J. S., Sutherland, B., Patel, A.,Stewart, C. and Bamber, C., Direct Measurement of the Quantum Wavefunction, *Nature*, **474** (2011) 188-191.
[26] Nelson, E.,*Quantum Fluctuations*, Princeton University Press, Princeton, N.J. (1985).
[27] Nelson, E., Derivation of Schrödinger's Equation from Newtonian Mechanics, *Phys. Rev.*, **150**, (1966), 1079-1085.
[28] Moyal, J. E. Quantum Mechanics as a Statistical Theory, *Proc. Camb. Phil. Soc.*, **45**, (1949), 99-123.
[29] Wiseman. H. M., Grounding Bohmian mechanics in weak values and Bayesianism, *New J.Phys.* **9** (2007) 165. DOI: 10.1088/1367-2630/9/6/165.
[30] Zachos, C., (2002), Deformation Quantization: Quantum Mechanics lives and works in phase space, *Int. J. Mod. Phys.*, **A17** (2002) 297-316.

EmerQuM 11: Emergent Quantum Mechanics 2011 IOP Publishing
Journal of Physics: Conference Series **361** (2012) 012015 doi:10.1088/1742-6596/361/1/012015

A Pseudo-Quantum Triad: Schrödinger's Equation, the Uncertainty Principle, and the Heisenberg Group

Maurice A. de Gosson

University of Vienna, Faculty of Mathematics, NuHAG, Nordbergstraße 15, A-1090 Vienna AUSTRIA

E-mail: maurice.de.gosson@univie.ac.at

Abstract. We show that the paradigmatic quantum triad "Schrödinger equation–Uncertainty principle–Heisenberg group" emerges mathematically from classical mechanics. In the case of the Schrödinger equation, this is done by extending the metaplectic representation of linear Hamiltonian flows to arbitrary flows; for the Heisenberg group this follows from a careful analysis of the notion of phase of a Lagrangian manifold, and for the uncertainty principle it suffices to use tools from multivariate statistics together with the theory of John's minimum volume ellipsoid. Thus, the mathematical structure needed to make quantum mechanics emerge already exists in classical mechanics.

1. Introduction

The Schrödinger equation, the uncertainty principle, and the Heisenberg group are three basic paradigms of Quantum Mechanics. The aim of this contribution is to show that this "triad" can actually be rigorously (i.e. mathematically) constructed within classical mechanics in its Hamiltonian formulation; Planck's constant h then appears as a scaling parameter whose dimension is that of an action. This fact is actually very much in accordance with Mackey's [31] statement following which quantum mechanics is a refinement of classical mechanics. Quantum Mechanics can thus only emerge if one can give a physical meaning to this triad, justifying the need for Planck's constant h. We will not attempt to do this in the present work.

Here is a short description of what we are going to do:

- **The Schrödinger Equation**. According to conventional wisdom, Schrödinger's equation cannot be derived from classical considerations; to witness Feynman's declaration: *"Where did that [the Schrödinger equation] come from? Nowhere. It came out of the mind of Schrödinger, invented in his struggle to find an understanding of the experimental observations in the real world.* However, at the time when Feynman made this declaration (1965) it was already pretty well-known among specialists working in harmonic analysis, that the Schrödinger equation is implicit in the metaplectic representation of the symplectic group; in fact it was already known that Schrödinger's equation could be rigorously derived for all Hamiltonians which are quadratic in the position and momentum variables. We will show that this equation can actually be derived for all Hamiltonian functions, not only for those of the classical type "kinetic energy plus potential"
- **The Uncertainty Principle.** In Classical Mechanics (CM) it is tacitly assumed that the physical quantities that are measured have an exact value, that can be *in principle*

determined. In practice, the numerical results obtained from our measurements will cluster around this objective exact value, and one then uses statistical methods to obtain approximations to the exact value. In Quantum Mechanics (QM) the situation is different; of course there are also measurements errors, but in the limiting case where perfect measurements is assumed, the results of identical experiments performed on identically prepared systems are generally not yielding identical results (this is well illustrated by the discussion in Peres [33], §4-3, of a spin experiment). This quantum indeterminacy is traditionally expressed in terms of the Uncertainty Principle of QM. We will show that the Heisenberg inequalities (or their refinement, the Robertson–Schrödinger inequalities) are –at least formally– perfectly classical.

- **The Heisenberg Group**. The theory of the Heisenberg group is of a slightly more abstract nature. Its "Schrödinger representation" is supposed to be a key to quantization, as explained in Refs. [23, 24]. We will see that the Heisenberg group and the associated Heisenberg–Weyl operators have a very simple interpretation in terms of the notion of phase of a Lagrangian manifold, using the properties of the Poincaré–Cartan invariant familiar from the theory of the Hamilton–Jacobi equation from classical mechanics.

Acknowledgement
This work has been financed by the Austrian Research Agency FWF (Projektnummer P 23902-N13). The author wishes to thank the organizers of the Heinz von Foerster Congress in Vienna.

2. Symplectic Mechanics

We gather in this preliminary section the tools from symplectic geometry and Hamiltonian mechanics we will need; for details and proofs see Refs. [1, 4, 13, 15, 18]; in [15] we have given a detailed study of the metaplectic group.

2.1. The symplectic group

A symplectic matrix is a real matrix S of size $2n$ satisfying anyone of the equivalent relations $S^TJS = J$ or $SJS^T = J$; here S^T is the transpose of S and J is the skew-symmetric matrix $\begin{pmatrix} 0 & I \\ -I & 0 \end{pmatrix}$ where 0 and I are, respectively, the $n \times n$ zero and identity matrices. In particular a 2×2 real matrix $\begin{pmatrix} a & b \\ c & d \end{pmatrix}$ is symplectic if and only if its determinant $ad - bc$ is equal to one; but it is not true in dimension $n > 1$ that every unimodular matrix is symplectic. Using the identity $S^TJS = J$ (or $SJS^T = J$) one easily checks that the symplectic matrices form a group, which is denoted by $\mathrm{Sp}(2n, \mathbb{R})$.

The symplectic group $\mathrm{Sp}(2n, \mathbb{R})$ is generated by the matrix J together with all matrices

$$M_L = \begin{pmatrix} L^{-1} & 0 \\ 0 & L^T \end{pmatrix} \ , \ V_{-P} = \begin{pmatrix} I & 0 \\ P & I \end{pmatrix} \tag{1}$$

where $\det L \neq 0$ and $P = P^T$ (L and P having size n); the matrices M_L, and V_{-P} are trivially symplectic. Since the determinants of J, M_L, and V_{-P} are equal to one, this shows at the same time that the determinant of a symplectic matrix always is equal to one. This fact implies Liouville's theorem on the conservation of phase space volume under Hamiltonian flows, see Refs. [4, 12].

2.2. The metaplectic representation

The symplectic group is connected and has covering spaces of all orders. Among all these, the two-fold covering group plays a very special role, because it can be faithfully represented by a group of unitary operators acting on the square integrable functions. This operator group is called the metaplectic group; we will denote it by $\mathrm{Mp}(2n,\mathbb{R})$. The easiest way to describe $\mathrm{Mp}(2n,\mathbb{R})$ is to use generators. Since $\mathrm{Mp}(2n,\mathbb{R})$ is a two-fold covering of $\mathrm{Sp}(2n,\mathbb{R})$ to each of the generators J, M_L, V_{-P} correspond two elements of $\mathrm{Mp}(2n,\mathbb{R})$, differing by a sign. These operators $\widehat{J}$, $\widehat{M}_L$, $\widehat{V}_{-P}$ are defined by

$$\widehat{J}\psi(x) = \left(\tfrac{1}{2\pi i}\right)^{n/2}\int e^{ix\cdot y}\psi(y)d^n y \tag{2}$$

$$\widehat{M}_L\psi(x) = \sqrt{\det L}\,\psi(Lx) \tag{3}$$

$$\widehat{V}_{-P}\psi(x) = e^{\frac{i}{2}Px\cdot x}\psi(x); \tag{4}$$

the argument of $\det L$ in formula (3) is chosen to be 0 or 2π if $\det L > 0$, and π or 3π if $\det L < 0$. The covering projection $\Pi : \mathrm{Mp}(2n,\mathbb{R}) \longrightarrow \mathrm{Sp}(2n,\mathbb{R})$ is determined by the equalities

$$\Pi(\widehat{J}) = J \ , \ \Pi(\widehat{M}_L) = M_L \ , \ \Pi(\widehat{V}_{-P}) = V_{-P}. \tag{5}$$

However, one obtains another perfectly honest covering projection by using inner automorphisms of the base group $\mathrm{Sp}(2n,\mathbb{R})$ (see e.g. de Gosson [13], §6.4.2, for a simple description of this conjugation relation). This means that given an arbitrary element S_0 of $\mathrm{Sp}(2n,\mathbb{R})$ the mapping Π_0 defined by $\Pi_0(\widehat{S}) = S_0\Pi(\widehat{S})S_0^{-1}$ is also a bona fide covering projection. Notice that the definition of Π_0 can be rewritten in the form

$$\Pi_0(\widehat{S}) = \Pi(\widehat{S}_0\widehat{S}\widehat{S}_0^{-1}) \ \ , \ \ \Pi_0(\widehat{S}_0) = S_0 \tag{6}$$

since Π is a group homomorphism. Choose now $S_0 = M_{1/\sqrt{\varepsilon}}$ (it is the symplectic matrix M_L with $L = (1/\sqrt{\varepsilon})I$) and define the projection

$$\Pi^\varepsilon(\widehat{S}) = M_{1/\sqrt{\varepsilon}}\Pi(\widehat{S})M_{\sqrt{\varepsilon}} = \Pi(\widehat{M}_{1/\sqrt{\varepsilon}}\widehat{S}\widehat{M}_{\sqrt{\varepsilon}}). \tag{7}$$

One immediately verifies that if we use Π^ε instead of Π, the projections formulas (5) become

$$\Pi^\varepsilon(\widehat{J}^\varepsilon) = J \ , \ \Pi^\varepsilon(\widehat{M}_L^\varepsilon) = M_L \ , \ \Pi^\varepsilon(\widehat{V}_{-P}^\varepsilon) = V_{-P}. \tag{8}$$

the metaplectic operators $\widehat{J}^\varepsilon$, $\widehat{M}_L^\varepsilon$, $\widehat{V}_{-P}^\varepsilon$ being defined by

$$\widehat{J}^\varepsilon\psi(x) = \left(\tfrac{1}{2\pi i\varepsilon}\right)^{n/2}\int e^{\frac{i}{\varepsilon}x\cdot y}\psi(y)d^n y \tag{9}$$

$$\widehat{M}_L^\varepsilon\psi(x) = \widehat{M}_L\psi(x) = \sqrt{\det L}\,\psi(Lx) \tag{10}$$

$$\widehat{V}_{-P}^\varepsilon\psi(x) = e^{\frac{i}{2\varepsilon}Px\cdot x}\psi(x). \tag{11}$$

In QM one chooses $\varepsilon = \hbar$.

2.3. Hamiltonian flows

Let $H = H(z)$ (the "Hamiltonian") be a smooth function of the variables x, p. Hamilton's equations of motion $\dot{x} = \nabla_p H(x,p)$ $\dot{p} = -\nabla_x H(x,p)$ can be written in compact form as

$$\dot{z} = J\nabla_z H(z). \tag{12}$$

The presence of the matrix J in this formula indicates that the symplectic group might lurk behind Hamiltonian mechanics. Let us examine this "educated guess" somewhat more in detail.

We denote by (f_t^H) the associated phase flow: f_t^H is the mapping which takes an initial point $z_0 = (x_0, p_0)$ at time $t = 0$ to the point $z = (x, p)$ at time t along the solution curve to (12) through z_0. That is, $z(t) = f_t^H(z_0)$ is the solution of (12) with initial condition $z(0) = z_0$. The importance of the symplectic group in Hamiltonian mechanics comes from the following fundamental property of the phase flows (f_t^H): each mapping f_t^H is a canonical transformation; see [1, 4, 12, 15, 18]. This means that for every $z = (x, p)$ the Jacobian matrix of f_t^H calculated at $z = (x, p)$ is symplectic:

$$Df_t^H(z) = \frac{\partial(x(t), p(t))}{\partial(x_0, p_0)} \in \mathrm{Sp}(2n, \mathbb{R}).$$

In particular, when the Hamiltonian function is quadratic in the x and p variables, that is of the type

$$H(z) = \frac{1}{2}Mz \cdot z = \frac{1}{2}(x, p)M(x, p)^T, \tag{13}$$

the flow determined by Hamilton's equations is linear, and consists of the symplectic matrices $S_t^H = e^{tJM}$. As we are going to see below, the study of linear flows is the first step towards a general derivation of Schrödinger's equation.

3. Schrödinger's Equation

We now proceed to show how the Schrödinger equation emerges from Hamiltonian mechanics. We assume that the Hamiltonians are time-independent, but everything in this Section can be extended to the time-dependent case at the price of some technical difficulties (see de Gosson [18], de Gosson and Hiley [19]).

3.1. Quadratic Hamiltonians

Consider first a quadratic Hamiltonian of the type (13). As we observed above, the phase flow can easily be calculated, and consists of symplectic matrices S_t^H. We have of course $S_t^H S_{t'}^H = S_{t+t'}^H$ so the flow is a one-parameter subgroup of $\mathrm{Sp}(2n, \mathbb{R})$. When t varies these matrices describe a curve in $\mathrm{Sp}(2n, \mathbb{R})$ which passes by the identity at the initial time $t = 0$. Now, the unique path lifting theorem in the theory of covering groups says that this curve can be lifted to any of the covering groups of $\mathrm{Sp}(2n, \mathbb{R})$ and that this can be made in an unique way if one specifies a point through which the lifting should pass. Choosing the metaplectic group $\mathrm{Mp}(2n, \mathbb{R})$ as the covering group, with a projection $\Pi^\varepsilon : \mathrm{Mp}(2n, \mathbb{R}) \longrightarrow \mathrm{Sp}(2n, \mathbb{R})$, and imposing to the lifting to pass through the identity of $\mathrm{Mp}(2n, \mathbb{R})$ at time $t = 0$ we thus obtain a curve of metaplectic operators $\widehat{S}_t^H$ such that $\Pi^\varepsilon(\widehat{S}_t^H) = S_t^H$; one moreover verifies that the group property $S_t^H S_{t'}^H = S_{t+t'}^H$ carries over to these operators: $\widehat{S}_t^H \widehat{S}_{t'}^H = \widehat{S}_{t+t'}^H$. The operators $\widehat{S}_t^H$ are unitary and act on the square integrable functions on $\mathbb{R}^n$. For such a function ψ_0 we set $\psi(x, t) = \widehat{S}_t^H \psi_0$ and ask the question "*what is the partial differential equation satisfied by* $\psi(x, t)$*?*" The answer is: if the initial function ψ_0 is at least twice continuously differentiable in the x variables, then

$$i\varepsilon \frac{\partial \psi}{\partial t}(x, t) = \widehat{H}(x, -i\varepsilon \nabla_x)\psi(x, t) \tag{14}$$

where $\widehat{H}(x, -i\varepsilon\nabla_x)$ is obtained from the Hamiltonian function (13) by the formal substitution $(x, p) \longrightarrow (x, -i\hbar\nabla_x)$. Writing $M = \begin{pmatrix} H_{xx} & H_{xp} \\ H_{px} & H_{pp} \end{pmatrix}$ this is

$$\widehat{H}(x, -i\varepsilon\nabla_x) = -\frac{\varepsilon^2}{2}H_{pp}x^2 - i\varepsilon H_{px}x\nabla_x + \frac{1}{2}H_{xx}x^2 - \frac{i}{2}\mathrm{Tr}(H_{px}) \tag{15}$$

where $\mathrm{Tr}(H_{px})$ (the trace) is the sum of the diagonal elements of H_{px}.

Thus, if *we choose* $\varepsilon = \hbar$ we obtain Schrödinger's equation; but this particular choice of parameter has to be motivated by physical considerations once a physical meaning has been given to the solution ψ. What must be remembered is that the derivation of (14) heavily depends on the choice of projection Π^{ε} of $\mathrm{Mp}(2n, \mathbb{R})$ onto $\mathrm{Sp}(2n, \mathbb{R})$: each choice of ε is a priori equally good.

We have thus established the existence, for each $\varepsilon > 0$, of a one-to-one and onto correspondence

$$C_{\varepsilon} : (S_t^H) \leftrightarrows (\widehat{S}_t^H) \tag{16}$$

between phase flows arising from quadratic Hamiltonian functions and unitary evolution groups $\widehat{S}_t^H = e^{-it\widehat{H}/\varepsilon}$, $\widehat{H} = \widehat{H}(x, -i\varepsilon\nabla_x)$. We now make a fundamental observation. Assume that we make a symplectic change of variables in the Hamiltonian $H(z)$, i.e. that we replace $H(z)$ with $H(S^{-1}z)$ where S is in $\mathrm{Sp}(2n, \mathbb{R})$. Then, the phase flow $(f_t^{H \circ S^{-1}})$ of $H(S^{-1}z)$ is obtained from the flow (f_t^H) by conjugation with S:

$$f_t^{H \circ S^{-1}} = S f_t^H S^{-1} \tag{17}$$

(see e.g. Refs. [4, 15, 18, 12, 34]). In the case of quadratic Hamiltonians we have $S_t^{H \circ S^{-1}} = S S_t^H S^{-1}$ and the group of unitary operators $\widehat{S}_t^H$ is replaced with $\widehat{S}_t^{H \circ S^{-1}} = \widehat{S}\widehat{S}_t^H \widehat{S}^{-1}$. It follows that the correspondence C_{ε} satisfies the covariance formula

$$C_{\varepsilon}(S S_t^H S^{-1}) = (\widehat{S}\widehat{S}_t^H \widehat{S}^{-1}). \tag{18}$$

3.2. The general case

We begin by recalling two results from functional analysis. The first is Stone's theorem [45] about one-parameter groups of operators (see Refs. [1, 35] for "modern" proofs):

> **Stone's Theorem:** *(i) For every strongly continuous one-parameter group (F_t) of unitary operators on a Hilbert space $L^2(\mathbb{R}^n)$ there exists a self-adjoint operator A on $L^2(\mathbb{R}^n)$ such that $F_t = e^{itA/\hbar}$; in particular A is closed and densely defined in $\mathcal{H}$. (ii) Conversely, if A is a self-adjoint operator on $L^2(\mathbb{R}^n)$ then there exists a unique one-parameter unitary group (F_t) whose infinitesimal generator is A, that is $F_t = e^{itA/\hbar}$.*

The second result says that every continuous operator (in a sense we will be precise) has a distributional kernel:

> **Schwartz's kernel theorem:** *Let A be a linear operator $\mathcal{S}(\mathbb{R}^n) \longrightarrow \mathcal{S}'(\mathbb{R}^n)$. We assume that A is continuous in the sense that if (ψ_k) is a sequence of functions such that $\psi_k \to \psi$ for $k \to \infty$ in $\mathcal{S}(\mathbb{R}^n)$ then $A\psi_k \to A\psi$ in $\mathcal{S}(\mathbb{R}^n)$. Then there exists a distribution K_A (the kernel of A) such that*
>
> $$A\psi(x) = \int K_A(x, y)\psi(y)d^n y \tag{19}$$
>
> *for all functions ψin $\mathcal{S}(\mathbb{R}^n)$ [the integral in the right hand-side of (19) is interpreted as a distributional bracket].*

Let now H be an arbitrary Hamiltonian function; we assume that the phase flow (f_t^H) generated by the Hamilton equations for H exist for all times t (this is just a technical requirement, which can be alleviated; see de Gosson and Hiley [19]). We are going to show that the functor (16) can be extended in a unique way to arbitrary Hamiltonian flows if one makes the following requirement:

- *The symplectic covariance relation* (18):

$$C_\varepsilon(Sf_t^H S^{-1}) = (\widehat{S}\widehat{F}_t^H \widehat{S}^{-1}) \tag{20}$$

should hold for each S *in* $\mathrm{Sp}(2n,\mathbb{R})$.

We have again $Sf_t^H S^{-1} = f_t^{H\circ S^{-1}}$ in view of formula (17), and $(\widehat{F}_t^H)$ is a strongly continuous one-parameter group of unitary operators on $L^2(\mathbb{R}^n)$ which has to be determined.

To the phase flow (f_t^H) (that is, equivalently, to the Hamiltonian function H) we associate the Weyl operator

$$\widehat{H} = \left(\tfrac{1}{2\pi i\varepsilon}\right)^{n/2} \int H_\sigma^\varepsilon(z)\widehat{T}^\varepsilon(z)d^{2n}z$$

where $\widehat{T}^\varepsilon(z) = e^{-i\sigma(\hat{z},z)/\varepsilon}$ is the ε-Heisenberg–Weyl operator (cf. formula (38) and

$$H_\sigma^\varepsilon(z_0) = \left(\tfrac{1}{2\pi i\varepsilon}\right)^n \int e^{-\frac{i}{\varepsilon}\sigma(z,z')}H(z')d^{2n}z'$$

is the ε-symplectic Fourier transform of H. Since H is real it follows from the standard theory of Weyl operators (see Refs. [15, 18, 29]) that the operator $\widehat{H}$ is self adjoint and densely defined on the Schwartz space of rapidly decreasing functions. In view of part (ii) of Stone's theorem there exists a unique one-parameter group of unitary operators $\widehat{F}_t^H = e^{-it\widehat{H}/\varepsilon}$, so we define

$$C_\varepsilon(f_t^H) = (\widehat{F}_t^H) = (e^{-it\widehat{H}/\varepsilon}). \tag{21}$$

To show that this is the desired extension of the functor C_ε we have to check that it satisfies the symplectic covariance property (20) and that its restriction to linear flows (S_t^H) is given by (16). The symplectic covariance property is proven as follows: since $Sf_t^H S^{-1} = f_t^{H\circ S^{-1}}$ the functor C_ε associates to $(Sf_t^H S^{-1})$ the one-parameter group $(e^{-it\widehat{H\circ S^{-1}}/\varepsilon})$. In view of the symplectic covariance property of Weyl operators we have $\widehat{H\circ S^{-1}} = \widehat{S}\widehat{H}\widehat{S}^{-1}$ and hence

$$e^{-it\widehat{H\circ S^{-1}}/\varepsilon} = e^{-it\widehat{S}\widehat{H}\widehat{S}^{-1}/\varepsilon} = \widehat{S}e^{-it\widehat{H}/\varepsilon}\widehat{S}^{-1}$$

so that $C_\varepsilon(Sf_t^H S^{-1}) = \widehat{S}C_\varepsilon(f_t^H)\widehat{S}^{-1}$ which is precisely formula (20) we set out to prove. Let us now show that definition (21) of the functor C_ε implies that we have $C_\varepsilon(S_t^H) = (\widehat{S}_t^H)$. We must thus show that the operators $\widehat{F}_t^H = e^{-it\widehat{H}/\varepsilon}$ are metaplectic when H is a quadratic Hamiltonian. The infinitesimal generator $\widehat{H}$ of $\widehat{F}_t^H$ satisfies the Schrödinger equation

$$i\hbar\frac{d}{dt}\widehat{F}_t^H = \widehat{H}\widehat{F}_t^H$$

where $\widehat{H} \overset{\text{Weyl}}{\longleftrightarrow} H$. Now, the operator $\widehat{H}(x, -i\varepsilon\nabla_x)$ defined by formula (15) is precisely the Weyl quantization of the quadratic Hamiltonian function H hence the $\widehat{F}_t^H$ are precisely the metaplectic operators $\widehat{S}_t^H$ corresponding to the lift of the linear flow (S_t^H) to $\mathrm{Mp}(2n,\mathbb{R})$.

We next have to show that, conversely, for every strongly-continuous one-parameter group $\widehat{F}_t$ of unitary operators on $L^2(\mathbb{R}^n)$ there exists a Hamiltonian function H such that $\widehat{F}_t = C_\varepsilon(f_t^H)$. This amounts to show that there exists a Weyl operator $\widehat{H} \overset{\text{Weyl}}{\longleftrightarrow} H$ such that $\widehat{F}_t = e^{-it\widehat{H}/\varepsilon}$. We begin by remarking that in view of part (i) of Stone's theorem there exists a unique self-adjoint operator $\widehat{H}$ such that $\widehat{F}_t = e^{-it\widehat{H}/\varepsilon}$ for all times t, so all we have to do is to prove that $\widehat{H}$ is indeed a Weyl operator; the Weyl correspondence $\widehat{H} \overset{\text{Weyl}}{\longleftrightarrow} H$ will then identify the Hamiltonian function such that $\widehat{F}_t = C_\varepsilon(f_t^H)$. The restriction of $\widehat{H}$ to the dense subspace $\mathcal{S}(\mathbb{R}^n)$ is continuous $\mathcal{S}(\mathbb{R}^n) \longrightarrow \mathcal{S}'(\mathbb{R}^n)$ because $\widehat{H}$ is essentially self-adjoint; it follows that Schwartz's kernel theorem applies, and we can thus find K such that

$$\widehat{H}\psi(x) = \int K(x,y)\psi(y)d^n y. \tag{22}$$

Let now τ be an arbitrary real number and define the "τ-symbol" H_τ of $\widehat{H}$ by the Fourier integral

$$H_\tau(z) = \int e^{-\frac{i}{\varepsilon}py}K(x+\tau y, x-(1-\tau)y)d^n y. \tag{23}$$

One shows (de Gosson [18], Shubin [44]) that the operator $\widehat{H}$ can be expressed in terms of H_τ by the formula

$$\widehat{H}\psi(x) = \left(\tfrac{1}{2\pi i\varepsilon}\right)^n \iint e^{-\frac{i}{\varepsilon}py}H_\tau((1-\tau)x+\tau y, p)\psi(y)d^n y d^n p \tag{24}$$

which is the "pseudo-differential representation of $\widehat{H}$ in terms of its τ-symbol". We next observe that if the operator is multiplication by a function $a(x)$ then we must have $K(x,y) = a(x)\delta(x-y)$ and hence, by (23),

$$H_\tau(z) = a(x)\int e^{-\frac{i}{\varepsilon}py}\delta(y)d^n y = a(x).$$

This property, together with the symplectic covariance requirement (20) implies (see Refs. [43, 50]), that this is only possible if we choose $\tau = \frac{1}{2}$, in which case formula (24) becomes

$$\widehat{H}\psi(x) = \left(\tfrac{1}{2\pi i\varepsilon}\right)^n \iint e^{-\frac{i}{\varepsilon}py}H_{1/2}(\tfrac{1}{2}(x+y), p)\psi(y)d^n y d^n p.$$

But this is exactly the integral representation of the Weyl operator $\widehat{H} \overset{\text{Weyl}}{\longleftrightarrow} H = H_{1/2}$, and we are done.

3.3. The Groenewold and Van Hove "no-go" theorem

It is sometimes claimed that one cannot "quantize classically" arbitrary non-quadratic observables; to sustain this claim an old (and famous) series of results going back to Groenewold [21] and van Hove [26, 27] is invoked. But this objection cannot be retained in the present case: a careful mathematical reading of the Groenewold-van Hove obstruction results reveals that what is actually proven is the impossibility to quantize observables in such a way that Dirac's [9] prescription that Poisson brackets should correspond to commutators is preserved. And, indeed, such a quantization is a chimera. This state of affairs is very lucidly analysed by Giulini's [11] who reminds us that Dirac [9] himself had noted that

> ...The strong analogy between quantum Poisson brackets [i.e. commutators] and classical Poisson brackets leads us to make the assumption that the quantum Poisson brackets, *or at any rate the simpler ones of them*, have the same values as the corresponding classical Poisson brackets.

The italics (added) show that Dirac himself was very aware of the fact that it was not reasonable to expect that the Poisson bracket/commutators correspondence should hold for all pairs of observables H and K. The Groenewold–van Hove obstructions thus have nothing to do with the impossibility of a derivation of Schrödinger's equation. They only show that one has to go beyond the metaplectic representation if one wants to quantize non-quadratic Hamiltonians. This is precisely what we have done by extending the correspondence C_ε to arbitrary Hamiltonian flows.

4. The Uncertainty Principle

We are following in this section the discussion in de Gosson [16, 17].

4.1. Uncertainty inequalities

Contrarily to what is often believed Heisenberg's uncertainty principle $\Delta p_j \Delta x_j \geq \frac{1}{2}\hbar$ is not a statement about the accuracy of our measurement instruments; its derivation assumes on the contrary perfect instruments. The correct interpretation of Heisenberg's inequalities is the following (see e.g. Peres [33], p.93): if the same preparation procedure is repeated a large number of times, and is followed by either by a measurement of x_j, or by a measurement of p_j, the results obtained have standard deviations Δx_j and Δp_j satisfying the Heisenberg inequalities. The same interpretation is of course true for the stronger Robertson–Schrödinger [36, 40] inequalities

$$(\Delta p_j)^2(\Delta x_j)^2 \geq \Delta(x_j, p_j)^2 + \tfrac{1}{4}\hbar^2 \tag{25}$$

to which the Heisenberg inequalities reduce if one neglects the covariances $\Delta(x_j, p_j)^2$; they are complemented by the trivial inequalities

$$\Delta p_j \Delta x_k \geq 0 \text{ if } j \neq k, \quad \Delta p_j \Delta p_k \geq 0 \ , \quad \Delta x_j \Delta x_k \geq 0. \tag{26}$$

where $\theta_{jk}^2 = \theta_{kj}^2$ and $\eta_{jk}^2 = \eta_{kj}^2$ (see Dias et al. [8] and the references therein). In *classical* statistical mechanics the situation is somewhat different: due to the inherent inaccuracy of the measurement apparatus there are uncertainties for all pairs of variables, conjugate or not. This is actually reminiscent of what happens in noncommutative mechanics (NCQM) where the second and third inequalities (26) are replaced with

$$\Delta p_j^2 \Delta p_k^2 \geq \Delta(p_j, p_k)^2 + \tfrac{1}{4}\theta_{jk}^2 \ , \ \Delta x_j^2 \Delta x_k^2 \geq \Delta(x_j, x_k)^2 + \tfrac{1}{4}\eta_{jk}^2. \tag{27}$$

4.2. The Minimum Volume Ellipsoid method

Performing position and momentum measurements on a large number K of identical copies of a system of particles, we get a cloud $\mathcal{S} = \{z_1, z_2, ..., z_N\}$, $N = nK$, of points in $\mathbb{R}^{2n}$. An efficient method for studying that cloud consists in using the minimum volume ellipsoid (MVE) method (Rousseeuw [37], Van Aelst and Rousseeuw [42]). Geometrically speaking this method is an application of John's theorem [28] which says that a convex set is contained in a unique ellipsoid with smallest volume. It works as follows: consider a subset $\{z_{i_1}, z_{i_2}, ..., z_{i_k}\}$ of $\mathcal{S}$ of points in general position. This condition is sufficient and necessary for any ellipsoid containing these points to have positive volume. The points $z_{i_1}, z_{i_2}, ..., z_{i_k}$ determine a polyhedron in $\mathbb{R}^{2n}$; we denote by $\mathcal{K}$ the convex hull of that polyhedron (it is the smallest subset of $\mathbb{R}^{2n}$ containing $\{z_{i_1}, z_{i_2}, ..., z_{i_k}\}$). John's theorem ensures us that there exists a unique ellipsoid $\mathcal{J}$ in $\mathbb{R}^{2n}$ containing $\mathcal{K}$ and having minimum volume among all the ellipsoids having this property. Repeating this process for all subsets of the cloud $\mathcal{S}$ having k elements in general position we get a family of ellipsoids; by definition the MVE is the one with the smallest volume. To determine that ellipsoid one proceeds as follows: choose an integer k between $[N/2]+1$ and N ($[N/2]$ the integer part of $N/2$); this constant k determines the robustness of the resulting estimator; a common choice is to take $k = [(N + 2n + 1)/2]$. The first step of the procedure is to minimize the determinant of the matrices M subject to the condition

$$\#\{j : (z_j - \bar{z})^T M^{-1}(z_j - \bar{z}) \leq m^2\} \geq k \tag{28}$$

where the minimization is over all $\bar{z} \in \mathbb{R}^{2n}$ and all positive definite symmetric matrices M of size $2n$. Here m is a fixed constant, chosen so that the MVE estimator is a consistent estimator for of the covariance matrix for data coming from a multivariate normal distribution, that is

$m = \sqrt{\chi^2_{2n,\alpha}}$, $\alpha = k/N$ where $\chi^2_{2n,\alpha}$ is a chi-square distribution with $2n$ degrees of freedom (see Lopuhäa and Rousseeuw [30]). Once the pair $(M, \bar{z})$ is determined, the minimum volume ellipsoid (MVE) is the set of all z in $\mathbb{R}^{2n}$ such that

$$(z - \bar{z})^T M^{-1}(z - \bar{z}) \leq m^2. \tag{29}$$

The second step consists in associating to the MVE $\mathcal{J}$ a covariance matrix. For this one has to choose an adequate value m_0 for m; denoting the corresponding matrix M by Σ the MVE is the ellipsoid

$$\mathcal{C} : (z - \bar{z})^T \Sigma^{-1}(z - \bar{z}) \leq m_0^2 \tag{30}$$

and Σ is then precisely the covariance matrix. We will write that matrix in the form

$$\Sigma = \begin{pmatrix} \Delta(x,x) & \Delta(x,p) \\ \Delta(p,x) & \Delta(p,p) \end{pmatrix} \tag{31}$$

where $\Delta(x,x) = ((\Delta(x_i, x_j))_{1 \leq i,j \leq n}$ and so on; writing as is customary $\Delta(x_i, x_i) = \Delta x_i^2$ and $\Delta(p_i, p_i) = \Delta p_i^2$ we have $\Delta(x_i, x_j) = \Delta(x_j, x_i)$, $\Delta(p_i, p_j) = \Delta(p_j, p_i)$, $\Delta(x_i, p_j) = \Delta(p_j, x_i)$ (observe that Σ is symmetric).

4.3. A condition on the covariance matrix

Let $A = (a_{jk})_{1 \leq j,k \leq n}$ and $C = (c_{jk})_{1 \leq j,k \leq n}$ be two real antisymmetric matrices, and $B = (b_{jk})_{1 \leq j,k \leq n}$ a real symmetric matrix. To A, B, C we associate the $2n \times 2n$ antisymmetric matrix

$$\Omega = \begin{pmatrix} A & B \\ -B & C \end{pmatrix}$$

which is the most general form an antisymmetric of size $2n$ can have. Assuming Ω invertible, the bilinear form ω on $\mathbb{R}^{2n}$ defined by $\omega(z, z') = -(z')^T \Omega^{-1} z$ is a symplectic form on $\mathbb{R}^{2n}$. Notice that when $A = C = 0$ and $B = I$ we have $\Omega = J = \begin{pmatrix} 0 & I \\ -I & 0 \end{pmatrix}$. The general case is reduced to the standard case by noting that there exists a matrix F such that $\Omega = F^T J F$; equivalently F is a linear canonical transformation $(\mathbb{R}^{2n}, \sigma) \longrightarrow (\mathbb{R}^{2n}, \omega)$.

Let now Σ be the covariance matrix (31). The matrix

$$\Sigma + i\Omega = \begin{pmatrix} \Delta(x,x) + iA & \Delta(x,p) + iB \\ \Delta(p,x) - iB & \Delta(p,p) + iC \end{pmatrix} \tag{32}$$

is Hermitian since Σ is symmetric and $(i\Omega)^* = i\Omega$, hence the eigenvalues of $\Sigma + i\Omega$ are real. Assuming in addition that these eigenvalues are nonnegative, $\Sigma + i\Omega$ is semi-definite positive, which we write $\Sigma + i\Omega \geq 0$. It follows that

$$\Delta(x,x) + iA \geq 0 \quad , \quad \Delta(p,p) + iC \geq 0. \tag{33}$$

When $n = 1$ the covariance matrix is just

$$\Sigma = \begin{pmatrix} \Delta x^2 & \Delta(x,p) \\ \Delta(p,x) & \Delta p^2 \end{pmatrix}$$

and the antisymmetric matrices Θ and N are zero so that $\Omega = aJ = \begin{pmatrix} 0 & a \\ -a & 0 \end{pmatrix}$. The condition $\Sigma + i\Omega = \Sigma + iaJ \geq 0$ is in this case equivalent to

$$\begin{pmatrix} \Delta x^2 + ia & \Delta(x,p) \\ \Delta(p,x) & \Delta p^2 - ia \end{pmatrix} \geq 0$$

that is, to the single inequality

$$\Delta x^2 \Delta p^2 \geq \Delta(x,p)^2 + a^2$$

which formally reduces to the Robertson–Schrödinger inequality (25) if one chooses $\varepsilon = \hbar/2$.

In higher dimensions one shows, using Sylvester's criterion which says that a Hermitian matrix is positive semidefinite if and only if all of its principal minors are nonnegative, that the condition $\Sigma + i\Omega \geq 0$ implies that

$$\Delta x_j^2 \Delta x_k^2 \geq \Delta(x_j, x_k)^2 + a_{jk}^2 \tag{34}$$

$$\Delta p_j^2 \Delta p_k^2 \geq \Delta(p_j, p_k)^2 + c_{jk}^2 \tag{35}$$

$$\Delta x_j^2 \Delta p_k^2 \geq \Delta(x_j, p_k)^2 + b_{jk}^2. \tag{36}$$

In particular, if $\Omega = \varepsilon J$, $\varepsilon > 0$, these conditions again reduce to the Robertson–Schrödinger type inequalities

$$\Delta x_j^2 \Delta p_k^2 \geq \Delta(x_j, p_k)^2 + \varepsilon^2.$$

In [16, 17] we have given an interpretation of the uncertainty principle (both quantum, and classical) in terms of the topological notion of symplectic capacity (also see the review paper by de Gosson and Luef [20]). This approach seems promising since it allows to recast the Robertson–Schrödinger inequalities in a topological form, invariant even under non-linear canonical transformations. This is an extension of the fact that the Robertson–Schrödinger inequalities are invariant under linear symplectic transformations.

5. Heisenberg Group and Operators

In this Section we show that the Heisenberg group is essentially a classical object associated to the notion of phase of a Lagrangian manifold (see de Gosson [14, 15]).

5.1. *The Heisenberg group*

The Heisenberg group $\mathbb{H}_n$ is the space $\mathbb{R}^{2n+1}$ equipped with the product

$$(z_1, t_1) \circledast (z_0, t_0) = (z_1 + z_0, t_0 + t_1 + \tfrac{1}{2}\sigma(z_0, z_1)). \tag{37}$$

It is intimately related to the Heisenberg–Weyl operators

$$\widehat{T}(z_0) = e^{-\frac{i}{\hbar}\sigma(\hat{z}, z_0)} \quad , \quad \sigma(\hat{z}, z_0) = p_0 x - x_0(-i\hbar\nabla_x) \tag{38}$$

whose explicit action on a function $\psi(x)$ is given by

$$\widehat{T}(z_0)\psi(x) = e^{\frac{i}{\hbar}(p_0 \cdot x - \frac{1}{2}p_0 \cdot x_0)}\psi(x - x_0) \tag{39}$$

(see e.g. de Gosson [15, 18], Littlejohn [29]). The Heisenberg–Weyl operators do not commute:

$$\widehat{T}(z_1)\widehat{T}(z_0) = e^{-\frac{i}{\hbar}\sigma(z_0, z_1)}\widehat{T}(z_0)\widehat{T}(z_1) \tag{40}$$

as immediately follows from the property

$$\widehat{T}(z_0 + z_1) = e^{-\frac{i}{2\hbar}\sigma(z_0, z_1)}\widehat{T}(z_0)\widehat{T}(z_1) \tag{41}$$

which leads to the definition of the Heisenberg group (see Refs. [23, 24]). This traditional construction gives the impression that some strange algebraic structure which is supposed to be the quintessence of QM abruptly emerges! We will see that this is not the case, by showing that formula (39) has a perfectly classical interpretation in terms of the notion of *phase*.

5.2. The phase of a Lagrangian manifold

A venerable topic in CM is the Hamilton–Jacobi equation

$$\frac{\partial\varphi}{\partial t} + H(x, \nabla_x\phi) = 0 \quad , \quad \varphi(x,0) = \varphi_0(x); \tag{42}$$

for instance, if H is of the classical type kinetic energy plus potential it takes the more familiar form (see Refs. [4, 12]):

$$\frac{\partial\varphi}{\partial t} + \sum_{j=1}^{n}\frac{1}{2m}(\nabla_x\phi)^2 + U(x) = 0 \quad , \quad \varphi(x,0) = \varphi_0(x).$$

The Hamilton–Jacobi equation allows, among other things, to solve explicitly Hamilton's equations for a wide class of systems. The solution (which we assume exists for t in some open interval surrounding the initial value $t = 0$), is constructed as follows. Let us denote by V_0 the graph of the equation $p = \nabla_x\varphi_0(x)$ (i.e. the set of all points $(x, \nabla_x\varphi(x))$ where x takes its values in the domain of φ). V_0 is a n-dimensional submanifold of phase space, in fact a Lagrangian manifold. This means that the symplectic product of any two tangent vectors Z, Z' at a same point z of V_0 is equal to zero. In [14] we have called the function φ_0 the *phase* of the manifold V_0; it is uniquely determined up to a constant. Lagrangian manifolds are the fundamental geometric objects of classical mechanics in its Hamiltonian formulation (see Refs [1, 4, 47, 48]; for instance, the invariant tori associated with completely integrable systems are Lagrangian manifolds (albeit not of the type V_0 above).

Consider now the time-dependent flow (f_t^H) determined by the Hamilton equations for H, and set $V_t = f_t^H V_0$. If $|t|$ is sufficiently small, the manifold V_t will again be a graph, that is, there exists a function $\varphi(x,t)$ (the phase of V_t) such that the equation $p = \nabla_x\varphi(x,t)$ represents V_t (if t is too large, bending of the original manifold V_0 may lead to the appearance of several points on V_t having same projection on x-space). The phase $\varphi = \varphi(x,t)$ (which is precisely the solution of the Hamilton–Jacobi equation (42)) is explicitly calculated as follows: since V_t is a graph, for each value x in the domain of φ there exits a unique p such that $p = \nabla_x\varphi(x,t)$, and the phase space point (x,p) is obtained from a unique point $(x(0), p(0))$ in V_0 by the flow (f_t^H). Denoting by Γ_t the arc of phase space trajectory joining $(x(0), p(0))$ in V_0 to (x,p) in V_t the phase φ of V_t satisfies

$$\varphi(x,t) = \varphi_0(x(0)) + \int_{\Gamma_t}\alpha_H \tag{43}$$

where

$$\alpha_H = pdx - Hdt \tag{44}$$

is the Poincaré–Cartan invariant (see Refs. [1, 4, 15]). We now remark that since both V_0 and V_t are graphs, the datum of $x(0)$ uniquely determines $x(t)$ and hence the formula $(x,p) = f_t^H(x(0), p(0))$ can be inverted, expressing $x(0)$ as a function of $x(0)$; we will write this as

$$x(0) = (f_t^H)_X^{-1}(x) \tag{45}$$

hence the solution $\varphi(x,t)$ is given by

$$\varphi(x,t) = \varphi_0((f_t^H)_X^{-1}(x)) + \int_{\Gamma_t}\alpha_H. \tag{46}$$

It is customary in theoretical statistical mechanics to identify a physical system with a pair (ρ, V) where V is a Lagrangian manifold and ρ a measure (probability density) carried by that manifold. At time $t = 0$ we thus have a pair (ρ_0, V_0) where the Lagrangian manifold V_0 is the graph of $p = \nabla_x \varphi_0(x)$. Since V_0 is uniquely determined by its phase φ_0 up to an inessential additive constant, we can without any loss of generality identify that system with the pair (ρ_0, φ_0). Under Hamiltonian evolution (ρ_0, φ_0) will become (ρ, φ) where ρ is obtained by "push-forward" of ρ_0, that is $\rho(z) = \rho_0((f_t^{z_0})^{-1}z)$ and $\varphi = \varphi(x, t)$ is given by formula (46). We now remark that since the variable $z = (x, p)$ in $\rho(z)$ is constrained by the condition $p = \nabla_x \varphi$ the measure ρ is actually only a function of x so we may write $\rho_0(z) = \rho_{0,X}(x)$ and

$$\rho(z, t) = \rho_X(x, t) = \rho_{0,X}\left[(f_t^H)_X^{-1} x\right]. \tag{47}$$

We next remark that the datum of the pair of real functions (ρ_X, φ) defined on $\mathbb{R}^n$ is mathematically equivalent to the datum of the complex function

$$\mu(x) = \rho_X(x) e^{i\phi(x)} \tag{48}$$

and the transformation formulas (45), (47) can then collectively be written in the compact form:

$$\mu(x, t) = e^{i\int_{\Gamma_t} \alpha_H} \mu_0\left[(f_t^H)_X^{-1}(x)\right] \tag{49}$$

5.3. Displacement operators

Let us now calculate (ρ, φ) in the particular case where H is the Hamiltonian

$$H^{z_0}(z) = \sigma(z, z_0) = p x_0 - p_0 x; \tag{50}$$

here $z_0 = (x_0, p_0)$ will play the role of a translation vector. The function H^{z_0} is called the "displacement Hamiltonian", because the solutions of the corresponding Hamilton equations $\dot{x} = x_0$, $\dot{p} = p_0$ are

$$x = x(t) = x(0) + t x_0 \ \ , \ \ p = p(t) = x(0) + t p_0 \tag{51}$$

so the flow $(f_{t,0}^{z_0})$ determined by H^{z_0} is just the phase space translation with vector $t z_0 = (t x_0, t p_0)$. Let us calculate $\varphi(x, t)$. The arc of trajectory Γ_t is here the line segment joining $(x(0), p(0))$ to $(x(0) + t x_0, p(0) + t p_0)$ and the Hamiltonian is $\sigma(z(0) + t z_0, z_0) = \sigma(z(0), z_0)$ along Γ_t so that

$$\alpha_H = \left[(p(0) + t p_0) x_0 - \sigma(z(0), z_0)\right] dt$$

A straightforward calculation then shows that we have

$$\int_{\Gamma_t} p dx - H^{z_0} dt = t p_0 x(0) + \tfrac{1}{2} t^2 p_0 x_0 \tag{52}$$

and hence, applying formula (46), the phase $\varphi(x, t)$ is given by

$$\varphi(x, t) = \varphi_0(x(0)) + t p_0 x(0) + \tfrac{1}{2} t^2 p_0 x_0$$

that is, using the relation $x = x(0) + t x_0$,

$$\varphi(x, t) = \varphi_0(x - t x_0) + t p_0 x - \tfrac{1}{2} t^2 p_0 x_0. \tag{53}$$

The measure $\rho_t(z) = \rho_0((f_t^{z_0})^{-1} z)$ is also easily calculated: since $f_t^{z_0}$ is the translation $z \longrightarrow z + t z_0$ we immediately get

$$\rho_t(z) = \rho_0(z - t z_0). \tag{54}$$

Let us now choose $t = 1$ and set $T(z_0) = f_1^{z_0}$: it is the phase space translation with vector $z_0 = (x_0, p_0)$. Under the action of $T(z_0)$ the complex function $\mu_0(x) = \rho_{0,X}(x) e^{i\phi_0(x)}$ becomes a new function

$$\widetilde{T}(z_0)\mu_0(x) = \mu_0(x - x_0) e^{i\int_{\Gamma_1} \alpha_H}$$

where $\Gamma_1 = \Gamma_{t=1}$. In view of Eqn. (53) we have

$$\int_{\Gamma_1} \alpha_H = p_0 x - \tfrac{1}{2} p_0 x_0 \tag{55}$$

and hence

$$\widetilde{T}(z_0)\mu_0(x) = \mu_0(x - x_0)e^{i(p_0 x - \frac{1}{2}p_0 x_0)} \tag{56}$$

and this is exactly formula (39) giving the action of the Heisenberg–Weyl operator with $\hbar = 1$ on the function $\psi = \mu_0$! From this we immediately deduce without any calculation at all that the operators $\widetilde{T}(z_0)$ satisfy the relations

$$\widetilde{T}(z_1)\widetilde{T}(z_0) = e^{-i\sigma(z_0,z_1)}\widetilde{T}(z_0)\widetilde{T}(z_1) \tag{57}$$

$$\widetilde{T}(z_0 + z_1) = e^{-\frac{i}{2}\sigma(z_0,z_1)}\widetilde{T}(z_0)\widetilde{T}(z_1). \tag{58}$$

Let us however check, for the sake of completeness, the relation (57) by a calculation of classical phases (the relation (58) will immediately follow). Let z_1 be a second fixed point in phase space; we set out to compare compare the states

$$\widetilde{T}(z_0 + z_1)(\rho_0, \varphi_0) \text{ and } \widetilde{T}(z_0)\widetilde{T}(z_1)(\rho_0, \varphi_0).$$

Both Lagrangian manifolds $\widetilde{T}(z_0 + z_1)V_0$ and $\widetilde{T}(z_0)\widetilde{T}(z_1)V_0$ are identical, being both obtained by a translation with vector $z_0 + z_1$, but these two states are not the same! First, replacing (x_0, p_0) in Eqn. (55) with $(x_0 + x_1, p_0 + p_1)$ the phase of $T(z_0 + z_1)V_0$ is

$$\varphi_{(z_0+z_1)}(x) = \varphi_0(x) + (p_0 + p_1)[x + \tfrac{1}{2}(x_0 + x_1)]. \tag{59}$$

The phase $\varphi_{(z_1,z_0)}(x)$ of $\widetilde{T}(z_1)\left[\widetilde{T}(z_0)V_0\right]$ is obtained by adding to the expression (55) of $\varphi_{(z_0)}(x)$ the integral of the Poincaré form $pdx - \sigma(z, z_1)dt$ along the line segment Γ_1' joining $(x+x_0, p+p_0)$ to $(x + x_0 + x_1, p + p_0 + p_1)$; this is

$$\int_{\Gamma_1'} pdx - \sigma(z, z_1)dt = p_1(x + x_0) + \tfrac{1}{2}p_1 x_1$$

and hence

$$\varphi_{(z_1,z_0)}(x) = \varphi_0(x) + p_0 x + p_1(x + x_0) + \tfrac{1}{2}(p_0 x_0 + p_1 x_1) \tag{60}$$

A straightforward calculation shows that the phases given by Eqns. (59) and (60) differ by the quantity

$$\varphi_{(z_1,z_0)}(x) - \varphi_{(z_0+z_1)}(x) = -\tfrac{1}{2}\sigma(z_0, z_1) \tag{61}$$

(it is, up to the sign, the area of the triangle with sides the vectors z_0, and z_1; see Fig. 3 in Littlejohn [29]); this immediately implies formula (57).

The important thing to remember is that everything in our construction is entirely classical, without any reference to the canonical commutation relations which are historically at the heart of the definition of the Heisenberg–Weyl operators and of the Heisenberg group.

6. Discussion and (Unconcluding) Remarks

Here are a few facts that could be used to illustrate the topics developed in this paper. It is usually assumed that the quantum wave-particle duality can have no counterpart in classical physics. It is also said that the probabilistic aspects of quantum dynamics are intrinsic and to have no possible relation with underlying unresolved dynamical phenomena. However, Yves Couder and his collaborators (Couder et al. [7]) have recently investigated the properties of a droplet bouncing on a vibrated bath. In this system, a propagative wave-particle association appears when the droplet couples to the surface wave it excites. Surprisingly, both a form of uncertainty and a form of quantization are observed. In these experiments you see textbook quantum mechanics, but everything is perfectly classical! It would be interesting to interpret Couder's results using our constructs. We mention that in a recent work [5] Bartlett et al. study the reconstruction of a version of QM from classical considerations; this work where the notion of *epistemic restriction* plays a key role is certainly closely related to our ideas. In [46] Gerard 't Hooft contends that "...Quantum mechanics is 'emergent' if a statistical treatment of large scale phenomena in a locally deterministic theory requires the use of quantum operators". The fact that Schrödinger's equation is (mathematically speaking) a classical equation contradicts in some sense this statement, since we are able to associate operators to classical observables, and these operators have a perfectly classical meaning unless one decides, a posteriori, to give their characteristics (e.g., their eigenvalues) a precise physical meaning.

To conclude, let us cite Stephen Adler [2]: "...quantum theory is not a complete, final theory, but is in fact an emergent phenomenon arising from a deeper level of dynamics..." (also see Bateson [6] who makes similar claims).

References

[1] Abraham R and Marsden J E 1978 Foundations of Mechanics. The Benjamin/Cummings Publishing Company, 2nd edition
[2] Adler S L 2004 Quantum Theory as an Emergent Phenomenon. The Statistical Mechanics of Matrix Models as the Precursor of Quantum Field Theory. Cambridge University Press
[3] Anderson P W 4 August 1972 More Is Different. Science, Vol. **177** No. 4047 393–396
[4] Arnold V I 1989 Mathematical Methods of Classical Mechanics. Graduate Texts in Mathematics, 2nd edition, Springer-Verlag
[5] Bartlett S D , Rudolph T and Spekkens R W Reconstruction of Gaussian quantum mechanics from Liouville mechanics with an epistemic restriction. arXiv:**1111.5057**
[6] Bateson R D 2010 Causal space-Time. Emergent Physics from Causality. CreateSpace ; extended version to be published by World Scientific.
[7] Couder Y, Boudaoud A, Protière S, Moukhtarb and Fort J E Walking droplets: a form of wave-particle duality at macroscopic level? , DOI: 10.1051/epn/2010101
[8] Dias N C, de Gosson M, Luef F and Prata J N 2010 A deformation quantization theory for noncommutative quantum mechanics, *J. Math. Phys.* **51** 072101
[9] Dirac P A M 1999 The Principles of Quantum Mechanics. *Oxford Science Publications*, 4th revised edition
[10] Fort E, Eddi A , Boudaoud A, Moukhtar J and Couder Y 2010 Path-memory induced quantization of classical orbits. *PNAS*, **107**(41), 17515–17520
[11] Giulini D 2003 That Strange Procedure Called Quantisation. *Lect. Notes Phys.* **361** 17–40
[12] Goldstein H 1959 Classical Mechanics. Addison–Wesley (1950), 2nd edition, (1980), 3d edition (2002)
[13] de Gosson M 2001 The Principles of Newtonian and Quantum Mechanics: The need for Planck's constant, h. With a foreword by Basil Hiley. Imperial College Press
[14] de Gosson M 2004 On the notion of phase in mechanics, *J. Phys. A*: Math. Gen., **37**(29), 7297–7314
[15] de Gosson M 2006 Symplectic Geometry and Quantum Mechanics. Birkhäuser, Basel, series "Operator Theory: Advances and Applications" (subseries: "Advances in Partial Differential Equations"), Vol. **166**
[16] de Gosson M 2009 The Symplectic Camel and the Uncertainty Principle: The Tip of an Iceberg? *Found. Phys.* **99** 194–214
[17] de Gosson M 2010 On the Use of Minimum Volume Ellipsoids and Symplectic Capacities for Studying Classical Uncertainties for Joint Position-Momentum Measurements. *J. Stat. Mech.* P11005
[18] de Gosson M 2011 Symplectic Methods in Harmonic Analysis and in Mathematical Physics. Birkhäuser, Basel

[19] de Gosson M and Hiley B J 2011 Imprints of the Quantum World in Classical Mechanics. *Found. Phys.* **41** 1415–1436
[20] de Gosson M and Luef F 2009 Symplectic Capacities and the Geometry of Uncertainty: the Irruption of Symplectic Topology in Classical and Quantum Mechanics. *Physics Reports* **484**, 131–179
[21] Groenewold H J 1946 On the principles of elementary quantum mechanics. *Physics* **12** 405–460
[22] Gromov M 1985 Pseudoholomorphic curves in symplectic manifolds. *Invent. Math.* **82** 307–47 (1985)
[23] Guillemin V and Sternberg S 1978 Geometric Asymptotics. *Math. Surveys Monographs* **14**, *Amer. Math. Soc.*, Providence R.I.
[24] Guillemin V and Sternberg S 1984 Symplectic Techniques in Physics. Cambridge University Press, Cambridge, Mass.
[25] Hall M.J.W and Reginatto M 2002 Schrödinger equation from an exact uncertainty principle. *J. Phys. A: Math.* Gen. 35 3289
[26] van Hove L 1951 Sur le Problème des Relations entre les Transformations Unitaires de la Mécanique Quantique et les Transformations Canoniques de la Mécanique Classique. *Mém. Acad. Roy. Belg.* 26:610
[27] van Hove L1952 Sur certaines représentations unitaires d'un groupe fini de transformations. *Mém. Acad. Roy. Belg.* Classe des Sci. **26**(6)
[28] John F 1948 Extremum problems with inequalities as subsidiary conditions, Studies and Essays Presented to R. Courant on his 60th Birthday, January 8, Interscience Publishers, Inc., New York, N.Y. 187–204
[29] Littlejohn R G 1986 The semiclassical evolution of wave packets. *Physics Reports* **138**(4–5), 193–291
[30] Lopuhäa H P and Rousseeuw P J 1991 Breakdown points of affine equivariant estimators of multivariate location and covariance matrices, *The Annals of Statistics* **19**(1) 229–248
[31] Mackey GW 1998 The Relationship Between Classical and Quantum Mechanics. In Contemporary Mathematics **214**, *Amer. Math. Soc.*, Providence, RI
[32] Nelson E 1966 Derivation of the Schrödinger Equation from Newtonian Mechanics. *Phys. Rev. A*, **150**(4):6, 1079–1085
[33] Peres A 1993 Quantum Theory: Concepts and Methods. Kluwer Academic Publishers
[34] Polterovich L 2001 The Geometry of the Group of Symplectic Diffeomorphisms. Lectures in Mathematics, Birkhäuser
[35] Reed M., Simon B 1972 Methods of Modern Mathematical Physics. Academic Press, New York
[36] Robertson H P 1929 The uncertainty principle, *Phys. Rev.* **34** 163–164
[37] Rousseeuw P J 1985 Multivariate estimation with high breakdown point. In Grossmann, W, Pflug, G, Vicenze, I & Wertz, (Eds.), Mathematical Statistics and Applications, Vol. B., Riedel Publishing, Dordrecht the Netherlands 283–297
[38] Rousseeuw P J and Leroy A M, 1987 Robust regression and outlier detection, John Wiley & Sons, New York
[39] Scheeres D J, Hsiao F Y, Park R S, Villac B F, and Maruskin J M 2006 Fundamental Limits on Spacecraft Orbit Uncertainty and Distribution Propagation, *Journal of the Astronautical Sciences* **54** 505–523
[40] Schrödinger E 1930 Zum Heisenbergschen Unschärfeprinzip, Berliner Berichte 296–303 [English translation: Angelow, A., Batoni, M.C.: About Heisenberg Uncertainty Relation. *Bulg. Journal of Physics*, **26**, nos.5/6, 193–203 (1999), and http://arxiv.org/abs/quant-ph/9903100]
[41] Schrödinger E 1926 Quantisierung als Eigenwertproblem, *Ann. der Physik*,**384**, 361–376
[42] Van Aelst S and Rousseeuw P J, 2009 Minimum volume ellipsoid, Wiley Interdisciplinary Reviews: *Computational Statistics* **1**(1) 71–82
[43] Stein EM 1993 Harmonic Analysis: Real Variable Methods, Orthogonality, and Oscillatory Integrals, Princeton University Press
[44] Shubin M A 1978 Pseudodifferential Operators and Spectral Theory, Springer-Verlag, (1987) [original Russian edition in Nauka, Moskva]
[45] Stone M H 1930 Linear transformations in Hilbert space, III: operational methods and group theory. *Proc. Nat. Acad. Sci.* U.S.A, 172–175
[46] 't Hooft G 2007 Emergent Quantum Mechanics and Emergent Symmetries. *AIP Conf. Proc.* **957**, 154-163 ; http://arxiv.org/abs/0707.4568
[47] Tulczyjew W 1976 *C.R. Acad. Sci.* Paris **283**, 15–18
[48] Tulczyjew W M 1977 The Legendre transformation, *Ann. Inst. H. Poincaré, Sect. A*, **27**(1), 101–114
[49] Williamson J 1936 On the algebraic problem concerning the normal forms of linear dynamical systems, *Amer. J. of Math.* **58** 141–163.
[50] Wong M W 1998 Weyl Transforms. Springer
[51] Woodhouse N M 1991 *Geometric Quantization,* second edition. Oxford Science Publications.

Quantumness beyond quantum mechanics

Ángel S. Sanz

Instituto de Física Fundamental (IFF–CSIC), Serrano 123, 28006 - Madrid, Spain

E-mail: asanz@iff.csic.es

Abstract. Bohmian mechanics allows us to understand quantum systems in the light of other quantum traits than the well-known ones (coherence, diffraction, interference, tunnelling, discreteness, entanglement, etc.). Here the discussion focusses precisely on two of these interesting aspects, which arise when quantum mechanics is thought within this theoretical framework: the *non-crossing property*, which allows for distinguishability without erasing interference patterns, and the possibility to define *quantum probability tubes*, along which the probability remains constant all the way. Furthermore, taking into account this hydrodynamic-like description as a link, it is also shown how this knowledge (concepts and ideas) can be straightforwardly transferred to other fields of physics (for example, the transmission of light along waveguides).

1. Introduction

At present, there is no doubt that quantum mechanics can be considered the most successful theory ever devised to explain the physical world. Its applicability ranges from very fundamental physical problems to the high-technology applications that are nowadays an important part of our daily life. This theory, though, still constitutes a veiled mystery at a deeper level of understanding, for there is a lack of a clear interpretation of the physics underlying quantum systems. This is somehow connected to its widely accepted interpretation, namely the Copenhagen interpretation [1], which not only does not allow us to think of quantum systems as we do of classical ones, but it just forbids such a thing.

A feasible way to surmount this drawback (although surely not the final one) comes through Bohmian mechanics [2–5]. According to this approach to the quantum world, systems are described in terms of trajectories evolving in configuration space. Rather than constituting a step backwards towards the classical paradigm, this quantum-mechanical formulation puts at the same descriptive level both classical and quantum systems, though the latter's behaviour differs from the former's one through the dynamical particularities induced by the quantum laws of motion. For instance, these laws give rise to motions constrained to the configuration space, uncertainty principles, interference phenomena, discreteness, and other features which are absent in classical systems. Of course, in the statistical limit, i.e., when an appropriate sampling over many of such Bohmian trajectories is carried out, the results of the standard quantum mechanics are recovered.

On the other hand, we find that in principle there are no restrictions to export the core idea in Bohmian mechanics, namely the possibility to reformulate quantum mechanics as a hydrodynamic-like theory, as well as the ideas emerging from it to other physical contexts where waves or distributions are the primary descriptor. This is somehow the other way around that

Madelung [6] considered in 1926, when he tried to provide a hydrodynamical interpretation of quantum processes based on the analogy between the latter and classical hydrodynamical flows. Within this approach, Bohmian trajectories were just considered to be the "tracks" displayed by some hypothetical *tracer particles* left on top of the quantum fluid [7], something very similar to what can be found in modern studies of flows, where the trajectories displayed by a number of tracer particles is recorded and subsequently analysed in order to extract information about the flow. By means of this method, for example, studies revealed anomalous diffusion and Lévy flights in two-dimensional rotating flows [8, 9]. However, closer in spirit to the present discussion, we find the recent observations by Couder and coworkers [10–14] as well as by Bush [15], where a seemingly classical analogue of Bohmian motion is found for droplets bouncing on a vertically vibrating bath of the same fluid.

The purpose of the present work is to pose an analysis and discussion of some interesting properties of quantum systems and their consequences, which arise when they are investigated from a Bohmian perspective. Furthermore, the link that can be established between the Bohmian viewpoint and its application or extension to other non-quantum wave theories will also be discussed. In order to make clear the arguments exposed, the work has been organized as follows. In Section 2, the essential elements/concepts of Bohmian mechanics are introduced as well as some important consequences implied by this theory. The transfer of knowledge from Bohmian mechanics to other classical wave theories, such as electromagnetism, is discussed in Section 3 in the context of light waveguiding. To conclude, a series of final remarks are highlighted in Section 4.

2. What can be learnt from Bohmian mechanics?

2.1. Some elementary background

In the standard Bohmian approach [2], one usually starts considering the wave function in polar form,

$$\Psi(\mathbf{r}, t) = R(\mathbf{r}, t) e^{iS(\mathbf{r},t)/\hbar}, \tag{1}$$

which is substituted into the (single-particle) non relativistic, time-dependent Schrödinger equation,

$$i\hbar \frac{\partial \Psi}{\partial t} = \left(-\frac{\hbar^2}{2m} \nabla^2 + V \right) \Psi, \tag{2}$$

to yield the system of real coupled equations

$$\frac{\partial R^2}{\partial t} + \nabla \cdot \left(R^2 \frac{\nabla S}{m} \right) = 0, \tag{3}$$

$$\frac{\partial S}{\partial t} + \frac{(\nabla S)^2}{2m} + V - \frac{\hbar^2}{2m} \frac{\nabla^2 R}{R} = 0. \tag{4}$$

In brief, the continuity (or conservation) equation (3) rules the ensemble dynamics, i.e., the number of particles described by the probability density has to remain constant; the quantum Hamilton-Jacobi equation (4), on the other hand, describes the time-evolution of the phase field, which governs the motion of quantum particles through the guidance equation

$$\mathbf{v}_B = \dot{\mathbf{r}} = \frac{\nabla S}{m}. \tag{5}$$

In this way, the simple transformation described by (1) allows us to go a step beyond standard quantum mechanics. In the latter, Ψ is all what is needed to describe and interpret the experimental outcomes; in Bohmian mechanics, on the other hand, the evolution of a *tracer particle* [7] according to the information supplied by Ψ (the guiding field) is also considered to further understand the quantum system dynamics.

Now, the same description can be formally inferred by simply assuming that the evolution of any quantum system can be identified with the diffusion of a (quantum) fluid throughout configuration space, as formerly proposed by Madelung [6]. Therefore, as in the case of classical fluids, the elementary quantities of this approach would be the (probability) density of the quantum fluid and its associated probability current density, $\mathbf{J}$, both related through the conservation (or continuity) equation

$$\frac{\partial \rho}{\partial t} + \nabla \cdot \mathbf{J} = 0, \tag{6}$$

where $\mathbf{J} = \rho \mathbf{v}_H$ and $\mathbf{v}_H$ is a local hydrodynamic velocity field. The probability density and the probability current density are related to the wave function (1) by the transformation relations

$$\rho = R^2, \qquad \mathbf{J} = R^2 \, \frac{\nabla S}{m}, \tag{7}$$

with the local velocity field being

$$\mathbf{v}_H = \frac{\mathbf{J}}{\rho} = \frac{\nabla S}{m}, \tag{8}$$

which is equivalent to the Bohmian one, $\mathbf{v}_B$, given by (5) (hence, from now on, both velocity fields will be denoted by $\mathbf{v}$). The integration in time of (8) generates a family of streamlines or paths (for each wave function considered) along which the quantum fluid propagates, just as in the case of a classical fluid. As it can be inferred, the first equality goes beyond Bohmian mechanics and allows to define streamlines in any system characterized by a certain density and a vector that transports it through the corresponding configuration space, regardless whether such a density describes a quantum system or not.

2.2. The non-crossing rule

One of the most relevant properties revealed by Bohmian mechanics is that quantum fluxes cannot cross in configuration space. That is, at a certain time t, each point on configuration space has a uniquely-defined value of the local velocity field, except for regions where the wave function displays a node (i.e., $\Psi = 0$). Therefore, two or more Bohmian trajectories cannot cross through such a point at the same time t or, equivalently, *quantum systems cannot reach the same configuration at a given time following different quantum streamlines.* In terms of the standard quantum mechanics, this just means that paths can be labelled, although they cannot be experimentally *observed* because this implies changing the process and therefore generating new, alternative paths. Obviously, this changes dramatically our physical view of the superposition principle, which states that any linear combination of solutions of Schrödinger's equation is also a formally valid solution to the same equation. Accordingly, given a coherent superposition of two counter-propagating wave packets, they can be evolved in time separately and then recombined at any time to determine when and where an interference pattern. Formally, there is no problem in proceeding in this way. However, from a (hydro) dynamical viewpoint the evolution of both wave packets has to be accounted for together for the interference pattern to be observable (which is what actually happens in an experiment), this implying that interpretations of physical phenomena cannot directly rely upon the superposition principle. Bohmian mechanics provides us an answer for this based precisely on the non-crossing rule.

In order to make clear such a statement, consider the wave packets above are represented by two free expanding Gaussian wave packets,

$$\Psi(x,t) = \left(\frac{1}{2\pi\tilde{\sigma}_t^2}\right)^{1/4} e^{-(x-x_{\rm cl})^2/4\tilde{\sigma}_t\sigma_0 + ip_0(x-x_{\rm cl})/\hbar + iEt/\hbar}, \tag{9}$$

where the time-dependence of their spreading is given by

$$\sigma_t = |\tilde{\sigma}_t| = \sigma_0 \sqrt{1 + \left(\frac{\hbar t}{2m\sigma_0^2}\right)^2}. \tag{10}$$

The wave packet (9) propagates along the classical trajectory $x_{\rm cl} = x_0 + v_0 t$ (with $v_0 = p_0/m$), while the associated Bohmian trajectories are given by

$$x(t) = x_{\rm cl} + \frac{\sigma_t}{\sigma_0}\left[x(0) - x_0\right], \tag{11}$$

with $x(0)$ denoting the initial position of the corresponding Bohmian trajectory. As can be noticed, there are two contributions to the evolution of the tracer particles, one coming from the classical drift and another one from a purely quantum origin [16, 17]. The former makes the particle to evolve according to only the laws of classical mechanics; the latter makes it to deviate in such a way that its motion accommodates to the evolution of the quantum fluid. The time-scale that separates both types of motion (to some extent) is given in this particular case by the characteristic time $\tau \equiv 2m\sigma_0^2/\hbar$ [16, 17]. Depending on the relative value of t when compared with τ, three regimes can then be found (in analogy with the regimes of optical physics [7]):

- *Ehrenfest or Huygens* ($\tau \equiv 2m\sigma_0^2/\hbar \ggg t$): Bohmian trajectories follow classical ones (in agreement with Ehrenfest's theorem).
- *Near-field or Fresnel* ($\tau \equiv 2m\sigma_0^2/\hbar > t$): Bohmian trajectories start to diverge from their classical counterparts.
- *Far-field or Fraunhofer* ($\tau \equiv 2m\sigma_0^2/\hbar \ll t$): Bohmian trajectories display a sort of rectilinear, uniform evolution due to the stationarity of the wave function far from any interaction potential.

Therefore, by inspecting the topology displayed by the quantum trajectories, one readily notices that the quantum flow evolves from an initially confined fluid to a linearly expanding one, undergoing at times of the order of τ a sort of internal boosting which bursts it open. However, this does not provide any clue on the true dynamics.

In order to understand the importance of not disentangling the joint evolution of both wave packets, as indicated by the superposition principle, consider the full wave function,

$$\Psi(\mathbf{r}, t) = \psi_1(\mathbf{r}, t) + \psi_2(\mathbf{r}, t), \tag{12}$$

where the $\psi_i = \rho_i^{1/2} e^{iS_i/\hbar}$ are properly normalized and have the shape indicated by (9), with $x_{2,0} = -x_{1,0} = x_0 > 0$ and $v_{1,0} = -v_{2,0} = v_0 > 0$. It is straightforward to notice that

$$\rho = \rho_1 + \rho_2 + 2\sqrt{\rho_1\rho_2}\cos\varphi, \tag{13}$$

$$\mathbf{J} = \frac{1}{m}\left[\rho_1 \nabla S_1 + \rho_2 \nabla S_2 + \sqrt{\rho_1\rho_2}\nabla(S_1 + S_2)\cos\varphi + \hbar\left(\rho_1^{1/2}\nabla\rho_2^{1/2} - \rho_2^{1/2}\nabla\rho_1^{1/2}\right)\sin\varphi\right], \tag{14}$$

with $\varphi = (S_2 - S_1)/\hbar$. Substituting these expressions into (5) leads to the equation of motion

$$\mathbf{r} = \frac{1}{m}\frac{\rho_1 \nabla S_1 + \rho_2 \nabla S_2 + \sqrt{\rho_1\rho_2}\nabla(S_1 + S_2)\cos\varphi}{\rho_1 + \rho_2 + 2\sqrt{\rho_1\rho_2}\sin\varphi} + \frac{\hbar}{m}\frac{\left(\rho_1^{1/2}\nabla\rho_2^{1/2} - \rho_2^{1/2}\nabla\rho_1^{1/2}\right)\sin\varphi}{\rho_1 + \rho_2 + 2\sqrt{\rho_1\rho_2}\cos\varphi}. \tag{15}$$

The crossed term in (13) and (14) arises from the interference of the probability densities and current densities associated with ψ_1 and ψ_2 after assuming the coherent superposition (12). However, as seen from (15), the local velocity field keeps a more complicated form, which cannot

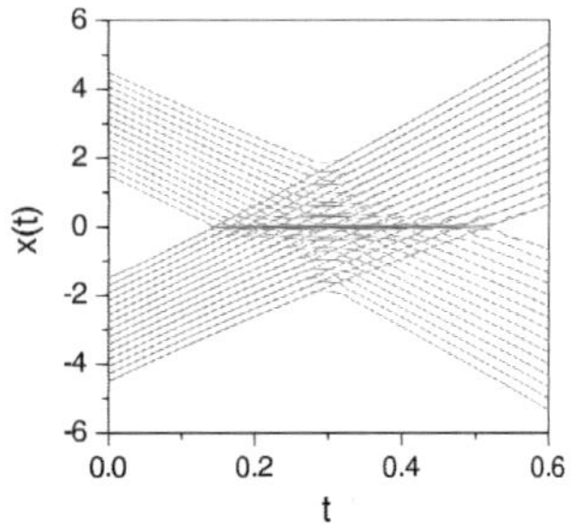

Figure 1. Bohmian trajectories associated with a coherent superposition of two Gaussian wave packets (gray lines) and with a single Gaussian wave packet (red lines).

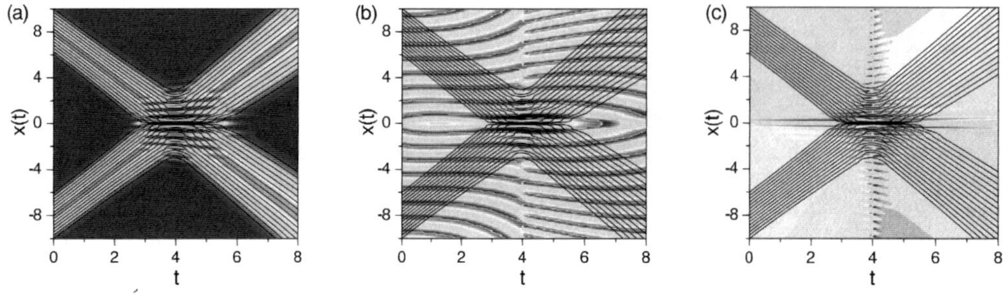

Figure 2. Contour-plots representing the evolution in time of the probability density (a), the phase field (b) and the local velocity field (c) of an initial coherent superposition of two Gaussian wave packets. The Bohmian trajectories (black lines) associated with this problem are also superimposed to each graph.

be expressed as a simple superposition, thus indicating that the dynamics is going to be far more complex. This is illustrated in Fig. 1, where the single wave-packet Bohmian trajectories (red) diverge from the behaviour displayed by the two wave-packet ones as they approach interference region, ending up eventually in the wrong place. This can be readily understood by looking at the three graphs displayed in Fig. 2: although the probability density (part (a)) does not provide any clue, both the phase field (part (b)) and the local velocity field (part (c)) immediately make apparent the dynamical consequences of the superposition.

It is worth stressing that the non-crossing takes place regardless of the presence of the quantum potential. Rather it is an intrinsic or inherent property associated with the dynamics generated through a guidance condition. In classical mechanics we can also observe something similar, but in phase space, where phase-space orbits cannot cross through the same point at the same time. In such a case, the "guiding field" is the classical action (this was actually the idea that led Schrödinger in the derivation of his wave theory of quanta [18]). Nevertheless, in the two wave-packet problem presented above, the non-crossing can still be well represented by the action of an effective potential acting on a single wave packet, with the form of a repulsive wall and an attractive region, both depending on time [17].

2.3. Quantum probability tubes

The possibility to study quantum processes in terms of Bohmian trajectories brings in another interesting property, namely the possibility to define *quantum probability tubes* [19], i.e., tubes in configuration space along which the integral of the probability density at a time t remains constant regardless of changes in their profile or in the shape of the enclosed sections of ρ. In order to show this, consider the so-called *restricted probability*, defined as

$$\mathcal{P}(t) \equiv \int_{\Omega} \rho(\mathbf{r}, t) d\mathbf{r}, \tag{16}$$

which gives the probability to find the system inside a certain region of interest Ω of the corresponding configuration space at a time t [20–22]. From standard quantum mechanics (and making use of the divergence or Gauss-Ostrogradsky theorem), the variation with time of $\mathcal{P}(t)$ can be expressed as

$$\frac{d\mathcal{P}(t)}{dt} = \int_{\Omega} \frac{\partial \rho}{\partial t}\, d\mathbf{r} = -\int_{\Omega} (\nabla \cdot \mathbf{J})\, d\mathbf{r} = -\int_{\Sigma} \mathbf{J} \cdot d\mathbf{S}, \tag{17}$$

where Σ is the boundary of Ω. That is, the losses/gains of $\mathcal{P}(t)$ inside Ω are described by the outgoing/ingoing probability flow through Σ, which is accounted for by the quantum current density $\mathbf{J}$. Defining an arbitrary region Ω in standard quantum mechanics is rather simple. However, how can this region be propagated in time so that $\mathcal{P}(t)$ keeps the same value along time?

Although the answer to the question posed above can be quickly answered in classical mechanics by using arguments based on the Liouvillian structure of this theory [23], the same cannot be found in standard quantum mechanics, but in Bohmian mechanics and, more specifically, in its non-crossing property. As shown elsewhere [19], if the surface Σ is the geometric place formed by a (discrete or continuous) set of initial conditions, because of the non-crossing property, any probability enclosed inside such a set will remain constant at any subsequent time. This is based on the fact that at any given time t no trajectory inside or outside the boundary defined by the set of *separatrix* trajectories will be able to penetrate into or exit from Ω respectively. Therefore, if the probability $P(t)$ is measured in terms of the accumulation of trajectories embraced by Σ, i.e.,

$$\mathcal{P}(t) \approx \sum_{i=1}^{N} \delta(\mathbf{r}_i(t) \in \Sigma(t)), \tag{18}$$

this number must remain constant at any time provided $\Sigma(t)$ is a causal map of $\Sigma(0)$ at time t. An important consequence of this result is that *any restricted probability can be determined directly from the initial state if the end points of the associated separatrix trajectories are known.* This means that, in principle, one could determine (or, at least, estimate) final probabilities without carrying out the full calculation [24], but directly from the particular region covered by the initial wave function causally connected with the feature of interest (e.g., a transmitted amplitude or a diffraction peak), since

$$\mathcal{P}_{\infty} = \int_{\Sigma_{\infty}} \rho(\mathbf{r}_{\infty}) d\mathbf{r}_{\infty} = \lim_{t\to\infty} \int_{\mathbf{r}(t)} \rho[\mathbf{r}(t)] d\mathbf{r}(t) = \lim_{t\to\infty} \mathcal{P}(t) = \int_{\mathbf{r}_0} \rho[\mathbf{r}_0] d\mathbf{r}_0 = \mathcal{P}_0. \tag{19}$$

Here, $\rho(\mathbf{r}_{\infty})d\mathbf{r}_{\infty}$ is the probability to find the system confined within the layer of configuration space defined by the separatrix trajectories ending at $\mathbf{r}_{\infty}$ and those ending at $\mathbf{r}'_{\infty} = [\mathbf{r} + d\mathbf{r}]_{\infty}$. Analogously, $\rho[\mathbf{r}(t)]d\mathbf{r}(t)$ is the probability to find the system confined within the layer defined by the sets of separatrix trajectories $\mathbf{r}(t)$ and $\mathbf{r}'(t) = [\mathbf{r} + d\mathbf{r}](t)$ at a time t. Note that, since the probability density is being evaluated along quantum trajectories, it is preferably denoted as $\rho[\mathbf{r}(t)]$ instead of $\rho(\mathbf{r}, t)$, with $\mathbf{r}_{\infty} = \mathbf{r}(t \to \infty)$.

In order to illustrate this result, consider the scattering of a wave packet with a rather arbitrary shape by a barrier, as displayed in Fig. 3(a) (details on this calculation can be found in [19]). The initial and final probability densities are denoted by the black and red solid lines, respectively, while the barrier is indicated by the blue shadowed region. The propagation goes on until the probability in the region covered by the barrier (intra-barrier resonance) becomes negligible, which can be better seen in Fig. 3(b), where the transmission (green dashed line), reflection (blue dash-dotted line) and intra-barrier resonance (red dotted line) probabilities are

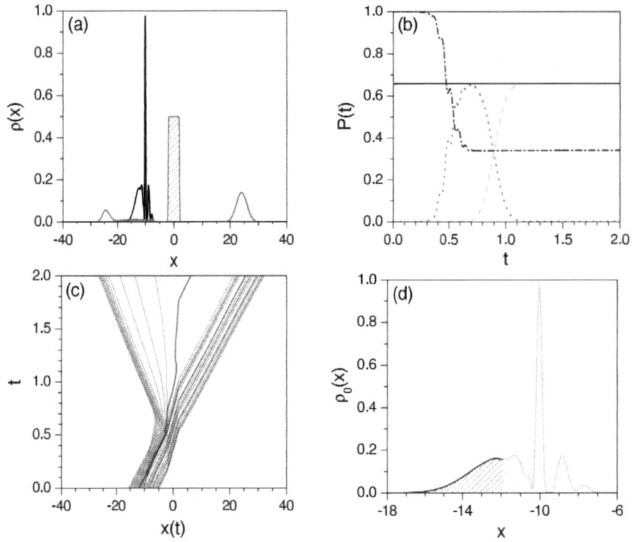

Figure 3. Scattering of a general wave packet by an almost square potential barrier. (a) Initial (black line) and final (red line) probability densities; the barrier is denoted with the blue shadowed region. (b) Time-dependence of different restricted probabilities: transmission (green dashed line), reflection (blue dash-dotted line) and intra-barrier resonance (red dotted line) probabilities. The black solid line denotes the transmission probability directly computed from the initial state (see text for details). (c) Bohmian trajectories illustrating the tunnelling dynamics. The separatrix trajectory is displayed with the red thicker line. (d) Initial probability density split according to the separatrix initial condition: initial conditions in the green shadowed region contribute to transmission and those from the blue shadowed one to reflection.

recorded along time. These are the three restricted probabilities considered in this example, with $\Omega_{\rm trans}$ being the region beyond the right edge of the barrier, $\Omega_{\rm res}$ the region bound between the two edges of the barrier, and $\Omega_{\rm refl}$ the region to the left edge of the barrier. After $t \approx 1.15$, $\mathcal{P}_{\rm res} \approx 0$ and $\mathcal{P}_{\rm trans}$ reaches its maximum, asymptotic value; $\mathcal{P}_{\rm refl}$, on the contrary, reaches its minimum earlier, at $t \approx 0.75$. This means than $\mathcal{P}_{\infty,\rm trans}$ can be computed by simply choosing as the separatrix trajectory a Bohmian trajectory that reaches the left tail of the transmitted wave packet in part (a) and propagating it backwards in time, as seen in part (c). Integrating the probability density between the position of this trajectory at each time and infinity (for practical purposes), it is then found that $\mathcal{P}_{\infty,\rm trans}$ remains constant all the way, as the black straight line shows in part (b). Regarding the initial set (probability density), it is found (see part (d)) that any initial condition started to the right of the separatrix initial condition will get into the transmitted region, while if it starts to its left, the trajectory becomes reflected.

3. Where can this knowledge be transferred to?

It is very common to separate quantum physics from other wave theories based on the difference in the treatment of probability amplitudes with respect to other types of waves (which are considered as perturbations propagating through a given medium or vacuum). However, the way how these wave theories operate is in essence the same, for both include the same type of concepts (e.g., coherence, interference, diffraction, tunnelling, discreteness, etc.) and principles (e.g., uncertainty, superposition). Taking this into account, it is possible to establish a sort of feedback regarding the understanding of both wave phenomena, in general, and quantum phenomena, in particular, through the hydrodynamic (Bohmian) approach considered above.

To briefly show how this transfer of knowledge can be done, consider the case of electromagnetic fields, for example. Bohmian mechanics was formerly formulated to describe massive particles, resulting very appealing to explain event-to-event experiments like those carried by Merli *et al* [25], Tonomura *et al.* [26], or Shimizu *et al* [27]. Now, there are also event-to-event experiments with light, such as those performed by Dimitrova and Weis [28], which claim for a treatment on equal footing. As shown by Prosser [29], this is actually possible by directly considering Maxwell's equations (which put electromagnetism at the same level of Schrödinger's wave mechanics [30]). In this case, given an electromagnetic field[1] defined by an electric field $\mathbf{E}(\mathbf{r})$ and a magnetic field $\mathbf{H}(\mathbf{r})$, the two key elements necessary to define the corresponding streamlines are the electromagnetic energy density and the Poynting vector (electromagnetic current density), i.e.,

$$U(\mathbf{r}) = \frac{1}{4}\left[\epsilon_0 \mathbf{E}(\mathbf{r})\cdot\mathbf{E}^*(\mathbf{r}) + \mu_0 \mathbf{H}(\mathbf{r})\cdot\mathbf{H}^*(\mathbf{r})\right]. \tag{20}$$

$$\mathbf{S}(\mathbf{r}) = \frac{1}{2}\,\mathrm{Re}\left[\mathbf{E}(\mathbf{r})\times\mathbf{H}^*(\mathbf{r})\right], \tag{21}$$

respectively. Since the electromagnetic energy density is transported through space in the form of the Poynting vector, a local velocity field can be defined [31] in analogy to (8), which reads as

$$\mathbf{S}(\mathbf{r}) = U(\mathbf{r})\mathbf{v}, \tag{22}$$

from which electromagnetic energy flow lines or photon paths are obtained by integrating the equation

$$\frac{d\mathbf{r}}{ds} = \frac{1}{c}\frac{\mathbf{S}(\mathbf{r})}{U(\mathbf{r})}, \tag{23}$$

along the arc-length coordinate s (which can be referred to a proper time $\tau = s/c$, with c being the speed of light). By means of this procedure, the interference of diffracted polarized beams was studied [32, 33], showing a good agreement with experiments carried out later on to infer the average paths displayed by photons [34].

The previous discussion focussed on Maxwell's equations. However, it is not necessary to stay at such a level in order to define hydrodynamic streamlines. As mentioned above, this can be done at any level where the elementary descriptive tool is a wave equation. Thus, consider, for example, the transmission of a light pulse through a waveguide within the small-angle or paraxial approximation [31]. Assuming that the optical axis is oriented along the z-axis and the electromagnetic field passing through the waveguide is time-harmonic, the field can be approximated by a plane wave along the z-direction modulated by a certain complex-valued amplitude,

$$\Phi(\mathbf{r}) = \phi(\mathbf{r})e^{ik_z z}, \tag{24}$$

where $k_z = n_0 k$, $k = 2\pi/\lambda$, λ is the light wavelength in vacuum and n_0 is the bulk refractive index. Substituting this expression into Helmholtz's equation,

$$\nabla^2\Phi + n^2k^2\Phi = 0, \tag{25}$$

with n being the refractive index inside the waveguide, one readily finds

$$2ik_z\frac{\partial\phi}{\partial z} + \frac{\partial^2\phi}{\partial z^2} = -\nabla_\perp^2\phi + (k_z^2 - n^2k^2)\phi, \tag{26}$$

[1] Time-independent electromagnetic fields will be considered here for simplicity, although time-dependent ones should be used, in general.

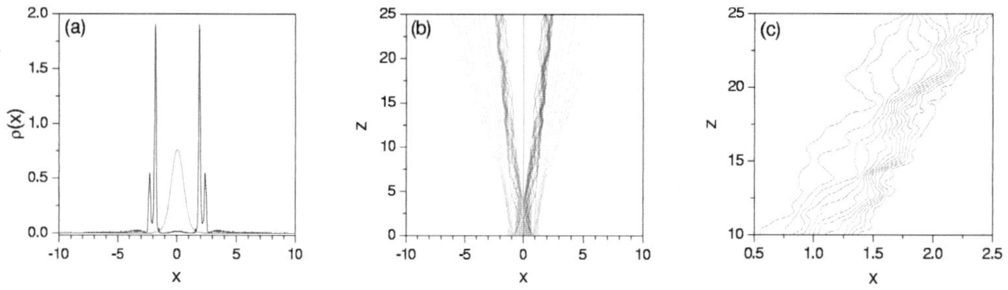

Figure 4. Light transport through a waveguide with the shape of a Y-junction. (a) Initial (black) and final (red) probability densities. (b) Optical streamlines illustrating the dynamics inside the waveguide. (c) Enlargement of part (b).

where $\nabla_\perp^2$ is the transverse part of the Laplacian (for example, in Cartesian coordinates, it reads as $\nabla_\perp^2 = \partial^2/\partial x^2 + \partial^2/\partial y^2$). If, apart from the paraxial approximation, it is also assumed that the slowly varying envelope approximation holds (i.e., the amplitude ϕ varies slowly in space compared to $2\pi/k_z$), the highest-order derivative in z can be neglected (i.e., $\partial^2\phi/\partial z^2 \approx 0$) and Eq. (26) can be recast as

$$i\frac{\partial\phi}{\partial z} = -\frac{\nabla_\perp^2\phi}{2k_z} + \frac{k}{2n_0}\left(n_0^2 - n^2\right)\phi. \tag{27}$$

This equation is, indeed, isomorphic to Schrödinger's one, with z playing the role of the evolution parameter (rather than the time t) and $k_z = n_0 k$ the role of an "optical mass". Equation (27) has been used, for example, to study the design of waveguides with optimal conditions of light transmission [35–37]. However, regarding our discussion here, it also brings the possibility to define a family of *optical streamlines* if the pulse $\phi(\mathbf{r})$ is expressed in polar from, as $\phi(\mathbf{R}, z) = \rho^{1/2}(\mathbf{R}, z)e^{iS(\mathbf{R},z)}$, where $\mathbf{R}$ denotes the vector perpendicular to the propagation direction (z). In this case, the local velocity field reads as

$$\mathbf{v} = \frac{d\mathbf{R}}{dz} = \frac{\nabla_\perp S}{k_z}, \tag{28}$$

which gives rise to such streamlines when integration over z is carried out.

In Fig. 4 we have an example of how these trajectories look like in the particular case of a Y-junction [38], which splits the initial pulse (red line) into two emerging pulses (black line), as seen in part (a). As it can be noticed in part (b), also here the electromagnetic flux follows the non-crossing rule, which allows to distinguish between different paths (and therefore to also define density tubes) without disturbing the experiment, also in agreement with the experiment performed by Kocsis *et al* [34]. Actually, in the enlargement of part (c), a behavior mimicking that of the frustrated internal reflection (inside the waveguide) can be observed, although contrary to geometric rays, the optical streamlines follow the wavy behaviour dictated by the pulse inside the wave, thus being affected by diffractions and interferences.

4. Final remarks

We are used to consider that quantum mechanics is about small objects. However, this is just a misunderstanding, for quantum mechanical behaviours arise whenever objects are simple enough, i.e., whenever the action of any internal or external surrounding degrees of freedom does not play any relevant role in their dynamics. As an example, just consider the interference patterns that can be obtained from large organic molecules [39] or Bose-Einstein condensates [40, 41]. Such a misunderstanding comes from technological lacks in the past, when quantum mechanics was developing, as well as *black-box* interpretations, like the Copenhagen one. The possibility to carry out interference diffraction of large objects, particle by particle, requires new viewpoints and interpretations that may go beyond the blind ones available at present.

In this sense, the role of Bohmian mechanics as a theory of quantum motion that describes the evolution of tracer particles when travelling along quantum fluids (i.e., the fluid associated with the diffusion of the wave function in a configuration space) is highly needed. Even if at a statistical level predictions are the same as those already provided by standard quantum mechanics, the fact that hydrodynamic-like analyses can be carried out, elucidating properties such as the flux non-crossing, results as very appealing, for it allows us to understand the (quantum) physical world in a different manner, without conceptual restrictions.

Furthermore, such approaches should also be extended to other fields of physics, for they could help us to get different insights of already well-known and established processes and phenomena. This is the case, as it has been seen, of the so-called classical electromagnetism. In this case, not only theoretical descriptions have been developed to better understand the evolution of the electromagnetic energy through space, but also experiments have been motivated that have corroborated the models proposed [34]. Of course, there is also no doubt about the important role these approaches present from a pedagogical viewpoint [42] by allowing us to connect the distribution of energy (probability) through space with its diffusion (current densities).

Finally, I would like to close this work with a simple, but striking reflection. Usually, one is always tempted to think that only those physical problems that are directly connected with a particular practical application are of interest and, therefore, worth spending time and efforts to investigate them. Of course, this is a legitimate point of view regarding where efforts and resources should be targeted to. However, this strategy cannot be used to get deeper into the physical theories and approaches behind such problems in order to get a better and deeper understanding of them. This is precisely what happens with quantum mechanics and, by extension, with wave theories. Nowadays nobody questions their use, but there are less and less researchers facing the problem of the understanding of the quantum world and the paradoxes it brings in to our intuition, which in spite of all is *shaped* by a classical world.

Acknowledgments

The author would like to thank Gerhard Grössing for his kind invitation to participate at the Conference "Emergent Quantum Mechanics", within the 5th Heinz von Foerster Congress, held in Vienna, as well as for his hospitality (and that of Siegfried Fussy, Johannes Mesa Pascasio and Herbert Schwabl) and interesting discussions during the event. Moreover, he also wishes to acknowledge Profs. Basil Hiley and Helmut Rauch for stimulating discussions and interesting questions. This work has been supported by the Ministerio de Economía y Competitividad (Spain) under Projects FIS2010-18132, FIS2010-22082 and FIS2011-29596-C02-01, as well as by the COST Action MP1006 (*Fundamental Problems in Quantum Physics*). The author is also grateful to the Ministerio de Economía y Competitividad for a "Ramón y Cajal" Research Fellowship.

References

[1] Greenberger D, Hentschel K and Weinert F (Eds) 2009 *Compendium of Quantum Physics: Concepts, Experiments, History and Philosophy* (Berlin: Springer)
[2] Bohm D 1952 *Phys. Rev.* **85** 166
[3] Bohm D 1952 *Phys. Rev.* **85** 180
[4] Holland P R 1993 *The Quantum Theory of Motion* (Cambridge: Cambridge University Press)
[5] Bohm D and Hiley B J 1993 *The Undivided Universe* (New York: Routledge)
[6] Madelung E 1926 *Z. Phys.* **40** 322
[7] Sanz A S and Miret-Artés S 2012 *Am. J. Phys.* (at press);

2011 Quantum phase analysis with quantum trajectories: A step towards the creation of a Bohmian thinking *Preprint* arXiv:1104.1296
[8] Solomon T H, Weeks E R and Swinney H L 1993 *Phys. Rev. Lett.* **71** 3975
[9] Klafter J, Shlesinger M F and Zumofen G 1996 *Phys. Today* **49** 33
[10] Couder Y, Protière S, Fort E and Boudaoud A 2005 *Nature* **437** 208
[11] Protière S, Boudaoud A and Couder Y 2006 *J. Fluid. Mech.* **554** 85
[12] Couder Y and Fort E 2006 *Phys. Rev. Lett.* **97** 154101
[13] Eddi A, Fort E, Moisy F and Couder Y 2009 *Phys. Rev. Lett.* **102** 240401
[14] Fort E, Eddi A, Boudaoud A, Moukhtar J and Couder Y 2010 *Proc. Natl. Acad. Sci. USA* **108** 17515
[15] Bush J W M 2010 *Proc. Natl. Acad. Sci. USA* **107** 17455
[16] Sanz A S and Miret-Artés S 2007 *Chem. Phys. Lett.* **445** 350
[17] Sanz A S and Miret-Artés S 2008 *J. Phys. A: Math. Theor.* **41** 435303
[18] Sanz A S and Miret-Artés S 2012 *A Trajectory Description of Quantum Processes. I. Fundamentals* (Berlin: Springer)
[19] Sanz A S and Miret-Artés S 2011 Determining final probabilities directly from the initial state *Preprint* arXiv:1104.1296
[20] Sanz A S, Giménez X, Bofill J M and Miret-Artés S 2009 *Chem. Phys. Lett.* **478** 89
[21] Sanz A S, Giménez X, Bofill J M and Miret-Artés S 2010 *Chem. Phys. Lett.* **488** 235
[22] Sanz A S, López-Durán D and González-Lezana T 2012 *Chem. Phys.* doi:10.1016/j.chemphys.2011.07.017
[23] Gutzwiller M C 1990 *Chaos in Classical and Quantum Mechanics* (New York: Springer-Verlag)
[24] Sanz A S and Miret-Artés S 2011 *J. Phys. A: Math. Theor.* **44** 485301
[25] Merli P G, Missiroli G F and Pozzi G 1976 *Am. J. Phys.* **44** 306
[26] Tonomura A, Endo J, Matsuda T, Kawasaki T and Ezawa 1989 *Am. J. Phys.* **57** 117
[27] Shimizu F, Shimizu K and Takuma H 1992 *Phys. Rev. A* **46** R17
[28] Dimitrova T L and Weis A 2008 *Am. J. Phys.* **76** 137
[29] Prosser R D 1976 *Int. J. Theor. Phys.* **15** 169
[30] Scully M O and Zubairy S 1997 *Quantum Optics* (Cambridge: Cambridge University Press)
[31] Born M and Wolf E 1999 *Principles of Optics. Electromagnetic Theory of Propagation, Interference and Diffraction of Light* 7th ed (Cambridge: Cambridge University Press)
[32] Sanz A S, Davidović M, Božić M and Miret-Artés S 2010 *Ann. Phys.* **325** 763
[33] Božić M, Davidović M, Dimitrova T L, Miret-Artés S, Sanz A S and Weis A 2010 *J. Russ. Laser Res.* **31** 117
[34] Kocsis S, Braverman B, Ravets S, Stevens M J, Mirin R P, Shalm L K and Steinberg A M 2011 *Science* **332** 1170
[35] Pant D K, Coalson R D, Hernández M I and Campos-Martínez J 1998 *J. Light. Tech.* **16** 292
[36] Pant D K, Coalson R D, Hernández M I and Campos-Martínez J 1999 *Appl. Opt.* **38** 3917
[37] Campos-Martínez J and Coalson R D 2003 *Appl. Opt.* **42** 1732
[38] Sanz A S, Campos-Martínez J and Miret-Artés S 2012 *J. Opt. Am. Soc. A* (at press);
2011 Transmission properties in waveguides: An optical streamline analysis *Preprint* arXiv:1108.4964
[39] Gerlich S, Eibenberger S, Tomandl M, Nimmrichter S, Hornberger K, Fagan P J, Tüxen J, Mayor M and Arndt M 2011 *Nat. Commun.* **2** 263
[40] Andrews M R, Townsend C G, Miesner H J, Durfee D S, Kurn D M and Ketterle W 1997 *Science* **275** 637
[41] Schumm T, Hofferberth S, Andersson L M, Wildermuth S, Groth S, Bar-Joseph I, Schmiedmayer J and Krüger P 2005 *Nature Phys.* **1** 57
[42] Wünscher T, Hauptmann H and Hermann F 2002 *Am. J. Phys.* **70** 599

Non-equilibrium in Stochastic Mechanics

Guido Bacciagaluppi[1]
Department of Philosophy, University of Aberdeen
Institut d'Histoire et de Philosophie des Sciences et des Techniques (CNRS, Paris 1, ENS)
E-mail: g.bacciagaluppi@abdn.ac.uk

Abstract. The notion of non-equilibrium, in the sense of a particle distribution other than $\rho = |\psi|^2$, is imported into Nelson's stochastic mechanics, and described in terms of effective wavefunctions obeying non-linear equations. These techniques are applied to the discussion of non-locality in non-linear Schrödinger equations.

1. Introduction

The ideal of quantum mechanics as an emergent theory is well represented by Nelson's stochastic mechanics, which aims at recovering quantum mechanics from an underlying stochastic process in configuration space. More precisely, as we sketch in Section 2, Nelson starts from a time-reversible description of a diffusion process in configuration space, and then introduces some (time-symmetric) dynamical conditions on the process, leading to the Madelung equations for two real functions R and S, which are implied by the Schrödinger equation for $\psi = Re^{iS/\hbar}$.

The resulting theory has a number of similarities with de Broglie and Bohm's pilot-wave theory (although in the latter ψ is interpreted as a fundamental quantity). Indeed, particle trajectories in Nelson's stochastic mechanics can be intuitively thought of as de Broglie–Bohm trajectories with a superimposed white noise; and several concepts and techniques from pilot-wave theory can be easily and usefully imported into stochastic mechanics (although this is not usually discussed explicitly). One of these concepts, that of non-equilibrium, is the topic of the present paper.

Non-equilibrium in pilot-wave theory is defined as a situation in which the particle distribution is not equal to $\rho = R^2 = |\psi|^2$. In Section 3 we discuss non-equilibrium in pilot-wave theory and how it makes sense even in stochastic mechanics.

In Section 4 we develop a formalism for describing non-equilibrium in stochastic mechanics in terms of effective wavefunctions and effective non-linear Schrödinger equations related to but different from the ones describing equilibrium.

In Section 5, this formalism is applied to the discussion of non-locality in the context of non-linear Schrödinger equations. Specifically, we shall find a class of entangled solutions of the equation

$$i\hbar\frac{\partial\psi}{\partial t} = -\frac{\hbar^2}{2m}\Delta\psi + V\psi + \frac{h^2}{m}\frac{\Delta|\psi|}{|\psi|}\psi \tag{1}$$

(with no interaction terms in V), possessing a *local* hidden variables model.

[1] Address for correspondence: Department of Philosophy, University of Aberdeen, The Old Brewery, High Street, Aberdeen AB24 3UB, Scotland, U.K.

Some open questions conclude the paper (Section 6). One should note in particular that the theory of non-equilibrium in stochastic mechanics is potentially richer that the corresponding one in pilot-wave theory. The techniques developed in this paper, however, do not yet allow to discuss the genuinely novel cases (which, among other things, might be useful for the study of causally symmetric models of quantum mechanics).

2. Nelson's stochastic mechanics

In this section, we give a brief summary of Nelson's theory (largely after Nelson 1966 and Davidson 1979), with particular reference to aspects we shall need later. We consider only one particle for simplicity (the general case is analogous).[2]

Nelson (1966) considers the following stochastic differential equations,

$$d\mathbf{x}(t) = \mathbf{b}(\mathbf{x}(t),t)dt + d\mathbf{w}(t) \ , \tag{2}$$

$$d\mathbf{x}(t) = \mathbf{b}_*(\mathbf{x}(t),t)dt + d\mathbf{w}_*(t) \ , \tag{3}$$

where $\mathbf{b}(\mathbf{x}(t),t)$ is the mean forward velocity and $\mathbf{b}_*(\mathbf{x}(t),t)$ the mean backward velocity of the particle, and $\mathbf{w}(t)$ and $\mathbf{w}(t)_*$ are suitable Wiener processes with mean square fixed by the diffusion coefficient $\nu > 0$.

These equations provide a time-symmetric kinematics for describing diffusion processes. In pilot-wave terminology they can be thought of as forward and backward stochastic 'guidance equations' for the particle.

One has

$$D\mathbf{x} = \mathbf{b} \quad \text{and} \quad D_*\mathbf{x} = \mathbf{b}_* \ , \tag{4}$$

with the forward and backward stochastic derivatives D and D_* given by

$$D\mathbf{x}(t)\Big|_{\mathbf{x}(t)=\mathbf{x}} = \lim_{\varepsilon\to 0^+} E_t\left[\frac{\mathbf{x}(t+\varepsilon)-\mathbf{x}(t)}{\varepsilon}\Big|\mathbf{x}(t)=\mathbf{x}\right] \tag{5}$$

and

$$D_*\mathbf{x}(t)\Big|_{\mathbf{x}(t)=\mathbf{x}} = \lim_{\varepsilon\to 0^+} E_t\left[\frac{\mathbf{x}(t-\varepsilon)-\mathbf{x}(t)}{-\varepsilon}\Big|\mathbf{x}(t)=\mathbf{x}\right] \ . \tag{6}$$

($E_t[\,.\,|\mathbf{x}(t)=\mathbf{x}]$ is the expectation value at time t conditional on the value of the process being $\mathbf{x}$.[3])

Now take an arbitrary solution ρ of the forward Fokker–Planck (FP) equation

$$\frac{\partial\rho}{\partial t} = -\mathrm{div}(\mathbf{b}\rho) + \nu\Delta\rho \ . \tag{7}$$

If we define an osmotic velocity

$$\mathbf{u}_\rho := \nu\frac{\boldsymbol{\nabla}\rho}{\rho} \ , \tag{8}$$

and current velocity

$$\mathbf{v}_\rho := \mathbf{b} - \mathbf{u}_\rho \ , \tag{9}$$

ρ satisfies a continuity equation:

$$\frac{\partial\rho}{\partial t} = -\mathrm{div}\left[\left(\mathbf{v}_\rho + \nu\frac{\boldsymbol{\nabla}\rho}{\rho}\right)\rho\right] + \nu\Delta\rho = -\mathrm{div}(\mathbf{v}_\rho\rho) \ . \tag{10}$$

[2] For a much more comprehensive overview, see Nelson (1985). An introduction with an eye to conceptual questions is given by Bacciagaluppi (2005).

[3] The definition of $D\mathbf{x}(t)$ thus involves the forward transition probabilities from time t to times $t+\varepsilon$, while the definition of $D_*\mathbf{x}(t)$ involves the backward transition probabilities from time t to times $t-\varepsilon$.

If it is further the case that

$$\mathbf{b}_* = \mathbf{v}_\rho - \mathbf{u}_\rho \ , \tag{11}$$

the same ρ satisfies also the analogous backward equations:

$$\frac{\partial \rho}{\partial t} = -\mathrm{div}(\mathbf{b}_*\rho) - \nu\Delta\rho = -\mathrm{div}\left[\left(\mathbf{v}_\rho + \nu\frac{\boldsymbol{\nabla}\rho}{\rho}\right)\rho\right] - \nu\Delta\rho = -\mathrm{div}(\mathbf{v}_\rho\rho) \ . \tag{12}$$

Note that if only (2) is given, the time reversal (3) is not unique. From *any* solution ρ of the forward FP equation, one could define a backward mean velocity $\mathbf{b}_* = \mathbf{v}_\rho - \mathbf{u}_\rho$ and construct a time reversal of (2). (We shall use this in Section 4.) But if $\mathbf{b}_*$ is given, the single-time distribution ρ of the process will need to solve both FP equations and will thus be unique, because the corresponding current and osmotic velocities are fixed by

$$\frac{1}{2}(\mathbf{b} + \mathbf{b}_*) = \mathbf{v}_\rho \quad \text{and} \quad \frac{1}{2}(\mathbf{b} - \mathbf{b}_*) = \mathbf{u}_\rho = \nu\frac{\boldsymbol{\nabla}\rho}{\rho} \ , \tag{13}$$

and ρ is normalised.

As yet, $\mathbf{b}$ and $\mathbf{b}_*$ (which define the dynamics of the process) are left unspecified. Nelson's aim is to find natural constraints that will yield the Madelung equations for R and S, with R^2 defining the distribution of the process and with

$$\mathbf{v} = \frac{1}{m}\boldsymbol{\nabla}S \ . \tag{14}$$

The latter can be justified for instance from the variational approach of Guerra and Morato (1983), further motivated in Nelson (1985). Alternatively, note that $\mathbf{v}$ being a gradient implies

$$D\mathbf{b} = D_*\mathbf{b}_* \ . \tag{15}$$

This was pointed out already by de la Peña and Cetto (1982, eq. (16)).[4] Thus we can also justify setting $\mathbf{v}$ equal to a gradient as a simple way of enforcing the time-symmetry condition (15) — which we now see is an identity in Nelson's stochastic mechanics.

Given (14) and writing $R^2 := \rho$ we obtain the first Madelung equation (the continuity equation),

$$\frac{\partial R^2}{\partial t} = -\mathrm{div}\left(\frac{1}{m}(\boldsymbol{\nabla}S)R^2\right) = -\frac{1}{m}(\Delta S)R^2 - \frac{1}{m}\boldsymbol{\nabla}S\boldsymbol{\nabla}R^2 \ . \tag{16}$$

By defining further the *mean stochastic acceleration* as

$$\mathbf{a} := \frac{1}{2}\left(D\mathbf{b}_* + D_*\mathbf{b}\right) \ , \tag{17}$$

and imposing 'Newton's law',

$$m\mathbf{a} = -\boldsymbol{\nabla}V \ , \tag{18}$$

Nelson obtains also the Hamilton–Jacobi–Madelung (HJM) equation

$$\frac{\partial S}{\partial t} = -\frac{1}{2m}(\boldsymbol{\nabla}S)^2 - V + \frac{\hbar^2}{2m}\frac{\Delta R}{R} \ . \tag{19}$$

[4] We give an explicit proof in the Appendix.

As originally shown by Madelung (1926a,b), equations (16) and (19) are implied by Schrödinger's equation for $\psi = Re^{iS/\hbar}$. However, as pointed out by Wallstrom (1994), the converse is not true unless S has the right multi-valuedness behaviour.[5]

More generally (Davidson 1979), we can define the mean acceleration as

$$\frac{\alpha}{2}(D\mathbf{b}_* + D_*\mathbf{b}) + \frac{\beta}{2}(D\mathbf{b} + D_*\mathbf{b}_*) \ , \tag{20}$$

with $\alpha, \beta \geq 0$, $\alpha + \beta = 1$. Imposing Newton's law with the acceleration defined by (20),[6] we now obtain

$$\frac{\partial S}{\partial t} = -\frac{1}{2m}(\boldsymbol{\nabla} S)^2 - V + 2m\nu^2(\alpha - \beta)\frac{\Delta R}{R} \ . \tag{21}$$

As is easy to see, this HJM equation is implied by a generally *non-linear* Schrödinger equation:

$$i\hbar\frac{\partial\psi}{\partial t} = -\frac{\hbar^2}{2m}\Delta\psi + V\psi + \left(\frac{\hbar^2}{2m} - 2m\nu^2(\alpha - \beta)\right)\frac{\Delta|\psi|}{|\psi|}\psi \ . \tag{22}$$

We can distinguish three canonical cases:

(1) $\nu \neq 0$ and $\alpha > \beta$: one obtains the linear Schrödinger equation by choosing

$$\nu = \frac{\hbar}{2m\sqrt{\alpha - \beta}} \ ; \tag{23}$$

(2a) $\nu = 0$ (deterministic case), or **(2b)** $\alpha = \beta$: one has

$$i\hbar\frac{\partial\psi}{\partial t} = -\frac{\hbar^2}{2m}\Delta\psi + V\psi + \frac{\hbar^2}{2m}\frac{\Delta|\psi|}{|\psi|}\psi \ , \tag{24}$$

corresponding to the classical Hamilton–Jacobi equation and thus called the 'Schrödinger equation of classical mechanics' (Holland 1993, Sect. 2.6 and references therein);[7]

(3) $\nu \neq 0$ and $\alpha < \beta$: one can choose

$$\nu = \frac{\hbar}{2m\sqrt{\beta - \alpha}} \tag{25}$$

to obtain the non-linear equation

$$i\hbar\frac{\partial\psi}{\partial t} = -\frac{\hbar^2}{2m}\Delta\psi + V\psi + \frac{\hbar^2}{m}\frac{\Delta|\psi|}{|\psi|}\psi \ . \tag{26}$$

[5] Possibly the most promising strategy for overcoming Wallstrom's objection is the suggestion by L. Smolin (personal communication, Waterloo, Ontario, July 2005) to exploit the fact that quantum mechanical ground states will be typically nodeless under appropriate conditions. Instead, the line of argument suggested in Smolin (2006, Sect. IV) does not go through; see Valentini (2010, footnote 3) and Schmelzer (2011, Sect. 2.5). For two recent approaches to overcoming Wallstrom's objection from the point of view of slightly different theories, see de la Peña *et al.* (2011, Sect. 5.3) and the detailed proposal by Schmelzer (2011).

[6] This is equivalent to the procedure by de la Peña (equation (25) in his (1969)) and de la Peña and Cetto (1975). Note that in their treatment the case $\beta = 1$ is shown to correspond to (classical) Brownian motion.

[7] Note that the description of ensemble motions provided in the case $\nu = 0$ corresponds to the special case of classical mechanics in which an initial momentum field is fixed by the choice of the initial Hamilton–Jacobi function. For a careful discussion of these and related issues, see Holland (1993, Chap. 2, esp. Sects. 2.5 and 2.6). The case $\alpha = \beta$ instead is physically very different, as we shall note again in Section 4.

3. Pilot-wave theory and non-equilibrium

The pilot-wave theory by de Broglie (1928) and Bohm (1952) describes deterministic particle trajectories in configuration space, where the particle velocity at any time is given by the 'guidance equation'

$$\mathbf{v} = \frac{1}{m}\nabla S \tag{27}$$

(or $\mathbf{v}_i = \frac{1}{m}\nabla_i S$ for several particles), S being the phase of Schrödinger's wavefunction. If one considers an ensemble of particles guided by identical wavefunctions and distributed according to $\rho = |\psi|^2$, then this relation will be preserved by the Schrödinger equation (a fact already known to de Broglie). Thus the theory describes very easily any diffraction and interference phenomena with material particles.

In order to account for measurements more general than simple detections, in particular of observables other than position, one has to include also the measuring apparatus into the description. The wavefunction of the combined system will develop into a superposition of components that are and — provided no macroscopic reinterference takes place — remain (approximately) non-overlapping in the combined configuration space, and the system and apparatus will thus be trapped inside one of these components, the others being 'empty waves'. This component will be solely responsible for guiding the future motion of the combined system. Thus, the theory recovers an *effective collapse* of the wavefunction.

This analysis of measurement was Bohm's decisive contribution, and can be extended to more general cases of decoherence by the environment, thus arguably allowing pilot-wave theory to reproduce the classical regime of quantum mechanics as understood in the theory of decoherence (see the discussion in Bacciagaluppi 2003).

Since the current velocity $\mathbf{v}$ in stochastic mechanics has the same form as the particle velocity in de Broglie–Bohm theory, many of the results and techniques developed in the context of pilot-wave theory can be straightforwardly imported into stochastic mechanics, even though in the latter theory the wavefunction is not considered a fundamental quantity.[8]

We shall now consider the notion of (quantum, or sub-quantum) non-equilibrium as discussed in pilot-wave theory, and suggest it should also be imported into stochastic mechanics.

In pilot-wave theory the (independently postulated) wavefunction ψ has two roles: it guides the particle via $\mathbf{v} = \frac{1}{m}\nabla S$, and it defines the particle distribution via $\rho = |\psi|^2$. At first, it may seem puzzling how the wavefunction can play these two roles simultaneously. Indeed, if one thinks of the wavefunction principally as defining the particle distribution (in a maybe hypothetical ensemble), it will be puzzling to say it also guides the motion of the (actual) particle.[9]

If, however, one thinks of the wavefunction principally in its dynamical role as guiding the particle's motion, then it is perfectly natural to expect the particle distribution to depend on the dynamics — namely, if one understands the distribution as an *equilibrium* distribution. By the same token, non-equilibrium distributions, in which the statistical distribution in the ensemble is not equal to $|\psi|^2$, become perfectly intelligible.

The notion of non-equilibrium in pilot-wave theory has been discussed by several authors over the years. Immediately following his initial papers, Bohm (1953) discussed relaxation of the particle distribution from initial non-equilibrium towards $|\psi|^2$ via random external influences.

[8] This is true in particular of the notion of effective collapse. This is an obvious move, but usually not considered explicitly in treatments of stochastic mechanics. I conjecture it would, among other things, resolve the puzzle of two-time correlations raised by Nelson (2006).

[9] Indeed, this was precisely the reason for Schrödinger's dismissal of Bohm's 1952 theory, as expressed in a letter to Einstein (Schrödinger to Einstein, [after 18 but before 31 January] 1953, *Archive for the History of Quantum Physics*, microfilm no. 37, sect. 005-012 (manuscript) and 005-013 (carbon copy) (in German)) reproduced in the collection of Schrödinger's correspondence edited by von Meyenn (2011, vol. 2., pp. 673–675).

The next year, Bohm and Vigier (1954) hypothesised relaxation through an intrinsic random noise. This theory was discussed in more detail by Bohm and Hiley (1993, Chap. 9), and leads to trajectories identical to those of stochastic mechanics (!). The differences between this stochastic variant of pilot-wave theory and stochastic mechanics proper are, first and obviously, that Bohm and co-workers take the wavefunction to be fundamental, and, second, that they disregard the backward equations and thus the time-symmetric formulation of the theory. As a matter of fact, they are explicitly concerned with the relaxation of non-equilibrium distributions towards (future) equilibrium, and thus with much the same situation as we shall discuss in the next section, of particles evolving under Nelson's stochastic guidance equation (say, the forward one), but subject to a (say, initial) *non-equilibrium constraint.*

Further important work has focused on the analogy between $|\psi|^2$ in pilot-wave theory and equilibrium in classical statistical mechanics. Major examples are Valentini's (1991a,b) sub-quantum H-theorem, the analysis of equilibrium and of how it relates to uncertainty by Dürr, Goldstein and Zanghì (1992), various studies of relaxation behaviour, e.g. the recent work by Towler, Russell and Valentini (2011) (see also references therein), and various studies of the consequences, signatures and possible residues of non-equilibrium by Valentini (see e.g. Valentini 2007, 2010). The last type of investigations is particularly exciting, because it opens up the possibility of new empirical predictions.[10]

The situation is now perfectly analogous in Nelson's theory. The (independently postulated) vector fields $\mathbf{b}$ and $\mathbf{b}_*$ have two roles: they guide the particle via $\mathbf{v} = \frac{1}{2}(\mathbf{b}+\mathbf{b}_*)$ and $\mathbf{u} = \frac{1}{2}(\mathbf{b}-\mathbf{b}_*)$, and they define the particle distribution via $\mathbf{u} = \nu\frac{\boldsymbol{\nabla}\rho}{\rho}$. The guidance equations are law-like, and so is ρ, but this is unsurprising if we interpret it as the equilibrium distribution of the process, while the actual distribution $\tilde{\rho}$ is contingent and might be different from ρ. Perhaps, however, I am labouring an obvious point. A stochastic process (as in Nelson's theory) is a probability measure over a space of trajectories, and *actual frequencies* neither are nor need always match the probabilities (see the more extensive discussion in Bacciagaluppi 2010).

4. Description of non-equilibrium

Suppose we impose a constraint on the particle distribution at some time t_0,

$$\tilde{\rho}(t_0) \neq |\psi(t_0)|^2 \ . \tag{28}$$

We shall now describe the constrained process for $t \geq t_0$. (The same result can be extended to all times t.[11])

In this case, the time development of the distribution still satisfies the forward FP equation for $t > t_0$,

$$\frac{\partial\tilde{\rho}}{\partial t} = -\mathrm{div}(\mathbf{b}\tilde{\rho}) + \nu\Delta\tilde{\rho} \ , \tag{29}$$

but (always for $t > t_0$) it *no longer* satisfies the backward equation,

$$\frac{\partial\tilde{\rho}}{\partial t} = -\mathrm{div}(\mathbf{b}_*\tilde{\rho}) - \nu\Delta\tilde{\rho} \ . \tag{30}$$

Since ρ does satisfy the forward equation, however, we can define:

$$\tilde{\mathbf{u}} := \mathbf{u}_{\tilde{\rho}} = \nu\frac{\boldsymbol{\nabla}\tilde{\rho}}{\tilde{\rho}} \ , \tag{31}$$

[10] Dürr and co-workers instead take it as an advantage of pilot-wave theory over classical statistical mechanics that there is no obvious non-equilibrium to be observed that would require explanation.

[11] This can be done by carrying out the analogous proof for $t \leq t_0$, or — more economically — by exploiting the time-symmetry properties of the resulting equations.

$$\tilde{\mathbf{v}} := \mathbf{b} - \tilde{\mathbf{u}} = \mathbf{v} + \mathbf{u} - \tilde{\mathbf{u}} \ , \tag{32}$$

and

$$\tilde{\mathbf{b}}_* := \tilde{\mathbf{v}} - \tilde{\mathbf{u}} = \mathbf{v} + \mathbf{u} - 2\tilde{\mathbf{u}} \ . \tag{33}$$

As discussed in Section 2, $\tilde{\rho}$ will now satisfy a new effective backward FP equation:

$$\frac{\partial \tilde{\rho}}{\partial t} = -\mathrm{div}(\tilde{\mathbf{b}}_* \tilde{\rho}) - \nu \Delta \tilde{\rho} \ . \tag{34}$$

(Note that if $\mathbf{v}$ is a gradient, given that $\mathbf{u}$ and $\tilde{\mathbf{u}}$ are also gradients, so is $\tilde{\mathbf{v}}$.)

We can now define

$$\tilde{R} := \sqrt{\tilde{\rho}} \ , \tag{35}$$

and

$$\tilde{S} := S + \nu m (\log R^2 - \log \tilde{R}^2) = S + 2\nu m (\log R - \log \tilde{R}) \ , \tag{36}$$

so that

$$\frac{1}{m}\boldsymbol{\nabla}\tilde{S} = \frac{1}{m}\boldsymbol{\nabla} S + \nu \frac{\boldsymbol{\nabla} R^2}{R^2} - \nu \frac{\boldsymbol{\nabla} \tilde{R}^2}{\tilde{R}^2} \ , \tag{37}$$

and a new effective wavefunction $\tilde{\psi} := \tilde{R} e^{i\tilde{S}/\hbar}$. (Note also that S and $\tilde{S}$ have the same multi-valuedness behaviour.)

By construction, $\tilde{R}^2$ satisfies the continuity equation with current velocity $\frac{1}{m}\boldsymbol{\nabla}\tilde{S}$:

$$\frac{\partial \tilde{R}^2}{\partial t} = -\mathrm{div}\left(\frac{1}{m}(\boldsymbol{\nabla}\tilde{S})\tilde{R}^2\right) = -\frac{1}{m}(\Delta \tilde{S})\tilde{R}^2 - \frac{1}{m}\boldsymbol{\nabla}\tilde{S}\boldsymbol{\nabla}\tilde{R}^2 \ . \tag{38}$$

We shall now obtain also an effective HJM equation. Partial differentiation of (36) gives

$$\frac{\partial \tilde{S}}{\partial t} = \frac{\partial S}{\partial t} + \nu m \left(\frac{1}{R^2}\frac{\partial R^2}{\partial t} - \frac{1}{\tilde{R}^2}\frac{\partial \tilde{R}^2}{\partial t}\right) \ . \tag{39}$$

And substituting (21), (16) and (38) into (39) yields

$$\begin{aligned}\frac{\partial \tilde{S}}{\partial t} = &-\frac{1}{2m}(\boldsymbol{\nabla} S)^2 - V + 2m\nu^2(\alpha - \beta)\frac{\Delta R}{R} \\ &- \nu \Delta S - \nu \boldsymbol{\nabla} S \frac{\boldsymbol{\nabla} R^2}{R^2} + \nu \Delta \tilde{S} + \nu \boldsymbol{\nabla}\tilde{S}\frac{\boldsymbol{\nabla}\tilde{R}^2}{\tilde{R}^2} \ .\end{aligned} \tag{40}$$

Using further the fact that

$$\Delta S = \Delta \tilde{S} - 2\nu m \frac{\Delta R}{R} + 2\nu m \frac{(\boldsymbol{\nabla} R)^2}{R^2} + 2\nu m \frac{\Delta \tilde{R}}{\tilde{R}} - 2\nu m \frac{(\boldsymbol{\nabla}\tilde{R})^2}{\tilde{R}^2} \ , \tag{41}$$

and

$$\boldsymbol{\nabla} S = \boldsymbol{\nabla}\tilde{S} - \nu m \frac{\boldsymbol{\nabla} R^2}{R^2} + \nu m \frac{\boldsymbol{\nabla}\tilde{R}^2}{\tilde{R}^2} \ , \tag{42}$$

and hence also

$$\begin{aligned}(\boldsymbol{\nabla} S)^2 = &(\boldsymbol{\nabla}\tilde{S})^2 - 2\nu m \boldsymbol{\nabla}\tilde{S}\frac{\boldsymbol{\nabla} R^2}{R^2} + 2\nu m \boldsymbol{\nabla}\tilde{S}\frac{\boldsymbol{\nabla}\tilde{R}^2}{\tilde{R}^2} \\ &+ \nu^2 m^2 \left(\frac{\boldsymbol{\nabla} R^2}{R^2}\right)^2 - 2\nu^2 m^2 \frac{\boldsymbol{\nabla} R^2}{R^2}\frac{\boldsymbol{\nabla}\tilde{R}^2}{\tilde{R}^2} + \nu^2 m^2 \left(\frac{\boldsymbol{\nabla}\tilde{R}^2}{\tilde{R}^2}\right)^2 \ ,\end{aligned} \tag{43}$$

after suitable simplification we finally obtain

$$\frac{\partial \tilde{S}}{\partial t} = -\frac{1}{2m}(\nabla \tilde{S})^2 - V + 2m\nu^2(\alpha - \beta + 1)\frac{\Delta R}{R} - 2m\nu^2 \frac{\Delta \tilde{R}}{\tilde{R}} \ . \tag{44}$$

We see that the equation for $\tilde{S}$ is not the same as that for S, but has acquired the extra term

$$2m\nu^2(\alpha - \beta + 1)\left(\frac{\Delta R}{R} - \frac{\Delta \tilde{R}}{\tilde{R}}\right) \ . \tag{45}$$

And the corresponding effective Schrödinger equation for $\tilde{\psi}$ is

$$i\hbar\frac{\partial \tilde{\psi}}{\partial t} = -\frac{\hbar^2}{2m}\Delta\tilde{\psi} + V\tilde{\psi} + 2m\nu^2(\alpha - \beta + 1)\frac{\Delta|\psi|}{|\psi|}\tilde{\psi} + \left(\frac{\hbar^2}{2m} + 2m\nu^2\right)\frac{\Delta|\tilde{\psi}|}{|\tilde{\psi}|}\tilde{\psi} \ . \tag{46}$$

Thus, the *non-equilibrium* process can be equally well described as an *equilibrium* process with a different wavefunction and Schrödinger equation. In general this equation is *not* the same as the one satisfied by ψ (and it is not an autonomous equation for $\tilde{\psi}$, because of the presence of the 'old' quantum potential). In special cases, however, the extra term in the HJM equation vanishes, and ψ and $\tilde{\psi}$ will both satisfy the same (autonomous) equation. This happens iff $2m\nu^2(\alpha - \beta + 1) = 0$, i.e.

(a) if $\nu = 0$ (but *not* if $\alpha = \beta$, although in both cases the Schrödinger equation is the 'Schrödinger equation of classical mechanics'), or

(b) if $\beta = 1$, $\alpha = 0$ and ν is arbitrary.

Case **(b)** corresponds to the non-linear Schrödinger equation

$$i\hbar\frac{\partial \psi}{\partial t} = -\frac{\hbar^2}{2m}\Delta\psi + V\psi + \left(\frac{\hbar^2}{2m} + 2m\nu^2\right)\frac{\Delta|\psi|}{|\psi|}\psi \ , \tag{47}$$

and a non-equilibrium distribution in this case is simply an equilibrium distribution corresponding to a *different* wavefunction. For this special case thus, there is no privileged equilibrium distribution (and a case to be made that the wavefunction is to a large extent epistemic).

5. Non-linearity and non-locality

We now apply the results of the previous section to the discussion of non-locality in the context of theories with non-linear Schrödinger equations.

Rigorous results are available on this issue, specifically the well-known paper by Gisin (1989), to which we shall return below. What these results, however, do *not* show is that one automatically has superluminal signalling whenever one has entangled states obeying non-linear equations. Indeed, we shall now construct a counterexample to this proposition.

First of all, note that for product wavefunctions, and only for product wavefunctions,

$$S(\mathbf{x}_1, \mathbf{x}_2) = S_1(\mathbf{x}_1) + S_2(\mathbf{x}_2) \tag{48}$$

and

$$R(\mathbf{x}_1, \mathbf{x}_2) = R_1(\mathbf{x}_1)R_2(\mathbf{x}_2) \ . \tag{49}$$

In this case, and only in this case,

$$\mathbf{v}_i = \frac{1}{m_i}\nabla_i S(\mathbf{x}_1, \mathbf{x}_2) = \frac{1}{m_i}\nabla_i S_i(\mathbf{x}_i) \tag{50}$$

$(i = 1, 2)$ and

$$\mathbf{u}_i = \nu \boldsymbol{\nabla}_i \log R^2(\mathbf{x}_1, \mathbf{x}_2) = \nu \boldsymbol{\nabla}_i \log R_i^2(\mathbf{x}_i) \ . \tag{51}$$

In particular, if the wavefunction has product form, the Nelsonian dynamics decomposes into independent dynamics for the two particles, and is thus local.

Now take any Schrödinger equation of the form (22), i.e.

$$i\hbar \frac{\partial \psi}{\partial t} = -\frac{\hbar^2}{2m} \Delta \psi + V\psi + \left(\frac{\hbar^2}{2m} - 2m\nu^2(\alpha - \beta) \right) \frac{\Delta |\psi|}{|\psi|} \psi \ . \tag{52}$$

If there are no interaction terms in the potential, this equation preserves product wavefunctions. We can see this by noting that, if the potential decomposes as $V_1(\mathbf{x}_1) + V_2(\mathbf{x}_2)$, the linear Schrödinger equation preserves products. But the quantum potential

$$\left(\frac{\hbar^2}{2m} - 2m\nu^2(\alpha - \beta) \right) \frac{\Delta R}{R} \tag{53}$$

just *is* a potential that decomposes in this way:

$$\frac{\Delta R}{R} = \frac{(\Delta_1 + \Delta_2) R_1 R_2}{R_1 R_2} = \frac{\Delta_1 (R_1 R_2)}{R_1 R_2} + \frac{\Delta_2 (R_1 R_2)}{R_1 R_2} = \frac{\Delta_1 R_1}{R_1} + \frac{\Delta_2 R_2}{R_2} \ . \tag{54}$$

To simplify the construction of our counterexample, we now specialise to one of the invariant non-linear Schrödinger equations derived above, e.g.

$$i\hbar \frac{\partial \psi}{\partial t} = -\frac{\hbar^2}{2m} \Delta \psi + V\psi + \frac{\hbar^2}{m} \frac{\Delta |\psi|}{|\psi|} \psi \tag{55}$$

(case of $\alpha = 0$, $\beta = 1$, $\nu = \frac{\hbar}{2m}$).[12]

We take an initial product wavefunction $\psi(\mathbf{x}_1, \mathbf{x}_2, t_0)$. From the above, it follows that the corresponding stochastic mechanics for this system is *local* (and similarly if the two particles separately interact locally with measurement apparatuses).

Now we impose *classical correlations* on the initial distribution:

$$\tilde{\rho}(t_0) \neq \tilde{\rho}_1(t_0) \tilde{\rho}_2(t_0) \ . \tag{56}$$

The resulting effective wavefunction $\tilde{\psi}$ is *entangled*, because the functions

$$\tilde{R}(t_0) = \sqrt{\tilde{\rho}(t_0)} \tag{57}$$

and

$$\tilde{S}(t_0) = S(t_0) + \hbar \log R(t_0) - \hbar \log \tilde{R}(t_0) \tag{58}$$

fail to decompose appropriately. The constrained process is thus described by an entangled $\tilde{\psi}$ satisfying a non-linear Schrödinger equation. But, by construction, *it is a local process with added classical correlations.* This refutes the 'folk' claim that entanglement and non-linearity together imply superluminal signalling.

What about Gisin's result, however? What Gisin (1989) has shown (illustrated even more strikingly in Gisin (1990)) is that *if* a theory with a non-linear Schrödinger equation is such that it reproduces the usual phenomenology of collapse, *then* one can use non-linearity in conjunction

[12] With slight modifications, the argument will cover all the non-linear Schrödinger equations (22).

with entanglement to obtain signalling.[13] As Gisin is very well aware (private communicaton, Arolla, Switzerland, July 2009), this is a substantive assumption, and his theorem is not meant to apply more generally. Our counterexample thus shows (perhaps unsurprisingly) that the family of non-linear equations we have considered are not good candidates for reproducing the standard collapse phenomenology of quantum mechanics!

6. Open questions

To conclude, let us briefly return to the comparison between stochastic mechanics and pilot-wave theory, and to a few open questions that suggest themselves.

Specifically in the context of non-linear Schrödinger equations, we have seen that an apparent threat of non-locality arises from the entanglement of the effective wavefunction describing the non-equilibrium situation, i.e. from the fact that $\tilde{\mathbf{v}}$ and $\tilde{\mathbf{u}}$ do not decompose in terms of the velocities of the single particles. Locality, however, is *preserved* because their sum $\mathbf{b} = \mathbf{v} + \mathbf{u} = \tilde{\mathbf{v}} + \tilde{\mathbf{u}}$ decomposes appropriately. On the other hand, in a pilot-wave model of the same situation (wavefunction $\tilde{\psi}$ satisfying the given non-linear equation) the motion of the particles depends only on $\tilde{\mathbf{v}}$. Thus, despite the fact that the particle distribution in both models would be the same, it would appear that the de Broglie–Bohm model of the situation would be more non-local than the Nelsonian model.[14]

Further, whether one considers linear or non-linear Schrödinger equations, non-equilibrium stochastic mechanics is a potentially richer theory (including a potentially richer range of novel phenomena) than non-equilibrium pilot-wave theory. This is so because the theory is stochastic. Indeed, in a deterministic theory such as pilot-wave theory one can only consider constraints at a single time: the distribution of trajectories is fixed by the particle distribution at a single instant. By contrast, in a stochastic theory any additional constraint on the distribution of trajectories merely subselects further from the ensemble of all possible trajectories that defines the stochastic process; thus, independent multi-time constraints make perfect sense. In particular, one can impose initial and final constraints on a process and might thereby obtain some qualitatively new phenomena (for instance genuine non-locality even if the unconstrained process is local). Such time-symmetrically constrained processes could provide good test cases for the ideas put forward in particular by Price (1996) about retrocausal hidden variables models for quantum mechanics. The techniques developed in this paper, however, cannot be straightforwardly extended to the case of multi-time constraints.

Acknowledgments

I would like to thank Elise Crull, Thomas Durt, Chris Timpson, Sonja Smets, Rob Spekkens and Antony Valentini for correspondence, discussion and other precious help, and Rafael Sorkin for very perceptive comments on a previous talk I gave on this material. I would also like to thank Mark Davidson, Edward Nelson, Ángel Sanz and other members of the audience at Vienna for questions, comments and correspondence, in particular Ana Maria Cetto. Previous versions of this talk were given at the Universities of Aberdeen, Firenze, Oxford, Pavia, Sydney and Utrecht and at Perimeter Institute. Finally, a warm thanks goes to the organisers of the Vienna conference, especially to Gerhard Grössing.

[13] Imagine for simplicity an EPR setup with a singlet state. If Alice by measuring spin in direction x or y can create on Bob's side equal up-down mixtures of spin eigenstates in direction x or y, respectively, Bob can then use the non-linearity of the evolution to discriminate between these two different mixtures. Thus, Alice can signal arbitrarily fast to Bob using an ensemble of sufficiently separated EPR pairs.

[14] This also suggests the question of whether entangled wavefunctions with decomposable $\frac{1}{m}\nabla S + \nu\frac{\nabla R^2}{R^2}$ are somehow less intrinsically non-local than generic ones.

Appendix

We now show explicitly that requiring that $\mathbf{v}$ be a gradient is a simple sufficient condition for the time-symmetry condition (15), i.e.

$$D\mathbf{b} = D_*\mathbf{b}_* \ . \tag{A.1}$$

Note first of all that applying the stochastic derivatives D or D_* to an arbitrary random variable $f(\mathbf{x}(t), t)$ one obtains

$$Df(\mathbf{x}(t), t)\Big|_{\mathbf{x}(t)=\mathbf{x}} = \left[\frac{\partial}{\partial t} + \mathbf{b}(\mathbf{x}, t) \cdot \boldsymbol{\nabla} + \nu\Delta\right] f(\mathbf{x}, t) \tag{A.2}$$

and

$$D_*f(\mathbf{x}(t), t)\Big|_{\mathbf{x}(t)=\mathbf{x}} = \left[\frac{\partial}{\partial t} + \mathbf{b}_*(\mathbf{x}, t) \cdot \boldsymbol{\nabla} - \nu\Delta\right] f(\mathbf{x}, t) \ . \tag{A.3}$$

By (A.2) and (A.3), condition (A.1) is equivalent to

$$\frac{\partial \mathbf{u}}{\partial t} + (\mathbf{v} \cdot \boldsymbol{\nabla})\mathbf{u} + (\mathbf{u} \cdot \boldsymbol{\nabla})\mathbf{v} + \nu\Delta\mathbf{v} = 0 \ . \tag{A.4}$$

Now, by the definition of $\mathbf{u}$, we have

$$\frac{\partial \mathbf{u}}{\partial t} = \frac{\partial}{\partial t}\nu\boldsymbol{\nabla} \ln \rho = \nu\boldsymbol{\nabla}\Big(\frac{1}{\rho}\frac{\partial \rho}{\partial t}\Big) \ . \tag{A.5}$$

And using the continuity equation (10),

$$\begin{aligned} \frac{\partial \mathbf{u}}{\partial t} &= -\nu\boldsymbol{\nabla}\Big(\frac{1}{\rho}\boldsymbol{\nabla} \cdot (\mathbf{v}\rho)\Big) = -\nu\boldsymbol{\nabla}\Big(\frac{1}{\rho}(\boldsymbol{\nabla} \cdot \mathbf{v})\rho + \frac{1}{\rho}\mathbf{v} \cdot \boldsymbol{\nabla}\rho\Big) = \\ &= -\nu\boldsymbol{\nabla}\Big(\boldsymbol{\nabla} \cdot \mathbf{v} + \mathbf{v} \cdot \boldsymbol{\nabla} \ln \rho\Big) = -\nu\boldsymbol{\nabla}(\boldsymbol{\nabla} \cdot \mathbf{v}) - \boldsymbol{\nabla}(\mathbf{v} \cdot \mathbf{u}) \ . \end{aligned} \tag{A.6}$$

Therefore, (A.4) becomes

$$\nu\Big(\boldsymbol{\nabla}(\boldsymbol{\nabla} \cdot \mathbf{v}) - \Delta\mathbf{v}\Big) + \boldsymbol{\nabla}(\mathbf{v} \cdot \mathbf{u}) - (\mathbf{v} \cdot \boldsymbol{\nabla})\mathbf{u} - (\mathbf{u} \cdot \boldsymbol{\nabla})\mathbf{v} = 0 \ . \tag{A.7}$$

From the identity

$$\boldsymbol{\nabla}(\mathbf{a} \cdot \mathbf{b}) = (\mathbf{a} \cdot \boldsymbol{\nabla})\mathbf{b} + (\mathbf{b} \cdot \boldsymbol{\nabla})\mathbf{a} + \mathbf{a} \times (\boldsymbol{\nabla} \times \mathbf{b}) + \mathbf{b} \times (\boldsymbol{\nabla} \times \mathbf{a}) \ , \tag{A.8}$$

and the fact that

$$\boldsymbol{\nabla}(\boldsymbol{\nabla} \cdot \mathbf{v}) = \boldsymbol{\nabla} \times (\boldsymbol{\nabla} \times \mathbf{v}) + \Delta\mathbf{v} \ , \tag{A.9}$$

we can further transform (A.7) to

$$\nu\boldsymbol{\nabla} \times (\boldsymbol{\nabla} \times \mathbf{v}) + \mathbf{u} \times (\boldsymbol{\nabla} \times \mathbf{v}) + \mathbf{v} \times (\boldsymbol{\nabla} \times \mathbf{u}) = 0 \ . \tag{A.10}$$

Finally, since $\mathbf{u}$ is a gradient, we obtain the equivalence of (A.1) and

$$\nu\boldsymbol{\nabla} \times (\boldsymbol{\nabla} \times \mathbf{v}) + \mathbf{u} \times (\boldsymbol{\nabla} \times \mathbf{v}) = 0 \ , \tag{A.11}$$

for which $\mathbf{v}$ being a gradient is obviously a sufficient condition.

If we further impose (18) or some similar dynamical law, $\mathbf{v}$ being a gradient is presumably also a necessary condition for (A.1), since all time derivatives of (A.11) need to vanish, too, and they will generally depend on the arbitrary external potential V.[15]

[15] The condition $D\mathbf{b}_* = D_*\mathbf{b}$ is similarly equivalent to $\frac{\partial \mathbf{u}}{\partial t} + (\mathbf{v} \cdot \nabla)\mathbf{u} - (\mathbf{u} \cdot \nabla)\mathbf{v} - \nu\Delta\mathbf{v} = 0$, which assuming (A.1) simplifies to $\nu\Delta\mathbf{v} + (\mathbf{u} \cdot \nabla)\mathbf{v} = 0$.

References

Bacciagaluppi G 2003 The role of decoherence in quantum theory *The Stanford Encyclopedia of Philosophy*, ed E N Zalta http://plato.stanford.edu/entries/qm-decoherence/ [next major update 2012]

Bacciagaluppi G 2005 A conceptual introduction to Nelson's mechanics *Endophysics, Time, Quantum and the Subjective*, eds R Buccheri *et al* (Singapore: World Scientific) pp 367–88 *Preprint (with corrections)* http://philsci-archive.pitt.edu/8853/

Bacciagaluppi G 2010 Probability and time symmetry in classical Markov processes *Probabilities, Causes and Propensities in Physics* Synthese Library vol 347, ed M Suárez (Berlin and Heidelberg: Springer) pp 41–60 *Preprint* http://philsci-archive.pitt.edu/3534/

Bohm D 1952 A suggested interpretation of the quantum theory in terms of 'hidden' variables, I. and II. *Phys. Rev.* **85** 166–79 and 180–93

Bohm D 1953 Proof that probability density approaches $|\psi|^2$ in causal interpretation of the quantum theory *Phys. Rev.* **89** 458–66

Bohm D and Hiley B 1993 *The Undivided Universe: an Ontological Interpretation of Quantum Theory* (London: Routledge)

Bohm D and Vigier J P 1954 Model of the causal interpretation of quantum theory in terms of a fluid with irregular fluctuations *Phys. Rev.* **96** 208–16

de Broglie L 1928 La nouvelle dynamique des quanta *Électrons et Photons: Rapports et Discussions du Cinquième Conseil de Physique Solvay* [ed H A Lorentz] (Paris: Gauthiers-Villars) pp 105–32 Translated in G Bacciagaluppi and A Valentini *Quantum Theory at the Crossroads: Reconsidering the 1927 Solvay Conference* (Cambridge: Cambridge University Press 2009) pp 374–407

Davidson M 1979 A generalization of the Fényes–Nelson stochastic model of quantum mechanics *Lett. Math. Phys.* **3** 271–7

Dürr D, Goldstein S and Zanghì N 1992 Quantum equilibrium and the origin of absolute uncertainty *J. Stat. Phys.* **67** 843–907

Gisin N 1989 Stochastic quantum dynamics and relativity *Helv. Phys. Acta* **62** 363–71

Gisin N 1990 Weinberg's non-linear quantum mechanics and supraluminal communications *Phys. Lett.* A **143** 1–2

Guerra F and Morato L 1983 Quantization of dynamical systems and stochastic control theory *Phys. Rev.* D **27** 1774–86

Holland P R 1993 *The Quantum Theory of Motion: an Account of the de Broglie–Bohm Causal Interpretation of Quantum Mechanics* (Cambridge: Cambridge University Press)

Madelung E 1926a Eine anschauliche Deutung der Gleichung von Schrödinger *Die Naturwissenschaften* **14** 1004

Madelung E 1926b Quantentheorie in hydrodynamischer Form *Zeitschrift für Physik* **40** 322–6

von Meyenn K (ed) 2011 *Eine Entdeckung von ganz außerordentlicher Tragweite: Schrödingers Briefwechsel zur Wellenmechanik und zum Katzenparadoxon* 2 vols (Berlin and Heidelberg: Springer)

Nelson E 1966 Derivation of the Schrödinger equation from Newtonian mechanics *Phys. Rev.* **150** 1079–85

Nelson E 1985 *Quantum Fluctuations* (Princeton: Princeton University Press)

Nelson E 2006 Afterword *Diffusion, Quantum Theory, and Radically Elementary Mathematics* Mathematical Notes vol 47, ed W G Faris (Princeton: Princeton University Press) pp 227–30

de la Peña L 1969 New formulation of stochastic theory and quantum mechanics *J. Math. Phys.* **10** 1620–30

de la Peña L and Cetto A M 1975 Stochastic theory for classical and quantum mechanical systems *Found. Phys.* **5** 355–70

de la Peña L and Cetto A M 1982 Does quantum mechanics accept a stochastic support? *Found. Phys.* **12** 1017–37

de la Peña L, Cetto A M, Valdés-Hernández A and França H M 2011 Origin and meaning of quantum nonlocality *Preprint* arXiv:1110.4641v1 [quant-ph]

Price H 1996 *Time's Arrow and Archimedes' Point: New Directions for the Physics of Time* (New York: Oxford University Press)

Schmelzer I 2011 An answer to the Wallstrom objection against Nelsonian stochastics *Preprint* arXiv:1101.5774v3 [quant-ph]

Smolin L 2006 Could quantum mechanics be an approximation to another theory? *Preprint* arXiv:quant-ph/0609109v1

Towler M D, Russell N and Valentini A 2011 Timescales for dynamical relaxation to the Born rule *Preprint* arXiv:1103.1589v2 [quant-ph]

Valentini A 1991a,b Signal-locality, uncertainty, and the subquantum H-theorem, I. and II. *Phys. Lett.* A **156** 5–11 and **158** 1–8

Valentini A 2007 Astrophysical and cosmological tests of quantum theory *J. Phys. A: Math. Theor.* **40** 3285–303

Valentini A 2010 Inflationary cosmology as a probe of primordial quantum mechanics *Phys. Rev.* D **82** 063513

Wallstrom T 1994 Inequivalence between the Schrödinger equation and the Madelung hydrodynamic equations *Phys. Rev.* A **49** 1613–7

Understanding quantization: a hidden variable model

Agung Budiyono

Institute for the Physical and Chemical Research, RIKEN, 2-1 Hirosawa, Wako-shi, Saitama 351-0198, Japan

E-mail: agungby@riken.jp

Abstract. We argue that to solve the foundational problems of quantum theory one has to first understand what it means to quantize a classical system. We then propose a quantization method based on replacement of deterministic c-numbers by stochastically-parametrized c-numbers. Unlike canonical quantization, the method is free from operator ordering ambiguity and the resulting quantum system has a straightforward interpretation as statistical modification of an ensemble of classical trajectories. We then develop measurement without wave function collapse à la pilot-wave theory and point out new testable predictions.

1. Introduction

While pragmatically quantum theory has shown spectacular successes, its foundation with respect to quantum-classical correspondence and measurement problem, despite many attempts, still resist unambiguous explanation. We believe that these difficulties originate from the (physical) ambiguity of the quantization scheme through which many successful quantum systems are obtained from the corresponding classical systems. In other words, to solve the foundational problems of quantum theory, we have to first understand what is meant by quantizing a classical system. In the canonical quantization, one first writes the classical dynamical equation in Cartesian coordinate. Quantization is then done by promoting the pair of canonical conjugate variables into the corresponding Hermitian operators, $\circ \mapsto \hat{\circ}$, and Poisson bracket is replaced by the commutator: $\{\circ, \star\} \mapsto [\hat{\circ}, \hat{\star}]/i\hbar$, where $\hbar$ is the reduced Planck constant. This procedure is usually said of as replacement of commuting c-number (classical number) by non-commuting q-number (quantum number or Hermitian operators).

Such direct substitution rule however implies formal, conceptual, and foundational problems. First, given a classical quantity, the above rule in general leads to infinitely many different alternatives of Hermitian operators due to operator ordering ambiguity. Second, the physical meaning of the resulting quantum systems can not be deduced "directly" from the quantization processes: the quantization does not tell us the physical meaning of the Schrödinger equation, nor offers it an explanation to the physical origin of Planck constant. In other words, the quantum-classical correspondence is not physically transparent. Hence, unlike the original classical system, we need to further introduce a physical interpretation to the resulting quantum system. Unfortunately, there are many physical interpretations with the same empirical prediction: standard Copenhagen interpretation, pilot-wave theory, many worlds etc. There is then the foundational problem that some of the existing interpretations suffer from the infamous measurement problem [1]. In the present paper, we shall discuss a new method of quantization which is free from the above mentioned problems.

2. Quantum fluctuations as statistical modification of classical ensemble parametrized by unbiased hidden random variable

2.1. General formalism

Let us consider a classical dynamics with N degrees of freedom whose configuration coordinate is $q = (q_1, \ldots, q_N)$. Generalization to infinite degrees of freedom is straightforward at least formally. In classical mechanics, all the dynamical information is contained in the Lagrangian function $\underline{L} = \underline{L}(q, \dot{q})$, where $\dot{q} \doteq dq/dt$ is the velocity. For simplicity, let us assume that the Lagrangian is non-singular. From the Hamilton principle, one then has the Euler-Lagrange equation $(d/dt)(\partial\underline{L}/\partial\dot{q}_i) - \partial\underline{L}/\partial q_i = 0$, $i = 1, \ldots, N$. To go to the Hamiltonian formalism, first, the canonical momentum is defined as $\underline{p}_i \doteq \partial\underline{L}/\partial\dot{q}_i$, $i = 1, \ldots, N$, which, due to non-singularity of the Lagrangian, can be inverted to give $\dot{q}_i = \dot{q}_i(q, \underline{p}; t)$, $i = 1, \ldots, N$. The classical Hamiltonian is defined as $\underline{H}(q, \underline{p}) \doteq \underline{p}\dot{q}(q, \underline{p}; t) - \underline{L}(q, \underline{p}; t)$. The Euler-Lagrange equation can then be put into the Hamilton equation

$$\dot{q}_i = \frac{\partial\underline{H}}{\partial\underline{p}_i}, \quad \dot{\underline{p}}_i = -\frac{\partial\underline{H}}{\partial q_i}, \quad i = 1, \ldots, N. \tag{1}$$

Now let us consider a congruence of curves, $\{q(t; u), p(t; v)\}$, satisfying the Hamilton equation, where $u = \{u_1, \ldots, u_N\}$ and $v = \{v_1, \ldots, v_N\}$ are parameters of the congruences. Namely, each pair of the value of (u, v) corresponds to a single member of the congruence. Let us then assume that there is a one-parameter family of hypersurfaces $\underline{S}(q; t) = \tau$, where $\underline{S}$ is a real-valued function and τ is a parameter, satisfying the following relation:

$$p_i = \partial_{q_i}\underline{S}. \tag{2}$$

Then, it can be shown that $S(q; t)$ also satisfies the following Hamilton-Jacobi equation [2]:

$$\partial_t\underline{S} + \underline{H}(q, \partial_q\underline{S}) = 0. \tag{3}$$

Conversely, defining $\underline{p}$ as in Eq. (2), the Hamilton-Jacobi equation of (3) can be shown to lead to the Hamilton equation of (1).

Now let us consider an ensemble of copies of the system. Then, if the probability density of the position of the ensemble of trajectories in configuration space is denoted by $\underline{\rho}(q; t)$, it must also satisfy the following continuity equation:

$$\partial_t\underline{\rho} + \partial_q \cdot (\dot{q}(\underline{S})\underline{\rho}) = 0, \tag{4}$$

where the functional form of the velocity $\dot{q}$ with respect to $\underline{S}$ is determined by substituting Eq. (2) into the left equation of (1)

$$\dot{q}_i(\underline{S}) = \frac{\partial\underline{H}}{\partial\underline{p}_i}\Big|_{\underline{p}=\partial_q\underline{S}}, \quad i = 1, \ldots, N. \tag{5}$$

Hence, the dynamics and statistics of the ensemble of trajectories is given by solving Eqs. (3), (4) and (5).

Let us develop a general scheme to modify the above classical dynamics of the ensemble of trajectories [3]. To do this, let us introduce two real-valued functions $\Omega(q, \lambda; t)$ and $S(q, \lambda; t)$ where λ is a non-vanishing hidden random variable, $\lambda \neq 0$. $\Omega(q, \lambda; t)$ is the joint-probability density of the fluctuations of q and λ so that the marginal probability densities are given by

$$\rho(q; t) \doteq \int d\lambda\Omega, \quad P(\lambda) \doteq \int dq\Omega, \tag{6}$$

where we have assumed that the probability density of λ is stationary.

Now let us postulate the following rule of replacement of (deterministic) c-number by (stochastic) c-number to be applied to Eqs. (3) and (4) governing the dynamics of the ensemble of trajectories:

$$\underline{\rho} \mapsto \Omega, \quad \partial_q \underline{S} \mapsto \partial_q S + \frac{\lambda}{2}\frac{\partial_q \Omega}{\Omega}, \quad \partial_t \underline{S} \mapsto \partial_t S + \frac{\lambda}{2}\frac{\partial_t \Omega}{\Omega} + \frac{\lambda}{2}\partial_q \cdot \dot{q}(S), \tag{7}$$

where the functional form of $\dot{q}(S)$ in the third line is determined by the classical Hamiltonian as in Eq. (5). That is, we just substitute $\underline{S}$ in Eq. (5) with S. Let us show that the above rule of replacement has a consistent classical correspondence. First, expanding as $\Delta F = \partial_t F \Delta t + \partial_q F \cdot \Delta q$, using the last two equations of (7) one gets $\Delta \underline{S} \mapsto \Delta S + (\lambda/2)(\Delta\Omega/\Omega + \partial_q \cdot \dot{q}(S)\Delta t)$. In the limit of $S \to \underline{S}$, the second term on right hand side has to be vanishing to give $\Delta\Omega + \Omega\partial_q \cdot \dot{q}(\underline{S})\Delta t = 0$. This is just the classical continuity equation so that one has $\rho = \int d\lambda\Omega \to \underline{\rho}$.

The next question is then what is the statistics of λ. We shall show in the next subsection by taking a concrete example that the above modification of classical dynamics of an ensemble of trajectories leads to the Schrödinger equation with unique quantum Hamiltonian if the probability density of λ is given by

$$P(\lambda) = \frac{1}{2}\delta(\lambda - \hbar) + \frac{1}{2}\delta(\lambda + \hbar). \tag{8}$$

Namely λ is a binary unbiased random variable which can take values $\pm\hbar$. Extension to a more general continuous hidden random variable satisfying the unbiased condition

$$P(\lambda) = P(-\lambda), \tag{9}$$

will be given in Section 4, suggesting a very small yet finite correction to the prediction of quantum mechanics.

2.2. Particle in external potentials

Let us apply the above general formalism to a single particle subjected to external potentials. The Lagrangian is given by $\underline{L} = \frac{1}{2}g_{ij}(q)\dot{q}^i\dot{q}^j + A \cdot \dot{q} - V(q)$, where $A = (A_1, A_2, A_3)$ and V are the vector and scalar potentials, respectively, $g_{ij}(q)$ is an invertible matrix which might depend on q (for example in the case of particle with position-dependent mass widely used in solid state physics), and summation over repeated indices is implied. Writing the inverse of g_{ij} as g^{ij} so that $g_{ik}g^{kl} = \delta_i^l$, the momentum is related to the velocity as

$$\dot{q}^i = g^{ij}(\underline{p}_j - A_j). \tag{10}$$

so that the classical Hamiltonian takes the following form:

$$\underline{H}(q, \underline{p}) = \frac{g^{ij}(q)}{2}(\underline{p}_i - A_i)(\underline{p}_j - A_j) + V. \tag{11}$$

The Hamilton-Jacobi equation of (3) then reads

$$\partial_t \underline{S} + \frac{g^{ij}}{2}(\partial_{q_i}\underline{S} - A_i)(\partial_{q_j}\underline{S} - A_j) + V = 0. \tag{12}$$

On the other hand, putting Eq. (2) into Eq. (10), the functional relation between $\dot{q}$ and $\underline{S}$ is given by

$$\dot{q}^i(\underline{S}) = g^{ij}(\partial_{q_j}\underline{S} - A_j). \tag{13}$$

Inserting this into Eq. (4), one thus has

$$\partial_t \underline{\rho} + \partial_{q_i}\Big((g^{ij}(\partial_{q_j}\underline{S} - A_j))\underline{\rho}\Big) = 0. \tag{14}$$

Hence, the dynamics and statistics of the classical trajectories are determined by solving Eqs. (12), (13) and (14). On the other hand, using Eq. (13), Eq. (7) becomes

$$\underline{\rho} \mapsto \Omega, \ \ \partial_q \underline{S} \mapsto \partial_q S + \frac{\lambda}{2}\frac{\partial_q \Omega}{\Omega}, \ \ \partial_t \underline{S} \mapsto \partial_t S + \frac{\lambda}{2}\frac{\partial_t \Omega}{\Omega} + \frac{\lambda}{2}\partial_{q_i} g^{ij}(\partial_{q_j} S - A_j), \tag{15}$$

Now let us apply the rule of replacement of Eq. (15) to Eqs. (12) and (14). First, imposing the first two equations of Eq. (15) into Eq. (14), one has

$$\partial_t \Omega + \partial_{q_i}\Big(g^{ij}(\partial_{q_j} S - A_j)\Omega\Big) + \frac{\lambda}{2}\partial_{q_i}(g^{ij}\partial_{q_j}\Omega) = 0. \tag{16}$$

On the other hand, inserting the last two equations of (15) into Eq. (12), one has, after arrangement,

$$\begin{aligned}\partial_t S + \frac{g^{ij}}{2}(\partial_{q_i} S - A_i)(\partial_{q_j} S - A_j) + V - \frac{\lambda^2}{2}\Big(g^{ij}\frac{\partial_{q_i}\partial_{q_j} R}{R} + \partial_{q_i} g^{ij}\frac{\partial_{q_j} R}{R}\Big) \\ + \frac{\lambda}{2\Omega}\Big(\partial_t \Omega + \partial_{q_i}\Big(g^{ij}(\partial_{q_j} S - A_j)\Omega\Big) + \frac{\lambda}{2m}\partial_{q_i}(g^{ij}\partial_{q_j}\Omega)\Big) = 0,\end{aligned} \tag{17}$$

where we have defined $R \doteq \sqrt{\Omega}$ and used the following identity

$$\frac{1}{4}\frac{\partial_{q_i}\Omega\partial_{q_j}\Omega}{\Omega^2} = \frac{1}{2}\frac{\partial_{q_i}\partial_{q_j}\Omega}{\Omega} - \frac{\partial_{q_i}\partial_{q_j} R}{R}. \tag{18}$$

Inserting Eq. (16) into Eq. (17), the second line is vanishing to give

$$\partial_t S + \frac{g^{ij}}{2}(\partial_{q_i} S - A_i)(\partial_{q_j} S - A_j) + V - \frac{\lambda^2}{2}\Big(g^{ij}\frac{\partial_{q_i}\partial_{q_j} R}{R} + \partial_{q_i} g^{ij}\frac{\partial_{q_j} R}{R}\Big) = 0. \tag{19}$$

We have thus pair of coupled equations (16) and (19) which are parametrized by λ.

Now let us assume that $\Omega(q, \lambda; t)$ has the following symmetry:

$$\Omega(q, \lambda; t) = \Omega(q, -\lambda; t), \tag{20}$$

so that the probability density of λ is unbiased, $P(\lambda) = \int dq\Omega(q, \lambda; t) = P(-\lambda)$. In this case, $S(q, \lambda; t)$ and $S(q, -\lambda; t)$ satisfy the same differential equation of (19): namely, the last term of Eq. (19) is not sensitive to the signs of λ. Hence, if initially one has $S(q, \lambda; 0) = S(q, -\lambda; 0)$, then the symmetry will be preserved for all the time

$$S(q, \lambda; t) = S(q, -\lambda; t). \tag{21}$$

These properties can then be used to eliminate the last term of Eq. (16): taking the case when λ is positive, add to it the case when λ is negative and divided by two, one gets

$$\partial_t \Omega + \partial_{q_i}\Big(g^{ij}(\partial_{q_j} S - A_j)\Omega\Big) = 0. \tag{22}$$

We have thus a pair of coupled equations (19) and (22) which are still parametrized by λ.

Next, since λ is assumed to be non-vanishing, one can define the following complex-valued function:

$$\Psi(q,\lambda;t) \doteq R\exp\Big(\frac{i}{|\lambda|}S\Big). \tag{23}$$

It differs from the Madelung transformation in which S is divided by $|\lambda|$ instead of $\hbar$. Equations (19) and (22) can thus be recast into the following modified Schrödinger equation:

$$i|\lambda|\partial_t\Psi = \frac{1}{2}(-i|\lambda|\partial_{q_i} - A_i)g^{ij}(q)(-i|\lambda|\partial_{q_j} - A_j)\Psi + V\Psi. \tag{24}$$

Here we have assumed that the fluctuations of λ in space and time are ignorable as compared to that of S. Let us proceed to assume that Ω is separable $\Omega(q,\lambda;t) = \rho(q;t)P(\lambda)$, where $P(\lambda)$ takes the form given by Eq. (8). In this case, Eq. (24) reduces to the celebrated Schrödinger equation

$$i\hbar\partial_t\Psi_Q(q;t) = \hat{H}\Psi_Q(q;t), \text{ with } \Psi_Q(q;t) = \sqrt{\rho}e^{\frac{i}{\hbar}S_Q(q;t)}, \text{ and } S_Q(q;t) = S(q,\pm\hbar;t), \tag{25}$$

where the quantum Hamiltonian $\hat{H}$ is given by

$$\hat{H} = \frac{1}{2}(\hat{p}_i - A_i)g^{ij}(q)(\hat{p}_j - A_j) + V, \text{ with } \hat{p}_i \doteq -i\hbar\partial_{q_i}. \tag{26}$$

From Eq. (25) we know that the Born's statistical interpretation of wave function is valid by construction for all time, $\rho(q;t) = |\Psi_Q(q;t)|^2$.

Let us mention that there are many approaches to derive Schrödinger equation with quantum Hamiltonian of the type of Eq. (26) [4]. The advantage of our approach is that it can be applied directly as soon as the classical Hamiltonian (or Lagrangian) is given, if a solution exists. Our method of deriving the Schrödinger equation can thus be regarded as to provide a method of quantization of a classical Hamiltonian. For the case $g_{ij} = m\delta_{ij}$ where m is constant so that $g^{ij} = \delta_{ij}/m$, we regain the result of canonical quantization. For the case where g^{ij} depends on q, then in contrast to canonical quantization which leads to infinite different alternatives with different ordering of operators, our model imposes a unique ordering. The same ordering is also obtained in Ref. [5]. However in contrast to the method reported there which can only be applied to classical Hamiltonian without linear term in classical momentum, our method can be applied to such a case.

Moreover, in this specific case where Ω is separable $\Omega(q,\lambda;t) = \rho(q;t)P(\lambda)$ and $P(\lambda)$ is given by Eq. (8), Eq. (22) becomes

$$\partial_t\rho + \partial_{q_i}\Big(g^{ij}(\partial_{q_j}S_Q - A_j)\rho\Big) = 0. \tag{27}$$

We can then identify an "effective" velocity field given by

$$v_e^i = g^{ij}(\partial_{q_j}S_Q - A_j). \tag{28}$$

Since $S_Q(q;t)$ is just the phase of the Schrödinger wave function, it turns out that the above effective velocity field is equal to the "actual" velocity of particle (beable) in pilot-wave theory [6]. This allows us to draw a conclusion that our model will reproduce the prediction of pilot-wave theory on statistical wave-like interference pattern in slits experiment and tunnelling over potential barrier [7]. However unlike pilot-wave theory, in our model, the dynamics is strictly stochastic, the wave function is not physically real-field and the Born's statistical interpretation of wave function is valid for all time by construction.

3. Measurement without wave function collapse and external observer

Now let us apply the quantization method developed in the previous Section to a class of von Neumann measurement model. To do this, let us consider the dynamics of two interacting particles, the first particle with coordinate q_1 represents the system whose properties being measured, and the second particle with coordinate q_2 represents the measuring apparatus. Let us first discuss how classical dynamics describes the whole system. Let us suppose that one wants to measure a quantity $\underline{A}_1 = \underline{A}_1(q_1, \underline{p}_1)$. To do this, let us choose the following classical measurement-interaction Hamiltonian:

$$\underline{H} = g\underline{A}_1(q_1, \underline{p}_1)\underline{p}_2, \tag{29}$$

where g is a coupling constant. Let us further assume that the interaction is impulsive so that the individual free Hamiltonians of the particles are ignorable.

In classical mechanics we first have $d\underline{A}_1/dt = \{\underline{A}_1, \underline{H}\} = 0$. $\underline{A}_1$ is thus conserved during the measurement-interaction. The idea is then to correlate the value of $\underline{A}_1(q_1, \underline{p}_1)$ with the classical momentum of the apparatus $\underline{p}_2$ while keeping the value of $\underline{A}_1(q_1, \underline{p}_1)$ unchanged. On the other hand, one also has $dq_2/dt = \{q_2, \underline{H}\} = g\underline{A}_1$, which can be integrated to give

$$q_2(T) = q_2(0) + g\underline{A}_1 T, \tag{30}$$

where T is the time span of the measurement-interaction. The value of $\underline{A}_1$ prior to the measurement can thus be inferred from the observation of the initial and final values of q_2. In other words, the position of the second particle plays the role of the pointer of the measurement apparatus. The above model is of course far from realistic. Especially, our model excludes the irreversibility of the registration process which can only be done by realistic apparatus plus bath with large degree of freedom. See for example Ref. [8] for a realistic model of measurement. Here we shall focus on the issue of wave function collapse.

Now let us consider an ensemble of identically prepared system whose classical Hamiltonian is given by Eq. (29), and quantize it as in the previous section. For concreteness, without loosing generality, let us consider measurement of angular momentum. To make explicit the three dimensional nature of the problem, let us put $q_1 = (x_1, y_1, z_1)$. First let us consider the measurement of $z-$part angular momentum of the first particle

$$\underline{L}_{z1} = x_1\underline{p}_{y_1} - y_1\underline{p}_{x_1}, \tag{31}$$

where $\underline{p}_{x1}$ is the conjugate momentum of x_1 and so on. The measurement-interaction classical Hamiltonian of Eq. (29) thus reads

$$\underline{H}_l = g\underline{L}_{z_1}\underline{p}_2 = g(x_1\underline{p}_{y_1} - y_1\underline{p}_{x_1})\underline{p}_2. \tag{32}$$

The Hamilton-Jacobi equation of (3) then becomes

$$\partial_t\underline{S} + g(x_1\partial_{y_1}\underline{S} - y_1\partial_{x_1}\underline{S})\partial_{q_2}\underline{S} = 0. \tag{33}$$

On the other hand, substituting Eq. (32) into Eq. (5), the classical velocity field is given by

$$\dot{x}_1(\underline{S}) = -gy_1\partial_{q_2}\underline{S},\ \ \dot{y}_1(\underline{S}) = gx_1\partial_{q_2}\underline{S},\ \ \dot{z}_1(\underline{S}) = 0,\ \ \dot{q}_2(\underline{S}) = g(x_1\partial_{y_1}\underline{S} - y_1\partial_{x_1}\underline{S}). \tag{34}$$

The continuity equation of (4) then becomes

$$\partial_t\underline{\rho} - gy_1\partial_{x_1}(\underline{\rho}\partial_{q_2}\underline{S}) + gx_1\partial_{y_1}(\underline{\rho}\partial_{q_2}\underline{S}) + gx_1\partial_{q_2}(\underline{\rho}\partial_{y_1}\underline{S}) - gy_1\partial_{q_2}(\underline{\rho}\partial_{x_1}\underline{S}) = 0. \tag{35}$$

Next, keeping in mind Eq. (34), the rule of replacement of Eq. (7) thus becomes

$$\underline{\rho} \mapsto \Omega,\ \ \partial_{x_1}\underline{S} \mapsto \partial_{x_1}S + \frac{\lambda}{2}\frac{\partial_{x_1}\Omega}{\Omega},\ \ \partial_{y_1}\underline{S} \mapsto \partial_{y_1}S + \frac{\lambda}{2}\frac{\partial_{y_1}\Omega}{\Omega},\ \ \partial_{q_2}\underline{S} \mapsto \partial_{q_2}S + \frac{\lambda}{2}\frac{\partial_{q_2}\Omega}{\Omega},$$
$$\partial_t\underline{S} \mapsto \partial_t S + \frac{\lambda}{2}\frac{\partial_t\Omega}{\Omega} + g\lambda(x_1\partial_{y_1}\partial_{q_2}S - y_1\partial_{x_1}\partial_{q_2}S). \quad (36)$$

Let us proceed to see how Eq. (36) modifies Eqs. (33) and (35). Imposing the first four equations of (36) into Eq. (35) one obtains, after a simple calculation

$$\partial_t\Omega - gy_1\partial_{x_1}(\Omega\partial_{q_2}S) + gx_1\partial_{y_1}(\Omega\partial_{q_2}S) + gx_1\partial_{q_2}(\Omega\partial_{y_1}S)$$
$$-gy_1\partial_{q_2}(\Omega\partial_{x_1}S) - g\lambda(y_1\partial_{x_1}\partial_{q_2}\Omega - x_1\partial_{y_1}\partial_{q_2}\Omega) = 0. \quad (37)$$

On the other hand, imposing the last four equations of (36) into Eq. (33), one has, after an arrangement

$$\partial_t S + g(x_1\partial_{y_1}S - y_1\partial_{x_1}S)\partial_{q_2}S - g\lambda^2\Big(x_1\frac{\partial_{y_1}\partial_{q_2}R}{R} - y_1\frac{\partial_{x_1}\partial_{q_2}R}{R}\Big)$$
$$+\frac{\lambda}{2\Omega}\Big(\partial_t\Omega - gy_1\partial_{x_1}(\Omega\partial_{q_2}S) + gx_1\partial_{y_1}(\Omega\partial_{q_2}S) + gx_1\partial_{q_2}(\Omega\partial_{y_1}S) - gy_1\partial_{q_2}(\Omega\partial_{x_1}S)$$
$$-g\lambda(y_1\partial_{x_1}\partial_{q_2}\Omega - x_1\partial_{y_1}\partial_{q_2}\Omega)\Big) = 0, \quad (38)$$

where we have again defined a real-valued function $R \doteq \sqrt{\Omega}$ and used the identity of Eq. (18). Substituting Eq. (37) into Eq. (38), the last term in the bracket vanishes to give

$$\partial_t S + g(x_1\partial_{y_1}S - y_1\partial_{x_1}S)\partial_{q_2}S - g\lambda^2\Big(x_1\frac{\partial_{y_1}\partial_{q_2}R}{R} - y_1\frac{\partial_{x_1}\partial_{q_2}R}{R}\Big) = 0. \quad (39)$$

The dynamics of ensemble of trajectories is then determined by pair of coupled Eqs. (37) and (39) which are parametrized by the hidden random variable λ.

Again, assuming the symmetry of Eq. (20), one can see that $S(q,\lambda;t)$ and $S(q,-\lambda;t)$ satisfy the same differential equation of (39). Hence, assuming that initially $S(q,\lambda;0) = S(q,-\lambda;0)$, one obtains $S(q,\lambda;t) = S(q,-\lambda;t)$. As in the previous section, this can then be used to eliminate the last term on the left hand side of Eq. (37) to give

$$\partial_t\Omega - gy_1\partial_{x_1}(\Omega\partial_{q_2}S) + gx_1\partial_{y_1}(\Omega\partial_{q_2}S) + gx_1\partial_{q_2}(\Omega\partial_{y_1}S) - gy_1\partial_{q_2}(\Omega\partial_{x_1}S) = 0. \quad (40)$$

Further, defining complex-valued wave function as in Eq. (23), Eqs. (39) and (40) can then be rewritten into the following modified Schrödinger equation:

$$i|\lambda|\partial_t\Psi = -g\lambda^2(x_1\partial_{y_1} - y_1\partial_{x_1})\partial_{q_2}\Psi = g\frac{\lambda^2}{\hbar^2}\hat{L}_{z_1}\hat{p}_2\Psi, \quad (41)$$

where $\hat{L}_{z_1} \doteq -i\hbar(x_1\partial_{y_1} - y_1\partial_{x_1})$ and $\hat{p}_2 \doteq -i\hbar\partial_{q_2}$ are the quantum mechanical $z-$angular momentum and linear momentum operators pertaining to the wave functions of the first and second particle, respectively, and again we have assumed that the spatiotemporal fluctuations of λ is ignorable as compared to that of S. Finally, assuming that $\Omega(q,\lambda;t) = \rho(q;t)P(\lambda)$ is separable and taking the case when $P(\lambda)$ is given by Eq. (8), the above modified Schrödinger equation reduces to

$$i\hbar\partial_t\Psi_Q = \hat{H}_I\Psi_Q,\ \ \text{with}\ \ \hat{H}_I \doteq g\hat{L}_{z_1}\hat{p}_2, \quad (42)$$

where $\Psi_Q(q;t)$ is the Schrödinger wave function defined as in Eq. (25). This result can be extended to the measurement of angular momentum along the $x-$ and $y-$ directions by cyclic permutation of (x, y, z). In this case, $\hat{L}_{z_1}$ in Eq. (42) is replaced by $\hat{L}_{x_1}$ and $\hat{L}_{y_1}$, the quantum mechanical angular momentum operators along the $x-$ and $y-$ directions, respectively. We have thus reproduced the results of canonical quantization as a specific case of our hidden variable model. Further, in this case, Eq. (40) becomes

$$\partial_t \rho - g y_1 \partial_{x_1}(\rho \partial_{q_2} S_Q) + g x_1 \partial_{y_1}(\rho \partial_{q_2} S_Q) + g x_1 \partial_{q_2}(\rho \partial_{y_1} S_Q) - g y_1 \partial_{q_2}(\rho \partial_{x_1} S_Q) = 0, \tag{43}$$

from which we can extract an effective velocity field which is equal to the actual velocity field of particles in pilot-wave theory. This then allows us to follow all the argumentation of pilot-wave theory to describe measurement without wave function collapse and external observer reproducing the prediction of quantum mechanics.

As is discussed in Ref. [3], the above method can also be applied to measurement of position and momentum. Let us remark however that unlike measurement of linear and angular momentum, the measurement of position is special. Namely, we can show that in this case the pair of classical quantities $\{\underline{\rho}, \underline{S}\}$ and the corresponding quantum quantities $\{\rho, S_Q\}$ satisfy the same Schrödinger equation with quantum Hamiltonian $\hat{H} = g q_1 \hat{p}_2$. Hence, there is no quantum correction to the original classical equations. In other words, unlike measurement of linear and angular momentum, the measurement of position can reveal the value of the position of the particle prior to measurement. This implies important consequence. Recall that the results of measurement of linear and angular momentum (or measurement of any physical quantities) are inferred from the position of the second particle, namely the apparatus pointer (in reality, any model of measurement should be reducible to position measurement). Then one might argue that one needs another particle, the third particle, as the second apparatus to probe the position of the second particle (the first apparatus). Proceeding in this way thus will lead to infinite regression: one will further need the forth particle (the third apparatus) to probe the position of the third particle (the second apparatus) and so on. In our model, however, since the quantum treatment of the position measurement is equivalent to the classical treatment revealing the position of the particle prior-measurement, then the second measurement on the position of the second particle (the first apparatus) is in principle not necessary. Namely, the results of position measurement by the second, third, forth apparatuses and so on are all equal to each other.

4. Possible correction to quantum mechanical prediction

We have shown in the previous two Sections that the prediction of quantum mechanics is reproduced corresponding to a specific distribution of hidden random variable λ given by Eq. (8). It is then imperative to see the implications if one allows $|\lambda|$ to fluctuate around $\hbar$ with small yet finite width. In this case, instead of using Eqs. (25) or (42), one has to work with Eqs. (24) or (41), and regard them as the natural generalization of the Schrödinger equation.

Let us first discuss measurement of angular momentum in an ensemble of identically prepared systems so that the initial wave function of the system (first particle) $\psi(q_1)$ is given by one of the eigenfunctions of the angular momentum operator $\psi(q_1) = \phi_l(q_1)$, $\hat{L}_{z_1}\phi_l = l\phi_l$, where l is the eigenvalue. Let us denote the initial wave function of the apparatus (second particle) by $\varphi_0(q_2)$, assumed to be sufficiently localized. The total initial wave function of the system-apparatus is thus given by

$$\Psi(q; 0) = \phi_l(q_1)\varphi_0(q_2). \tag{44}$$

We have thus made an idealization that the initial wave function is independent of λ. Recall that in this case, according to the standard quantum mechanics, each single measurement event will give outcome l with certainty (probability one). This is one of the postulates of quantum mechanics in addition to the Schrödinger equation.

Let us solve Eq. (41) with the initial condition given by Eq. (44). To do this, let us assume that after time-span t of measurement-interaction, the wave function can be written as $\Psi(q,\lambda;t)=\phi_l(q_1)\varphi(q_2,\lambda;t)$. Inserting this into Eq. (41) and keeping in mind that $\hat{L}_{z_1}\phi_l=l\phi_l$, one has $\partial_t\varphi+gl'\partial_{q_2}\varphi=0$, with $l'(\lambda)=\frac{|\lambda|}{\hbar}l$. It can then be directly integrated with the initial condition $\varphi(q_2,\lambda;0)=\varphi_0(q_2)$ to give $\varphi(q_2,\lambda;t)=\varphi_0(q_2-gl't)$. One thus finally has $\Psi(q,\lambda;t)=\phi_l(q_1)\varphi_0(q_2-g|\lambda|lt/\hbar)$. Hence, in each single measurement event, the wave function of the apparatus becomes correlated to the initial state of the system and is shifted by an amount of $gl'(\lambda)t$. This means that at the end of each single measurement event, the initial position of the second particle (the apparatus pointer) is shifted uniformly and randomly as

$$q_2(t,\lambda)=q_2(0)+gl'(\lambda)t. \tag{45}$$

Now let us interpret the above formalism in a similar way as with classical measurement. As discussed in the previous section, in the latter case, after time-span of measurement-interaction t, the position of the apparatus-particle is shifted as $q_2(t)=q_2(0)+g\underline{L}_{z_1}t$. From this, one infers the result of measurement to be given by $\underline{L}_{z_1}$. Similarly, it is natural to interpret Eq. (45) that the outcome of each single measurement event is given by $l'(\lambda)=|\lambda|l/\hbar$. Hence, instead of obtaining a sharp value l as postulated by the standard quantum mechanics, one obtains a random value $l'(\lambda)$ which depends on the value of the hidden variable λ. One can also see that when the distribution of λ is given by Eq. (8) so that $\lambda=\pm\hbar$, then the randomness of the outcome of single measurement disappears and one regains the prediction of quantum mechanics: $l'(\pm\hbar)=l$ with probability one. For general distribution of λ satisfying Eq. (9), we have thus a random correction to the prediction of quantum mechanics: even when the initial wave function of the system is given by one of the eigenfunctions of the angular momentum operator, the result of each single measurement will still be random with statistical properties determined by the distribution of λ. This observation in turn leads to a finite broadening of the spectral line purely induced by the hidden random variable. Detailed elaboration of these observations is reported somewhere else [9].

Next, let us notice that the generalized Schrödinger equation of Eqs. (24) or (41) is still linear. Hence, given two solutions $\Psi_i(q,\lambda;t)=\sqrt{\Omega}\exp(iS/|\lambda|)$, $i=1,2$, one can construct new solution through linear superposition $\Psi_{12}=a\Psi_1+b\Psi_2$, where a and b are complex numbers. Calculating the probability density of the position, assuming that Ω_i is separable $\Omega_i(q,\lambda;t)=\rho_i(q;t)P(\lambda)$, the interference term then takes the following form

$$I_{12}(q;t)\sim\sqrt{\rho_1\rho_2}\int d\lambda P(\lambda)\exp\Big(\frac{iS_1}{|\lambda|}-\frac{iS_2}{|\lambda|}\Big), \tag{46}$$

which, for general type of $P(\lambda)=P(-\lambda)$ and $S_i(q,\lambda;t)$, will give results different from the quantum mechanics. In particular if $P(\lambda)$ takes a form of symmetric log-normal function then there will be Gaussian suppression of the quantum mechanical interference even in a closed system [10].

5. Conclusion and discussion

We have proposed a quantization method based on replacement of c-number by another c-number parametrized with hidden random variable. The result of canonical quantization is reproduced if the hidden random variable λ can only take discrete values $\pm\hbar$ with equal probability. Of course λ can be a function of other continuous hidden random variables. Unlike canonical quantization, the method is free from operator ordering ambiguity and has a straightforward interpretation as statistical modification of classical dynamics of ensemble of trajectories. Hence, the quantum-classical correspondence is kept physically transparent in the quantization processes. The classical limit of the Schrödinger equation is given by the classical dynamics of an ensemble of trajectories.

In particular, for all the systems considered, we can identity an "effective" velocity field which turns out to be equal to the "actual" velocity of the particle in pilot-wave theory. This then allows us to conclude that the model reproduces the statistical wave-like interference pattern in slits experiments; and further we can borrow all argumentation of pilot-wave theory on measurement without wave function collapse and external classical observer. However unlike pilot-wave theory, the model is stochastic, the wave function is not physically real and the Born's statistics is valid for all time by construction. Moreover, the construction is unique given the classical Lagrangian or Hamiltonian. Finally, assuming that $|\lambda|$ fluctuates around $\hbar$ with a very small yet finite width, the model predicts a small correction to the prediction of quantum mechanics. This might lead to precision tests of quantum mechanics against our hidden variable model.

It is then imperative to ask how our model will deal with Bell's no-go theory. Since our model reproduces the prediction of quantum mechanics for specific distribution of λ, then for this case, it must violate Bell inequality which implies that it is non-local in the sense of Bell [11], or there is no global Kolmogorovian space which covers all the probability spaces of the incompatible measurement in EPR-type of experiments [12], or both. We believe that this question can be discussed only if we know the physical origin of the the general rules of replacement postulated in Eq. (7). To this end, a discussion on the derivation of the rules from Hamilton-Jacobi theory with random constraint will be given somewhere else [13].

References

[1] Isham C J 1955 *Lectures on Quantum Theory: Mathematical and Structural Foundation* (London: Imperial College Press)

[2] Rund H 1966 *The Hamilton-Jacobi theory in the calculus of variations: its role in mathematics and physics* (London: Van Nostrand)

[3] Budiyono A 2012 to appear in *Physica A*; arXiv:1103.4737

[4] Nelson E 1966 *Phys. Rev.* **150** 1079; de la Peña-Auerbach L 1969 *J. Math. Phys.* **10** 1620; de la Pena-Auerbach L and Cetto A. M 1982 *Found. Phys.* **12** 1017; Santamato E 1984 *Phys. Rev. D* **29** 216; Frieden B R 1989 *Am. J. Phys.* **57** 1004; Nagasawa M 1989 *Probab. Th. Rel. Fields* **82** 109; Garbaczewski P and Vigier J-P 1992 *Phys. Rev. A* **46** 4634; Nottale L 1993 *Fractal Space-Time and Microscopic physics: Towards a Theory of Scale Relativity* (Singapore: World Scientific); Reginatto M 1998 *Phys. Rev. A* **58** 1775; Olavo L S F 1999 *Physica A* **262** 197; Kaniadakis G 2002 *Physica A* **307** 172; Hall M J W and Reginatto M 2002 *J. Phys. A: Math. Gen.* **35** 3289; Fritsche L and Haugk M 2003 *Ann. Phys. (Leipzig)* **12** 371; Markopoulou F and Smolin L 2004 *Phys. Rev. D* **70** 124029; Parwani R 2005 *J. Phys. A: Math. Gen.* **38** 6231; Smolin L 2006 arXiv:quant-ph/0609109v1; Santos E 2006 *Phys. Lett. A* **352** 49; Brénig L 2007 *J. Phys. A: Math. Theor.* **40** 4567; Grössing G 2008 *Phys. Lett. A* **372** 4556; Rusov V D, Vlasenko D S and Mavrodiev S C 2009 *Ukr. J. Phys.* **54** 1131; Gogberashvili M 2011 arXiv:gr-gc/1008.2544v2

[5] Hall M J W, Kumar K and Reginatto M 2003 *J. Phys. A* **36** 9779

[6] Bohm D 1952 *Phys. Rev.* **85** 166

[7] Philippidis C, Dewdney C and Hiley B 1979 *Nuovo Cimento* **52B** 15

[8] Allahverdyan A E, Balian R and Nieuwenhuizen T M 2003 *Europhys. Lett.* **61** 452

[9] Budiyono A 2012 to appear in *Physica A*; arXiv:1201.4554

[10] Budiyono A 2012 submitted for publication

[11] Bell J S 1988 *Speakable and Unspeakable in Quantum Mechanics* (Cambridge: Cambridge Univ. Press); Genovese M 2005 *Physics Rep.* **413** 319

[12] Accardi L 1981 *Phys. Rep* **77** 169; Fine A 1982 *Phys. Rev. Lett.* **48** 291; Pitowsky I 1982 *Phys. Rev. Lett.* **48** 1299; Rastal P 1983 *Found. Phys* **13** 555; Kupczynski M 1986 *Phys. Lett. A* **116** 417; Garola C and Solombrino L 1996 *Found. Phys.* **26** 1121; Cetto A M, Brody T and de la Peña L 1997 *Lett. Nuovo Cimento* **5** 177; Hess K and Philipp W 2001 *PNAS* **98** 14224; Zhao S, de Raedt H and Michielsen K 2008 *Found. Phys.* **38**, 322; Khrennikov A Y 2002 *Found. Phys* **32** 1159; Adenier G 2007 *J. Phys. B* **40** 131; Nieuwenhuizen T M 2011 *Found. Phys.* **41** 580.

[13] Budiyono A 2012 in preparation

Journal of Physics: Conference Series **361** (2012) 012019 doi:10.1088/1742-6596/361/1/012019

Particle and/or wave features in neutron interferometry

Helmut Rauch

Atominstitut, Vienna University of Technology, 1020 Wien, Austria

Abstract. Neutron interferometry provides a powerful tool to investigate particle and wave features in quantum physics. Single particle interference phenomena can be observed with neutrons and the entanglement of degrees of freedom, i.e., contextuality can be verified and used in further experiments. Entanglement of two photons, or atoms, is analogous to a double slit diffraction of a single photon, neutron or atom. Neutrons are proper tools for testing quantum mechanics because they are massive, they couple to electromagnetic fields due to their magnetic moment, they are subject to all basic interactions, and they are sensitive to topological effects, as well. The 4π-symmetry of spinor wave functions, the spin-superposition law and many topological phenomena can be made visible, thus showing interesting intrinsic features of quantum physics. Related experiments will be discussed. Deterministic and stochastic partial absorption experiments can be described by Bell-type inequalities. Neutron interferometry experiments based on post-selection methods renewed the discussion about quantum non-locality and the quantum measuring process. It has been shown that interference phenomena can be revived even when the overall interference pattern has lost its contrast. This indicates a persisting coupling in phase space even in cases of spatially separated Schrödinger cat-like situations. These states are extremely fragile and sensitive against any kind of fluctuations and other decoherence processes. More complete quantum experiments also show that a complete retrieval of quantum states behind an interaction volume becomes impossible in principle, but where and when a collapse of the wave-field occurs depends on the level of experiment.

1. Basic Relations

Neutrons are usually considered as particles but throughout this article they are considered as waves underlying the duality features of quantum physics. Experiments reported in this article have been performed with monochromatic low energy neutrons from a research reactor and with neutron interferometers based on wave-front and amplitude division [1, 2, 3, 4]. The most frequently used neutron interferometer is the perfect crystal interferometer first tested in 1974 at our 250 kW TRIGA reactor in Vienna. The wide beam separation of several centimeters and the relatively high intensity make it useful for fundamental-, nuclear- and solid-state physics [5] (Fig. 1). This kind of neutron interferometry is based on the undisturbed arrangement of atoms in a monolithic perfect silicon crystal [6, 2]. An incident beam is split coherently at the first crystal plate, reflected at the middle plate and coherently superposed at the third plate. From general symmetry considerations follows immediately that the wave functions in both beam paths, which compose the beam in the forward direction behind the interferometer, are equal ($\psi_0^{\mathrm{I}} = \psi_0^{\mathrm{II}}$), because they are transmitted-reflected-reflected (TRR) and reflected-reflected-transmitted (RRT), respectively. The theoretical treatment of the diffraction process from the perfect crystal is described by the dynamical diffraction theory [7, 8]. To preserve the interference

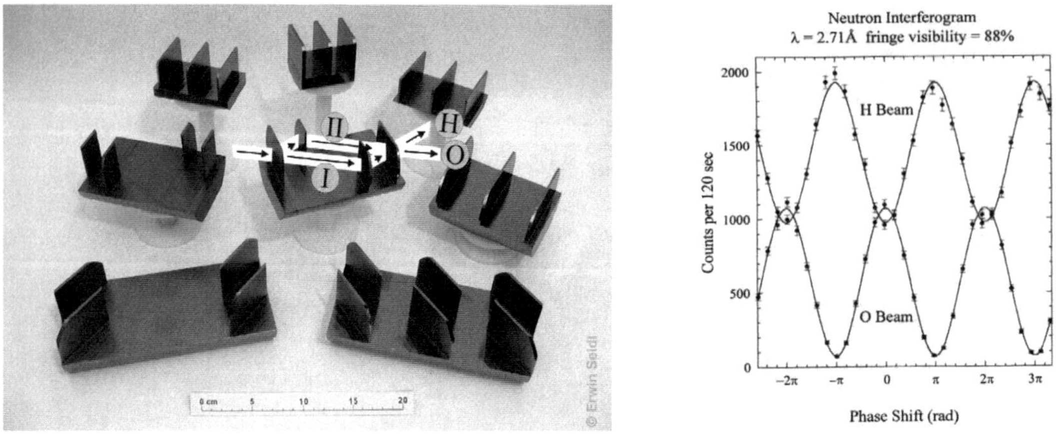

Figure 1. Various perfect crystal neutron interferometers and a typical interference pattern.

properties over the length of the interferometer, the lattice planes have to be parallel within one lattice constant and the dimensions of the monolithic system have to be accurate on a scale comparable to the so-called Pendellösung length (~50 μm). The whole interferometer crystal has to be placed on a stable goniometer table under conditions avoiding temperature gradients and vibrations. A phase shift between the two coherent beams can be produced by nuclear, magnetic or gravitational interactions. In the first case, the phase shift for non-absorbing and weakly absorbing materials is most easily calculated using the index of refraction [9, 10]:

$$n = \frac{k_{\rm in}}{k_0} = 1 - \frac{\lambda^2 N}{2\pi}\sqrt{b_{\rm c}^2 - \frac{\sigma_{\rm r}^2}{2\lambda}} + i\frac{\sigma_{\rm r} N\lambda}{2\pi} \simeq 1 - \lambda^2\frac{Nb_{\rm c}}{2\pi}\,, \tag{1}$$

where $b_{\rm c}$ is the coherent scattering length, $\sigma_{\rm r}$ the attenuation cross section, and N is the particle density of the phase shifting material. The different k-vector inside the phase shifter of thickness D causes a spatial shift of the wave packet which depends on the orientation of the sample surface $\hat{s}$ and which is related to the scalar phase shift χ by

$$\psi \to \psi_0 {\rm e}^{{\rm i}\vec{\Delta}\cdot\vec{k}} = \psi_0 {\rm e}^{{\rm i}(k_{\rm in}-k_0)D} = \psi_0 {\rm e}^{-{\rm i}Nb_{\rm c}\lambda D} = \psi_0 {\rm e}^{{\rm i}\chi}\,, \tag{2}$$

where χ can be written as a path integral of the canonical momentum $\mathrm{k_c}$ along the beam paths $\chi = \oint \vec{k}_{\rm c}\,{\rm d}\vec{s}$ [11]. Therefore, the intensity behind the interferometer becomes

$$\mathrm{I}_0 \propto \left|\psi_0^{\rm I} + \psi_0^{\rm II}\right|^2 \propto (1+\cos\chi)\,. \tag{3}$$

The intensity of the beam in the deviated direction $I_{\rm H}$ follows from particle conservation ($I_0 + I_{\rm H}$ = const.). Thus, the intensities behind the interferometer vary as a function of the thickness D of the phase shifter, the particle density N and the neutron wavelength λ.

Neutron optics is a part of quantum optics and many phenomena can be described properly in that terminology where the coherence function plays an important role [12, 13]

$$\Gamma(\vec{\Delta}) = <\psi(0)\,\psi(\vec{\Delta})>, \tag{4}$$

which is the autocorrelation function of the wave function. Using a wave packet description for the wave functions (amplitude $a(\vec{k})$)

$$\psi(\vec{x}) \propto \int a(\vec{k}) e^{{\rm i}\vec{k}\cdot\vec{x}}\,{\rm d}k, \tag{5}$$

one obtains

$$I_0(\vec{\Delta}) \quad \propto \quad \left|\psi_0^{\rm I} + \psi_0^{\rm II}\right|^2 \propto 1 + \left|\Gamma(\vec{\Delta})\right| \cos\chi_0 = 1 + |\Gamma(\Delta)|\,\cos(\vec{\Delta}_0\cdot\vec{k_0}), \tag{6}$$

where $\vec{\Delta}$ and χ_0 denote the phase shifts at the mean momentum $\vec{k}_0$. This gives:

$$\left|\Gamma(\vec{\Delta})\right| \propto \left|\int g(\vec{k})e^{i\vec{k}\cdot\vec{\Delta}}\,d^3\vec{k}\right| . \tag{7}$$

$\Gamma(\Delta)$ depends on the phase shift and therefore each interference fringe is slightly different from any other and this shows that each interference fringe has a distinct identity. The absolute value of the coherence function can be obtained from the fringe visibility $\left|\Gamma(\vec{\Delta})\right| = (I_{\mathrm{Max}} - I_{\mathrm{Min}})/(I_{\mathrm{Max}} + I_{\mathrm{Min}})$ or as the Fourier transform of the momentum distribution $g(\vec{k}) = \left|a(\vec{k})\right|^2$. The mean square distance related to $\left|\Gamma(\vec{\Delta})\right|$ defines the coherence length Δ_i^c which is for Gaussian distribution functions directly related to the minimum uncertainty relation ($\Delta_i^c \delta k_i = \frac{1}{2}$). Similar relations can be obtained for time-dependent phenomena where the spectral distribution $g(w)$ and the temporal coherence function come into play.

Any experimental device deviates from the idealized situations: the perfect crystal can have slight deviations from its perfectness, and its dimensions may vary slightly; the phase shifter contributes to such deviations by variations in its thickness and due to its inhomogeneities; and even the neutron beam itself contributes to a deviation from the idealized situation because of its momentum spread δk. Therefore, the experimental interference patterns have to be described by a generalized relation

$$\mathrm{I} \propto A + B\left|\Gamma(\vec{\Delta})\right| \cos(\chi + \Phi) . \tag{8}$$

where A, B and Φ are characteristic parameters of a certain set-up. It should be mentioned, however, that the idealized behavior described by Eq. (3) can be approached by a well balanced set-up (Fig. 1). Phase shifts can be applied in the longitudinal, transverse and vertical directions and the related coherence properties can be measured [14]. In the transverse direction the phase shift becomes wavelength independent ($\chi_T = -2d_{\mathrm{hkl}}Nb_cD_0$; d_{hkl} ... reflecting lattice plane distance), which implies a much larger coherence length in that direction.

All the results of interferometric measurements obtained up to now can be explained well in terms of the wave picture of quantum mechanics and the complementarity principle of standard quantum mechanics. Nevertheless, one should bear in mind that neutrons also carry well defined particle properties, which have to be transferred through the interferometer. These properties are summarized in Table 1 together with a formulation in the wave picture. Both particle and wave properties are well established and, therefore, neutrons seem to be a proper tool for testing quantum mechanics with massive particles, where the wave-particle dualism becomes very obvious.

All neutron interferometric experiments pertain to the case of self-interference, where during a certain time interval, only one neutron is inside the interferometer, if at all. Usually, at that time the next neutron has not yet been born and is still contained in the uranium nuclei of the reactor fuel. Although there is no interaction between different neutrons, they have a certain common history within predetermined limits which are defined, e.g., by the neutron moderation process, by their movement along the neutron guide tubes, by the monochromator crystal and by the special interferometer set-up. Therefore, any interferometer pattern contains single particle and ensemble properties together.

2. CLASSIC NEUTRON INTERFERENCE EXPERIMENTS

Here only short comments on these experiments are given since at least some of them provide the basis of the more recent investigation described in the following chapters. More details can be found in a related book [5].

Journal of Physics: Conference Series **361** (2012) 012019 doi:10.1088/1742-6596/361/1/012019

PARTICLE PROPERTIES		**WAVE PROPERTIES**
$m = 1.674928(1) \cdot 10^{-27}\,\mathrm{kg}$	CONNECTION	$\lambda_c = \frac{h}{m \cdot c} = 1.319695(20) \cdot 10^{-15}\,\mathrm{m}$
$s = \frac{1}{2}\hbar$ $\mu = -9.6491783(18) \cdot 10^{-27}\,\mathrm{J/T}$	de Broglie $\lambda_B = \frac{h}{m.v}$	for thermal neutrons: $\lambda = 1.8\,\mathrm{A}$, $v = 2200\,\mathrm{m/s}$
$\tau = 887(2)\,\mathrm{s}$ $R = 0.7\,\mathrm{fm}$ $\alpha = 12.0(2.5) \cdot 10^{-4}\,\mathrm{fm}^3$ u - d - d - quark structure	Schrödinger $H_\psi(\vec{r}, t) = i\hbar\frac{\delta\psi(\vec{r},r)}{\delta t}$ & boundary conditions	$\lambda_B = \frac{h}{m.v} = 1.8 \cdot 10^{-10}\,\mathrm{m}$ $\Delta_c = \frac{1}{2\delta k} \simeq 10^{-8}\,\mathrm{m}$ $\Delta_p = v \cdot \Delta t \simeq 10^{-2}\,\mathrm{m}$ $\Delta_d = v \cdot t = 1.942(5) \cdot 10^6\,\mathrm{m}$ $0 \leq \chi \leq 2\pi(4\pi)$
m mass, s spin, μ magnetic moment, τ β-decay lifetime, R (magnetic) confinement radius, α electric polarizability; all other measured quantities like electric charge, magnetic monopole, and magnetic dipole moment are compatible with zero.	$-\mu B$ ——— ↕ two level system μB ———	λ_c Compton wavelength, λ_B de Broglie wavelength, Δ_c coherence length, Δ_p packet length, δk momentum width, Δt chopper opening time, v group velocity, χ phase.

Table 1. Properties of neutrons.

2.1. 4π-Spinor symmetry

This is probably one of the most intensively discussed interference experiments done with matter waves. Based on elementary principles of quantum mechanics, the propagation of a wave function can be described by a unitary transformation, given by the relevant Hamiltonian. For magnetic interaction, $H_m = -\vec{\mu}\,\vec{B}$, the propagation of the two-component *spinor wave function*, which describes the neutron as a fermion, can be represented as follows:

$$\psi(t) = e^{\mathrm{i}Ht/\hbar}\,\psi(0) = e^{-\mathrm{i}\vec{\mu}\vec{B}t/\hbar}\,\psi(0) = e^{-\mathrm{i}\vec{\sigma}\vec{\alpha}/2}\,\psi(0) = \psi(\alpha), \tag{9}$$

where α means the *Larmor precession angle*

$$|\alpha| = \frac{2\mu}{\hbar}\int B dt = \frac{2\mu}{\hbar v}\int B ds. \tag{10}$$

When inserting the Pauli spin operators, one can easily show that $\psi(\alpha)$ has a 4π-*symmetry*, and not the 2π-symmetry which we are used to with respect to expectation values and within the scope of classical physics,

$$\begin{array}{l} \psi(2\pi) = -\psi(0) \\ \psi(4\pi) = \psi(0) \end{array} . \tag{11}$$

These facts, which were not previously regarded as verifiable, can be elucidated very easily with neutron interferometry by observing the intensity modulations, while one of the coherent beams

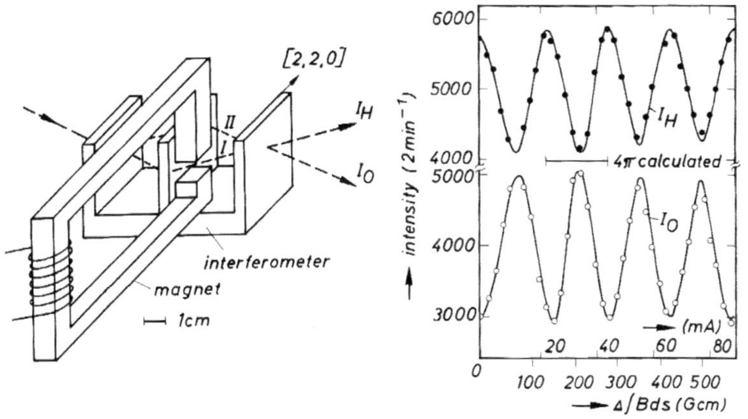

Figure 2. Results of the neutron interferometric 4π experiment [20].

passes through a magnetic field,

$$I_0 = |\psi_0(0) + \psi_0(\alpha)|^2 \propto \left(1 + \cos\frac{\alpha}{2}\right) . \tag{12}$$

The above relations are valid for polarized as well as for unpolarized neutrons, which points to the inherent symmetry properties of fermions. From Eqs. ((13)) and ((14)) one recognizes that only for $\alpha = 4\pi$ the original state is reproduced. This was verified, nearly simultaneously, in measurements by Rauch *et al.* [15] and by Werner *et al.* [16]. Afterwards, this effect was also proven through several other methods and for a series of other fermion systems.
A distinction between dynamical and topological phases will be discussed in Chapter 4.

2.2. Spin-superposition

Spin superposition is a frequently used principle of quantum mechanics. Its curiosity value has been stressed by Wigner [17]. The wave function of both coherent beams is originally polarized in $|z>$-direction. One beam is then inverted to a polarization in $|-z>$-direction, whereas the other remains unchanged. Both beams are then superimposed. This spin flip can be produced, for example, by Larmor precession around a magnetic field perpendicular to z-direction. The result of superposition of these two beams can be obtained by applying the rotation operator to the spin-flipped beam for a rotation of 180° in y-direction (Equ. (11)). If we also allow for a nuclear phase shift, one gets

$$\psi(\chi, \pi) = e^{i\chi}\, e^{-i\sigma_y \pi/2}\, |+z> \; = -i\pi\sigma_y\, e^{i\chi}\, |+z> = e^{i\chi}\, |-z> . \tag{13}$$

The total wave function $\psi = |+z> + e^{i\chi}\, |-z>$ leads to the following intensity and polarization of the out-going beam

$$I = \text{const.}, \;\; \vec{p} = \frac{\psi * \vec{\sigma}\psi}{\psi * \psi} = \begin{pmatrix} \cos\chi \\ \sin\chi \\ 0 \end{pmatrix} . \tag{14}$$

Thus, the intensity does not show any dependence on the phase shift but the polarization shows a marked χ dependence where the polarization vector lies in the x,y-plane, and is perpendicular to the polarizations of the two superimposed, coherent beams. The results of a related experiment are shown in Fig. 4. This implies that a pure quantum state in $|z>$-direction, e.g. for $c = 0$, has

been transformed into a quantum state in $|x>$-direction, and, in the sense of self-interference, which definitively applies here, it seems that each neutron has information about the physical situation in both of the widely separated coherent beams. The experiment by Summhammer *et al.* [18] has fully confirmed this process. Intensity modulations appear only when the polarization analysis is done in the x,y-plane.
The experiment mentioned above has been repeated with a *Rabi resonance flipper*, where an energy exchange of $\hbar w_r = 2mB_0$ occurs which causes a Larmor rotation of the polarization vector behind the interferometer [19].

In connection with these results, the obvious question arises whether the measurement of the energy transfer makes a determination of the beam path possible. One can, however, show that this is impossible, because interference vanishes in the presence of a measurable energy shift (i.e., larger than the energy width of the beam), and because the measurement of the energy change of the flip-field is impossible due to the *photon number-phase uncertainty relationship* ($\mathrm{D}\phi\mathrm{D}N > 1$).

2.3. Stochastic versus deterministic beam path detection
A certain beam attenuation can be achieved either by a semi-transparent material or by a proper chopper or slit system. The transmission probability in the first case is defined by the attenuation cross section σa of the phase shifting material $[a = I/I_0 = \exp(-\sigma_a ND)]$. The change of the wave function is obtained directly from the complex index of refraction (Equ. (1)):

$$\psi \rightarrow \psi_0 \mathrm{e}^{\mathrm{i}(n-1)kD} = \psi_0 \mathrm{e}^{\mathrm{i}\chi} \mathrm{e}^{-\sigma_a ND/2} = \psi_0 \mathrm{e}^{\mathrm{i}\chi} \sqrt{\underline{a}}\, \psi\,, \tag{15}$$

Therefore, the beam modulation behind the interferometer is obtained in the following form

$$I_0 \propto \left|\psi_0^{\mathrm{I}} + \psi_0^{\mathrm{II}}\right|^2 \propto [(1-\underline{a}) + 2\sqrt{\underline{a}}\, \cos\chi]\,. \tag{16}$$

On the other hand, the transmission probability of a chopper wheel or another shutter system is given by the open to closed ratio, $t = t_{\mathrm{open}}/(t_{\mathrm{open}} + t_{\mathrm{closed}})$, and one obtains after straightforward calculations

$$I \propto \left[(1-\underline{a})\left|\psi_0\right|^2 + \underline{a}\left|\psi_0^{\mathrm{I}} + \psi_0^{\mathrm{II}}\right|^2\right] \propto [(1-\underline{a}) + 2\underline{a}\cos\chi]\,. \tag{17}$$

i.e., the contrast of the interference pattern is proportional to $\sqrt{\underline{a}}$, in the first case, and proportional to $\underline{\underline{a}}$ in the second case, although the same number of neutrons are absorbed in both cases. The absorption represents a measuring process in both cases, i.e., a beam path detection, because compound nuclei are produced with an excitation energy of several MeV, which are usually de-excited by capture gamma rays. The measured contrast verifies the "stochastic" and "deterministi" predictions (Eqs. (16) and (17)) [20, 21]. The different contrast becomes especially obvious for low transmission probabilities. The discrepancy diverges for $\underline{a}$ but it has been shown that in this regime the variations of the transmission due to variations of the thickness or of the density of the absorber plate have to be taken into account which shifts the points below the $\sqrt{\underline{a}}$-("stochastic") curve [22]. The region between the linear and the square root behavior can be reached by very narrow chopper slits or by narrow transmission lattices, where one starts to loose information about which individual slit the neutron went through. This is exactly the region which shows the transition between a deterministic and a stochastic situation, i.e., between a particle-like and a wave-like behavior.

The stochastic limit corresponds to the quantum limit when one does not know anymore through which individual slit the neutron went. Which situation is given depends on how the slit widths l compare to the coherence lengths in the related direction. In case that the slit widths become comparable to the coherence lengths, the wave functions behind the slits show distinct diffraction peaks which correspond to new quantum states ($n \neq 0$). The creation of the new quantum states means that those labeled neutrons carry information about the chosen beam path and, therefore, do not contribute to the interference amplitude [23] (Fig. 3). A related

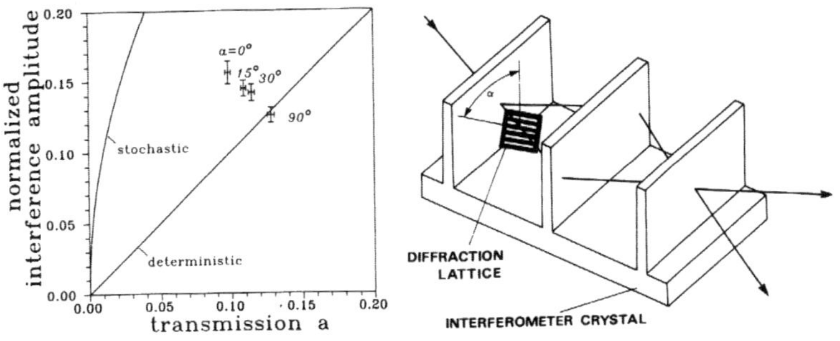

Figure 3. Lattice absorber in the interferometer approaching the classical limit when the slits are oriented horizontally and the quantum limit when they are oriented vertically [30].

experiment has been carried out by rotating an absorption lattice around the beam axis where one changes from $l \ll \Delta_x^c$(vertical slits) to $l \gg \Delta_y^c$ (horizontal slits). Thus, the attenuation factor $\underline{\underline{a}}$ has to be generalized including not only nuclear absorption and scattering processes but also lattice diffraction effects if they remove neutrons from the original phase space. The partial absorption and coherence experiments are closely connected to the quantum duality principle which states that the observation of an interference pattern and the acquisition of which-way information are mutually exclusive. Various inequalities have been formulated to describe this mutual exclusion principle [24, 25, 26]. The most concise formulation reads as

$$V^2 + P^2 \leq 1 \tag{18}$$

where V denotes the fringe visibility (Eq. (6)) and P is the predictability of the path through the interferometer, which is a quantitative measure of the *a priori* which-way knowledge.

3. Post-selection Experiments

Various post-selection measurements in neutron interferometry have shown that interference features can be restored by proper filtering methods even in cases when the overall beam does not exhibit any interference fringes due to spatial phase shifts larger than the coherence lengths of the interfering beams [27, 28]. Post-selection procedures can be applied to various parameters of an experiment:

(i) spatial post-selection
(ii) momentum post-selection
(iii) counting statistic post-selection
(iv) phase post-selection
(v) topology post-selection

In each case more information about the quantum system can be extracted than without post-selection. Fig. 4 shows some of them schematically. Here we discuss momentum post-selection and phase phase-echo experiments and for other methods we refer the reader to the literature [29, 30].

3.1. Post-selection of momentum states

The experimental arrangement with an indication of the wave packets at different parts of the interference experiment is shown in Fig. 4. An additional monochromatization is applied behind

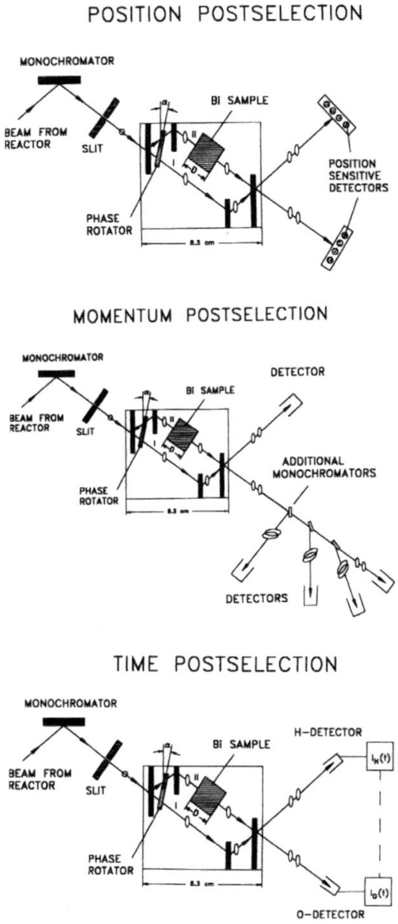

Figure 4. Sketch of various post-selection procedures.

the interferometer by means of Bragg diffraction from single crystals or by time-of-flight systems. The momentum-dependent intensity for Gaussian momentum distributions reads as:

$$I_0(k) = \exp[-(k-k_0)^2/2\delta k^{2]}\left\{1+\cos\left(\chi_0\frac{k_0}{k}\right)\right\} \tag{19}$$

The spatial phase shift-dependent intensity is given by Equ. (6). The formula show that the overall interference fringes disappear for spatial phase shifts much larger than the coherence lengths $[\Delta_i \geq \Delta_i^c = 1/(2\delta k_i)]$. The surprising feature is that $I_0(k)$ becomes oscillatory for large phase shifts where the interference fringes disappear ([27]). This indicates that interference in phase space has to be considered [31]. The amplitude function of the packets arising from beam paths I and II determines the spatial shape of the packets behind the interferometer

$$I_0(x) = |\psi(x) + \psi(x+\Delta)|^2\,, \tag{20}$$

which separates for large phase shifts ($\Delta \gg \Delta^c$), into two peaks. The related state can be interpreted as a superposition state of two macroscopically distinguishable states, that is a stationary Schrödinger cat-like state [31, 32] - here, for the first time, for massive particles. These states - separated in ordinary space and oscillating in momentum space - seem to be notoriously fragile and sensitive to dephasing and decoherence effects [32, 33, 34, 35, 36, 37].

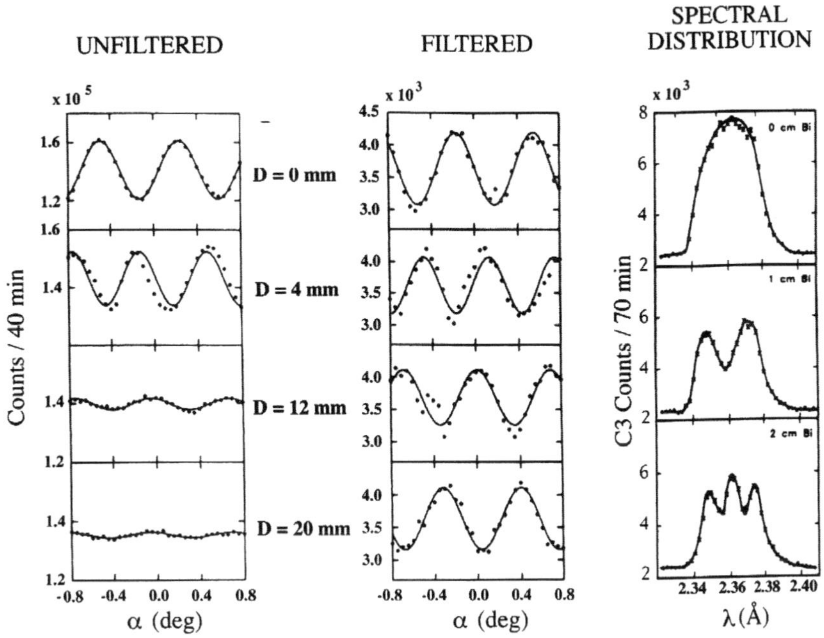

Figure 5. Interference pattern of the unfiltered overall beam ($dk/k_0 = 0.012$ middle) and the filtered beam reflected from a nearly perfect crystal analyzer in the anti-parallel position ($dk'/k_0 = 0.0003$, left), and the observed spectral modulation (right) of the outgoing beam for different phase shifter thicknesses [28]. These measurements have been done with a set-up shown in the middle of Fig. 4.

Measurements of the wavelength (momentum) spectrum were made with an additional silicon crystal placed behind the interferometer with a rather narrow mosaic spread (high resolution) which reflects in the parallel position a rather narrow band of neutrons only ($\delta k'/k_0 \approx 0.0003$) causing a restored visibility even at large phase shifts ([28], Fig. 5). This feature shows that an interference pattern can be revived even behind the interferometer by means of a proper post-selection procedure. In this case the overall beam does not show interference fringes anymore and the wave packets originating from the two different beam paths do not overlap. The momentum distribution has been measured by scanning the analyzer crystal through the Bragg-position. These results clearly demonstrate that the predicted spectral modulation (Eq. (19)) appears when the interference fringes of the overall beam disappear. The modulation is somehow smeared out due to averaging processes across the beam due to various imperfections, unavoidably existing in any experimental arrangement. The contrast of the empty interferometer was 60%.

It should be mentioned that momentum post-selection in typical Bell experiments (EPR-experiments) with entangled photons may also provide a less mystic view about these experiments since more information can be extracted when a momentum post-selection is added [27].

General conclusions about wave function properties should only be drawn if all accessible information about it is included. Thus the completeness of a quantum experiment has to be seen in a new light. The non-locality phenomenon of quantum mechanics can be understood as the far reaching action of the plane wave components of the wave function as well.

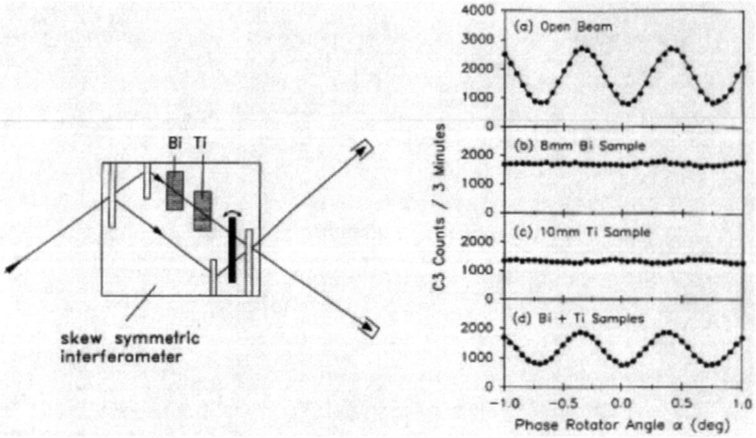

Figure 6. Loss of contrast at high interference and its retrieval by an opposite phase shifter inserted into the same beam [38].

3.2. Contrast retrieval by phase-echo

A large phase shift ($\Delta > \Delta^c$) can be applied in one arm of the interferometer, which can be compensated by a negative phase shift acting in the same arm or by the same phase shift applied to the second beam path [38]. Because the phase shift is additive, the coherence function depends on the net phase shift only. Thus, the interference pattern can be restored as it is shown in form of an experimental example in Fig. 6. The phase-echo method can also be applied behind the interferometer loop when multi-plate interferometers are used [30]. In this case, the situation becomes even more similar to the situation discussed in the previous section. The experimental results completely confirmed that behavior. Phase echo is a similar technique to spin echo [3], which is routinely used in neutron spectroscopy and which represents an interference experiment as well.

Nevertheless it should be mentioned that a complete retrieval seems to be impossible due to theoretical and practical limitations [39, 40].

4. Topological Effects

Topological and geometrical effects appear in the solution of the Schrödinger equation due to special geometric forms of the interaction [41, 42, 43]. Thus they are part of quantum mechanics but they are easily overlooked by pure intensity experiments. It also shows that a wave-function often carries more information than those extracted in a standard experiment. A typical example is the spin superposition experiment discussed in Chap. 2.3, where the result also depends around which axis the spin has been rotated into the opposite direction. In this respect the action of a Hamiltonian can be separated into a part related to its strength (dynamical) and its geometry, which results from the sum of state changes along the excursion in phase space

$$\phi = -\frac{1}{\hbar}\int_0^T <\psi(t)\,|\mathrm{H}|\,\psi(t)> \mathrm{d}t + \mathrm{i}\int_0^T <\phi(t)\,\frac{\mathrm{d}}{\mathrm{d}t}\,|\phi\,(t)> \mathrm{d}t = \alpha + \phi_{\mathrm{g}}, \tag{21}$$

with: $|\phi(t)>= e^{\mathrm{i}\phi}\,|\psi(t)>$.

Wagh *et al.* [44] did recently a related experiment and showed clearly the existence of the topological phase. In a similar sense the scalar and the vector *Aharonov-Bohm effects* of neutrons have been verified by neutron interferometric methods [45, 46]. In the case of an adiabatic excursion the geometrical phase becomes half the solid angle of the excursion seen on the Bloch

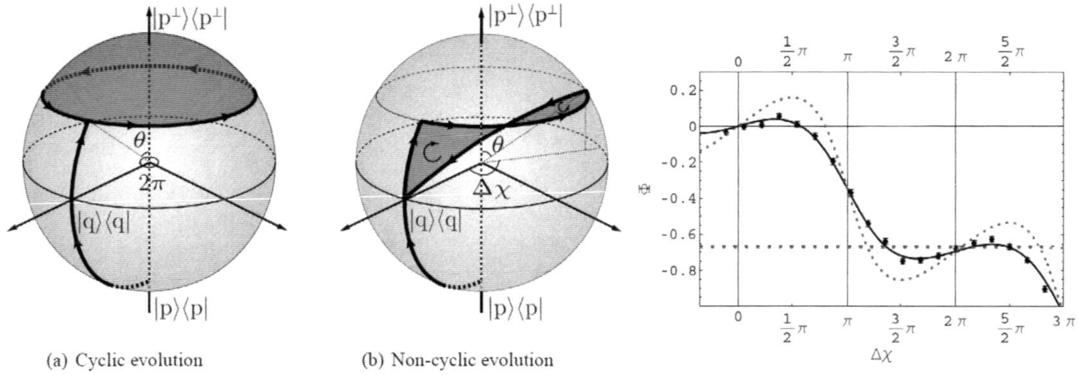

(a) Cyclic evolution (b) Non-cyclic evolution

Figure 7. Diagonal and off-diagonal geometric phases drawn on Poincaré spheres and results for a non-adabatic and non-cyclic excursion measurement [51].

sphere

$$\phi_g = -\frac{\Omega}{2}, \tag{22}$$

This has been verified recently with a high accuracy with ultra-cold neutrons guided by slowly varying magnetic fields $\phi_g = -0.51(1)\Omega$ [54].

Off-diagonal and non-adiabatic geometrical phases have been predicted as well [43, 48]. Detailed proposals and related experiments have been done [49, 50]. In a Poincaré representation diagonal phases are given by the solid angle opened up by the excursion line $|\psi_i>$ to $|\psi_f>$ and their geodesics to the pole, whereas off-diagonal phases are given by two excursion lines and their connection line in form of geodesics. In a related experiment non-adiabatic and non-cyclic phases have been verified with a double loop interferometer where two phase shifters (PS) and an absorber (A) permit quite peculiar state excursions as shown in Fig 7 [51].

It should be mentioned that just geometric phases show a high robustness against fluctuation and dissipative effects as predicted by DeChiara and Palma [59]. This has been verified experimentally by Filipp *et al.* [47], which may have remarkable consequences for quantum communication systems.

5. Quantum contextuality

A. Einstein, B. Podolsky and N. Rosen [53] argued that quantum mechanics may not be complete since non-local correlations between spatially separated systems are predicted, which stimulated the discussion about "hidden" variables and a more "realistic" theory. J. Bell [54] formulated inequalities which can decide between the quantum mechanical and the "realistic" view [55, 56]. Related experiments with entangled photons verified the non-local view of quantum mechanics [57, 58, 59, 60]. Entanglement does not only exist between two particles (photons), but also between different degrees of freedom of a single system (neutron). This yields to the concept of "contextuality", which states that independent measurements of independent observables are correlated. In our case the beam path through the interferometer and the spin states are taken as independent observables. In this case a Bell-like inequality can be formulated, which can be measured from the counting rates N at different values of the phase shift χ and the spin rotation angle a [50].

$$-2 \leq S \leq 2,$$

$$S = E(\alpha_1, \chi_1) + E(\alpha_1, \chi_2) - E(\alpha_2, \chi_1) + E(\alpha_2, \chi_2),$$

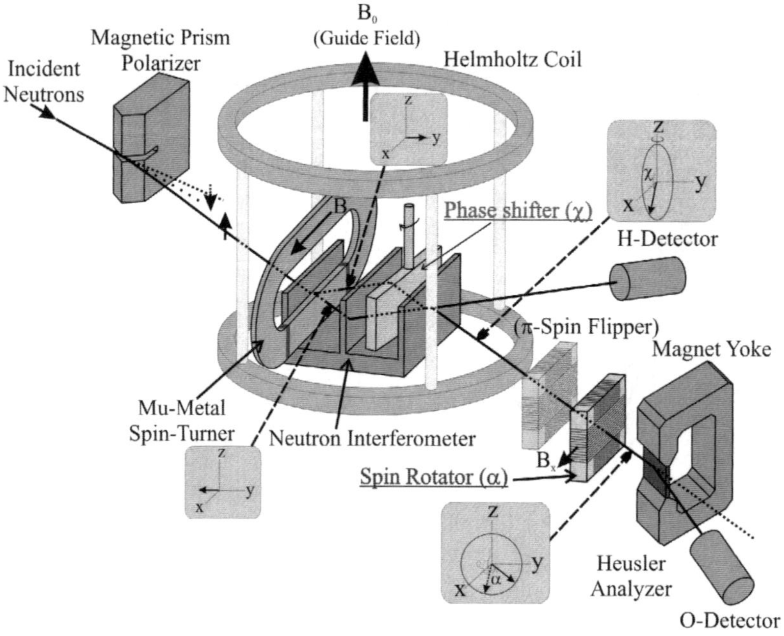

Figure 8. Sketch of the experimental setup for the contextuality experiment. The phase χ and the polarization rotation a could be varied independently [50].

$$E(\alpha,\chi) \;=\; \frac{N(\alpha,\chi)+N(\alpha+\pi,\chi+\pi)-N(\alpha,\chi+\pi)-N(\alpha+\pi,\chi)}{N(\alpha,\chi)+N(\alpha+\pi,\chi+\pi)+N(\alpha,\chi+\pi)+N(\alpha+\pi,\chi)}. \tag{23}$$

The maximal violation of this inequality due to quantum mechanics happens for the following parameters: $\alpha_1 = 0$, $\alpha_2 = \pi/2$, $\chi_1 = \pi/4$ and $\chi_2 = -\pi/4$ and amounts to $S = 2\sqrt{2} = 2.82$.

The measurement scheme is shown in Fig. 8. The entangled neutron state has been produced by rotating the neutron spin in beam path I into the |-y> and in beam path II into the |y> direction respectively. The precise determination of the related counting rates at the parameter values given above yielded a value for S of [50]:

$$S = 2.051 \pm 0.019,$$

which is by a 3s-limit above 2, verifying for the first time the contextuality principle of quantum mechanics. The maximal violation of $S = 2.82$ has not been achieved because the contrast of the interference pattern and the neutron polarization were below unity. In this kind of measurements these quantities play a similar role than the finite efficiency of the photon detectors in entangled photon experiments. In a subsequent and improved experiment a violation up to $S = 2.291 \pm 0.008$ has been measured [61].

The same set-up as shown in Fig. 8 has been used to perform experiments related to the Kochen-Specker theorem [62] and the Mermin inequalities [63], where even stronger violations of classical hidden variable theories can be verified. For neutron matter-waves a related proposal came from Basu *et al.* [49]. In this experiment the beam paths could be closed alternatively by means of an absorber sheet [64]. The measurement of the product observable $(\sigma_x^s\sigma_y^p)\cdot(\sigma_y^s\sigma_x^p)$ was done by measuring $(\sigma_z^s\sigma_z^p)$ and using *a priori* the non-contextuality relation. The measurable quantity is defined by a sum of product observables

$$C \;=\; \hat{I} - \sigma_x^s\sigma_x^p - \sigma_y^s\sigma_y^p - (\sigma_x^s\sigma_y^p)\cdot(\sigma_y^s\sigma_x^p). \tag{24}$$

In any experiment expectation values only can be measured. For non-contextual models the last term can be separated:

$$< (\sigma_x^s \sigma_y^p) >< (\sigma_y^s \sigma_y^p) > = < \sigma_x^s >< \sigma_y^p >< \sigma_y^s >< \sigma_x^p > \tag{25}$$

which gives

$$C_{nc} = \pm 2, \tag{26}$$

whereas quantum mechanics predicts

$$C_{qm} = 4. \tag{27}$$

The measured value was

$$C_{\text{exp}} = 3.138 \pm 0.0115, \tag{28}$$

which is well above the non-contextuality (classical) limit of 2 and provides an all-versus-nothing-type contradiction. It is also a Peres-Mermin proof of quantum-mechanics against non-contextual hidden variable theories.

A debate in literature [65, 66] criticized the *a priori* use of the non-contextuality relation $(\sigma_x^s \sigma_y^p) \cdot (\sigma_y^s \sigma_x^p) = (\sigma_z^s \sigma_z^p)$ and in this connection the use of an absorber to measure this quantity. In a follow-up proposal [67] and subsequent experiment [68] the previous result (Eq. 28) has been verified and an even stronger violation has been observed. In this case a quantum erasure has been used instead of an absorber and, therefore, all quantities required for Eq. 24 could be measured within the same context.

6. Discussion

It has been shown that more information about a quantum system can be extracted when more accessible parameters are measured, i.e., when post-selection methods are applied. It becomes obvious that a system may remain coupled in phase space even when it becomes separated in any other parameter space. Thus, interference properties can be shifted from one parameter space to another one and back again. Related bands of plane wave components which compose the wave packets may be considered as a responsible factor for the understanding of the coupling and non-locality phenomena in quantum mechanics. It looks like these plane wave components of the wave packets, i.e., narrow bands, interact over much larger distances than the sizes of the overall packets. This interaction guides neutrons of certain momentum bands to the 0- or H-beam, respectively. These phenomena throw a new light on the discussion on Schrödinger-cat-like situations in quantum mechanics [53, 55]. It may be considered as a contribution to speakable and unspeakable aspects of quantum mechanics [56]. Spatially separated packets remain entangled (correlated) in phase space and non-locality appears as a result of this entanglement. Since entanglement exists not only between objects but also between different degrees of freedom, Bell-type experiments can also be done in single particle experiments [56]. In this respect, contextuality experiments with neutrons (Chap. 5) may be of special value since they show that the experimental outcomes when measuring commuting observables (spin and beam path) are intrinsically correlated and quantum contextuality may be considered as an important feature of quantum physics. This also shows that quantum systems contain stronger correlations than classical ones and contextuality may cause an additional loophole for the deviation of Bell inequalities [69].

The summaries drawn for the different experimental situations discussed in this article are followed by statements that the retrieval of the interference properties by several post-selection procedures became increasingly more difficult the wider the separation in any parameter space of the quantum system happened before. This is caused by fluctuations, which are unavoidable due to residual quantum fluctuations inherent to any physical system.

Unavoidable fluctuations (even zero-point fluctuations) cause an irreversibility effect which becomes more influential for widely separated Schrödinger-cat like states. All these effects can be described by an increasing entropy inherently associated with any kind of interaction. This also supports the idea that irreversibility is a fundamental property of nature and reversibility an approximation only, as stated by several authors [70, 71, 72, 73].

All the results of the neutron interferometric experiments are well described by the formalism of quantum mechanics. According to the complementarity principle of the Copenhagen interpretation, the wave picture has to be used to describe the observed phenomena. The question of how the well-defined particle properties of the neutron are transferred through the interferometer is not a meaningful one within this interpretation, but from the physical point of view it should be an allowed one.

More complete quantum experiments show that a complete retrieval of all wave components behind an interaction the quantum system experienced becomes impossible, in principle. It also shows commuting variables are still correlated in the sense of quantum contextuality.

Acknowledgments

Most of the experimental work has been supported by projects of the Austrian Science Foundation (FWF), most recently by the project P 18943-N20. All coworkers mentioned in the self citations are gratefully acknowledged.

References

[1] Maier-Leibnitz H., and Springer T., Z. Physik 167, 386 (1962).
[2] Rauch H., Treimer W., Bonse U., Phys. Lett. A47, 369 (1974).
[3] Mezei F., Z. Physik 25, 146 (1972).
[4] Ioffe A.I., Zabiyankan V.S., Drabkin G.M., Phys. Lett. 111, 373 (1985).
[5] Rauch, H. and Werner S.A., "Neutron Interferometry", Clarendon Press, Oxford, 2000.
[6] Bonse U., Hart M., Appl. Phys. Lett. 6, 155 (1965).
[7] Bauspiess W., Bonse U., Graeff W., J. Appl. Cryst. 9, 68 (1976).
[8] Rauch H., and Petrascheck D., "Dynamical Neutron Diffraction and its Application", in Neutron Diffraction, edited by H. Dachs, Springer Verlag, StateBerlin, 1978, Chap. 9.
[9] Goldberger M.L., and Seitz F., Phys Rev. 71, 294 (1947).
[10] Sears V.F., "Neutron Optics", Oxford University Press, 1989.
[11] Feynman R.P., Leighton R.B., Sands M., "The Feynman Lectures on Physics", Vol. III, Addison-Wesley, 1965.
[12] Glauber R.J., Phys. Rev. 130, 2529 and 131, 2766 (1963).
[13] Mandel L., and Wolf, E., "Optical Coherence and Quantum Optics", Cambridge University Press, 1995.
[14] Rauch H., Wölwitsch H., Kaiser H., Clothier R., Werner S.A., Phys. Rev. A53, 902 (1996).
[15] Rauch H., Zeilinger A., Badurek G., Wilfing A., Bauspiess W., Bonse U., Phys. Lett. A54, 425 (1975).
[16] Werner S.A., Colella R., Overhauser, A.W., Eagen C.F., Phys. Rev. Lett. 35, 1053 (1975).
[17] Wigner E.P., Am. J. Phys. 31, 6 (1963).
[18] Summhammer J., Badurek G., Rauch H., Kischko U., Zeilinger A., Phys. Rev. A27, 2523 (1983).
[19] Badurek G., Rauch H., Summhammer J., Phys. Rev. Lett. 51, 1015 (1983).
[20] Summhammer J., Rauch H., Tuppinger D., Phys. Rev. A36, 4447 (1987).
[21] Rauch H., Summhammer J., Zawisky M., Jericha E., Phys. Rev. A42, 3726 (1990).
[22] Namiki M., Pascazio S., Phys. Lett. 147A, 430 (1990).
[23] Rauch H., and Summhammer J., Phys. Rev. 46, 7284 (1992).
[24] Wooters W.K., and Zurek W.H., Phys. Rev. D19, 473 (1979).
[25] Jaeger G., Shimony A., Vaidman L., Phys. Rev. A51, 54 (1995).
[26] Englert B.-G., Phys. Rev. Lett. 77, 2154 (1996).
[27] Rauch H. Phys. Lett. A173, 240 (1993).
[28] Jacobson D.L., Werner S.A., Rauch H., Phys. Rev. A49, 3196 (1994).
[29] Zawisky M., Rauch H., Hasegawa Y., Phys. Rev. A50, 5000 (1994).
[30] Heinrich M., Petrascheck D., Rauch H., Z. Physik B72, 357 (1988).
[31] Suda M., "Quantum interferometry in phase space: Theory and Applications", Springer, Berlin 2005.
[32] Schleich W., Pernigo M., Fam Le Kien Phys. Rev. A44, 2172 (1991).

[33] Walls D.F., Milburn G.J., Phys. Rev. A31, 2403 (1985).
[34] Zurek H., Physics Today, Oct. 1991, p.36.
[35] Blanchard P., and Jadczyk A., Phys. Lett. A175, 157 (1993).
[36] Buzek V., Keitel C.H., Knight P.L., Phys. Rev. A51, 2594 (1995).
[37] Jansky J., Vinogradov A.V., Phys. Rev. Lett. 64, 2771 (1990).
[38] Clothier R., Kaiser H., Werner S.A., Rauc H., Wölwitsch H., Phys. Rev. A44, 5357 (1991).
[39] Rauch H., J. Phys.: Conf.Series 36, 164 (2006).
[40] Rauch H., Phys.Scr. T135, 014027 (2009)
[41] Pancharatram S., Proc. Indian Acad. Sci. A44, 247 (1956).
[42] Berry M.V., Proc. Royal Soc. London A392, 45 (1984).
[43] Aharonov Y., Anandan J., Phys. Rev. Lett. 58, 1593 (1987).
[44] Wagh A.G., Rakhecha V.C., Summhammer J., Badurek G., Weinfurter H., Allman B.M., Kaiser H., Hamacher K., Jacobson D.L., Werner S.A., Phys. Rev. Lett. 78, 755 (1997).
[45] Allman B.E., Cimmino A., Klein A.G., Opat G.I., Kaiser H., Werner S.A., Phys. Rev. Lett. 68, 2409 (1992).
[46] Cimmino A., Opat G.I., Klein A.G., Kaiser, H. Werner S.A., Arif M., Clothier R., Phys. Rev. Lett. 63, 380 (1989).
[47] Filipp S., Klepp J., Hasegawa Y., Plonka-Spehr C., Schmidt U., Geltenbort P., Rauch H., Phys. Rev. Lett. 102, 030404 (2009)
[48] Samuel J., Bhandari R. Phys. Rev. Lett. 60, 2339 (1988).
[49] Basu S., Bandyopadhyay S., Kar G., Home D., Phys. Lett. A270, 281 (2001).
[50] Hasegawa Y., Loidl R., Badurek G., Baron M., Rauch H., Nature 425, 46 (2002).
[51] Filipp S., Hasegawa, Loidl R., Rauch H., Phys. Rev. A 72, 021602 (2005).
[52] DeChiara G., Palma G.M., Phys. Rev. Lett., 91, 090404 (2003).
[53] Einstein A., Podolsky B., Rosen N., Phys. Rev. 47, 777 (1935).
[54] Bell J., Physics 1, 195 (1955)
[55] Greenberger D.M., Horne M.A., Zeilinger A., "Going beyond Bell's Theorem", in "Bell's Theorem, Quantum Theory and Conceptions of the Universe" edited by M. Kafatos, Kluwer Publishing, Dordrecht, 1989, p.69.
[56] Bell S.J., "Speakable and Unspeakable in Quantum Mechanics", Cambridge University Press, Cambridge, 1987.
[57] Aspect A., Grangier P., Roger G., Phys. Rev. Lett. 49, 91 (1982).
[58] Ou Z.Y., Mandel L., Phys. Rev. Lett. 61, 50 (1988).
[59] Weihs G., Jennewein T., Simon C., Weinfurter H., Zeilinger A., Phys. Rev. Lett. 81, 5039 (1998).
[60] Zou X.Y., Grayson T.P., Mandel L., Phys. Rev. Lett. 69, 3041 (1992).
[61] Bartosik H., Klepp J., Schmitzer C., Sponar S., Cabello A., Rauch H., Hasegawa Y., Phys. Rev. Lett. 103, 040403 (2009).
[62] Kochen S., Specker F.P. J. Math. Mech. 17, 59 (1967).
[63] Mermin D., Phys. Rev. Lett. 65, 1838 (1990).
[64] Hasegawa Y., Loidl R., Badurek G., Baron M., Rauch H., Phys. Rev. Lett. 97, 230401 (2006).
[65] Simon C., Zukovski M., Weinfurter H., Zeilinger A., Phys. Rev. Lett. 85, 1783 (2000).
[66] Cinelli C., Barbieri M., Mataloni P., De Martini F., Phys. Rev. Lett. 95 240405 (2005).
[67] Cabello A., Filipp S., Rauch H., Hasegawa Y., Phys. Rev. Lett. 100, 130404 (2008).
[68] Sponar S., Klepp J., Loidl R., Filipp S., Badurek G., Rauch H., Phys. Rev. A78, 061604 (2008).
[69] Nieuwenhuizen T.M., Found.Phys. 41, 580 (2011).
[70] Lorentz H.A., "Theorie der Strahlung", Akademische Verlagsgesellschaft, Leipzig, 1927.
[71] Prigogine I., Proc. Ecol. Phys. Chem., Siena, Elsevier, Amsterdam, 1991, p. 8.
[72] Haag R., Comm. Math. Phys. 132, 245 (1990).
[73] Kono N., Machida K., Namiki M., Pascazio S., Phys. Rev. A54, 1064 (1996).

Nonlinear quantum mechanics, complex classical mechanics and conservation laws for closed and open systems

Dieter Schuch

Institut für Theoretische Physik, J.W. Goethe-Universität Frankfurt am Main, Max-von-Laue-Str. 1, D-60438 Frankfurt am Main, Germany

E-mail: Schuch@em.uni-frankfurt.de

Abstract. Phase and amplitude of the complex wave function are not independent of each other, but coupled, which becomes obvious looking at Madelung's hydrodynamic formulation of quantum mechanics. In the time-independent case, this leads to a kind of conservation law that allows for the reformulation of the linear Schrödinger equation in terms of a nonlinear Ermakov equation which is equivalent to a complex Riccati equation where the quadratic term in this equation explains the origin of the phase-amplitude coupling. A similar conservation law and corresponding nonlinear equations can also be found in the time-dependent case. The gain from the nonlinear formulations emerges when open systems with dissipation and irreversibility are considered. Describing this kind of systems by an effective nonlinear Schrödinger equation leads to a modification of the above-mentioned equations with new qualitative effects like Hopf bifurcations.

1. Introduction

This is my contribution to the proceedings of a conference held in Vienna in November 2011 in honour of the 100th birthday of Heinz von Foerster. Almost 25 years earlier, in March 1987, I attended a conference in London celebrating the centenary of the birth of another Austrian polymath, Erwin Schrödinger. One of the talks there was given by the Nobel Laureate Chen Ning Yang entitled "Square root of minus one, complex phases and Erwin Schrödinger" [1] in which he stated that with quantum mechanics, for the first time the imaginary unit enters physics in a fundamental way. So, he claims: "Complex numbers became a conceptual element of the very foundation of physics: the fundamental equations of matrix mechanics and wave mechanics,

$$p\,q - q\,p = -i\hbar \tag{1}$$

$$i\hbar\frac{\partial}{\partial t}\Psi = H\Psi \tag{2}$$

both explicitly contain the imaginary unit $i = \sqrt{-1}$. It is to be emphasized that the very meaning of these equations would be totally destroyed if one tries to get rid of i by writing (1) and (2) in terms of real and imaginary parts."

I totally agree with this statement, but want to go even a step further in this paper. The reason why it is not sufficient to simply write these equations in terms of two real equations shall be explained in the case of the Schrödinger equation (SE), Eq.(2). In Section 2, Madelung's hydrodynamic formulation [2] of Schrödinger's theory will be used, where the complex wave function is written in polar form to show that phase and amplitude (or real and imaginary parts), and hence the corresponding equations of motion are not independent of each other but uniquely coupled. First, for the time-independent (TI)SE, it will be shown that this coupling can be explained in terms of a conservation law that leads to a nonlinear (NL) formulation of quantum mechanics that can be rewritten in terms of a complex NL Riccati equation, where the origin of the coupling becomes obvious. Formally, a similar situation is also found in the case of the time-dependent (TD)SE, at least for quadratic Hamiltonians with Gaussian wave packet solutions. In this case, spatial derivatives are replaced by temporal ones, but the system of interrelated differential equations looks very much like in the TI case.

All realistic physical systems are in contact with some kind of environment, thus introducing phenomena like irreversibility and dissipation. There are different approaches in the literature for including these aspects in a quantum mechanical context. The ansatz that will be discussed in Section 3 is based on a modified equation for the probability density and leads to an effective NLSE with complex logarithmic nonlinearity. In the cases discussed in Section 2 where Gaussian wave packet solutions exist, there are also the same type of solutions for our NLSE. In Section 4, the modifications of the results found in Section 2 for the TD case will be discussed if dissipation and irreversibility are included and a particular example leading to Hopf bifurcation will be considered in detail. From the form of the modifications in the TD case one can draw conclusions for the TI case and find hints for the derivation of the corresponding equations. Section 5 will summarize the results and show possible further developments.

2. Complex Riccati equations and NL Ermakov equations in time-independent and time-dependent quantum mechanics

Already in the same year (1926) when Schrödinger's communications on quantum mechanics appeared, Madelung [2] published a paper where he used the polar form

$$\Psi(\mathbf{r},t) = \varrho^{1/2}(\mathbf{r},t)\exp\left(\frac{i}{\hbar}S(\mathbf{r},t)\right) \tag{3}$$

of the complex wave function that fulfils the TDSE

$$i\hbar\frac{\partial}{\partial t}\Psi(\mathbf{r},t) = \left(-\frac{\hbar^2}{2m}\Delta + V\right)\Psi(\mathbf{r},t) = H_L\Psi(\mathbf{r},t)\ . \tag{4}$$

Inserting (3) into (4), one obtains two real equations,

$$\frac{\partial}{\partial t}S + \frac{1}{2m}(\nabla S)^2 + V - \frac{\hbar^2}{2m}\frac{\Delta\varrho^{1/2}}{\varrho^{1/2}} = 0 \tag{5}$$

and

$$\frac{\partial}{\partial t}\varrho + \frac{1}{m}\nabla(\varrho\,\nabla S) = 0\ . \tag{6}$$

The first one essentially determines the evolution of the phase $S(\mathbf{r},t)$ and has almost the form of the Hamilton–Jacobi equation, only that there is an additional term, the last one on the rhs, that depends on the amplitude $\varrho^{1/2}$ of the wave function and is often deceptively called "quantum potential" (V_{qu}) although it is essentially related to the kinetic energy of the quantum system (see below). This equation reflects the classical mechanical (deterministic) trajectory aspect of the system however, via ϱ, also a probabilistic element enters. The quantity $\varrho = \Psi^*\Psi$ represents the probability density and fulfils the continuity equation (6). This equation that essentially determines the evolution of the amplitude of the wave function, contains the (gradient of) the phase in the probability current $\mathbf{j} = \varrho(\frac{1}{m}\nabla S)$, where the gradient of the action S is usually associated with the momentum of the system. So, the Hamilton–Jacobi equation (5) and the continuity equation (6) (or, phase and amplitude of the wave function Ψ) are coupled. This coupling, however, is not arbitrary but related to a conservation law which will be shown in the following for the TI case.

2.1. *Nonlinear formulations and invariants in time-independent quantum mechanics*

In 1994, G. Reinisch [3] presented a NL formulation of TI quantum mechanics. Since in this case $\frac{\partial}{\partial t}\varrho = 0$ and $\frac{\partial}{\partial t}S = -E$ are valid, the continuity equation (6) (we use now the notation $\varrho^{1/2} = |\Psi| = a$) turns into

$$\nabla(a^2\nabla S) \;=\; 0 \tag{7}$$

and the modified Hamilton-Jacobi equation into

$$-\frac{\hbar^2}{2m}\Delta a \;+\; (V-E)\,a \;=\; -\frac{1}{2m}(\nabla S)^2 a\;. \tag{8}$$

Equation (7) is certainly fulfilled for $\nabla S = 0$, turning (8) into the usual TISE for the real wave function $a = |\Psi|$ with position-independent phase $S(\mathbf{r})$. (NB: the kinetic energy term $-\frac{\hbar^2}{2m}\Delta a$ is just identical to $V_{qu}\ a$!)

However, Eq.(7) can also be fulfilled for $\nabla S \neq 0$ if only the conservation law

$$\nabla S \;=\; \frac{C}{a^2} \tag{9}$$

with constant (or, at least, position-independent) C is fulfilled. This relation now shows explicitly the coupling between phase and amplitude of the wave function. Inserting (9) into the rhs of Eq.(8) changes this into the real NL Ermakov equation

$$\Delta a \;+\; \frac{2m}{\hbar}(E-V)a \;=\; (\frac{1}{\hbar}\nabla S)^2\ a \;=\; \left(\frac{C}{\hbar}\right)^2\frac{1}{a^3} \tag{10}$$

with inverse cubic nonlinearity. (This equation had been studied by V. Ermakov [4] long before quantum mechanics was developed and we will come back to it in the TD case.) Knowing the solution of (10), via (9) the phase (up to an integration constant) can be determined and thus the complex wave function can be obtained.

For a better understanding of the origin of the coupling expressed by Eq.(9) it should be mentioned that the real NL Ermakov equation (10) is equivalent to a complex NL Riccati equation. In our case, this quadratic NL equation would read

$$\nabla\left(\frac{\nabla\Psi}{\Psi}\right) \;+\; \left(\frac{\nabla\Psi}{\Psi}\right)^2 \;+\; \frac{2m}{\hbar}(E-V) \;=\; 0 \tag{11}$$

with the complex variable $\left(\frac{\nabla\Psi}{\Psi}\right) = \frac{\nabla a}{a} + i\frac{1}{\hbar}\nabla S$.

First, it is straightforward to show that Eq.(11) can be linearised to yield the usual TISE,

$$-\frac{\hbar^2}{2m}\Delta a \;+\; V\ a \;=\; E\ a\;, \tag{12}$$

but inserting the definition of the complex variable into Eq.(11) and separating real and imaginary parts shows that the imaginary part leads to the conservation law (9), whereas the

Journal of Physics: Conference Series **361** (2012) 012020 doi:10.1088/1742-6596/361/1/012020

real part leads to the Ermakov equation (10). The origin of the coupling between phase and amplitude is obviously the quadratic nonlinearity since it introduces a term proportional to $(\nabla S)^2$ into the real part (Eq.(10)) and is responsible for the product term proportional to $(\nabla a)(\nabla S)$ in the imaginary part, leading to Eq.(9).

The same type of Riccati and Ermakov equations also occurs in the TD case, as will be shown in the following.

2.2. Nonlinear formulations and invariants in time-dependent quantum mechanics

The TDSE (4) always possesses for at most quadratic Hamiltonians exact analytic solutions in the form of Gaussians wave packets that can be written (in the following in one dimension) as

$$\Psi(x,t) = N(t) \exp\left\{ i \left[y(t)\tilde{x}^2 + \frac{1}{\hbar} < p > \tilde{x} + K(t) \right] \right\} \,, \tag{13}$$

where the maximum is at $\eta(t)$ which is identical with the mean value of the position, $< x >= \int_{-\infty}^{+\infty} \Psi^* x \Psi dx = \eta$, $\tilde{x} = x - \eta$ and the width of the WP is related with the imaginary part of the complex quantity $y(t)$ via $y_I = \frac{1}{4 < \tilde{x}^2 >}$ with $< \tilde{x}^2 > = < x^2 > - < x >^2$. ($N$ is a, possibly TD, normalization factor, $< p >$ the classical momentum and K a purely TD function that will not be relevant for the following.) Inserting wave packet (13) into the SE (4) leads to the equations of motion for $\eta(t)$ and $y(t)$. Particularly for the harmonic oscillator (i.e., $V = \frac{m}{2}\omega^2 x^2$ with possibly TD $\omega(t)$), one obtains the usual Newtonian equation of motion

$$\ddot{\eta} + \omega^2(t)\eta = 0 \tag{14}$$

for the maximum and a complex Riccati equation, but now for the TD variable $\left(\frac{2\hbar}{m} y\right)$,

$$\frac{2\hbar}{m}\dot{y} + \left(\frac{2\hbar}{m} y\right)^2 + \omega^2(t) \;=\; 0, \tag{15}$$

where overdots denote time-derivatives.

Introducing a new variable $\alpha_L(t)$ via $\frac{2\hbar}{m} y_I = \frac{\hbar}{2m < \tilde{x}^2 >_L} = \frac{1}{\alpha_L^2}$ allows one to determine the real part $\frac{2\hbar}{m} y_R$ from the imaginary part of Eq.(15) as $\frac{2\hbar}{m} y_R = \frac{\dot{\alpha}_L}{\alpha_L}$. Inserting this into the real part of (15) yields again an Ermakov equation,

$$\ddot{\alpha}_L + \omega^2(t)\alpha_L \;=\; \frac{1}{\alpha_L^3} \,. \tag{16}$$

Eliminating $\omega^2(t)$ from Eqs.(14) and (16) leads to the dynamical invariant

$$I_L \;=\; \frac{1}{2}\left[(\dot{\eta}\,\alpha_L - \eta\,\dot{\alpha}_L)^2 \;+\; \left(\frac{\eta}{\alpha_L}\right)^2 \right] \;=\; \text{constant} \tag{17}$$

that also exists for TD $\omega(t)$, whereas the corresponding Hamiltonian is no longer a dynamical invariant (for further details see, e.g., [4, 5]).

Also the NL Riccati equation (15) can be linearised via the ansatz $\left(\frac{2\hbar}{m} y\right) = \frac{\dot{\lambda}}{\lambda}$ with complex $\lambda(t)$ to yield the **complex** Newtonian equation

$$\ddot{\lambda} + \omega^2(t)\lambda = 0 \,. \tag{18}$$

With the polar form $\lambda = \alpha\, e^{i\varphi}$ the complex variable $\frac{2\hbar}{m} y$ can be written as $\frac{\dot{\lambda}}{\lambda} = \frac{\dot{\alpha}}{\alpha} + i\,\dot{\varphi}$. Inserting this into Eq.(15) shows that the imaginary part leads to the **conservation law**

$$\dot{\varphi} = \frac{1}{\alpha^2} \tag{19}$$

which represents a kind of conservation of angular momentum for the motion of $\lambda(t)$ in the complex plane [6]. From the real part of Eq.(15) one finally obtains the Ermakov equation (16) with $\alpha \equiv \alpha_L$.

3. Irreversibility, dissipation and effective NLSEs

Since realistic physical systems are always in contact with some kind of environment and this coupling usually introduces the phenomena irreversibility and dissipation (but not necessarily both simultaneously!), the question arises of how this can be taken into account particularly in a quantum mechanical context. On the classical level, one must go beyond the established formalism of canonical transformations to be able to include these aspects into the Hamiltonian formalism if one wants to avoid a rather unwieldy and computationally time-consuming inclusion of a large number of environmental degrees of freedom. There are, however, numerous approaches (see, e.g., [7, 8, 9] and references cited therein) that only take into account the effect of the environment on the system of interest by modifying the respective equation of motion without including the environmental degrees of freedom explicitly. In classical Newtonian mechanics, this can be done, e.g., by introducing a dissipative friction force. If it is assumed that this force depends linearly on the velocity (or momentum), this leads to

$$m\ddot{x} = m\dot{v} = -m\gamma v - \frac{\partial}{\partial x} V \tag{20}$$

which is the Langevin equation without stochastic contribution (γ is the friction coefficient).

This equation was used to describe Brownian motion from a trajectory point of view. An equivalent description of the phenomenon can also be given in terms of (classical) probability distributions ϱ_{cl} via the Fokker-Planck equations that contain irreversible diffusion terms. Particularly in position-space this can be written in form of the Smoluchowski equation [10] for $\varrho_{cl}(x,t)$,

$$\frac{\partial}{\partial t}\varrho_{cl} + \frac{\partial}{\partial x}\left(\frac{F(x)}{m\gamma}\varrho_{cl}\right) - \frac{\mathsf{kT}}{m\gamma}\frac{\partial^2}{\partial x^2}\varrho_{cl} = 0\ , \tag{21}$$

with the conservative force $F(x)$, Boltzmann's constant k and temperature T. Comparison with Einstein's theory of Brownian motion [11] shows that the coefficient of the diffusion term fulfils the Einstein relation $D = \frac{\mathsf{kT}}{m\gamma}$.

Equation (21) provided the basis for our modification of the SE in order to include irreversibility and dissipation. We started from the continuity equation (6) for the probability density $\varrho(\mathbf{r},t)$, for which Madelung [12] and Mrowka [13] had shown that with a bilinear ansatz for the probability density, $\varrho = \Psi^*\Psi$, and for the probability current density, $\mathbf{j} = \varrho\left(\frac{1}{m}\nabla S\right) = \varrho\mathbf{v}_- = \left(\frac{\hbar}{2mi}\right)(\Psi^*\nabla\Psi - \Psi\nabla\Psi^*)$, this can be separated into the TDSE and its complex conjugate, where the separation "constant" is proportional to the potential V. Keeping Eq.(21) in mind, we tried to break the time-symmetry by adding an irreversible diffusion term leading (in the following again in one dimension) to

$$\frac{\partial}{\partial t}\varrho + \frac{\partial}{\partial x}(\varrho\, v_-) - D\frac{\partial^2}{\partial x^2}\varrho = 0\ . \tag{22}$$

However, due to the diffusion term, Ψ- and Ψ^*-dependent terms can generally no longer be separated. But if one introduces an additional separation condition

$$-D\frac{\frac{\partial^2}{\partial x^2}\varrho}{\varrho} = \gamma\,(\ln\varrho - <\ln\varrho>)\ , \tag{23}$$

where $\ln\varrho = \ln\Psi + \ln\Psi^*$, separation is possible and leads to an additional complex logarithmic term in the SE [8] for $\Psi_{NL}(x,t)$,

$$i\hbar\frac{\partial}{\partial t}\Psi_{NL} = \{H_L + \gamma\,\frac{\hbar}{i}\,(\ln\Psi_{NL} - <\ln\Psi_{NL}>)\}\Psi_{NL} = \{H_L + W\}\Psi_{NL}\ , \tag{24}$$

where H_L is the usual linear Hamiltonian. Now the question arises, has this additional term W, that can be written as real and imaginary contributions in the form

$$W = W_R + W_I = \frac{\gamma}{2}\frac{\hbar}{i}\left(\ln\frac{\Psi_{NL}}{\Psi^*_{NL}} - <\ln\frac{\Psi_{NL}}{\Psi^*_{NL}}>\right) + \frac{\gamma}{2}\frac{\hbar}{i}\,(\ln\varrho_{NL} - <\ln\varrho_{NL}>)\ , \tag{25}$$

any physical meaning?

The real part was already know in the literature (see: Kostin [14]) and leads, according to Ehrenfest, to a linear velocity dependent term in the equation of motion for the mean value of position, thus turning Eq.(14) into a Langevin equation like (20).

Since the mean value of W_R (as well as of W_I) vanishes, the mean value of the energy is still given by $< H_L >$, but now calculated with Ψ_{NL}, the solution of the NLSE (24). Therefore, the energy is no longer constant but decreases in a way that would be expected from the classical counterpart. However, W_R alone does not completely describe the physics of this system, as has been criticized in comments concerning Kostin's work [7, 8]. In particular, W_R would still allow the solutions of the undamped harmonic oscillator to be also solutions of the damped one and, since W_R is real, the equation of motion for $\varrho_{NL} = \Psi^*_{NL}\Psi_{NL}$ would still be the reversible continuity equation (6). All these problems are eliminated when the imaginary part W_I is also taken into account.

A term formally like the imaginary part W_I (but in the context of density operators) was independently discussed around the same time by Beretta [15] to describe non-equilibrium systems (without dissipation). An interesting interpretation of W_I can be found if one identifies, according to Grössing et al [16], the Einstein diffusion coefficient with the quantum mechanical one (if the SE is considered a diffusion equation with imaginary diffusion coefficient), i.e., $D = \frac{\mathsf{kT}}{m\gamma} = \frac{\hbar}{2m}$. Then, W_I turns into

$$W_I = -\,i\mathsf{Tk}\,(\ln\varrho_{NL} - <\ln\varrho_{NL}>)\ , \tag{26}$$

where $-\mathsf{k} < \ln\varrho >= -\mathsf{k}\int_{-\infty}^{+\infty}\varrho\ln\varrho dx$ has a form like the definition of entropy, $\mathcal{S}$. So, the mean value of the linear Hamiltonian that still represents the energy of the system, $< H_L >= E$, together with the second term of (26) would look like $E - i\mathsf{T}\mathcal{S}$, i.e., it has similarity with an expression for the free energy only that here, again, the imaginary unit i turns up in the quantum mechanical context. This point still needs further investigation.

It should be emphasized that the imaginary part by itself does not introduce any dissipation of energy but a kind of non-unitarity in the time-evolution of the wave function, however, still guaranteeing normalizability (due to the subtraction of $< \ln\varrho >$).

The fact that both real and imaginary parts of W are necessary to describe correctly dissipation as well as irreversibility of the system, where W_R depends only on the phase of the wave function, while W_I only depends on its amplitude, shows again that phase and amplitude are not independent of each other, but coupled, as already shown in the conservative reversible case.

How including the environment modifies the NL relations, and thus the coupling between phase and amplitude that were discussed in Sections 2.1 and 2.2, will be shown in the following.

4. Modification of the complex Riccati and nonlinear Ermakov equations

4.1. Time-dependent case

In the cases where the TDSE (4) possesses exact Gaussian-type wave packet solutions of the form (13), the same also holds for the NLSE (24). Due to the logarithmic nonlinearity, the equations of motion for the wave packet maximum and the wave packet width are modified. The maximum now follows the classical trajectory obeying the Newton-type equation

$$\ddot{\eta} + \gamma\dot{\eta} + \omega^2\eta = 0 \tag{27}$$

with an additional friction force and the complex NL Riccati equation gains an additional linear term according to

$$\frac{2\hbar}{m}\dot{y} + \gamma\left(\frac{2\hbar}{m}y\right) + \left(\frac{2\hbar}{m}y\right)^2 + \omega^2(t) \;=\; 0 \tag{28}$$

(here again for the harmonic oscillator potential).

With a definition analogous to the one in the conservative case, a new variable α_{NL} is introduced via $\frac{2\hbar}{m}y_I = \frac{\hbar}{2m<\tilde{x}^2>_{NL}} = \frac{1}{\alpha_{NL}^2}$. Inserting this into Eq.(28) yields from its imaginary part $\frac{2\hbar}{m}y_R = \frac{\dot{\alpha}_{NL}}{\alpha_{NL}} - \frac{\gamma}{2}$ which, inserted into the real part of (28), turns it into an Ermakov equation with shifted frequency

$$\ddot{\alpha}_{NL} + \left(\omega^2(t) - \frac{\gamma^2}{4}\right)\alpha_{NL} = \frac{1}{\alpha_{NL}^3} \quad . \tag{29}$$

Also in this dissipative case, a dynamical Ermakov invariant can be obtained by eliminating ω^2 between Eqs.(27) and (29), leading to

$$I_{NL} \;=\; \frac{1}{2}e^{\gamma t}\left[\left(\dot{\eta}\,\alpha_{NL} - \left(\dot{\alpha}_{NL} - \frac{\gamma}{2}\alpha_{NL}\right)\eta\right)^2 + \left(\frac{\eta}{\alpha_{NL}}\right)^2\right] \;=\; \text{constant} \; . \tag{30}$$

An interesting qualitative change occurs when considering the modified Riccati equation (28), which shall be shown in the following for the damped free motion. Knowing a particular solution $\left(\frac{2\hbar}{m}\tilde{y}\right)$ of the inhomogeneous differential equation (28) allows it to be turned into the homogeneous Bernoulli equation

$$\frac{2\hbar}{m}\dot{v} + \left(\gamma + \frac{4\hbar}{m}\tilde{y}\right)\left(\frac{2\hbar}{m}v\right) + \left(\frac{2\hbar}{m}v\right)^2 \;=\; 0 \tag{31}$$

with $\gamma + \frac{4\hbar}{m}\tilde{y} = A$, where the general solution of the Riccati equation is obtained from $\frac{2\hbar}{m}y = \frac{2\hbar}{m}\tilde{y} + \frac{2\hbar}{m}v(t)$. For $\frac{2\hbar}{m}\tilde{y} =$ constant one finds

$$A \;=\; \pm 2\left(\frac{\gamma^2}{4} - \omega^2\right)^{1/2} \tag{32}$$

or, in our case for $\omega = 0$, $A = \pm\gamma$. In the conservative case, for $\gamma = 0$, only one solution exists with $A = 0$, but now there are two different values possible for A (like in a Hopf bifurcation) and, therefore, two different solutions which both correspond to different wave packets with different widths and different contributions to the energy. For comparison, the position uncertainty $< \tilde{x}^2 > (t)$ and the minimum energy that is proportional to the momentum uncertainty via $\tilde{E} = \frac{1}{2m} < \tilde{p}^2 >= \frac{1}{2m}(< p^2 > - < p >^2)$ shall be given for the conservative ($\gamma = 0$) and dissipative ($\gamma \neq 0$) case, respectively.

For $A = 0$, one obtains

$$< \tilde{x}^2 >_L \;=\; < \tilde{x}^2 >_0 \{1 + (\beta_0 t)^2\} \tag{33}$$

$$\tilde{E}_L = \frac{<\tilde{p}^2>_0}{2m} = \frac{\hbar}{4}\,\beta_0 \tag{34}$$

with the constant frequency-type quantity $\beta_0 = \frac{\hbar}{2m<\tilde{x}^2>_0}$, depending on the initial position uncertainty $<\tilde{x}^2>_0$.

For $A = +\gamma$, this changes into

$$<\tilde{x}^2>_+ = <\tilde{x}^2>_0 \left\{ e^{\gamma t} + \left(\frac{\beta_0}{\gamma/2}\right)^2 \sinh^2 \frac{\gamma}{2} t \right\} \tag{35}$$

$$\tilde{E}_+ = \frac{\hbar}{4}\,\beta_0\, e^{-\gamma t}\,, \tag{36}$$

for $A = -\gamma$ into

$$<\tilde{x}^2>_- = <\tilde{x}^2>_0 \left\{ e^{-\gamma t} + \left(\frac{\beta_0}{\gamma/2}\right)^2 \sinh^2 \frac{\gamma}{2} t \right\} \tag{37}$$

$$\tilde{E}_- = \frac{\hbar}{4}\,\beta_0 \left(1 + \left(\frac{\gamma}{\beta_0}\right)^2\right) e^{-\gamma t}\,. \tag{38}$$

It becomes obvious that the coupling to the environment in both cases causes a faster spreading of the wave packet width (due to a faster dephasing; for details see [17]) and an exponential decay of $\tilde{E}_\pm$ (nevertheless, the uncertainty product always fulfils the relation $U_\pm = <\tilde{x}^2>_\pm <\tilde{p}^2>_\pm \geq \frac{\hbar^2}{4}$).

To further compare these results with the conservative case, we consider the time $t = 0$ with $\tilde{E}_+(0) = \frac{\hbar}{4}\beta_0 = \tilde{E}_L$ and $\tilde{E}_-(0) = \frac{\hbar}{4}\beta_0\left(1 + \left(\frac{\gamma}{\beta_0}\right)^2\right) > \tilde{E}_L$. The "splitting" of energy level is given by

$$\Delta\tilde{E}_0 = \tilde{E}_-(0) - \tilde{E}_+(0) = \frac{\hbar\,\gamma^2}{4\,\beta_0} = \frac{m}{2}\gamma^2 <\tilde{x}^2>_0 \tag{39}$$

and is thus independent of $\hbar$, but only depending on γ, the parameter that characterizes the environment! From the Smoluchowski equation (22) and the NLSE (24) it follows that $D = \frac{\gamma}{2} <\tilde{x}^2>$ (in particular for time-dependent $<\tilde{x}^2>(t)$, the diffusion coefficient also becomes time-dependent). Assuming for $t = 0$ an equilibrium condition where the Einstein relation applies, we then obtain

$$\Delta\tilde{E}_0 = \mathsf{kT}\,. \tag{40}$$

So, the coupling to the environment (thermal bath) breaks the time-symmetry and removes a kind of "degeneracy" of the ground state energy, introducing a new state where this second state is initially kT higher in energy than the original quantum mechanical ground state and both energies, $\tilde{E}_+$ and $\tilde{E}_-$, decay exponentially.

Finally, it shall be mentioned that also the Riccati equation (28) can be linearised via $\left(\frac{2\hbar}{m}y\right) = \frac{\dot{\tilde{\lambda}}}{\tilde{\lambda}}$ with $\tilde{\lambda} = \lambda e^{-\gamma t/2} = \alpha_{NL} e^{-\gamma t/2 + i\varphi}$ to yield the complex Newtonian equation

$$\ddot{\tilde{\lambda}} + \gamma\dot{\tilde{\lambda}} + \omega^2\tilde{\lambda} = 0\,. \tag{41}$$

Also in this case, phase and amplitude of $\tilde{\lambda}$ are coupled via $\dot{\varphi} = \frac{1}{\alpha_{NL}^2}$ and α_{NL} can be determined by solving the Ermakov equation (29).

4.2. Time-independent case

Due to the formal similarities between the systems of equations that we discussed in Section 2 for the TDSE and TISE, the modifications we obtained in the TD case when a dissipative environment is included may provide a hint in which way the corresponding equations in the TI situation must be modified. From the Riccati equation (28) one can conclude that a term proportional to $\left(\frac{\nabla\Psi}{\Psi}\right)$ must be added to Eq.(11), or, from Eqs.(27) and (41), that a term proportional to $\nabla\Psi$ must be added to the TISE (12). Since Ψ is complex, from the imaginary part one would obtain an additional contribution to the continuity equation (6). In the following, only a short outline of the reasoning shall be given as to how the additional terms can be obtained; further details will be presented elsewhere [18]. Since the SE can be obtained from the continuity equation via separation, as well as the NLSE from the Smoluchowski equation (plus separation condition), the additional term in the continuity equation should take a separable form proportional to $\nabla\varrho = (\Psi^*\nabla\Psi + \Psi\nabla\Psi^*)$, leading to a $\nabla\Psi$-term with imaginary coefficient in the corresponding modified TISE. Furthermore, at least for the cases with Gaussian wave packet solutions, this additional term should have the same effect as the diffusion term or the $\ln\varrho$-term, i.e.,

$$-D\Delta_x\varrho \;=\; \gamma(\ln\varrho - <\ln\varrho>)\varrho \;=\; \frac{\gamma}{2}\left(1 \;-\; \frac{\tilde{x}^2}{<\tilde{x}^2>}\right)\varrho \qquad (42)$$

should be reproduced. Comparison shows that $\nabla_x\left(\frac{\gamma}{2}\tilde{x}\varrho\right)$ yields the desired result.

To compare the modified equations with the situation discussed in Section 2.1., we consider again the TI case with $\frac{\partial}{\partial t}\varrho = 0$ (and use again the notation $\varrho^{1/2} = |\Psi| = a$), where Eq.(7) is now replaced by

$$\nabla\left(a^2\left(\nabla S + m\frac{\gamma}{2}\tilde{x}\right)\right) \;=\; \nabla\left(a^2\nabla S'\right) \;=\; 0 \qquad (43)$$

with $S' = S + m\frac{\gamma}{4}\tilde{x}^2 + f(t)$. So now the conservation law (9) must be replaced by $\nabla S' = \frac{C}{a^2}$ for the modified action S', i.e. a change of the phase, corresponding to a gauge transformation, must be performed.

Consequently, also the modified Hamilton–Jacobi equation (8) must be changed which (with $\frac{\partial}{\partial t}S = -E$) finally replaces the Ermakov equation (10) by

$$\Delta a \;+\; \frac{2m}{\hbar}\left(E - \left(V - \frac{m}{2}\frac{\gamma^2}{4}\tilde{x}^2\right)\right)a \;=\; (\frac{1}{\hbar}\nabla S')^2\; a \;=\; \left(\frac{C}{\hbar}\right)^2\frac{1}{a^3}\,. \qquad (44)$$

5. Conclusions

We start out with the statement that a fundamental difference between classical physics and quantum mechanics exists because the latter is based essentially on complex quantities where the imaginary unit cannot simply be eliminated by writing down two equations for the real and imaginary parts of the complex quantities, or for the phase and amplitude, respectively. The reason for this already becomes evident in Madelung's hydrodynamic formulation of Schrödinger's wave equation using a polar form for the complex wave function. Obviously, there is a coupling between the two components that can be related to some kind of conservation law. The origin of this coupling can however be understood immediately when the TISE is rewritten as a complex NL Riccati equation. In particular, the quadratic nonlinearity shows right away how the mixing of phase and amplitude occurs.

In a further step it has been shown that similar formal relationships also occur in TD quantum mechanics, at least (but not only) in cases where Gaussian wave packet solutions exist. Here, the complex Riccati equation essentially describes the time-evolution of the quantum uncertainties and can again be rewritten in terms of a (real) NL Ermakov equation.

In this case, this equation, together with the corresponding Newtonian equation that describes the motion of the wave packet maximum, allows for the construction of a dynamical invariant that still exists in cases where the corresponding Hamiltonian does not have this property.

A different way of treating the Riccati equation is to linearise it to a Newtonian equation for a complex quantity ($\lambda(t)$) where, formally, the same relations between its phase and amplitude exist as in the TI case between phase and amplitude of the wave function. The corresponding conservation law can be interpreted as a kind of conservation of angular momentum for the motion of λ in the complex plane.

Next, the systems considered were coupled to a dissipative environment. In the TD case, this was studied using an effective NLSE with complex logarithmic nonlinearity. The changes for the afore-mentioned Riccati, Ermakov and Newtonian equations due to the coupling to the environment were discussed. Now the quadratic nonlinearity in the Riccati equation leads to new qualitative effects like Hopf bifurcations since two physically distinguishable solutions occur. A possible interpretation of this phenomenon has been given.

From the results obtained in the TD case, conclusions for corresponding modifications in the TI case have been drawn leading to the results given in Eqs.(43) and (44). One interesting aspect of this comparison is that the same type of Ermakov equation can be obtained either by changing the phase of the complex quantity (in the TI case, $S \to S' = S + m\frac{\gamma}{4}\tilde{x}^2$) or by changing the amplitude (in the TD case, $\alpha \to \alpha e^{-\gamma t/2}$). Consequences for the relations between gauge transformations and non-unitary transformations are still under investigation.

Acknowledgments
The author wishes to thank the organizers of EmerQuM 2011 for the invitation to participate in this really stimulating conference.

References

[1] Yang C N 1987 *Schrödinger – Centenary Celebration of a Polymath* ed C W Kilmister (Cambridge: Cambridge University Press) p 53
[2] Madelung E 1926 *Z. Physik* **40** 322
[3] Reinisch G 1994 *Physica* A **206** 229
[4] Ermakov V P 1880 *Univ. Jzv. Kiev* **20** No 9, 1
[5] Ray J R and Reid J L 1979 *Phys. Lett.* A **71** 317
[6] Schuch D 1992 *Int. J. Quantum Chem.* **42** 663
[7] Hasse R W 1975 *J. Math. Phys.* **16** 2005
[8] Schuch D, Chung K-M and Hartmann H 1983 *J. Math. Phys.* **24** 1652
[9] Schuch D 1999 *Int. J. Quantum Chem.* **72** 537
Schuch D 2007 *J. Math. Phys.* **48** 122701
[10] Kramers H A 1940 *Physica* **7** 284
[11] Einstein A 1905 *Ann. d. Phys.* **17** 549
[12] Madelung E 1950 *Die mathematischen Hilfsmittel des Physikers* (Berlin: Springer Verlag) p 432
[13] Mrowka B 1951 *Z. Phys.* **130** 164
[14] Kostin M D 1972 *J. Chem. Phys.* **57** 3589
[15] Beretta G P 1987 *Found. Phys.* **17** 365
Beretta G P 2006 *Phys. Rev.* E **73** 026113
[16] Grössing G, Mesa Pasasio J and Schwabl H 2011 *Found. Phys.* **41** 1437
[17] Schuch D, Chung K-M and Hartmann H 1984 *J. Math. Phys.* **25** 3086
[18] Schuch D 2012 *J. Phys.: Conf. Series* in preparation

Quantum features derived from the classical model of a bouncer-walker coupled to a zero-point field

H Schwabl[1], J Mesa Pascasio[1,2], S Fussy[1], G Grössing[1]

[1] Austrian Institute for Nonlinear Studies, Akademiehof, Friedrichstr. 10, 1010 Vienna, Austria
[2] Institute for Atomic and Subatomic Physics, Vienna University of Technology, Operng. 9, 1040 Vienna, Austria

E-mail: ains@chello.at

Abstract. In our bouncer-walker model a quantum is a nonequilibrium steady-state maintained by a permanent throughput of energy. Specifically, we consider a "particle" as a bouncer whose oscillations are phase-locked with those of the energy-momentum reservoir of the zero-point field (ZPF), and we combine this with the random-walk model of the walker, again driven by the ZPF. Starting with this classical toy model of the bouncer-walker we were able to derive fundamental elements of quantum theory [1]. Here this toy model is revisited with special emphasis on the mechanism of emergence. Especially the derivation of the total energy $\hbar\omega_0$ and the coupling to the ZPF are clarified. For this we make use of a sub-quantum equipartition theorem. It can further be shown that the couplings of both bouncer and walker to the ZPF are identical. Then we follow this path in accordance with Ref. [2], expanding the view from the particle in its rest frame to a particle in motion. The basic features of ballistic diffusion are derived, especially the diffusion constant D, thus providing a missing link between the different approaches of our previous works [1, 2].

1. Introduction

As explicated already in some of our previous papers [1, 2, 3, 4], we understand the quantum as a well-coordinated *emergent system*, where particle-like and wave-like phenomena are the result of both stochastic and regular dynamical processes which exchange energy with the surrounding "vacuum", i.e., the zero-point field (ZPF). Thus, a quantum is modeled as a nonequilibrium steady-state maintained by a permanent throughput of energy. Specifically, we consider a "particle" as a bouncer-walker whose combined movements are coupled to the energy-momentum reservoir of the ZPF. The notion of the bouncer-walker is derived from classical physics, as shown experimentally via the "bouncing droplets" of Couder's group [5, 6, 7, 8, 9].

The possible existence of such a corresponding, underlying "medium" is *a priori* independent of quantum theory. For similar views, compare, for example, the works of Hestenes [10], Khrennikov [11], Nelson [12], Nottale [13], and de la Peña and Cetto [14]. We want to underline that by using the expression "classical", we imply the "updated" present-day status of classical physics, i.e., including present-day statistical physics, nonequilibrium thermodynamics, and the like. Vacuum fluctuations in our terminology thus refer to the statistical mechanics of the ZPF, i.e., a "classical" sub-quantum medium.

In our picture the quantum emerges via phase coupling of the bouncer's oscillatory frequency ω with corresponding modes of the ZPF fluctuations, combined with the random-walk model of the walker, again driven by the ZPF. Starting with this toy model of the bouncer-walker, we were able to derive fundamental elements of quantum theory from such a classical approach. In Section 2 this toy model is revisited according to [1], but this time with special emphasis on the mechanism of emergence. Especially, the derivation of the total energy $\hbar\omega_0$ and the coupling to the ZPF are clarified. Section 3 follows this path, expanding the view from the particle in its rest frame to a particle in motion [2]. The basic features of ballistic diffusion are derived via a link to the individual bouncer-walker model.

2. Modeling the quantum: "bouncer" and "walker"

The bouncer is modeled as a classical oscillator with the following Newtonian equation

$$m\ddot{\mathsf{x}} = -m\omega_0^2\mathsf{x} - 2\gamma m\dot{\mathsf{x}} + F_0\cos\omega t. \tag{1}$$

Eq. (1) describes a forced oscillation of a mass m swinging around a center point along $\mathsf{x}(t)$ with amplitude A and resonant frequency ω_0. The damping factor γ, or friction, allows the oscillator to exchange energy to the ZPF, which in this case is modeled by a locally independent driving force $F(t) = F_0\cos\omega t$. In the center of mass frame, the system is characterized by a single degree of freedom (DOF).

The stationary solution of Eq. (1), i.e., for $t \gg \gamma^{-1}$, with the ansatz

$$\mathsf{x}(t) = A\cos(\omega t + \varphi), \tag{2}$$

yields for the phase shift between the forced oscillation and the forcing oscillation that

$$\tan\varphi = -\frac{2\gamma\omega}{\omega_0^2 - \omega^2}\,, \tag{3}$$

and for the amplitude of the forced oscillation

$$A(\omega) = \frac{F_0/m}{\sqrt{(\omega_0^2 - \omega^2)^2 + (2\gamma\omega)^2}}\,. \tag{4}$$

It is well known that the system is stable at the resonant frequency ω_0 of the free undamped oscillator

$$\omega = \omega_0. \tag{5}$$

With identity (5) we introduce the notations

$$\tau = \frac{2\pi}{\omega_0}\,, \quad r := A(\omega_0) = \frac{F_0}{2\gamma m\omega_0}\,. \tag{6}$$

We assume the net energy balance of the exchange oscillator–ZPF to be zero, i.e., the oscillator–ZPF system is in a steady state. For this, we analyze the energetic balance. By multiplying Eq. (1) with $\dot{\mathsf{x}}$ we obtain

$$m\ddot{\mathsf{x}}\dot{\mathsf{x}} + m\omega_0^2\mathsf{x}\dot{\mathsf{x}} = -2\gamma m\dot{\mathsf{x}}^2 + F_0\cos(\omega t)\dot{\mathsf{x}}. \tag{7}$$

Due to the friction term $-2\gamma m\dot{\mathsf{x}}^2$ the oscillator loses its energy to the ZPF bath, whereas the oscillator regains its power $F_0\cos(\omega t)\dot{\mathsf{x}}$ from the energy bath via the force $F(t)$. We can rewrite Eq. (7) as

$$\frac{\mathrm{d}}{\mathrm{d}t}\left(\frac{1}{2}m\dot{\mathsf{x}}^2 + \frac{1}{2}m\omega_0^2\mathsf{x}^2\right) = -2\gamma m\dot{\mathsf{x}}^2 + F_0\cos(\omega t)\dot{\mathsf{x}} = 0. \tag{8}$$

The term within the brackets is the Hamiltonian of the system. By inserting Eqs.(2) and (5) we see

$$\mathcal{H} = \frac{1}{2}m\dot{\mathrm{x}}^2 + \frac{1}{2}m\omega_0^2\mathrm{x}^2 = \frac{m\omega_0^2A^2}{2} = \text{const.}, \tag{9}$$

thus providing the vanishing of Eq. (8).

One can write down the net work-energy that is taken up by the bouncer during each period τ as

$$\begin{aligned} W_{\text{bouncer}} &= \int_\tau F_0\cos(\omega t)\dot{\mathrm{x}}\,\mathrm{d}t = \int_\tau 2\gamma m\dot{\mathrm{x}}^2\,\mathrm{d}t \\ &= 2\gamma m\omega^2A^2\int_\tau \sin^2(\omega t+\varphi)\,\mathrm{d}t \\ &= \gamma m\omega^2A^2\tau. \end{aligned} \tag{10}$$

With Eqs. (5) and (6) we obtain in the steady state

$$W_{\text{bouncer}} = W_{\text{bouncer}}(\omega_0) = \gamma m\omega_0^2r^2\tau = 2\pi\gamma m\omega_0r^2. \tag{11}$$

We now shift to polar coordinates, which allow us to model the system in continuous circular motion. If one introduces the angle $\theta(t) := \omega_0 t$ and substitutes Eq. (2) into Eq. (9), this yields, as is well known, the two equations

$$\ddot{r} - r\dot{\theta}^2 + \omega_0^2r = 0, \tag{12}$$

and

$$r\ddot{\theta} + 2\dot{r}\dot{\theta} = 0. \tag{13}$$

From Eq. (13), an invariant quantity is obtained: it is the angular momentum

$$L(t) = mr^2\dot{\theta}(t). \tag{14}$$

With $\theta(t) = \omega_0 t$, and thus $\dot{\theta} = \omega_0$, the quantity of Eq. (14) becomes a time-invariant expression, which we denote as

$$\hbar := mr^2\omega_0. \tag{15}$$

Note that $L(t)$ is an invariant even more generally, i.e., for $\theta(t) := \int\omega(t)\,\mathrm{d}t$. Still, for all cases where the time average $\langle\theta(t)\rangle = \omega_0 t$, one can again write down $\hbar$ in the form Eq. (15). Thus, we rewrite our result (11) as

$$W_{\text{bouncer}} = 2\pi\gamma\hbar. \tag{16}$$

Now let us revisit the energy (9) of the linear harmonic oscillator, which reads, together with Eqs. (6) and (15),

$$\mathcal{H} = \frac{m\omega_0^2r^2}{2} = \frac{\hbar\omega_0}{2} = E_{\text{bouncer}}. \tag{17}$$

In a second step, we model another dissipative system. The "walker" is a "particle" driven via a stochastic force, e.g., due to not just one regular, but due to different fluctuating wave-like configurations in the ZPF environment. The particle's motion, which will generally assume a Brownian-type character, is then described (in any one dimension) by a Langevin stochastic differential equation with velocity $u = \dot{x}$, driving force $f(t)$, and friction coefficient ζ,

$$m\dot{u} = -m\zeta u + f(t). \tag{18}$$

Again, "friction", earlier represented by γ and now by ζ, generally describes the coupling between the oscillator (or particle in motion) on the one hand, and the ZPF bath on the other hand. The time-dependent force $f(t)$ is stochastic, i.e., one has as usual for the time-averages

$$\langle f(t)\rangle = 0\,, \quad \langle f(t)f(t')\rangle = \phi(t-t'), \tag{19}$$

where $\phi(t)$ differs noticeably from zero only for $t < \zeta^{-1}$. The correlation time ζ^{-1} denotes the time during which the fluctuations of the stochastic force remain correlated.

One usually introduces a coefficient λ that measures the strength of the mean square deviation of the stochastic force, such that

$$\phi(t) = \lambda\delta(t). \tag{20}$$

Since friction increases in proportion to the frequency of the stochastic collisions, there must be a connection between λ and ζ. One solves the Langevin equation (18) in order to find this connection. Solutions of this equation are well known from the Ornstein-Uhlenbeck theory of Brownian motion [15, 16].

Since the dependence of $f(t)$ is known only statistically, one does not consider the average value of $u(t)$, but instead that of its square,

$$\begin{aligned}\overline{u^2(t)} &= \mathrm{e}^{-2\zeta t}\int_0^t \mathrm{d}\tau \int_0^t \mathrm{d}\tau' \mathrm{e}^{\zeta(\tau+\tau')}\phi(\tau-\tau')\frac{1}{m} + u_0^2\mathrm{e}^{-2\zeta t} \\ &= \frac{\lambda}{2\zeta m^2}\left(1-\mathrm{e}^{-2\zeta t}\right) + u_0^2\mathrm{e}^{-2\zeta t} \quad \xrightarrow{t\gg\zeta^{-1}} \quad \frac{\lambda}{2\zeta m^2}\,,\end{aligned} \tag{21}$$

with u_0 being the initial value of the velocity. For $t \gg \zeta^{-1}$, the term with u_0 becomes negligible, i.e., ζ^{-1} then plays the role of a relaxation time. We require that our particle attains thermal equilibrium [17, 18] after long times, so that due to the *equipartition theorem on the sub-quantum level* the average value of the kinetic energy becomes

$$\frac{1}{2}m\,\overline{u^2(t)} = \frac{\lambda}{4\zeta m} = E_{\mathrm{zp}} = \frac{1}{2}kT_0. \tag{22}$$

where we introduce the average kinetic Energy E_{zp} of the zero-point field, which is the sub-quantum analogon to the thermodynamical expression $k_{\mathrm{B}}T/2$, where k_{B} is Boltzmann's constant, and T the classical temperature, whereas T_0 in our scenario denotes the vacuum temperature. However, as we today neither know T_0 nor the constant k (unless it should turn out as identical to k_{B}), we shall mostly stick to formally using E_{zp}. In other words, we shall use the specification "kT_0" only occasionally, i.e., in order to point out the close analogy to the usual thermodynamical formalism, and as a reminder that E_{zp} is the "kinetic temperature" of the vacuum's heat reservoir.

From Eq. (22) one obtains an Einstein-type relation

$$\lambda = 4\zeta m E_{\mathrm{zp}}. \tag{23}$$

Similarly, we obtain the mean square displacement of $x(t)$ for $t \gg \zeta^{-1}$. Therefore, one integrates twice to obtain

$$\overline{x^2(t)} = \int_0^t \mathrm{d}\tau \int_0^t \mathrm{d}\tau' \frac{\lambda}{2\zeta m^2}\mathrm{e}^{-\zeta|\tau-\tau'|} \simeq \frac{\lambda}{\zeta^2 m^2}t = 2Dt, \tag{24}$$

with the diffusion constant turning out as

$$D = \frac{\lambda}{2\zeta^2 m^2} = \frac{2E_{\mathrm{zp}}}{\zeta m}\,. \tag{25}$$

Now we remind ourselves that we deal here with a steady-state system. Just as with the friction ζ there exists a flow of (kinetic) energy into the ZPF environment, there must also exist a work-energy flow back into our system of interest. For its calculation, we multiply Eq. (18) by $u = \dot{x}$ and obtain an energy-balance equation. With a natural number $n > 0$ chosen so that $n\tau$ is large enough to make all fluctuating contributions negligible, it yields for the duration of time $n\tau$ the net work-energy of the walker

$$W_{\text{walker}} = \int_{n\tau} m\zeta \, \overline{\dot{x}^2} \, \mathrm{d}t = m\zeta \int_{n\tau} \overline{u^2(t)} \, \mathrm{d}t. \tag{26}$$

Inserting Eq. (22), we obtain

$$W_{\text{walker}} = n\tau m\zeta \, \overline{u^2(t)} = 2n\tau\zeta E_{\text{zp}}. \tag{27}$$

We have so far analyzed two perspectives for our model of a single-particle quantum system:

(i) A harmonic oscillator is driven by the environment via a periodic force $F_0 \cos\omega_0 t$. However, in the N-dimensional reference frame of the laboratory, the oscillation is not fixed *a priori*. Rather, with $\hbar$ as angular momentum, there will be a free rotation in all N dimensions, and possible exchanges of energy will be equally distributed in a stochastic manner.

(ii) Concerning the latter, the flow of energy is on average distributed evenly via the friction γ in all N dimensions of the laboratory frame. It can thus also be considered as the stochastic source of the particle moving in N dimensions, each described by the Langevin equation (18).

In order to make the result comparable with Eq. (16), we choose $\tau = 2\pi/\omega_0$ to be identical with the period of Eq. (6). The work-energy for the particle undergoing Brownian motion can thus be written as

$$W_{\text{walker}} = n\frac{4\pi}{\omega_0}\zeta E_{\text{zp}}, \tag{28}$$

and, for the general case of N DOF

$$W_{\text{walker}} = n\frac{N4\pi}{\omega_0}\zeta E_{\text{zp}}. \tag{29}$$

Accordingly, the walker gains its energy from the heat bath via the oscillations of the bouncer-ZPF bath system in N dimensions: The bouncer, via the coupling γ, pumps energy to and from the heat bath. There is a continuous flow from the bath to the oscillator, and *vice versa*. Moreover, and most importantly, during that flow, for long enough times $n\tau$, this coupling of the bouncer can be assumed to be exactly identical with the coupling of the walker. For this reason, we directly compare the results of Eqs. (16) and (29),

$$nW_{\text{bouncer}} = W_{\text{walker}}. \tag{30}$$

With $n \gg 1$, since we have to take the mean over a large number of stochastic motions, we get

$$n2\pi\gamma\hbar = n\frac{N4\pi}{\omega_0}\zeta E_{\text{zp}}. \tag{31}$$

Now, one generally has that the total energy of a sinusoidal oscillator exactly equals twice its average kinetic energy. With Eq. (17) the bouncer model provided already

$$E_{\text{tot}} = 2E_{\text{bouncer}} = \hbar\omega_0, \tag{32}$$

and we compare this result with Eq. (31). We achieve the same result for the total energy only if both systems, the bouncer and the walker, are coupled with the same strength to the ZPF bath, i.e., the friction coefficient for both the bouncer and the walker must be identical, $\gamma = \zeta$. We have thus *derived* the total energy of our model for a quantum "particle", i.e., a driven steady-state oscillator system, as

$$E_{\text{tot}} = 2NE_{\text{zp}} = 2E_{\text{bouncer}} = \hbar\omega_0, \tag{33}$$

where $\hbar$, as defined in Eq. (15), can now be identified with Planck's reduced constant. Note that via (15), $\hbar$ is defined as angular momentum in exactly the same manner as it is obtained in an independent earlier derivation by Puthoff [19].

Now, with Boltzmann's relation $\Delta Q = 2\omega_0 \delta S$ between the heat applied to an oscillating system and a change in the action function $\delta S = \delta \int E_{\text{kin}}\, dt$, respectively, [17, 18] one has

$$\nabla Q = 2\omega_0 \nabla(\delta S). \tag{34}$$

δS relates to the momentum fluctuation via

$$\nabla(\delta S) = \delta \mathbf{p} =: m\mathbf{u} = -\frac{\hbar}{2}\frac{\nabla P}{P}, \tag{35}$$

and therefore, with $P = P_0 \mathrm{e}^{-\delta Q/kT_0}$ and (33),

$$m\mathbf{u} = \frac{\nabla Q}{2\omega_0}. \tag{36}$$

As the friction force in Eq. (18) is equal to the gradient of the heat flux,

$$m\zeta\mathbf{u} = \nabla Q, \tag{37}$$

comparison of (36) and (37) provides now a detailed expression for the coupling to the ZPF bath

$$\zeta = \gamma = 2\omega_0. \tag{38}$$

Note that with Eqs. (33) and (38) one obtains in one dimension the expression for the diffusion constant (25) as

$$D = \frac{2E_{\text{zp}}}{\zeta m} = \frac{\hbar}{2m}, \tag{39}$$

which is exactly the usual expression for D in the context of quantum mechanics.

Finally, we can also introduce the recently proposed concept of an "entropic force" [20, 21]. We revisit Eq. (33) and look at the total energy equaling a total work applied to the system. With S_{e} denoting the entropy one can write

$$\begin{aligned} E_{\text{tot}} &= 2\left\langle E_{\text{kin}} \right\rangle =: \mathbf{F}\cdot\Delta\mathbf{x} = T_0 \Delta S_{\text{e}} = \frac{1}{2\pi}\oint \nabla Q \cdot \mathrm{d}\mathbf{r} \\ &= \Delta Q\,(\text{circle}) = 2\left[\frac{\hbar\omega_0}{4} - \left(-\frac{\hbar\omega_0}{4}\right)\right] = \hbar\omega_0 \end{aligned} \tag{40}$$

which provides an "entropic" view of a harmonic oscillator in its thermal bath.

We know already the total energy of a simple harmonic oscillator E_{bouncer}, Eq. (17). The average kinetic energy of a harmonic oscillator is given by half of its total energy, i.e., by $\left\langle E_{\text{kin}} \right\rangle = E_{\text{bouncer}}/2 = \hbar\omega_0/4$, which — because of the local equilibrium — is both the average kinetic energy of the bath and that of the "bouncer" particle. As the latter during one oscillation varies between 0 and $\hbar\omega_0/2$, one has the following entropic scenario. When it is minimal, the tendency towards maximal entropy will provide an entropic force equivalent to the absorption of the heat quantity $\Delta Q = \hbar\omega_0/4$. Similarly, when it is maximal, the same tendency will now enforce that the heat $\Delta Q = \hbar\omega_0/4$ is given off again to the "thermostat" of the thermal bath. In sum, then, the total energy throughput E_{tot} along a full circle will equal, according to Eq. (40), $2\langle E_{\mathrm{kin}}\rangle(\mathrm{circle}) = 2\hbar\omega_0/2 = \hbar\omega_0$. In other words, the formula $E = \hbar\omega_0$ does not refer to a classical "object" oscillating with frequency ω_0, but rather to a process of a "fleeting constancy": due to entropic requirements, the energy exchange between bouncer and heat bath will constantly consist of absorbing and emitting heat quantities such that in sum the "total particle energy" emerges as $\hbar\omega_0$.

As was shown in [1] one can continue along these classical lines to express further features of quantum mechanics, like the energy spectrum of the harmonic oscillator, or spin, for example.

3. Gaussian particle behaviour

So far, we have described the emerging entity in its rest frame. Now we expand this model and describe the randomly moving walker, characterized by the diffusion constant D, Eq. (39). Here we follow the arguments presented in [2]. As shown above, the nonequilibrium steady-state is characterized by a permanent throughput of energy, or heat flow. First, let us reconsider the Brownian motion of the particle, but from another perspective as compared to the previous chapter. The Brownian motion is a form of kinetic energy provided by the ZPF interaction, and is of course different from the "ordinary" kinetic energy of the particle, which will be introduced later. The total energy of the whole system can be written as

$$E_{\mathrm{tot}} = \hbar\omega + \frac{(\delta p)^2}{2m}, \tag{41}$$

where $\hbar\omega$ is the generalized total energy of the particle and $\delta p := mu$ is said additional, fluctuating momentum of the particle of mass m [2]. Note that δp can take on an arbitrary value such that E_{tot} is generally variable.

Every bouncer-walker is a rapidly oscillating object, which itself is guided by the ZPF environment that also contributes some fluctuating momentum to the walker's propagation. In fact, the walker is the cause of the waves surrounding the particle, and the detailed structure of the wave configurations influences the walker's path, as the particle both absorbs heat from and emits heat into its environment, both cases of which can be described in terms of momentum fluctuations. Thus, if we imagine the bouncing of a walker in its "fluid" environment, the latter will become "excited" or "heated up" wherever the momentum fluctuations direct the particle to. After some time span (which can be rather short, considering the very rapid oscillations of elementary particles), a whole area of the particle's environment will be coherently heated up in this way.

Now we expand this further to a source of identical particles, which are prepared in such a way that each one ideally has an initial (classical) velocity $\mathbf{v}$, which is also called "convective" velocity. Similar arguments are presented by Grössing [in this volume] and [1, 2, 3, 4, 17, 18, 22], but here we want to focus on the interaction with the ZPF. The particles emerge from the source, one at a time only, with a Gaussian probability density P. This comes along with a heat distribution generated by the oscillating ("bouncing") particle(s), with a maximum at the center of the aperture $\boldsymbol{x}_0 = \mathbf{v}t$. In one dimension the corresponding solution of the heat equation is then

$$P(x,t) = \frac{1}{\sqrt{2\pi}\sigma} e^{-\frac{(x-x_0)^2}{2\sigma^2}}, \tag{42}$$

with the usual variance $\sigma^2 = \overline{(\Delta x)^2} = \overline{(x-x_0)^2}$, where we shall choose $x_0(t=0) = 0$. Note that from Eq. (41) one has for the averages over particle positions and fluctuations

$$\overline{E_{\text{tot}}} = \overline{\hbar\omega} + \frac{\overline{(\delta p)^2}}{2m} = \text{const.}, \tag{43}$$

with the mean values (generally defined in N-dimensional configuration space

$$\overline{(\delta p)^2} := \int P\,(\delta p)^2 \,\mathrm{d}^N x\,. \tag{44}$$

Eq. (43) is a statement of total average energy conservation, i.e., holding for all times t . Being a central argument, it is also the starting point for [2]. In Eq. (43), a variation in δp implies a varying "particle energy" $\hbar\omega$, and *vice versa*, such that each of the summands on the right hand side for itself is not conserved. As can be shown [2], there is an exchange of momentum between the two terms providing a net balance

$$m\delta v - m\delta u = 0 \tag{45}$$

where δv describes a change in the convective velocity v paralleled by the "diffusive" momentum fluctuation $\delta(\delta p) := m\delta u$ in the thermal environment.

Now let us look at the contributions of the diffusive and convective velocities to the total energy. As from Eq. (43) one has that $\frac{\partial}{\partial t}\overline{E_{\text{tot}}} = 0$ and thus also $\delta\overline{E_{\text{tot}}(t)} - \delta\overline{E_{\text{tot}}(0)} = 0$, and as only the kinetic energy varies, one obtains $\delta\overline{E_{\text{kin}}(t)} = \delta\overline{E_{\text{kin}}(0)} = \text{const.}$, which yields for any t, with (35) and (42),

$$\delta\overline{E_{\text{kin}}(t)} = \frac{m}{2}\overline{(\delta v)^2} + \frac{m}{2}\overline{u^2} = \frac{m}{2}\overline{(\delta v)^2} + \frac{\hbar^2}{8m\sigma^2}\,, \tag{46}$$

and thus at the initial time, where $v = 0$:

$$\delta\overline{E_{\text{kin}}(0)} = 0 + \frac{m}{2}\overline{u^2}\Big|_{t=0} =: \frac{m}{2}u_0^2 = \frac{\hbar^2}{8m\sigma_0^2}\,. \tag{47}$$

Now we again make use of the equipartition equation Eq. (22). Together with Eq. (25) we obtain

$$\frac{m}{2}u_0^2 = \frac{kT_0}{2} = \frac{D\zeta m}{2}\,. \tag{48}$$

At the time $t = 0$ the system is in the prepared state where the fluctuating kinetic energy term is solely determined by the initial value σ_0, whereas for later times t it decomposes into the term representing the particle's changed kinetic energy and the term including $\sigma(t)$. At $t = 0$ the velocity u_0 is determined by the mean displacement σ_0 and the relaxation time ζ^{-1} of the walker, Eq. (18). By using Eq. (48) we can write

$$u_0 = \zeta\sigma_0 = \frac{D}{\sigma_0}. \tag{49}$$

With the Gaussian distribution Eq. (42) being a solution of the diffusion equation, one has for the particle's drift the familiar relation

$$\overline{x^2(t)} = \overline{x^2}\Big|_{t=0} + 2Dt. \tag{50}$$

However, comparison with Eq. (24) indicates that (50) is only a solution for the particle in its rest frame. From Eqs.(46) and (47) we see that in order to account for a Gaussian dispersion as in (46), the diffusivity must become time dependent, too. Thus, one must rewrite Eq. (50) as

$$\overline{x^2} = \overline{x^2}\Big|_{t=0} + D(t)t. \tag{51}$$

In Refs. [2, 3] and [Grössing, this volume] we show that $D(t) = u_0^2 t$. We have thus connected the diffusion processes as described by $D(t)$, which we have found through our model of the bouncer-walker, to the movement of a particle following a Gaussian distribution. With this, we have derived the elements of ballistic diffusion from our classical bouncer-walker model and thus found a missing link between the different approaches of our previous work [1, 2]. We can now expand our thoughts towards interference and look deeper into the emergent properties of the quantum world, and beyond.

References

[1] G. Grössing, J. Mesa Pascasio, H. Schwabl, A Classical Explanation of Quantization, Found. Phys. 41 (2011) 1437–1453. `arXiv:0812.3561v4`.

[2] G. Grössing, S. Fussy, J. Mesa Pascasio, H. Schwabl, Emergence and collapse of quantum mechanical superposition: Orthogonality of reversible dynamics and irreversible diffusion, Physica A 389 (2010) 4473–4484. `arXiv:1004.4596v1`.

[3] G. Grössing, S. Fussy, J. Mesa Pascasio, H. Schwabl, Elements of sub-quantum thermodynamics: quantum motion as ballistic diffusion, J. Phys.: Conf. Ser. 306 (2011) 012046. `doi:10.1088/1742-6596/306/1/012046`.

[4] G. Grössing, S. Fussy, J. Mesa Pascasio, H. Schwabl, An explanation of interference effects in the double slit experiment: Classical trajectories plus ballistic diffusion caused by zero-point fluctuations, Ann. Phys. 327 (2012) 421–437. `arXiv:1106.5994v3`.

[5] Y. Couder, S. Protière, E. Fort, A. Boudaoud, Dynamical phenomena: Walking and orbiting droplets, Nature 437 (2005) 208–208.

[6] Y. Couder, E. Fort, Single-particle Diffraction and Interference at a macroscopic scale, Phys. Rev. Lett. 97 (2006) 154101.

[7] S. Protière, A. Boudaoud, Y. Couder, Particle-wave association on a fluid interface, J. Fluid Mech. 554 (2006) 85–108.

[8] A. Eddi, E. Fort, F. Moisy, Y. Couder, Unpredictable Tunneling of a Classical Wave-Particle Association, Phys. Rev. Lett. 102 (2009) 240401.

[9] E. Fort, A. Eddi, A. Boudaoud, J. Moukhtar, Y. Couder, Path-memory induced quantization of classical orbits, PNAS 107 (2010) 17515 –17520.

[10] D. Hestenes, Zitterbewegung in Quantum Mechanics, Found. Phys. 40 (2009) 1–54.

[11] A. Khrennikov, Prequantum classical statistical field theory: background field as a source of everything?, J. Phys.: Conf. Ser. 306 (2011) 012021. `doi:10.1088/1742-6596/306/1/012021`.

[12] E. Nelson, Quantum fluctuations, Princeton series in physics, Princeton Univ. Press, Princeton, NJ, 1985.

[13] L. Nottale, Scale Relativity and Fractal Space-Time, World Scientific, 2011.

[14] L. de la Peña, A. Valdés-Hernández, A. Cetto, H. França, Genesis of quantum nonlocality, Phys. Lett. A 375 (2011) 1720–1723.

[15] G. E. Uhlenbeck, L. S. Ornstein, On the Theory of the Brownian Motion, Phys. Rev. 36 (1930) 823–841.

[16] W. T. Coffey, Y. P. Kalmykov, J. T. Waldron, The Langevin equation: With applications to stochastic problems in physics, chemistry and electrical engineering, 2nd Edition, Vol. 14 of World Scientific series in contemporary chemical physics, World Scientific, Singapore, 2004.

[17] G. Grössing, The Vacuum Fluctuation Theorem: Exact Schrödinger Equation via Nonequilibrium Thermodynamics, Phys. Lett. A 372 (2008) 4556–4563. `arXiv:0711.4954v2`.

[18] G. Grössing, On the thermodynamic origin of the quantum potential, Physica A 388 (2009) 811–823. `arXiv:0808.3539v1`.

[19] H. E. Puthoff, Ground state of hydrogen as a zero-point-fluctuation-determined state, Phys. Rev. D 35 (1987) 3266–3269.

[20] E. Verlinde, On the origin of gravity and the laws of Newton, J. High Energy Phys. 2011. `arXiv:1001.0785v1`.

[21] T. Padmanabhan, Thermodynamical aspects of gravity: new insights, Rep. Prog. Phys. 73 (2010) 046901. `arXiv:0911.5004v2`.

[22] G. Grössing, Sub-Quantum Thermodynamics as a Basis of Emergent Quantum Mechanics, Entropy 12 (2010) 1975–2044. `doi:10.3390/e12091975`.

Particles as stable topological solitons

Manfried Faber

Institute for Atomic and Subatomic Physics, Vienna University of Technology, Operngasse 9, 1040 Wien, Austria

E-mail: faber@kph.tuwien.ac.at

Abstract. We discuss a model for charged particles and their fields where mass is field energy only, charge appears as a topological quantum number, photons are Goldstone bosons of spontaneous symmetry breaking, spin and chirality are described as topological charge of solitons interacting by Coulomb and Lorentz forces. The model is U(1) gauge invariant. We discuss open problems and add some speculations. The model is as simple as possible, with three degrees of freedom only describing the rotations of Dreibeins (triads) in space-time and dynamics defined by a Lorentz covariant Lagrangian.

1. Introduction

We can ask some naive questions:

- What is the origin of mass?
- Why do all electrons (protons) have the same mass?
- Why is charge quantised?
- What is spin?
- What is the origin of the Pauli principle?
- Where does gauge invariance emerge from?

We will discuss that there are models indicating answers to these questions.

There are several interesting relativistically covariant models with solutions which contain particle-like excitations. Five of them I would like to mention shortly since they are related to the model presented in this paper. The first one was formulated by Dirac, when he aimed to symmetrize electro-dynamics by introducing magnetic currents into the Maxwell equations [1, 2]. Dirac's magnetic monopoles are defined in 3+1D, in three space and one time dimension. They have two types of singularities, line-like singularities, Dirac strings, from the center of the monopole to an antimonopole or to infinity which are gauge dependent and therefore unphysical and point-like singularities in the center, well-known also from classical electron models. Wu and Yang [3, 4, 5] succeeded to formulate magnetic monopoles without the line-like singularities of the Dirac strings by using either a fibre-bundle construction with two different gauge fields, one for the northern and one for the southern hemisphere of the monopole, or by a normalised three-dimensional vector field $\vec{n}(x)$ in 3+1D. These Wu-Yang monopoles still suffer from the point-like singularities in the center. There are monopole solutions without any singularity in the Georgi-Glashow model [6], the 'tHooft-Polyakov monopoles [7, 8]. The Georgi-Glashow model, formulated in 3+1D, has 15 degrees of freedom, an adjoint Higgs field with three degrees of freedom and an SU(2) gauge field with $4 \cdot 3 = 12$ field components. Only one degree of freedom is needed for the Sine-Gordon model [9], a model in 1+1D. It addition to waves it has kink and anti-kink solutions which interact with each other. The kink-antikink configuration is attracting and the kink-kink configuration is repelling. The simplicity of the Sine-Gordon model inspired Skyrme [10, 11, 12] to a model in 3+1D with a scalar SU(2)-valued field, i.e. three degrees of freedom, with stable topological solitons with the properties of particles with a short range interaction, the Skyrmions. The model presented in this paper, also inspired by the simplicity and physical content of the Sine-Gordon model, the model of topological fermions, was first formulated in [13]. It has the same degrees of freedom as the Skyrme model but uses a different Lagrangian. Its relations to electrodynamics and symmetry breaking were discussed in [14, 15, 16, 17]. A python package for numerical calculations was written by Roman Bertle and actual calculations were performed in several diploma works at the Vienna University of Technology by Joachim Wabnig, Maria Hörndl, Julia Fornleitner, Xaver König, Roman Höllwieser and Josef Resch.

2. Formulation of the model

As basic field we use a scalar SO(3)-valued field in three space and one time dimension with coordinates $x^\mu = (ct, \mathbf{x})$, defining a rotation of a Dreibein in three-dimensional space. Below, we call it the soliton field. Objects not connected to the surrounding return after 2π-rotations to the original situation. One can easily show in a nice experiment performed at this conference that objects connected by wires to the surrounding behave differently [18]. Rotating such an object by 2π or 4π, the wires are completely entangled, but after 4π-rotations the wires can be disentangled without moving the object itself. In mathematics this is well known. SU(2) is the double (universal) covering group of SO(3), each pair of elements of SU(2) differing by a rotation by 2π can be mapped to an element of SO(3), and 4π-rotations are smoothly connected to no rotation. For the actual formulation of our basic SO(3)-field it is algebraically simpler to use an SU(2)-field $Q(x)$, keeping in mind that for differentiable fields every SO(3) field configuration corresponds to two SU(2)-field configuration differing by a multiplication with the non-trivial center element $z = -\mathbb{1}_2$ of SU(2). Active rotations Q by an angle $\omega = 2\alpha$ around the axis $\vec{n}$ can be represented by the quaternionic units $\mathbf{i}, \mathbf{j}, \mathbf{k}$ with the famous relation $\mathbf{i}^2 = \mathbf{j}^2 = \mathbf{k}^2 = \mathbf{ijk} = -1$ or by Pauli matrices σ_K with the correspondence $\mathbf{i} = -\mathrm{i}\sigma_1$, $\mathbf{j} = -\mathrm{i}\sigma_2$, $\mathbf{k} = -\mathrm{i}\sigma_3$,

$$Q = q_0 - \mathrm{i}\vec{\sigma}\vec{q} = \begin{pmatrix} q_0 - \mathrm{i}q_3 & -q_2 - \mathrm{i}q_1 \\ q_2 \mathrm{i} q_1 & q_0 + \mathrm{i}q_3 \end{pmatrix} \quad \text{with} \quad \vec{\sigma}\vec{q} = \sigma_K q_K,\ q_0 = \cos\alpha,\ \vec{q} = \vec{n}\sin\alpha \tag{1}$$

The SU(2)-manifold is isomorphic to S_3, to a three-dimensional sphere in four dimensions, see Fig. 1. We define an affine connection, a vector field $\vec{\Gamma}_\mu$ and a curvature tensor $\vec{R}_{\mu\nu}$

$$\partial_\mu Q Q^\dagger = -\mathrm{i}\vec{\Gamma}_\mu\vec{\sigma}, \quad \vec{R}_{\mu\nu} = \vec{\Gamma}_\mu \times \vec{\Gamma}_\nu. \tag{2}$$

with simple geometrical interpretations on S_3. From the condition for the function $Q(x)$ to be single-valued, the Schwarz integrability condition $\partial_\mu\partial_\nu Q(x) = \partial_\nu\partial_\mu Q(x)$, follows [13] the Maurer-Cartan equation $\partial_\mu\vec{\Gamma}_\nu - \partial_\nu\vec{\Gamma}_\mu = 2\vec{\Gamma}_\mu \times \vec{\Gamma}_\nu$, allowing to represent the curvature tensor $\vec{R}_{\mu\nu} = \vec{\Gamma}_\mu \times \vec{\Gamma}_\nu$ as $\vec{R}_{\mu\nu} = \frac{1}{2}(\partial_\mu\vec{\Gamma}_\nu - \partial_\nu\vec{\Gamma}_\mu)$ and in a form well-known from gravity and non-abelian gauge theories

$$\vec{R}_{\mu\nu} = \partial_\mu\vec{\Gamma}_\nu - \partial_\nu\vec{\Gamma}_\mu - \vec{\Gamma}_\mu \times \vec{\Gamma}_\nu. \tag{3}$$

The dynamics of our model is given by a Lagrangian density $\mathcal{L}$ which in generalisation of the Sine-Gordon model to 3+1D contains a curvature term $\vec{R}_{\mu\nu}\vec{R}^{\mu\nu}$ and a potential term $\Lambda(q_0)$

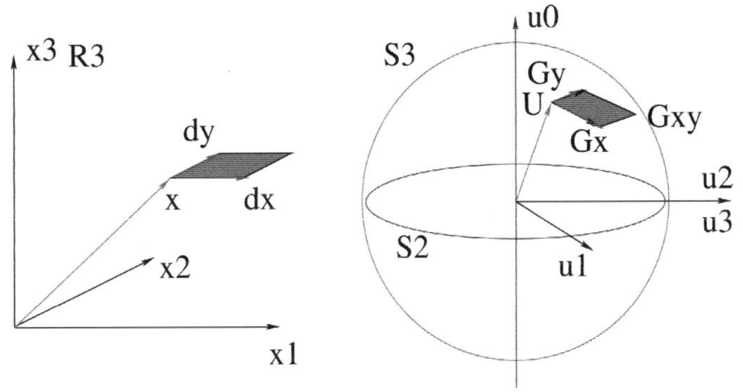

Figure 1. $Q(x)$, $\vec{\Gamma}_\mu \mathrm{d}x^\mu$ and $\vec{R}_{\mu\nu}\mathrm{d}x^\mu \mathrm{d}x^\nu$ represent points, line and area elements on S_3.

which is a function of q_0 only, with a minimum at $q_0 = 0$ and maxima at $|q_0| = 1$,

$$\mathcal{L} = -\frac{\alpha_f \hbar c}{4\pi}\left(\frac{1}{4}\vec{R}_{\mu\nu}\vec{R}^{\mu\nu} + \Lambda(q_0)\right) \quad \text{with} \quad \Lambda(q_0) = \frac{q_0^{2m}}{r_0^4}, \quad m = 1, 2, 3, \cdots . \tag{4}$$

The potential term has no derivatives and the curvature term four derivatives, this forces us to introduce a parameter r_0 for the length scale in Eq. (4). Since we want to compare the solitons emerging from this Lagrangian with electric charges, we have introduced Sommerfeld's fine-structure constant α_f. A relation to physics we get by the identification of the dual field-strength tensor with the curvature tensor ${}^*\vec{F}_{\mu\nu} = -\frac{e_0}{4\pi\varepsilon_0 c}\vec{R}_{\mu\nu}$.

3. Solitonic solutions

The Lagrangian (4) has solitonic solutions with the properties of electric charges. The static ansatz for a soliton, see left diagram in Fig. 2,

$$\vec{n}(x) = \frac{\mathbf{r}}{r}, \quad \alpha(x) = \alpha(r), \quad \text{with} \quad \mathbf{r} = (x_1, x_2, x_3) \tag{5}$$

and $\rho = \frac{r}{r_0}$ leads [13] to the energy functional

$$H = \frac{\alpha_f \hbar c}{r_0}\int_0^\infty \mathrm{d}\rho \left[\frac{\sin^4\alpha}{2\rho^2} + (\partial_\rho \cos\alpha)^2 + \rho^2 \cos^{2m}\alpha\right], \tag{6}$$

and by the variation of this functional to the non-linear second order differential equation

$$\partial_\rho^2 \cos\alpha + \frac{(1-\cos^2\alpha)\cos\alpha}{\rho^2} - m\rho^2\cos^{2m-1}\alpha = 0. \tag{7}$$

Solutions of this equation read for $m = 3$

$$\alpha(r) = \arctan\rho, \quad \sin\alpha(r) = \frac{\rho}{\sqrt{1+\rho^2}}, \quad \cos\alpha(r) = \frac{1}{\sqrt{1+\rho^2}}, \tag{8}$$

with a minimum value of the energy functional $H = \frac{\alpha_f \hbar c}{r_0}\frac{\pi}{4}$. Comparing this result with the mass of an electron, $m_e c^2 = 0.511$ MeV, we get $r_0 = 2.21$ fm, a value very close to the classical electron radius $\frac{\alpha_f \hbar c}{m_e c^2} = \alpha_f \frac{\hbar}{m_e c} = 2.82$ fm. For $m = 2$ we get the solution $q_0(\rho) = \frac{1}{1+\rho^2}$ of Eq. (7) by scaling the potential, $\Lambda(q_0) = \frac{7}{2}(\frac{q_0}{r_0})^4$, with the soliton energy $E_1 = \frac{\alpha_f \hbar c}{r_0}\frac{7\pi}{16}$. Implementing the factor $\frac{7}{2}$ in r_0 and again comparing the soliton energy with $m_e c^2$ leads to $\sqrt[4]{2/7}r_0 = 2.83$ fm. The three contributions (6) to the radial energy density are shown in the right diagram of Fig. 2.

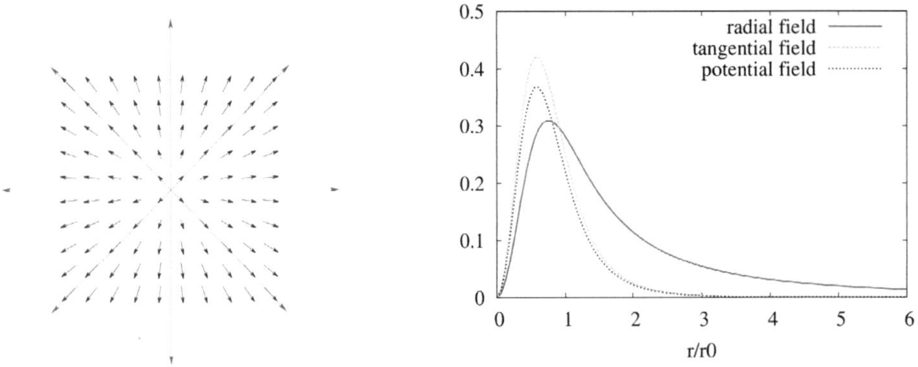

Figure 2. Left: The hedge-hog configuration according to Eq. (5) is depicted by arrows indicating the size of the imaginary part $\vec{q}(x)$ of the soliton field $Q(x)$. The lines connect $\vec{q}$-vectors with the same direction. Right: contributions to the radial energy density in Eq. (6) for $m = 3$ and $q_0(\rho) = \cos\alpha(\rho) = \frac{1}{1+\rho^2}$. For large distances the radial energy density is proportional to $1/\rho^2$.

4. Soliton mechanics

The variation of the soliton-field $Q(x)$ has to respect the unitarity of Q and can be done by $Q \to Q' = \mathrm{e}^{\mathrm{i}\vec{\sigma}\vec{\zeta}}Q$ with a three-component vector $\vec{\zeta}$ which can be considered as a generalised coordinate. The generalised momentum density follows to be $\pi^\mu = \frac{\partial\mathcal{L}}{\partial\partial_\mu\vec{\zeta}} = -\frac{\alpha_f\hbar c}{4\pi}\vec{\Gamma}_\nu \times \vec{R}^{\mu\nu}$ and can also be expressed by $\pi^\mu = \frac{\partial\mathcal{L}}{\partial\vec{\Gamma}_\mu}$. From the variation of $\vec{\zeta}$ follows the equation of motion

$$\partial_\mu[\vec{\Gamma}_\nu \times \vec{R}^{\mu\nu}] + \vec{q}\,\frac{\mathrm{d}\Lambda}{\mathrm{d}q_0} = 0. \tag{9}$$

It corresponds to Newton's second law of motion, where the derivative of the momentum is given by the variation of the potential.

From the Lagrangian (4) we get the canonical energy-momentum tensor by

$$\Theta^\mu{}_\nu(x) = \frac{\partial\mathcal{L}(x)}{\partial(\partial_\mu\alpha_k)}\partial_\nu\alpha_k - \mathcal{L}(x)\,\delta^\mu_\nu = \frac{\partial\mathcal{L}(x)}{\partial\vec{\Gamma}_\mu}\vec{\Gamma}_\nu - \mathcal{L}(x)\,\delta^\mu_\nu. \tag{10}$$

In Maxwell's electrodynamics the canonical energy-momentum tensor is asymmetric and has to be symmetrised by adding an appropriate total divergence [19]. The tensor (10) turns out to be symmetric from the very beginning

$$\Theta^\mu{}_\nu = -\frac{\alpha_f\hbar c}{4\pi}\left\{\left(\vec{\Gamma}_\nu \times \vec{\Gamma}_\sigma\right)\left(\vec{\Gamma}^\mu \times \vec{\Gamma}^\sigma\right)\right\} - \mathcal{L}(x)\,\delta^\mu_\nu. \tag{11}$$

For the trace of the energy-momentum tensor (11) we get

$$\Theta^\mu{}_\mu = 4\mathcal{H}_p. \tag{12}$$

It contributes in the presence of charges only.

The derivative of the energy-momentum tensor defines the force density. It vanishes [13] due to the equation of motion (9)

$$f_\nu = \partial_\mu\Theta^\mu{}_\nu = 0. \tag{13}$$

This is an obvious consequence of the unified description of charges and fields. Therefore, energy and momentum

$$P^\mu = \frac{1}{c}\int \Theta^{0\mu} \mathrm{d}^3x \tag{14}$$

are conserved quantities.

As expected the mass m_e of a single soliton increases with its velocity $\mathbf{v}$

$$P^0 = \gamma\, m_e c, \quad \mathbf{P} = \boldsymbol{\beta}\,\gamma\, m_e c, \quad \boldsymbol{\beta} = \mathbf{v}/c, \quad \gamma = 1/\sqrt{1-\beta^2}. \tag{15}$$

5. Electrodynamics, Coulomb and Lorentz forces

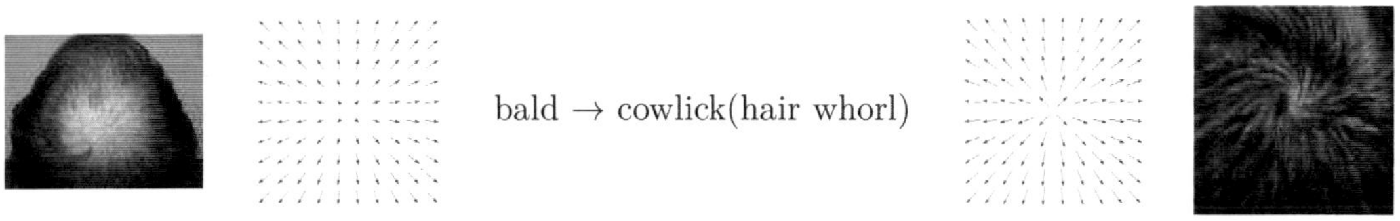

Figure 3. Left: Schematic picture for a regular soliton field $Q(x)$ with no singularity in the center. Right: Singular soliton field of a Wu-Yang monopole $Q(x) = -\mathrm{i}\vec{\sigma}\vec{n}$.

In order to derive Coulomb and Lorentz forces we have to artificially separate particles and their fields. In the limit

$$r_0 \to 0 \quad \Longleftrightarrow \quad q_0 = \cos\alpha = 0 \quad \Longleftrightarrow \quad \alpha = \frac{\pi}{2}, \tag{16}$$

see Fig. 3 for a descriptive picture, solitons transform to dual Dirac-monopoles in the Wu-Yang description [5]. Their remain only two degrees of freedom, a normalised three-dimensional vector field $\vec{n}(x)$ with

$$Q(x) = -\mathrm{i}\vec{\sigma}\vec{n}(x), \quad \vec{\Gamma}_\mu(x) = \vec{n}(x) \times \partial_\mu \vec{n}(x), \quad \vec{R}_{\mu\nu}(x) = \partial_\mu \vec{n}(x) \times \partial_\nu \vec{n}(x). \tag{17}$$

Since $\vec{R}_{\mu\nu}(x)$ points in $\vec{n}$-direction there is only one essential component left, $\vec{R}_{\mu\nu}\vec{n}$. The field-strength tensor and the Lagrangian reduce to

$${}^*F_{\mu\nu}(x) = -\frac{e_0}{4\pi\varepsilon_0 c}\vec{R}_{\mu\nu}\vec{n} = -\frac{e_0}{4\pi\varepsilon_0 c}\vec{n}(x)[\partial_\mu \vec{n}(x) \times \partial_\nu \vec{n}(x)], \quad \mathcal{L}_{\mathrm{ED}} = -\frac{1}{4\mu_0}{}^*F_{\mu\nu}(x)^*F^{\mu\nu}(x). \tag{18}$$

Since solitons can not move faster than light, the singularities can be enclosed in two-dimensional surfaces, e.g. parametrised by u and v, leading to the definition of electric charge

$$Q_{\mathrm{el}}(\mathcal{S}) = -\frac{e_0}{4\pi}\oint_{\mathcal{S}(u,v)} \mathrm{d}u\mathrm{d}v\, \vec{n}[\partial_u \vec{n} \times \partial_v \vec{n}]. \tag{19}$$

In Minkowski-space the singularities form closed world-lines, the electric charge is conserved,

$$j^\mu = -e_0 c \sum_{i=1}^{N} \int \mathrm{d}\tau_i \frac{\mathrm{d}X^\mu(\tau_i)}{\mathrm{d}\tau_i}\, \delta^4(x - X(\tau_i)) = (c\rho, \mathbf{j}). \tag{20}$$

After this artificial splitting of the soliton field in electric currents and electro-magnetic fields, we can also split the total force density (13) in two parts

$$f_\nu = \partial_\mu \Theta^\mu{}_\nu = f^\mu_{\text{charges}} + \partial^\nu T^\mu{}_\nu = 0, \tag{21}$$

Coulomb and Lorentz forces appear

$$f^0_{\text{charges}} = \frac{1}{c}\mathbf{j}\mathbf{E}, \quad \mathbf{f}_{\text{charges}} = \rho\mathbf{E} + \mathbf{j}\times\mathbf{B}. \tag{22}$$

It is well known from QCD that the tensor (3) is gauge-invariant. In the case of the soliton model this means invariant against rotations of the local coordinate systems on S^3. Since q_0 is also invariant, the Lagrangian (4) is invariant against SU(2)-gauge transformations. In the electrodynamic limit the soliton-field is restricted from the Q-field to the $\vec{n}$-field, from S^3 to S^2. The tangential spaces of S^2 are two-dimensional, the gauge-invariance reduces to a U(1)-invariance, well known from Maxwell's electrodynamics.

6. Topological charge

In addition to electric charge which turned out to be a topological quantum number explaining the quantisation of charge there is a further quantum number, the topological charge which counts the number of coverings of S^3

$$\mathcal{Q} = \frac{1}{V(S^3)}\int_0^\infty \mathrm{d}r \int_0^\pi \mathrm{d}\vartheta \int_0^{2\pi} \mathrm{d}\varphi\, \vec{\Gamma}_r(\vec{\Gamma}_\vartheta \times \vec{\Gamma}_\varphi), \quad V(S^3) = \int_{S^2} \mathrm{d}^2 n \int_0^\pi \mathrm{d}\alpha \sin^2\alpha = 2\pi^2. \tag{23}$$

The hedge-hog configuration (5) covers half of S^3, therefore it is characterised by $|\mathcal{Q}| = \frac{1}{2}$. It turns out that $\mathcal{Q}$ is a conserved quantity [13], $\frac{\mathrm{d}\mathcal{Q}(t)}{\mathrm{d}t} = 0$.

Transf.	1	z	Π_n	$z\Pi_n$
$\vec{n}$	$\vec{r}/r$	$-\vec{r}/r$	$-\vec{r}/r$	$\vec{r}/r$
q_0	≥ 0	≤ 0	≥ 0	≤ 0
Q_{el}/e_0	-1	1	1	-1
$\mathcal{Q} = \chi \times s$	$1/2$	$1/2$	$-1/2$	$-1/2$
diagram	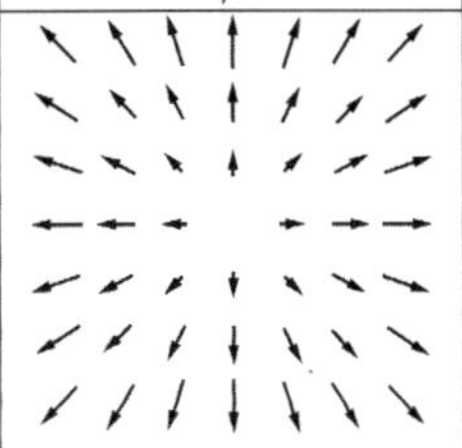	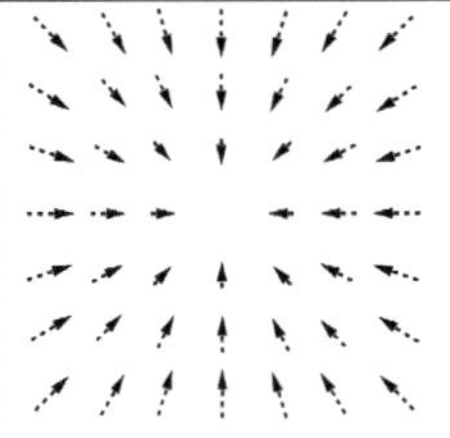	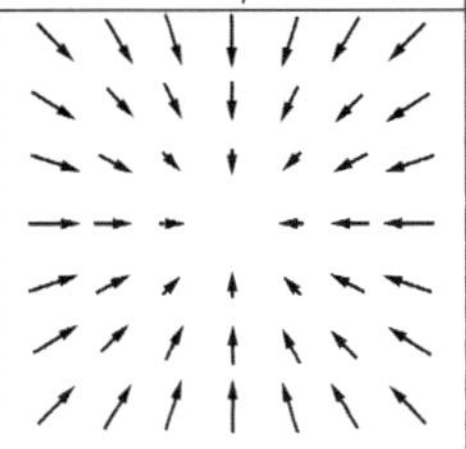	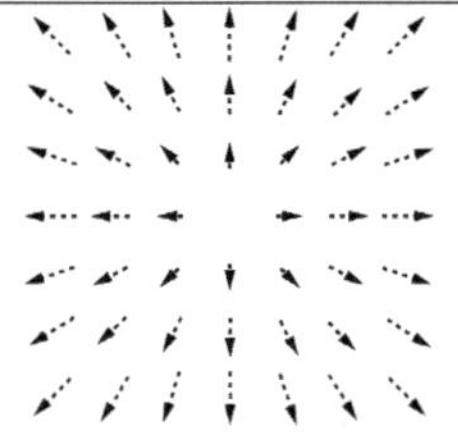

Table 1. Schematic pictures of the topologically different single-soliton configurations and their quantum numbers. Length and direction of arrows depict the value of the imaginary part $\vec{q}(x)$ of the soliton field $Q(x)$. The sign of q_0 is indicated by the line type of the arrow, full lines for $q_0 > 0$ and dashed lines for $q_0 < 0$.

Investigating the possible types of solitons, the physical meaning of the topological charge gets obvious. There are two transformations changing the homotopy type of the configuration characterised by the electric charge Q_{el} and the topological charge $\mathcal{Q}$. These are the transformation of the soliton field $Q(x)$ with the non-trivial center element $z = -\mathbb{1}_2$ and a reversal Π_n of the $\vec{n}$-direction. Applying these transformations, as shown in Table 1, we get four single-soliton configurations. It can be seen that arrows pointing outward (inward) define negative (positive) charges. Traversing the soliton, the arrows indicate a rotation by 2π. The topological charge reflects the chirality χ of this rotation, whether it is right or left-handed. Lines of constant $\vec{n}$-field, e.g. shown in the left part Fig. 2, connect solitons to the surroundings. Therefore, as argued in the beginning of section 2, solitons return only after a 4π-rotation to the original state. This is characteristic of particles with spin quantum number $s = 1/2$, of fermions. Since solitons are characterised by $|\mathcal{Q}| = \frac{1}{2}$ it is quite natural to interpret $\mathcal{Q}$ as the product of χ and s

$$\mathcal{Q} = \chi \times s. \tag{24}$$

This relation may answer a question, posed by T.D.Lee around 30 years ago, when he gave a talk at the University of Vienna: "Why does the mass violate chiral symmetry?"

7. Interacting solitons

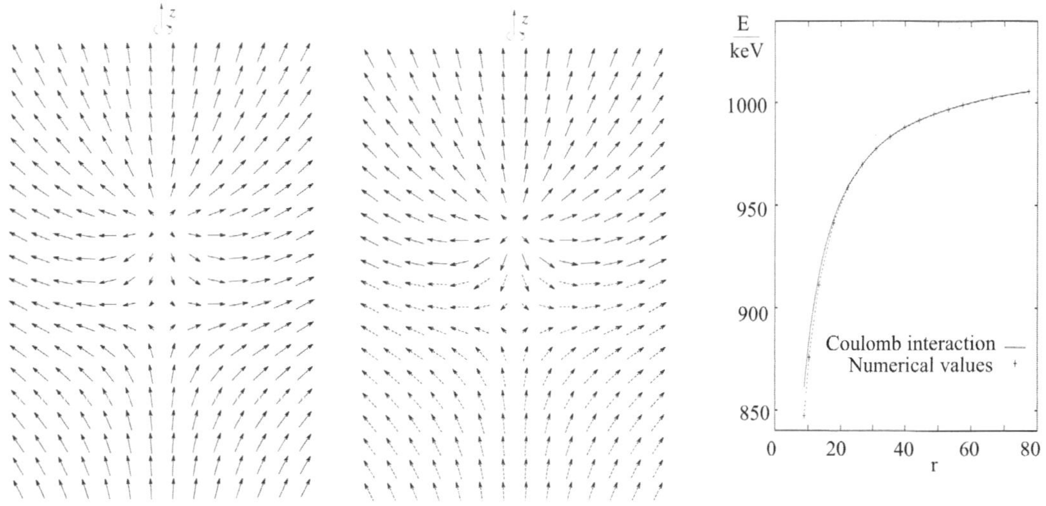

Figure 4. Soliton-antisoliton configurations, two opposite charges, left: $\mathcal{Q} = S = 0$, middle: $\mathcal{Q} = S = 1$, observe the dashed vectors in the lower half. Right: Total energy of an "electron-positron"-system in the $S = 0$-state for $m = 3$ in units of keV (crosses) as a function of distance r in lattice units $a = 3r_0$ for $r_0 = 2.21$ fm in comparison to the expectation for point-like charges (full line), from [20].

Solitons interact with electromagnetic forces. Therefore, static two-soliton configurations are in general not solutions of the equations of motion. Nevertheless, in a static approximation one can fix the soliton centers and determine the energy of the field configuration. In the left and middle diagram of Fig. 4 we show the two topological different soliton-antisoliton configurations, the left one corresponds to the two-particle spin quantum number $S = 0$, the middle diagram to $S = 1$. Both diagrams are rotationally symmetric around the z-axis. It should be noted that the antisolitons in the lower half of the diagrams are rotated by π around the symmetry-axis. In the left diagram there is no resistance when the two solitons approach each other, they attract each other and annihilate. In the middle diagram there is also attraction until the two "balds" in the center approach closely but can not annihilate since they belong to different half-spheres of S^3, indicated by the dashed arrows in the lower half. The right diagram in Fig. 4 is from the diploma work of Joachim Wabnig at the Vienna University of Technology [20] where he compared the energy of an "electron-positron" system for $m = 3$ and $S = 0$ in units of keV as a function of distance r in lattice units a with $r_0 = 3a = 2.21$ fm with the energy of point-like charges. It can be seen that the energy of the soliton-pair agrees with the Coulomb law for distances large compared to r_0 until the soliton cores of the two particles get close enough and increase the attraction. The energy density in a plane through the soliton centers is shown in Fig. 5. It is remarkable that the energy density is everywhere finite.

In scattering processes these solitons react almost like point-like particles do, they have no sub-structure which could be revealed. However, the strength of the interaction starts to change at distances of the order r_0. This behaviour is reminiscent of the running coupling in quantum field theory.

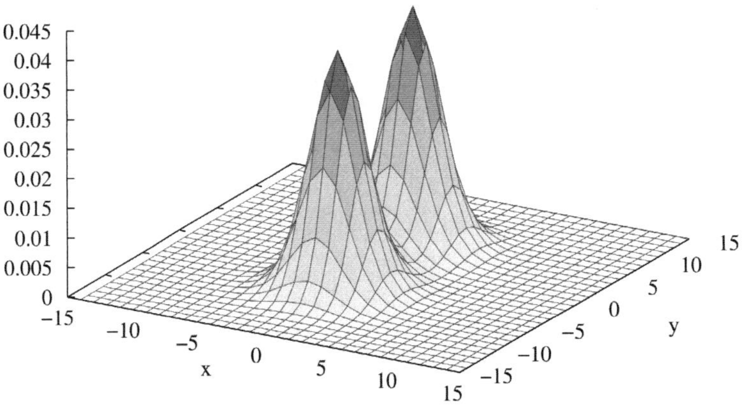

Figure 5. The energy density for a particle anti-particle solution

8. Conclusions and speculations

We can summarise our results:

- Mass is field energy
- Particles are topological solitons
- Charge and spin are topological quantum numbers
- Only integer multiples of elementary charge are possible
- Distinction between charges and their field is artificial
- There are two types of massless excitations, $\vec{n}$-waves
- Pauli principle has topological origin
- U(1) gauge invariance = invariance against rotations of a Zweibein (dyad) around $\vec{n}$
- Cosmological constant Λ $\rightarrow$ Cosmological Function $\Lambda(t, \vec{r})$

The above soliton model describes charged particles as topological solitons. The mass of these particles is field energy only. There is no Higgs particle necessary to give mass to particles. Due to their topological nature all free particles with the same topology have the same mass.

Charge is quantised in units of an elementary charge, it is a topological quantum number. There are only integer multiples of the elementary charge.

The spin quantum number is the absolute value of topological charge and can take only multiples of 1/2. The angular momentum properties of spin appear only when particles move around each other. Let's take the important example of a light negative charge rotating around a much heavier positive charge. By their common soliton field, see left and middle diagram of Fig. 4, the two charges are connected like two cogwheels of equal size. For a static heavy particle the light particle is forced to rotate twice around itself in a full orbit, as it should be for a gyromagnetic spin factor $g_s = 2$.

Charges and their fields are not distinguishable, they are build from the same field. The artificial separation between charges and fields as formulated in Maxwell's electrodynamics is a clever way to get a linear theory from an originally combined non-linear system.

The $\vec{n}$-field describing massless excitations has two degrees of freedom corresponding to the two polarisation directions of electromagnetic waves. Photons are realised as Goldstone bosons of the symmetry breaking of the vacuum described by a fixed direction of the $\vec{n}$-field.

A rudimentary form of the Pauli principle may be that two solitons in a $S = 1$ state can not be at the same position.

The local coordinate systems in the tangential space of S^3 are arbitrary. The area covered on S^3 by the Q-field is independent of the choice of this coordinate system. There is consequently an SU(2) gauge invariance of the field strength tensor. In the electrodynamic limit the elements $Q \in \mathrm{SU}(2)$ reduce to the normalised three dimensional vector field $\vec{n} \in \mathrm{S}^2$. There is only an invariance of the local coordinate system on S^2 left. This is a U(1) gauge-invariance also well known from QED.

The potential term $\Lambda(x)$ in the Lagrangian has some similarity to the cosmological constant Λ in the gravitation theory. In the presence of charged matter $\Lambda(x)$ is non-zero inside of solitons. It contributes with 25% to the soliton energy, see [13].

The model has also several properties which at first look differ from our present understanding of nature. The equations of motion allow for magnetic currents. These currents are non-topological [14] and act as sources of electric and magnetic fields. Magnetic currents are unknown in Maxwell theory. Further, there exist excitations in the α-degree of freedom. For $m = 1$ these excitations are realised as vibrational states of the solitons [17]. But for the electron such vibrational excitation are not known. For higher values of m there are no vibrational excitations, but α-waves with non-zero mass propagate in the vacuum. Such waves are also unknown. Since the potential term $\Lambda(x)$ in the Lagrangian has a dimension-full coupling constant the model is perturbatively non-renormalisable. Quantum theory can be included via a path integral formulation. Due to the smallness of the fine-structure constant α_f the quantum system should be in the strong coupling regime, at zero temperature it should be disordered. But there may exist a finite temperature phase transition from this disordered to an ordered phase. There is no definite answer to this question yet.

We can add some speculations about possible solutions of the above problems. To solve the problem on non-topological magnetic currents and α-waves we can claim that all particles which we measure in our detectors are topological solitons. Non-topological currents and waves escape our detectors. We measure them as electric and magnetic fields only. These non-topological fields could contribute to dark-matter. The potential term $\Lambda(x)$ in analogy to the cosmological constant Λ could be called a cosmological function and possibly contribute to dark energy. There are strong indications that in the early universe there was a period called inflation with a huge energy transfer and consequently fast expansion of the universe. One can speculate that the early universe was characterised by a vacuum with $Q(x) = 1$, $\alpha = 0$ and maximal potential energies, then a transition happened to a minimum of $\Lambda(x)$ with $\alpha = \pi/2$. Since this minimum is a two-dimensional manifold the vacuum is degenerate, spontaneous symmetry breaking happened and photons as Goldstone bosons appeared.

Finally, some words on the main topics of this conference, i.e., on the emergence of quantum mechanics. This model seems to favor a scenario where dark matter disturbs the motion of solitonic electrons when they orbit around nuclei. These statistical fluctuations are influenced by the density of dark matter and not by temperature, they induce zero-point motion and lead to probability distributions for the electrons.

These are some ideas which could be worked out and compared with experiment. The presented model is rather uncomplicated, it has three degrees of freedom only and gives a nice insight in many electro-dynamical phenomena and phenomena which are described by quantum field theory, like mass generation, particle-antiparticle annhihilation and the running of the coupling.

This model is a geometric model, the algebra is used as a tool only to describe geometry. It has a strong similarity with gravitation which can be formulated as translations of Vierbeins (tetrads). This model is based on local rotations of Dreibeins (triads). John A. Wheeler has formulated "Space-time tells matter how to move, matter tells space-time how to curve". We can add "··· and charge tells space how to rotate". This means the phenomenology of gravitation and electro-dynamics can possibly be understood by the properties of space and time only. For the description of strong and weak forces the model should probably be extended to include further dimensions, i.e. further degrees of freedom.

I think that physics is geometry and not algebra. We should use the algebra only to describe the geometry.

References

[1] P. A. M. Dirac, "Quantised singularities in the electromagnetic field," *Proc. Roy. Soc. Lond.* **A133** (1931) 60–72.
[2] P. A. M. Dirac, "The theory of magnetic poles," *Phys. Rev.* **74** (1948) 817–830.
[3] T. T. Wu and C.-N. Yang, "in properties of matter under unusual conditions,". edited by H. Mark and S. Fernbach (Interscience, New York 1969) p.349.
[4] T. T. Wu and C. N. Yang, "Concept of nonintegrable phase factors and global formulation of gauge fields," *Phys. Rev.* **D12** (1975) 3845–3857.
[5] T. T. Wu and C.-N. Yang, "Some remarks about unquantized nonabelian gauge fields," *Phys. Rev.* **D12** (1975) 3843–3844.
[6] H. Georgi and S. L. Glashow, "Unified weak and electromagnetic interactions without neutral currents," *Phys. Rev. Lett.* **28** (1972) 1494.
[7] G. 't Hooft, "MAGNETIC MONOPOLES IN UNIFIED GAUGE THEORIES," *Nucl. Phys.* **B79** (1974) 276–284.
[8] A. M. Polyakov, "Particle spectrum in quantum field theory," *JETP Lett.* **20** (1974) 194–195.
[9] M. Remoissenet, *Waves called solitons: concepts and experiments.* Springer, 1996.
[10] T. H. R. Skyrme, "A Nonlinear field theory," *Proc. Roy. Soc. Lond.* **A260** (1961) 127–138.
[11] T. H. R. Skyrme, "A Unified Field Theory of Mesons and Baryons," *Nucl. Phys.* **31** (1962) 556–569.
[12] V. Makhan'kov, Y. Rybakov, and V. Sanyuk, *The Skyrme model: Fundamentals, methods, applications.* Springer series in nuclear and particle physics. Springer-Verlag, 1993.
[13] M. Faber, "A model for topological fermions," *Few Body Syst.* **30** (2001) 149–186, `hep-th/9910221`.
[14] M. Faber, J. Wabnig, and A. N. Ivanov, "Solitons and spontaneous symmetry breaking in 2 and 4 dimensions," in *Confinement, Topology, and Other Non-Perturbative Aspects of QCD*, J. Greensite and Š. Olejník, eds., pp. 129–136. Kluwer Academic Publishers, Dordrecht, 2002. `http://arXiv.org/abs/hep-lat/0204031`.
[15] M. Faber and A. P. Kobushkin, "Electromagnetic field from the Skyrme term," *Phys. Rev.* **D69** (2004) 116002(13 pages), `hep-th/0207167`.
[16] D. Borisyuk, M. Faber, and A. Kobushkin, "Electro-Magnetic Waves within a Model for Charged Solitons," *J. Phys.* **A40** (2007) 525–531, `arXiv:0708.3173 [hep-th]`.
[17] M. Faber, A. Kobushkin, and M. Pitschmann, "Shape vibrations of topological fermions," *Adv. Stud. Theor. Phys.* **2** (2008) 11–22, `0812.4225`.
[18] C. Misner, K. Thorne, and J. Wheeler, *Gravitation*, Fig. 41.6, p. 1149. Physics Series, W. H. Freeman, 1973.
[19] J. Jackson, *Classical electrodynamics*, Wiley, 1999.
[20] J. Wabnig, "Interaction in the model of topological fermions," *Vienna University of Technology,* **diploma thesis** (2001).

An extended model of electrons: experimental evidence from high-resolution scanning tunneling microscopy.

Werner A. Hofer
Department of Physics, University of Liverpool
L69 3BX Liverpool, United Kingdom

Abstract. In a recent paper we introduced a model of extended electrons, which is fully compatible with quantum mechanics in the formulation of Schrödinger. However, it contradicts the current interpretation of electrons as point-particles. Here, we show by a statistical analysis of high-resolution scanning tunneling microscopy (STM) experiments, that the interpretation of electrons as point particles and, consequently, the interpretation of the density of electron charge as a statistical quantity will lead to a conflict with the Heisenberg uncertainty principle. Given the precision in these experiments we find that the uncertainty principle would be violated by close to two orders of magnitude, if this interpretation were correct. We are thus forced to conclude that the density of electron charge is a physically real, i.e. in principle precisely measurable quantity, as derived in a recent paper. Experimental evidence to the contrary, in particular high-energy scattering experiments, is briefly discussed. The finding is expected to have wide implications in condensed matter physics, chemistry, and biology, scientific disciplines which are based on the properties and interactions of electrons.

In density functional theory (DFT) a many-electron system is comprehensively described by the density of electron charge [1, 2]. However, the density itself, within the framework of second quantization, is thought to be a statistical quantity [3, 4]. In principle, this statement can be tested by a statistical analysis of high-resolution STM measurements including the uncertainty relations [5]. The reason that such an analysis at this point is necessary is the following. In a recent paper [6] it was found that an extended model of electrons fully recovers the formulation of quantum mechanics in terms of Schrödinger's equation [7]. These findings take us back more than eighty years to the original development of modern physics. At this point, two different approaches to atomic physics were proposed: one, by Heisenberg, Born and Jordan, based on matrices and classical mechanics; the other, by de Broglie and Schrödinger, based on wave properties and an adaptation of classical electrodynamics. The rift between these two approaches had repercussions in the development of physics for at least two generations. Today, both approaches are combined in our modern quantum theory. However, it was recently shown that a model of extended electrons, which is more in line with Schrödinger's than Heisenberg's concept, can remove some of the main obstacles to an understanding in atomic scale physics [6]. The model is sketched in Figure 1. It has the following advantages: (i) The self-energy, which would be infinite for a point-like electron, is calculated to be typically 8.16eV; it can be accounted for by a simple density-dependent cohesive potential. (ii) The spin $S = S\mathbf{e}_v$ of a free electron is a vector quantity directed either parallel or anti-parallel to the velocity vector $\mathbf{v}$, it changes in an external and time-dependent magnetic field B according to a modified Landau-Lifshitz equation:

$$\frac{d\mathbf{S}}{dt} = \alpha \cdot \mathbf{S} \times \mathbf{B} + \beta \cdot \mathbf{S} \times \left(\mathbf{v} \times \frac{d\mathbf{B}}{dt} \right) \tag{1}$$

where α and β are constants. A statistical manifold of spin-up and spin-down electrons in a constrained trajectory of the electron as e.g. within the Coulomb field of a nucleus in this case yields exactly two results in a Stern-Gerlach type experiment. (iii) The electron density is a real physical quantity, it varies with time and space in the wave-propagation of the electron and it can be shown how it changes due to accelerations. (iv) The wavefunction is not strictly a physical property of electrons, since it combines density components and field components in one (a multivector) structure. Therefore effects like the notorious spreading of the wavepacket for free electrons are mere artefacts within this new model and without physical reality. (v) While in standard theory the stability of hydrogen is referred to the uncertainty principle (which prevents that the electron falls into the nucleus), in the model of extended electrons it is the requirement that the energy density within one electron has to be constant, which prevents this outcome. (vi) While there is a strict divide between atomic-scale physics and macro physics in the standard framework, due to the necessary integrations over infinite space if one is to calculate a measured value, this divide vanishes in the new framework: a consistent theoretical framework can thus be constructed from the picometer to the meter scale [6].

This leads to a very fundamental issue in quantum mechanics. Originally, it was found by Born, Jordan and Schrödinger that the formulation of the theory in terms of the Schrödinger equation is equivalent to the formulation in terms of operators and matrices, as proposed by Heisenberg [8], implying the uncertainty relations. However, this equivalence depends on the interpretation of the electron as a point-particle. It is not a priori valid for an extended electron. For extended electrons the density of charge will be a physical property of a microsystem. The question then arises, whether the uncertainty relations are still meaningful in this context. This question can in principle be answered by an analysis of measurements of electron charge distributions. It should be noted that the interpretation of the electron wave as a physical wave makes the usual, purely mathematical, derivation of the Uncertainty Relations via Fourier transforms and a Gaussian wave packet untenable. If the electron is a real wave, then it does possess a physical momentum and it is essentially a plane wave confined to some region in space. In this case a composition of single electrons in terms of a Fourier series of partial waves with different momenta, even though mathematically possible, is physically not meaningful. The Uncertainty Relations are thus a statement about the physical properties of systems at the atomic scale [9]. If it turns out that even in our most precise measurements the interpretation of the density of electron charge as a statistical quantity is still possible, then the uncertainty relations have a meaning also in this context. However, if it turns out that standard statistical quantities, depending on the uncertainty of location and momentum, will be much smaller in the actual experiments than allowed for by a theoretical analysis, then the uncertainty relations, and by implication operator mechanics, cannot be considered an equivalent description. In this case, it should also be ultimately possible to measure any physical property of a microsystem, i.e. an atomic-scale system, with in principle infinite precision.

The statistical measure we shall use for our analysis is the standard deviation σ. The standard deviation for a measurement of position x, given N measured values x_i, is defined as:

$$\sigma_x = \sqrt{\frac{1}{N} \sum_{i=1,N} (x_i - \langle x \rangle)^2} \qquad \langle x \rangle = \frac{1}{N} \sum_{i=1,N} x_i \tag{2}$$

where $\langle x \rangle$ denotes the statistical average. We shall analyse the variation of the density of electron charge at a metal surface, and compare the local contrast in the experiments, which is given by the change of the density from one point of measurement to the next, with the

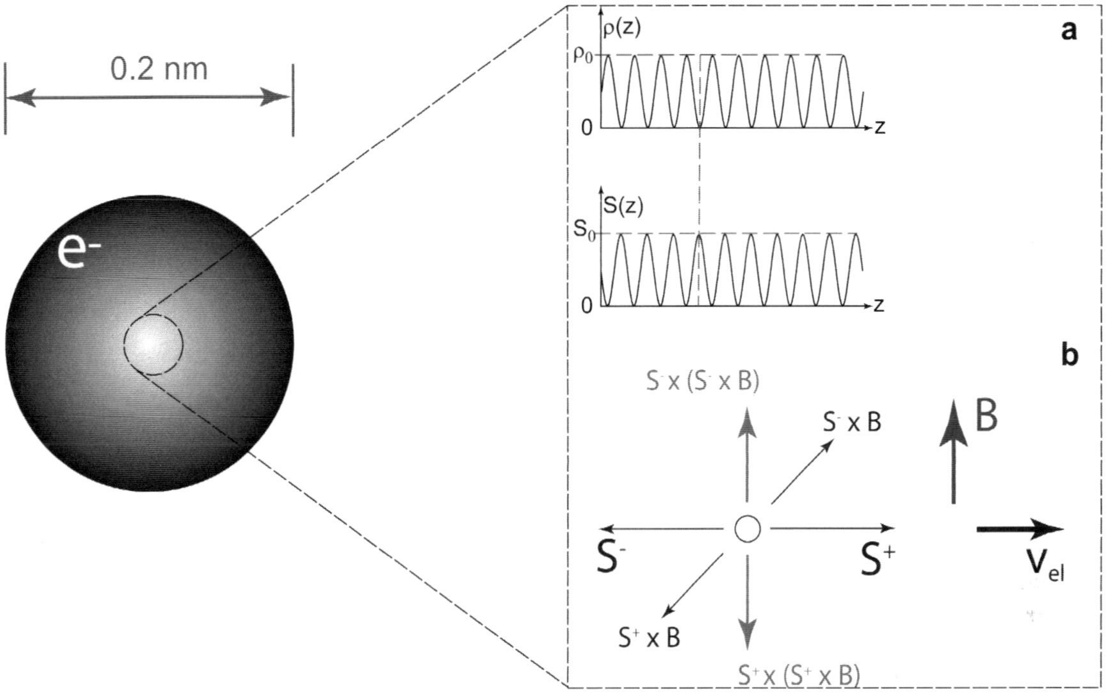

Figure 1. (online version in colour): Model of extended electrons. a The electron is an extended structure with a diameter of about 0.2 nm; the density of charge and the spin density S are oscillating functions, their wavelength complies with the de Broglie relations. b In a constrained trajectory magnetic fields will lead to a rotation of the spin density vector: the induced spin density is either parallel or antiparallel to the applied magnetic field.

achievable contrast on the basis of allowed energy, momentum, and position uncertainty. We shall explicitly treat noble metal (111) surfaces, where free electrons, confined to the surface layer, exist, which can be treated for all practical purposes as free electrons. They should thus comply with the following fundamental relation in quantum mechanics:

$$\Delta p \cdot \Delta x \geq \frac{\hbar}{2} \tag{3}$$

The relation between an uncertainty in momentum and an uncertainty in energy is described by:

$$\Delta E = \Delta \left(\frac{p^2}{2m} \right) \tag{4}$$

Today, STMs have reached a level of precision which is quite astonishing. While it was barely possible to resolve the positions of single atoms in early experiments [10, 11], it is now routine not only to resolve the atomic positions but also e.g. the standing wave pattern of surface state electrons [12] on metal surfaces, and even very subtle effects like inelastic excitations, Kondo resonances, or surface charging [13, 14, 15]. What the STM measures, in these experiments, is the current between a surface and a very sharp probe tip.

The current itself is in most cases proportional to the density of electron charge at the surface [16]. While one may dispute this claim for some special cases, and while it can be shown for specific situations that an explicit simulation of the scattering process does improve the agreement between experiment and theory (see, for example, [17]), in most measurements on metal surfaces the bias range is so low and the dominance of single electron-states at the tip so high, that the Tersoff-Hamann approximation [18], which assumes tunneling into a single tip-state with radial symmetry, is a very good approximation. If this approximation is valid, then the map of tunneling currents at a surface is, but for a constant, equal to the map of electron charge densities at the same surface. Then a standard deviation of the density of charge due to the uncertainty of position and momentum can be mapped identically onto a standard deviation of the tunneling current, which can immediately be compared to experimental results.

From the viewpoint of a physicist of the golden age of quantum mechanics, say the time around 1925, the precision of STM measurements, where the vertical resolution of the best instruments is close to fifty times smaller (about 0.05 pm [13]) than the Compton wavelength of an electron (or about 2.4 pm), must seem like magic. In Figure 2 we show a quantum corral of 51 silver atoms on an Ag(111) surface measured by Rieder et al. [19]. In some images the instrument was able to resolve not only the standing wave pattern of surface state electrons, but also the modulation due to the positions of single surface atoms. The measurement is quite spectacular, and it raises some very fundamental problems in quantum mechanics, which we shall explore in the following. It is irrelevant for this analysis, whether the remaining experimental imprecision is due to microscopic or macroscopic processes: it is the level of precision in these experiments itself, which is problematic.

It is difficult to define the exact uncertainty of momentum of a surface state electron in the experiments. From the energy resolution in today's best measurements $\Delta E = 1\text{meV}$ [20] and using Eq. (2) one could infer an uncertainty of momentum $\Delta p = p = 1.52 \times 10^{-26}$ kgms^{-1}, and a local uncertainty $\Delta x = 3.1 \times 10^{-9}$ m, or 3.1 nm. However, to establish that the uncertainty relations fail in this context, it is only necessary to provide an upper limit of the energy and momentum uncertainty. Thus, for an electron at the Fermi level of a Ag(111) surface the maximum available energy is given by the energy difference between the bottom of the parabolic surface state band and the energy at the Fermi level. This energy is about 80 meV. This corresponds also to the thermal energy at ambient conditions. As the experiments are performed at cryogenic temperatures of 5K, the thermal energy is much lower than the band energy. Hence the maximum uncertainty in the energy of the electron is also about 80 meV. From this value it is possible, via the energy uncertainty and the Heisenberg uncertainty relations (Eqs. (2)), to derive the uncertainty of the x-position: $\Delta x = 3.48 \times 10^{-10}$m or about 3.5Å. This value is about three times larger than the Wigner-Seitz radius (1.06Å), which describes the radius of a sphere with the typical electron density found in metals. The uncertainty Δx is also larger than the distance between two individual atoms on a silver surface, which is 2.9Å. The unit cell, the Wigner-Seitz cell, and the uncertainty are marked in Figure 2c.

Following the standard interpretation of quantum theory the electron is a point particle. Its appearance at a particular location is a statistical event governed by the Schrödinger equation and the probability density, which is equal to the square of the electron's wavefunction, or $|\psi|^2$. The same statement is also correct for alternative interpretations of quantum mechanics, e.g. based either on the pilot-wave theory [21, 22], or a theory of the Dirac electron based on Zitterbewegung [23]. In this case the precision of an experiment will depend on the statistical variation of the electron's location and its convolution over time. This, however, will also influence the statistical variation of e.g. the apparent height. If the apparent height can be measured with a precision of less than 0.05pm, as found in experiments [13], then the statistical probability of an electron to be found at a particular location can only have a certain standard deviation σ. Given the local uncertainty derived from the uncertainty relations, this deviation can, in principle, be linked either to a minimum number of events within the measurement interval of an STM, which is in the range of milliseconds, or to the error in the vertical distance measurement combined with the lateral resolution in the experiments. This argument differs

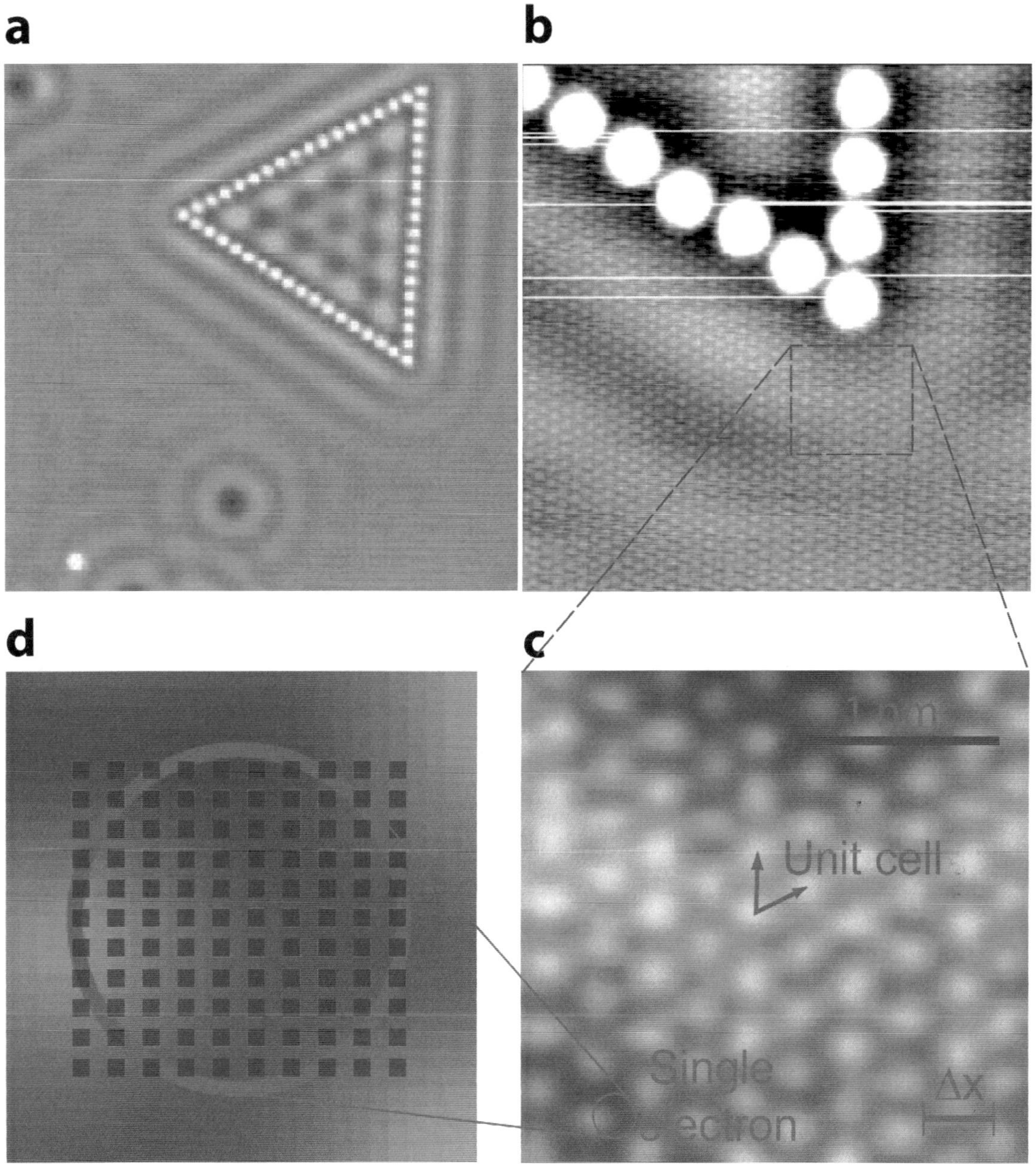

Figure 2. (online version in colour): Quantum corral and surface state electron on a Ag(111) surface. a The propagation of the surface state electrons leads to periodic variations of the density and the apparent height. b In high-precision measurements these variations are modulated by the electronic structure of surface atoms. c Small region of the central image blown up and annotated: the unit cell, the area of a single electron, and the local uncertainty are marked. d The Wigner-Seitz cell of a single electron contains at least 88 pixels of measurements, if the lateral resolution is assumed to be 20pm. From [19]

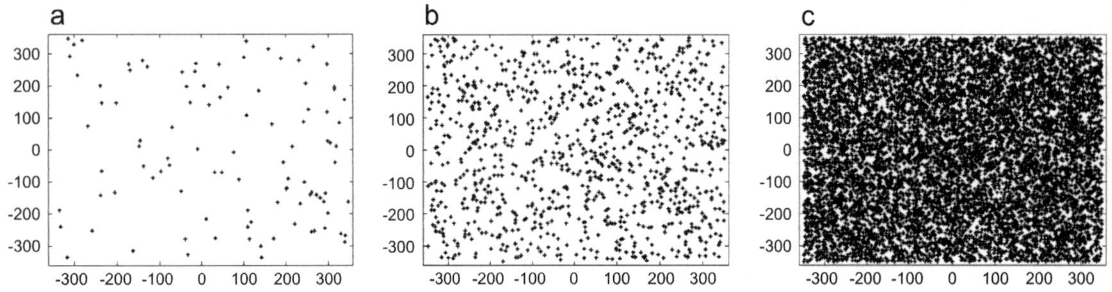

Figure 3. Random distribution of the location of a point-like electron on the surface. a 100 events, b 1000 events, c 10000 events. The standard deviation of the radius for each of the distributions is about 280 pm (random distribution taken from random.org).

substantially from previous analysis of STM measurements, which only considered the vertical precision: given the time resolution of the instrument and a tunneling current of 1nA about 10^7 electrons cross the vacuum barrier within the measurement resolution of the instrument. Then, assuming that the standard deviation decreases with the square root of the number of transitions $\sqrt{N}$ and based on a vertical uncertainty of 3.5Å, we obtain a resolution limit of 0.1 pm, which is about the precision in present experiments. However, this alone will not be sufficient to establish the statistical facts. Only the *lateral resolution* in combination with the *vertical precision* will be sufficient to analyse the statistics in these experiments comprehensively.

We have two possible ways in which the experimental values can be compared to the standard deviation inferred from the uncertainty relations. We may either find, that a threshold number of measurements N_0 is necessary to obtain the experimentally observed standard deviation. In this case we can estimate the minimum distance covered by the electron within the time-resolution limit of STM experiments $\Delta t(STM)$ (i.e. milliseconds), and derive a lower limit for the time-resolution of the instrument without a violation of the uncertainty relations. Such a limit is described by:

$$\Delta L = N_0 dl \quad dl \approx \Delta x \quad v(\Delta E)\Delta t(STM) \geq \Delta L$$
$$\Delta t(STM) \geq \frac{\Delta L}{\sqrt{2\Delta E/m}} \tag{5}$$

where ΔL is the total path during N_0 measurements, dl is the average path from one measurement to the next, and ΔE is the energy uncertainty. Or we may find that the calculated standard deviation is actually quite independent of the number of measurements. In this case the value of the standard deviation for an energy uncertainty ΔE can be directly compared to the maximum standard deviation of the tunneling current given the precision of the instrument. Here, the percentage of error in an STM measurement, calculated from ratio of the error in the distance measurement divided by the apparent height of a feature, can be used, in combination with the lateral resolution of the instrument, to determine the maximum standard deviation. For an error in the experiments of 0.3%, which is easily obtained in today's most precise experiments, we find then:

$$\frac{\Delta z}{z_0} \leq 0.3\% \quad \Rightarrow \quad \Delta x(STM) \geq 3\sigma(\Delta E) \tag{6}$$

where $\Delta x(STM)$ is the lateral resolution of an STM and ΔE the maximum energy uncertainty. If three times the standard deviation inferred from the energy uncertainty is substantially larger than the lateral experimental resolution $\Delta x(STM)$, then one of the original assumptions going into the analysis of the process must be flawed. We show that this is indeed the case.

In the following we start the discussion by analysing events at one particular pixel of the grid on the surface and analyse events from the viewpoint of one electron only. Subsequently, we shall determine how the presence of many electrons at the surface could alter the initial findings. The apparent height of a feature measured by STM is, for a metal surface, in the range of 10 to 200pm. For the sake of the argument, and to avoid being overly restrictive, we assume the vertical resolution to be 0.1pm, and the lateral resolution to be 20pm. In a Wigner-Seitz cell we then have about 88 individual pixels of STM measurements (see Figure 2d). If we assume that the feature height is 30pm, then the experimental precision is 0.3%. The apparent height at an individual pixel would be affected, however, if at a given moment during the measurement interval the electron were not found at the location of this pixel, but at a neighbouring pixel. Thus we may say that the standard deviation of the location measurement is such that the required percentage of measurements must be at this particular location. For a variation of less than 0.3% three times the standard deviation of the measured results must thus still be smaller than half the distance between the pixels (see Figure 2d), or about 10pm. The standard deviation must thus be smaller than 3.3pm. Given that the uncertainty $\Delta x = 348$pm, every single measurement event, at which the electron should be detected at the position of one pixel at $(x, y) = (0, 0)$, will lead to a statistically distributed measurement in the interval $-348 \leq x \leq +348$, and $-348 \leq y \leq 348$. We have simulated such a statistical distribution of events in Figure 3. It can be seen that the locations of the electron are indeed randomly distributed in the interval, as required by the uncertainty relations and quantum mechanics. We find a standard deviation of about 280pm for all sets of data. Increasing the number of measurements does not change the precision, since the precision is limited by the uncertainty relations themselves. The precision can only be increased by an increase of the electron's energy. Since σ is about one hundred times the required value, we may say that there is a clear contradiction between the observed precision of the experiments, and the calculated uncertainty for the location of the electron. It is also clear that the underlying assumption, leading to this contradiction, is the statistical nature of the measurement, encoded in the uncertainty relations. This contradiction, moreover, is not the result of a small variation of the values obtained, but a variation of two orders of magnitude.

It is straightforward to calculate the energy necessary to obtain the precision found in the experiments. If σ has to be around 3pm, then the necessary uncertainty will also be about 3pm. This means that the momentum uncertainty has to be around 1.76×10^{-23}kgms^{-1}, and the necessary energy will be more than 1000 eV. For a measurement taken at cryogenic temperatures of 5K, as routinely performed today, the probability of such energy, derived from the Fermi-Dirac distribution, is essentially zero; the precision found in the experiments is thus impossible to explain within the framework of standard quantum theory.

The result would not change substantially if the precision in the experiments were somewhat lower. However, so far we have assumed that the uncertainty leads to a loss of electron detection at a particular point. It is, at least in principle, also possible, that uncertainty, via the influx of adjacent electrons at this particular point, may lead to a gain in electron detection. We may calculate this gain from the normal distribution for the distance from adjacent pixels and the ratio of the diameter of a pixel, or 20pm and the circumference of a circle at this distance, so that the normalized gain g from a point at distance d will be (units in pm):

$$g(d) = \frac{20}{2\pi d} \frac{1}{\sqrt{2\pi\sigma^2}} \exp - \frac{d^2}{2\sigma^2} \tag{7}$$

Summing up all contributions from pixels in a distance d of less than 280 pm, we obtain a value of 0.65. For a contrast of 100, corresponding to an apparent height difference of 200pm, which is easily obtained in an STM, the gain at a particular pixel should not be larger than 0.01. Comparing with the number obtained from a normal distribution, based on a standard deviation of 280pm and a pixel width of 20pm, or 0.65, we see that such a contrast cannot be

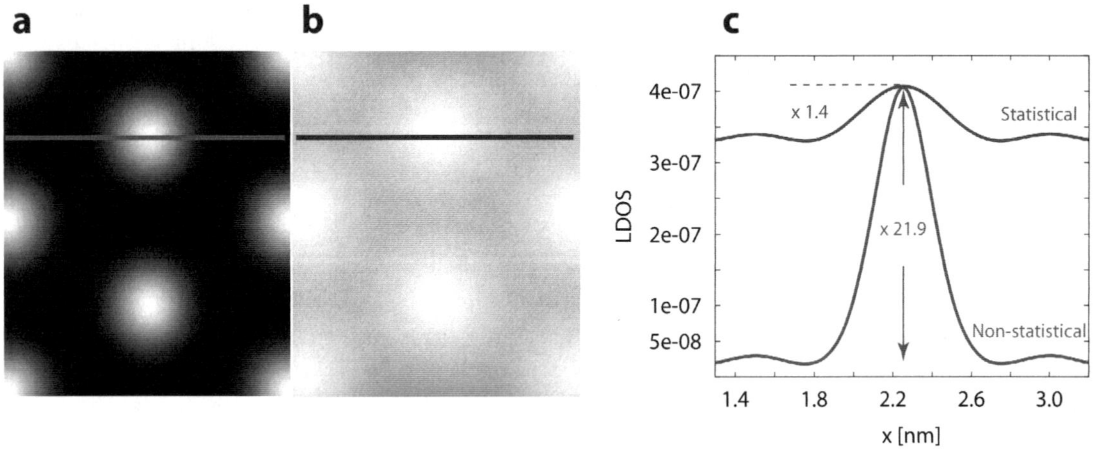

Figure 4. Simulated constant height contour of silver adatom. a Simulation based on the exact values of the local density of electron charge. The contour is similar to the experimental results shown in Fig. 1 and to measurements and simulations of a Pb adatom presented in Ref. 18. b Simulation based on a convolution of probabilities for a standard deviation of 280pm. c Linescan across the adatom for both simulations. The exact values of the density yield an apparent height which is more than one order of magnitude higher than the apparent height in the simulation based on quantum statistics.

obtained if the density of charge is assumed to be a statistical quantity complying with the uncertainty relations. Again, the value obtained from the statistical analysis differs by about two orders of magnitude from the value inferred from the experiments. So that from both perspectives, the loss due to the appearance of the electron at adjacent pixels, and the gain due to the appearance of the electron statistically deviating from adjacent pixels, we arrive at the same conclusion: the precision and contrast of today's STM experiments is well beyond the requirements of the uncertainty principle. It is thus safe to conclude that the density of charge cannot be a statistical quantity.

To show in detail, how a statistical density of electron charge complying with the uncertainty relations would affect the image of a single adatom at the silver surface, we have simulated the image of an adatom at very low bias voltage (20mV), in constant height mode at a distance of 500pm above the centre of the adatom. Figure 4 shows two possible images: Figure 4a reflects a non-statistical measurement, where the measured current is exactly proportional to the calculated local density of states (LDOS), this image is very close to experimental results (see Fig. 2a, and the measurements by Lian et al. [24]). Figure 4b shows a hypothetical measurement, where the LDOS value at every single point is affected by the random distribution of the electrons present at a given point. In this image we have summed up all contributions from the normal distributions of adjacent measurement points, where the amplitude is equal to the exact density of states at a given point, and its contribution to adjacent points is calculated from a normal distribution with a standard deviation of 280pm. We have rescaled the ensuing LDOS values to the maximum value for the density. Figure 4c shows a simulated linescan across a single adatom. Here, the numerical difference is more than one order of magnitude. The simulation allows the conclusion that measurements of adatoms would not be able to show an apparent height difference of more than about 10pm. This is, again, a clear contradiction with experimental results, where the apparent heights are typically larger than 100pm.

In the present context, what the STM will measure is the intrinsic density distribution of single surface electrons, which are spread-out into their lateral Wigner-Seitz cells. The measurements in this case can be arbitrarily precise: since the density is a physical, and not merely a statistical quantity, there is no limit to the precision in a measurement. Given the conceptual advantages of the model, and the fact that a statistical approach does violate the uncertainty principle, as shown, we think it is time to reconsider the famous Heisenberg-Schrödinger controversy.

The analysis seemingly contradicts high-energy scattering experiments with electrons insofar that no local extension of electrons is manifest in these experiments. This differs decisively from high-energy scattering of neutrons, protons, or nuclei, where a local extension of electron charge, both positive and negative, can be extracted from experimental data [25]. However, on closer analysis it turns out that the high-energy experiments do not establish a positive result, i.e. that the electron really *is* a point particle, but rather a negative one: no recurring local extension and shape of the electron's charge has ever been found. This is a crucial difference and it establishes only that the underlying assumption - a fixed local shape and extension of the electrons which becomes evident in the scattering cross sections - is not valid. In our view it points to a possible modification of the analysis of these experiments in the future.

In summary, we have shown that modern STM experiments violate the Heisenberg uncertainty relations by about two orders of magnitude. This indicates that the density of electron charge is not a statistical quantity, as currently believed.

Acknowledgments
Helpful discussions with Krisztian Palotas are gratefully acknowledged. The author thanks the Royal Society London for support.

References

[1] P. Hohenberg, W. Kohn, Phys. Rev. 136, B864 (1964).
[2] W. Kohn, L. J. Sham, Phys. Rev. 140, A1133 (1965).
[3] M. Born, Zeitschr. f. Phys. 37, 863 (1926).
[4] M. Born, Zeitschr. f. Phys. 38, 803 (1926).
[5] W. Heisenberg, Zeitschr. f. Phys. 43, 172 (1927).
[6] Werner A. Hofer, Found. Phys. 41, 754 (2011).
[7] E. Schrödinger, Phys. Rev. 28, 1049 (1926).
[8] W. Heisenberg, Zeitschr. F. Phys. 33, 879 (1925).
[9] James L. Park and Henry Margenau, Intl. J. Theor. Phys. 3, 211 (1968). See in particular page 213: "Many gedankenexperiments have been designed to illustrate Heisenberg's famous law; unfortunately, the false impression is often conveyed that his principle, which is actually a theorem about standard deviations in collectives of measurement results, imposes restrictions on *measurability*." [italics in the original text]
[10] G. Binnig, H. Rohrer, Ch. Gerber, E. Weibel, Phys. Rev. Lett. 49, 57 (1982).
[11] G. Binnig, H. Rohrer, Rev. Mod. Phys. 59, 615 (1987).
[12] E. J. Heller, M. F. Crommie, C. P. Lutz, D. M. Eigler, Nature 394, 464 (1994).
[13] H. Gawronski, M. Mehlhorn, K. Morgenstern, K. , Science 319, 930 (2008).
[14] N. Neel et al., Phys. Rev. Lett. 98, 016801 (2007).
[15] K. R. Harikumar et al., Nature Nanotechnology 3, 222 (2008).
[16] W. A. Hofer, A. S. Foster, A. L. Shluger, Rev. Mod. Phys. 75, 1287 (2003).
[17] Z. T. Deng, et al., Phys. Rev. Lett. 96, 156102 (2006).
[18] J. Tersoff, D. R. Hamann, Phys. Rev. B 31, 805 (1985).
[19] K.-H. Rieder et al., Phil. Trans. R. Soc. Lond. A 362, 1207 (2004).
[20] A. J. Heinrich, J. A. Gupta, C. P. Lutz, D. M. Eigler, Science 306, 466 (2004).
[21] David Bohm, Phys. Rev. 85, 166 (1952); 85, 180 (1952).
[22] L. De Broglie, Non-linear Wave Mechanics: A Causal Interpretation. Elsevier, Amsterdam (1960).
[23] David Hestenes, Found. Phys. 40, 1 (2010).
[24] J. C. Lian, et al. Phys. Rev. B 81, 195411 (2010).
[25] R. M. Littauer, H. F. Schopper, and R. R. Wilson, Phys. Rev. Lett. 7, 144 (1961).

Quantum Mechanics from Classical Logic

Gerard 't Hooft [1,2]

[1] Institute for Theoretical Physics, Utrecht University

[2] Spinoza Institute, Postbox 80.195, 3508 TD Utrecht, the Netherlands

E-mail: g.thooft@uu.nl http://www.phys.uu.nl/~thooft/

Abstract. Although quantum mechanics is generally considered to be fundamentally incompatible with classical logic, it is argued here that the gap is not as great as it seems. Any classical, discrete, time reversible system can be naturally described using a quantum Hilbert space, operators, and a Schrödinger equation. The quantum states generated this way resemble the ones in the real world so much that one wonders why this could not be used to interpret all of quantum mechanics this way. Indeed, such an interpretation leads to the most natural explanation as to why a wave function appears to "collapse" when a measurement is made, and why probabilities obey the Born rule. Because it is real quantum mechanics that we generate, Bell's inequalities should not be an obstacle.

1. Introduction

Quantum Mechanics is usually presented as being a totally new paradigm, complete with its own rules of logic. This, indeed, is how quantum mechanics works in practice. The rules of quantum logic, for instance, categorically disallow certain questions from being asked; in particular, one should not even try to ask "what is really going on?" when calculations are made concerning some quantum phenomenon. If a photon is sent through two tiny slits, and an interference pattern arises, one should never ask through which slit the photon is going, or whether the photon follows any particular path at all – or even what a photon "actually" is.

However, very occasionally, quantum mechanics is encountered in altogether different settings. Not always do we have to modify our "classical" sense of logic. Quantum mechanics can also be used as a tool to compute something. A question is then asked about a mathematical problem that itself has nothing to do with quantum mechanics. Nor does the answer to the question refer to quantum mechanics as a special doctrine. Only it so happens that the path from question to answer requires an intermediate stage where quantum technology, including its logical interpretation, is indispensable. My most cherished example is the Two-Dimensional Ising Problem. It is the following:

In statistical physics, one may consider atoms forming a rectangular grid, and each atom may have a spin pointing either up or down. Questions of the following form may then arise, although this particular question is a very special one:

Consider a two-dimensional rectangular grid with N rows and M columns, N and M

both being very large. For simplicity one may consider periodic boundary conditions. Fill the entire lattice with 1's and 0's. The total number of 1's and 0's are equal: $NM/2$. The total number of boundaries is $2NM$. Of these, a fraction x is such that a 0 is adjacent to a 1. The remaining boundaries, the fraction being $1-x$, are either between two 1's or between two 0's. Question: in how many different ways can we position the 1's and the 0's?

The one-dimensional version of this question is easy to solve for an undergraduate student, but the two-dimensional variant is quite hard. The solution was found by L. Onsager, and written down in two beautiful papers with his student, B. Kaufman[1]. In statistical physics, the problem can easily be rephrased in terms of a matrix (the transfer matrix) that describes the interactions between two adjacent rows. What has to be calculated is the trace of this matrix raised to the $N^{\rm th}$ power. For large N, only the lowest eigenvalue of the transfer matrix contributes. The transfer matrix can be rewritten as $e^{-\beta H}$, where β is related to the given fraction x, and H in every respect resembles a quantum mechanical hamiltonian.

What Onsager and Kaufman discovered is that this hamiltonian is the one of a quantum field theory describing fermions. These fermions indeed are anticommuting and have spin $\frac{1}{2}$. The zero eigenvalue belongs to the ground state of this system. So, they had to calculate the energy of the vacuum. It so happens that, in this special case, the fermions are free fermions – their interactions vanish. So, the energy of the vacuum could be computed exactly. Most notably, vacuum expectation values of spins exactly correspond to the 'average' values in the statistical system, and the correlation functions in the statistical system are exactly the connected Feynman diagrams of the quantum field theory.

This is what we now wish to emphasize: quantum mechanics was just being used as a tool, a smart tool indeed, but never in the argument was there any need to replace the logic of statistical physics into anything unconventional. There are no questions that are disallowed. Use was made of some good mathematics, and that's all there is to it.

In the world of atoms and sub-nuclear particles, we also have quantum mechanics, and their properties are contained in quantum field theories. Could it not be that these quantum field theories are also nothing but the answers to questions asked, about a system that itself is not quantum mechanical at all? We can confirm that the answer to the question indeed works correctly: it is what we call quantum mechanics. But now, what is the question? More precisely, what is the statistical system and what are its laws of nature, that also require the quantum mechanical method to arrive at an answer?

The Ising model is not quantum mechanical. The question could just as well have been asked about checker boards or beer crates. Similarly, when we consider dynamical laws, we can also use quantum notation to describe big, non quantum mechanical things. Take, for instance, the planets orbiting the sun, the prototype of a classical, deterministic system. It does not forbid the introduction of operators that look quantum mechanical. An example is the *Earth-Mars exchange operator*. It puts Earth where Mars was, and sticks Mars to Earth's previous position. The velocities of the two planets are exchanged accordingly. One can now ask the mathematically relevant question: how will this system continue to evolve? Will the planets, through their mutual interactions, settle for new stable orbits?

In the jargon of quantum mechanics, we can confirm that the position operators for Earth and Mars do not commute with this displacement operator. It is legitimate to ask how this operator will evolve with time. Its eigenvalues are easy to calculate: they are ± 1, but the time evolution of this operator is complicated.

There do seem to be features that quantum mechanics has, but planetary systems have not.

In planetary systems, operators such as the Earth-Mars exchange operators seem to be mostly useless[1]. In conventional quantum systems, however, we frequently see how observables such as positions of things, evolve and mix with typical exchange operators. For instance, magnetic fields can rotate the z-spin observable $\left(\begin{smallmatrix}1&0\\0&-1\end{smallmatrix}\right)$ into the z-spin exchange operator $\left(\begin{smallmatrix}0&1\\1&0\end{smallmatrix}\right)$. The reasons for this could be a very simple one:

> *In our quantum mechanical theories for the atomic scale, we have not yet identified those observables that truly correspond to the actual states of a system. All operators we use at the atomic scale are superpositions of primary observables* ('beables') *and typical permutation operators* ('changeables').

There is reason to suspect that, in the atomic and sub-atomic world, we are confusing the true 'beables' with the 'changeables' (such as the Earth-Mars exchange operator). It so happens that both of these evolve by the same rules: the linear Schrödinger equation.

2. Examples of quantum systems allowing for classical interpretations

One of the simplest examples of a quantum system that allows for a deterministic interpretation is the Zeeman atom, a spin 1 atom in a homogeneous magnetic field. The eigenvalues for the hamiltonian are $E = (\mu B\,,\ 0\,,\ -\mu B)$. Consider the time step $t_0 = \frac{2\pi}{3\mu B}$; the evolution operator for this time step, $U(t_0)$, obeys:

$$U(t_0) = e^{-it_0 H}\ , \qquad U(3t_0) = U(t_0)^3 = 1\ , \tag{2.1}$$

and in another basis (the discrete Fourier transform), this operator can be written as

$$U(t_0) = \begin{pmatrix} 0 & 0 & 1 \\ 1 & 0 & 0 \\ 0 & 1 & 0 \end{pmatrix}\ , \tag{2.2}$$

since it has the same eigenvalues. In this basis, U is simply a permutation operator, and it describes a completely deterministic system that permutes three states in every time step of length t_0.

In the limit where we have infinitely many energy eigenstates, all equally spaced: $E_n = (n+\frac{1}{2})\hbar\omega$, $n = 0, 1, \cdots$, we find that the time unit t_0 becomes infinitesimal. This is the quantum harmonic oscillator. The deterministic permutation operator now becomes an infinitesimal rotation over a circle. Thus, the quantum harmonic oscillator is seen to be mathematically mapped onto a *deterministic* particle running around in a circle, with a period T that equals the period of the oscillator.

A third example is that of *massless, chiral, second-quantized, non-interacting neutrinos*[2]. Performing a similar analysis, identifying the observables that commute at all times, one finds that these "neutrinos" can be regarded as infinite, totally flat membranes moving along a vector orthogonal to the membranes, with the speed of light. The two directions in which a membrane can travel correspond to the two values of the spin.

The most important example however is that of the cellular automaton. It is the first example where the dynamics is non-trivial. It is discussed in the next section.

[1] but they *do* qualify as observable operators just as in quantum mechanics!

3. The time-reversible cellular automaton

Consider a cellular automaton[3][4]. Space and time[5] are both discrete here: we have a D dimensional space, where positions are indicated by integers: $\vec{x} = (x^1, x^2, \cdots, x^D)$, with $x^i \in \mathbb{Z}$. Also time t will be indicated by integers, and time evolution takes place stepwise. The physical variables $F(\vec{x}, t)$ in the model could be assumed to take a variety of forms, but the most convenient choice is to take these to be integers modulo some integer $\mathbb{N}$. We now write down an explicit model[3], where these physical degrees of freedom are defined to be attached only to the even lattice sites:

$$\sum_{i=1}^{D} x^i + t = \text{even.} \tag{3.3}$$

Furthermore, the data can be chosen freely at two consecutive times, so for instance at $t = 0$, we can choose the initial data to be $\{F(\vec{x},\ t = 0),\ F(\vec{x},\ t = 1)\}$.

The dynamical equations of the model can be chosen in several ways, provided that they are time reversible. To be explicit, we choose them to be as follows:

$$\begin{aligned} F(\vec{x}, t+1) &= F(\vec{x}, t-1) + \\ & Q\left(F(x^1 \pm 1, x^2, \cdots, x^D, t), \cdots, F(x^1, \cdots, x^D \pm 1, t)\right) \text{ Mod } \mathbb{N}\,, \\ & \text{when } \textstyle\sum_i x^i + t \text{ is odd}\,, \end{aligned} \tag{3.4}$$

where the integer Q is some arbitrary given function of all variables indicated: all nearest neighbors of the site $\vec{x}$ at time t. This is time reversible because we can find $F(\vec{x}, t-1)$ back from $F(\vec{x}, t+1)$ and the neighbors at time t. Assuming Q to be a sufficiently irregular function, one generally obtains quite non-trivial cellular automata this way. Indeed, this category of models have been shown to contain examples that are computationally universal [6]. Models of this sort are often considered in computer animations.

We now discuss the mathematics of this model using Hilbert space notation. We switch from the Heisenberg picture, where states are fixed, but operators such as the beables $F(\vec{x}, t)$ are time dependent, to the Schrödinger picture. Here, we call the operators F on the even sites $X(\vec{x})$, and the ones on the odd sites $Y(\vec{x})$. As a function of time t, we alternatingly update $X(\vec{x})$ and $Y(\vec{x})$, so that we construct the evolution operator over two time steps. Keeping the time parameter t even:

$$U(t, t-2) = A \cdot B\,, \tag{3.5}$$

where A updates the data $X(\vec{x})$ and B updates the data $Y(\vec{x})$.

Updating the even sites only, is an operation that consists of many parts, each defined on an even space coordinate $\vec{x}$, and all commuting with one another:

$$A = \prod_{\vec{x}\ \text{even}} A(\vec{x})\,, \quad [A(\vec{x}), A(\vec{x}\,')] = 0\,, \tag{3.6}$$

whereas the B operator refers only to the odd sites,

$$B = \prod_{\vec{x}\ \text{odd}} B(\vec{x})\,, \quad [B(\vec{x}), B(\vec{x}\,')] = 0\,. \tag{3.7}$$

Note however, that the operators $A(\vec{x})$ and $B(\vec{x}\,')$ do not all commute. If $\vec{x}$ and $\vec{x}\,'$ are neighbors, then

$$\vec{x} - \vec{x}\,' = \vec{e}\,, \quad |\vec{e}| = 1 \quad \to \quad [A(\vec{x}), B(\vec{x}\,')] \neq 0\,. \tag{3.8}$$

It is important to observe here that both the operators $A(\vec{x})$ and $B(\vec{x})$ only act in finite subspaces of Hilbert space, and they are all unitary, so we can easily write them as follows:

$$A(\vec{x}) = e^{-i\mathfrak{a}(\vec{x})} , \qquad B(\vec{x}) = e^{-i\mathfrak{b}(\vec{x})} . \tag{3.9}$$

In general, $\mathfrak{a}(\vec{x})$ and $\mathfrak{b}(\vec{x})$ are hermitean. We can write

$$\mathfrak{a}(\vec{x}) = \mathcal{P}_x(\vec{x})\, Q(\{Y\}) , \quad \mathfrak{b}(\vec{x}) = \mathcal{P}_y(\vec{x})\, Q(\{X\}) , \tag{3.10}$$

where $\mathcal{P}_x(\vec{x})$ is the generator for a one-step displacement of $X(\vec{x})$:

$$e^{i\mathcal{P}_x(\vec{x})}|X(\vec{x})\rangle \stackrel{\text{def}}{=} |X(\vec{x}) - 1 \text{ Mod } \mathbb{N}\rangle , \tag{3.11}$$

and, similarly, $\mathcal{P}_y(\vec{x})$ generates one step displacement of the function $Y(\vec{x})$.

As an example, we give the matrix P for the case $\mathbb{N} = 5$. They are easily obtained by discrete Fourier transformation. Defining the numerical coefficients $\alpha = 2\sin(\pi/5) + \sin(2\pi/5)$ and $\beta = 2\sin(2\pi/5) - \sin(\pi/5)$, we have

$$P = \frac{4\pi i}{25}\begin{pmatrix} 0 & -\alpha & \beta & -\beta & \alpha \\ \alpha & 0 & -\alpha & \beta & -\beta \\ -\beta & \alpha & 0 & -\alpha & \beta \\ \beta & -\beta & \alpha & 0 & -\alpha \\ -\alpha & \beta & -\beta & \alpha & 0 \end{pmatrix} ; \qquad e^{iP} = \begin{pmatrix} 0 & 1 & 0 & 0 & 0 \\ 0 & 0 & 1 & 0 & 0 \\ 0 & 0 & 0 & 1 & 0 \\ 0 & 0 & 0 & 0 & 1 \\ 1 & 0 & 0 & 0 & 0 \end{pmatrix} . \tag{3.12}$$

We see that

$$[\mathfrak{a}(\vec{x}),\, \mathfrak{a}(\vec{x}\,')] = 0 , \quad [\mathfrak{b}(\vec{x}),\, \mathfrak{b}(\vec{x}\,')] = 0 , \quad \forall\, (\vec{x},\, \vec{x}\,') ; \tag{3.13}$$

$$[\mathfrak{a}(\vec{x}),\, \mathfrak{b}(\vec{x}\,')] = 0 \quad \text{only if} \quad |\vec{x} - \vec{x}\,'| > 1 . \tag{3.14}$$

A consequence of Eqs. (3.13) is that also the products A in Eq. (3.6) and B in Eq. (3.7) can be written as

$$A = e^{-i\sum_{\vec{x}\,\text{even}} \mathfrak{a}(\vec{x})} , \qquad B = e^{-i\sum_{\vec{x}\,\text{odd}} \mathfrak{b}(\vec{x})} . \tag{3.15}$$

However, now A and B do not commute. Nevertheless, we wish to compute the total evolution operator U for two consecutive time steps, writing it as

$$U = A \cdot B = e^{-i\mathfrak{a}}\, e^{-i\mathfrak{b}} = e^{-2iH} . \tag{3.16}$$

For this calculation, we could use the power expansion given by the Baker-Campbell-Hausdorff formula[7],

$$e^P\, e^Q = e^R ,$$
$$R = P + Q + \tfrac{1}{2}[P, Q] + \tfrac{1}{12}[P, [P, Q]] + \tfrac{1}{12}[[P, Q], Q] + \tfrac{1}{24}[[P, [P, Q]], Q] + \cdots , \tag{3.17}$$

a series that continues exclusively with commutators[7]. Replacing P by $-i\mathfrak{a}$, Q by $-i\mathfrak{b}$ and R by $-2iH$, we find a series for the 'hamiltonian' H in the form of an infinite sequence of commutators. Now note that the commutators of the local operators $\mathfrak{a}(\vec x)$ and $\mathfrak{b}(\vec x\,')$ are non-vanishing only if $\vec x$ and $\vec x\,'$ are neighbors, $|\vec x-\vec x\,'|=1$. Consequently, if we insert the sums (3.15) into Eq. (3.17), we obtain again a sum:

$$H=\sum_{\vec x}\mathcal{H}(\vec x)\ ,$$
$$\mathcal{H}(\vec x)=\tfrac12\mathfrak{a}(\vec x)+\tfrac12\mathfrak{b}(\vec x)+\mathcal{H}_2(\vec x)+\mathcal{H}_3(\vec x)+\cdots\ , \qquad (3.18)$$

where

$$\mathcal{H}_2(\vec x)\ =\ -\tfrac14 i\sum_{\vec y}[\mathfrak{a}(\vec x),\,\mathfrak{b}(\vec y)]\ ,$$
$$\mathcal{H}_3(\vec x)\ =\ -\tfrac{1}{24}\sum_{\vec y_1,\vec y_2}[\mathfrak{a}(\vec x)-\mathfrak{b}(\vec x)\,,\ [\mathfrak{a}(\vec y_1),\mathfrak{b}(\vec y_2)]]\ ,\quad \text{etc.} \qquad (3.19)$$

All these commutators are only non-vanishing if the coordinates $\vec y$, $\vec y_1$, $\vec y_2$, etc., are all neighbors of the coordinate $\vec x$. It is true that, in the higher order terms, next-to-nearest neighbors may enter, but still, one may observe that these operators are all local functions of 'field operators' $\Phi(\vec x,\,t)$, and thus we arrive at a hamiltonian H that can be regarded as the sum over D-dimensional space of a Hamilton density $\mathcal{H}(\vec x)$, which has the property that

$$[\mathcal{H}(\vec x),\,\mathcal{H}(\vec x\,')]=0\ ,\quad \text{if}\quad |\vec x-\vec x\,'|\gg 1\ . \qquad (3.20)$$

The $\gg$ symbol here means that at the nth order in the BCH series, $\vec x$ and $\vec x\,'$ must be further than n steps away from one another.

At every finite order of the series, the Hamilton density $\mathcal{H}(\vec x)$ is a finite-dimensional Hermitean matrix, and therefore, it will have a lowest eigenvalue h. In a large but finite volume V, the total hamiltonian H will therefore also have a lowest eigenvalue, obeying

$$E_0>hV\ . \qquad (3.21)$$

The associated eigenstate $|\,0\rangle$ might be identified with the 'vacuum'. This vacuum is stationary, even if the automaton itself may have no stationary solution. The next-to-lowest eigenstate may be a one-particle state. In a Heisenberg picture, the fields $F(\vec x,t)$ may create a one-particle state out of the vacuum. Thus, we arrive at something that resembles a genuine quantum field theory. The states are quantum states in complete accordance with a Copenhagen interpretation. The fields $\mathfrak{a}(\vec x,t)$ and $\mathfrak{b}(\vec x,t)$ should obey the Wightman axioms.

4. The real world

There are three ways, however, in which this theory differs from conventional quantum field theories. One is, of course, that space and time are discrete. Well, maybe there is an interesting 'continuum limit', in which the particle mass(es) is(are) considerably smaller than the inverse of the time quantum.

Actually, the existence of large scale difference between the region where our model is defined (the Planck scale) and the scale up to where its quantum behavior continues to dominate (the atomic scale), is one of Nature's greater mysteries, called the *Hierarchy problem.* This mystery is still completely unresolved. Be this as it may, Nature's degrees of freedom at the atomic scale relate to those of the Planck scale by a complicated renormalization group relation: the running parameters of the Standard Model undergo a complete metamorphosis, which may well explain why the beables and the changeables of the theory get thoroughly mixed at the atomic scale.

Secondly, no attempt has been made to arrive at Lorentz invariance, or even Galilei invariance. Thus, the dispersion relations for these particles, if they obey any at all, may be nothing resembling conventional physical particles. Do note, however, that no physical information can travel faster than velocity one in lattice units. This is an important constraint that the model still has in common with special relativity.

But the third difference is more profound. It was tacitly assumed that the Baker-Campbell-Hausdorff formula converges. This is often not the case. In Ref [3], it is argued that the series will converge well only if sandwiched between two eigenstates $|E_1\rangle$ and $|E_2\rangle$ of H, where E_1 and E_2 are the eigenvalues, that obey

$$2|E_1 - E_2| < 2\pi\hbar/\Delta t \ , \tag{4.22}$$

where Δt is the time unit of our clock, and the first factor 2 is the one in Eq. (3.16). ("Planck's constant", $\hbar$, has been inserted merely to give time and energy the usual physical dimensions.)

This may seem to be a severe restriction, but, first, one can argue that $2\pi\hbar/\Delta t$ here is the Planck energy, and in practice, when we do quantum mechanics, we only look at energies, or rather energy differences, that indeed are much smaller than the Planck energy.

This argument will not convince skeptics. We have not proven that the cellular automaton can generate theories closely resembling the Standard Model. But this is not the aim of this investigation. What we wish to demonstrate is that the mathematical machinery borrowed from real quantum mechanics applies to deterministic systems such as a cellular automaton. Its states evolve in accordance with a Schrödinger equation. The hamiltonian shows much resemblance with that of genuine quantum systems. If we try to diagonalize the states of the hamiltonian that we find, the original ontological states are rearranged in terms of eigenstates of a hamiltonian, and as such they may well exhibit delicate forms of quantum entanglement.

The mathematical description of the cellular automaton that we arrive at, will be practically indistinguishable from other quantum mechanical systems. However, its universe can be described in terms of a very specially chosen basis of Hilbert space, the 'ontological basis'. In terms of this basis, the wave function has the following properties:

- it is 1 for the 'true' state, and 0 elsewhere;
- it does not *spread*;
- it does not *collapse*, that is, in the ontological basis it is as collapsed as it ever can be;
- it describes *only one world*;
- it always obeys a linear Schrödinger equation.

We claim that our world could be of the same sort; the 'wave function of the universe' could be exactly as trivial as this one, but only if we manage to identify the ontological basis.

Actually, we suspect that the *macroscopic* states, such as the states that describe dead or live cats, can be distinguished by studying the statistical distributions of the data in the deterministic underlying automaton. This would mean that the wave function of the universe is always automatically collapsed when used to describe Schrödinger's cat. We return to this in Sections 6–8.

5. The Bell inequalities

One of the most difficult issues to understand in this context is the Bell inequalities[8][9]. These inequalities refer to gedanken experiments in which quantum mechanical objects are produced in some quantum entangled states, such as two photons in a state with total spin zero, and many possible variations of this theme[10].

As has been shown with abundant evidence, such states can indeed be produced in real experiments. Observers, separated by macroscopic distances away from one another, can choose which component(s) of the wave function to detect, and they can use "free will" to determine their choices. Bell's inequalities appear to imply that the correlations then found cannot possibly be reconciled with a deterministic hidden variable theory. In the hidden variable theories that one then has in mind, the quantum particles are, somehow, accompanied by classical hidden variables that decide what the outcome of any of the possible measurements will be, even if the measurement is not made.

Clearly, Bell has shown that such hidden variable theories are unrealistic. We must conclude that our cellular automaton (CA) theory cannot be of this particular type. Yet, we had a classical system, and we claim that it reproduces quantum mechanics, with probabilities generated by the squared norm of wave functions. Quantum states, in particular entangled quantum states, are perfectly legitimate to describe statistical distributions. But how exactly to understand why Bell's inequalities can be violated in spite of the fact that we do start off from a classical deterministic, discrete theory, (to be referred to as a CA), requires a more detailed explanation than the one that can be given here. At the risk of not being understood (see also Ref [11]), let me briefly summarize what we think the situation is.

Even though we work with wave functions that are quite complicated quantum superpositions of the "Standard Model" (SM) eigenstates, we also emphasized that quantum superpositions of "ontological states" themselves are not ontological. Thus, if in a Bell experiment one axis was chosen, say, for the measurement of a spin, any other axis where the spin variable would not commute with the previous one, is in principle forbidden. One would be tempted to conclude that Bell's inequalities should be obeyed, so an important question is raised: how can it be that experiments showing violations could be performed so easily, not only in our imagination but also in real experiments[12]?

It must be emphasized that all commonly employed wave functions associated to atoms, fields, etc., do not correspond to the ontological states of the CA. So, if one "considers" a decaying atom with total spin zero (such that two entangled photons emerge), this only describes the CA in a statistical sense. Many of the CA states qualify to describe the initial condition.

As will be explained in the next section, many of the atoms in the environment will therefore be entangled with the system that is being studied. We emphasize that this includes vacuum fluctuations, which are rarely included in the standard treatments of the Bell inequalities. These vacuum fluctuations generate spacelike correlations, which in principle could be held responsible for the apparent non-locality of systems that violate Bell.

The next notion that will have to be scrutinized is the concept of "free will"[13][14]. Clearly, an observer who uses "free will" to choose the direction of a spin to be measured (or any other, more general basis for a quantum measurement), actually makes his or her decision depending of the outcome of Nature's laws in his/her own system. This is inevitable in deterministic theories, so we do not have to worry about "free will" itself[15].

The most difficult scenario is the following. Suppose that the observer's decision is made to depend not on "free will" but on fluctuations of the light from a distant quasar. If Alice and Bob participate in the experiment, they both use quasars that are located diametrically opposite to one another. These quasars would both be spacelike separated from the device that produced an entangled particle state. The initial state of the entire system does not allow for quantum superpositions of these ontological states, but we were lead to the apparently inevitable conclusion that the atomic objects in the quantum measurement in question must be entangled with both quasars in a very delicate sense, even if all of them are spacelike separated.

The best way then to summarize this explanation is to emphasize that cellular automata *can* be described by having their elementary states form basis vectors in a linear Hilbert space. The evolution equation in this space is the linear Schrödinger equation. Then, the 'SM variables' (atoms, fields, ...) form a totally different basis in this Hilbert space, all of these elements being highly entangled. Included in this entanglement are the vacuum fluctuations, and it may generate entangled correlations over vast spacelike distances. This is why the initial state will be an entangled one, and all states at later times, such as the ones used in an EPR experiment[8], continue to produce results that exhibit quantum entanglement.

6. Collapsing quantum amplitudes

In this meeting, it was repeatedly claimed that the collapse of the wave function cannot be reconciled with Schrödinger's equation[16]–[20], and therefore has to be introduced as a separate axiom, while Schrödinger's equation is modified, for instance with non-linear "corrections", just in order to comply with this 'bizarre' axiom.

An important argument that can be brought up in favor of this position is the following: suppose that a system starting off in a quantum state $|A\rangle_0$ would, after some time t, lead to a collapsed state $|A\rangle_t$, while a system starting off as $|B\rangle_0$ would end up in the collapsed state $|B\rangle_t$. Then, where would the state $\lambda\,|A\rangle_0+\mu\,|B\rangle_0$ end up? If this isn't the state $\lambda\,|A\rangle_t+\mu\,|B\rangle_t$, would this not imply a violation of Schrödinger's equation?

Not only do we never explicitly observe that Schrödinger's equation is violated anywhere, but the claim is also at odds with models that this author brought up to explain quantum mechanics as being the realization of statistical features of an underlying deterministic theory (see also [21]–[27]).

Similar statements are encountered concerning the Born interpretation of the wave function as being a description of probabilities. The probability of finding a system described by a wave function $|\psi\rangle$ to be in a certain state $|x\rangle$ when a measurement is made, is exactly equal to the square of the norm of the inner product $\langle x|\psi\rangle$. This also appears to be a special, separate axiom. What, after all, do probabilities have to do with equations such as the Schrödinger equation?

It is important to clarify these issues. Once and for all? Some issues will not be completely settled with the arguments presented here, so that the discussions will doubtlessly continue. In particular, Bell's inequalities will continue to raise questions, but this author is convinced of the basic correctness of the presentation given below. As for the apparent breakdown of the superposition principle as mentioned above, the answer is simple: in a deterministic theory, a system can be in a state A or in a state B, but never in a superposition. More precisely, the *sub-microscopic* degrees of freedom that we suspect to be deterministic, may also serve to fix the *macroscopic* variables describing outcomes of measurements, in a classical statistical sense, which could be just a perfect explanation of why an apparent collapse takes place.

The degrees of freedom in terms of which we usually describe atoms, molecules, subatomic particles and their fields will be referred to as *microscopic* degrees of freedom. It is these that have to be described as superpositions of the sub-microscopic states, and in turn, the macroscopic states are superpositions of microscopically defined states. Perhaps the most accurate way to describe the situation is that the states we use to describe atoms, quantum fields, etc., is to say that these serve as *templates*. A particle in the state $|x\rangle$ or in the state $|p\rangle$ or whatever, will nearly always be a superposition of many of the sub-microscopic states; as such, they evolve *exactly* according to Schrödinger equations. In contrast, the sub-microscopic states evolve classically. The macroscopic states also evolve classically, but the details of their evolution laws are far too complicated to follow, which is what we need the microscopic template states for.

7. Wave functions

If a theory is deterministic, we can also choose the initial condition deterministically, that is, there is exactly one state that is realized at $t = 0$, and it describes "the universe". Being deterministic, such a theory describes a single world at all times – there is never any interference, *in terms of the* $\vec{q}$ *variables.* The wave function at $t = 0$ is $|\psi, 0\rangle$ and, in the continuum case, it could be written as

$$\langle\vec{q}\,|\psi, 0\rangle = \mathcal{N}\,\delta(\vec{q} - \vec{q}_0)\ , \qquad \langle\vec{q}\,|\psi, t\rangle = \mathcal{N}\,\delta(\vec{q} - \vec{q}_t)\ , \tag{7.23}$$

where $\mathcal{N}$ is a normalization constant, while in the discrete case,

$$\langle\vec{q}\,|\psi, t\rangle = \delta_{\vec{q},\,\vec{q}_t}\ . \tag{7.24}$$

Of course, such "wave functions" do not spread. At all times t, only one value of the 'ontological' variable $\vec{q}$ is realized. This wave function always takes the form of Eqs. (7.23) or (7.24). Note however that, even if it does neither spread nor collapse, it fully obeys the Schrödinger equation based on the hamiltonian constructed as in Eq. (3.18).

Now look at our world. At first sight, the wave functions we use to describe it look very different. But then we have to realize that we do not know the operators $\vec{q}(t)$. The operators that we do use, for instance in describing the Standard Model, actually refer to states close to the lowest eigenstates of H, so that they contain low-energy projection operators. Therefore, when expressed in terms of Standard Model (SM) operators, the observables $\vec{q}(t)$ always form non-trivial superpositions of eigenstates of SM operators, probably compounded by operators of as yet unknown particles and fields at higher energies (the "hidden" variables). Thus, the Standard Model describes the real world in terms of templates. We only have access to a small subset of all templates; those that describe objects beyond the Standard Model, are simply not yet known.

In terms of the eigenstates of SM operators, the eigenstates of $\vec{q}(t)$ must seem to be highly entangled. This is what happens in some of our models[3], and it may well be the reason why, in terms of microscopic variables, even the wave functions (7.23) and (7.24) may seem to be complicated entangled ones.

However, common sense gives us the suspicion that the macroscopic observables may be diagonal again in terms of the $\vec{q}(t)$ operators. It is an interesting assumption, which we will adhere to. It indeed implies that *the wave function of the universe* will be *collapsed* in terms of the macroscopic observables *at all times.* According to our deterministic theories, these wave functions should nevertheless obey the theory's Schrödinger equation. Now that we expressed our suspicion that wave functions can be constructed that always stay collapsed when macroscopic variables are considered, we can attempt to construct them more directly, starting from conventional theories of physics such as the Standard Model.

Journal of Physics: Conference Series **361** (2012) 012024 doi:10.1088/1742-6596/361/1/012024

8. Schrödinger's cat

The prototype example is the Schrödinger cat gedanken experiment[28]. Let us start with an over simplified description that runs as follows. At $t=0$, we have an unstable atom[2] in the initial state $|\psi(0)\rangle=|1\rangle$, and a cat, in a sealed box. A certain moment later, at $t=t_1$, there is, say, a 40% chance that the atom decayed into state $|2\rangle$, by emitting a photon γ. The wave function is then $|\psi(t_1)\rangle=\sqrt{0.6}|1\rangle+\sqrt{0.4}|2,\,\gamma\rangle$. If the atom at that moment has decayed, the cat is poisoned and dies, so, according to the simple argument, we then have a cat in the super-imposed state $\sqrt{0.6}\,|\,\text{live cat}\rangle+\sqrt{0.4}\,|\,\text{dead cat}\rangle$. Then, the box is opened for inspection; a "measurement" takes place.

Will the cat really be in a superimposed state? Of course not. Decoherence takes place[29], and we expect that the cat will be *either* in the live *or* in the dead state. The question usually asked is how the wave function really evolves. Does it "collapse"?

To do this right, we have to take into account all those physical degrees of freedom that might be responsible for decoherence. Billions of atoms interact very weakly with the decaying atom and the cat. Each of these atoms can be in dozens of states, so the total dimensionality of the vector space spanned by these atoms is a huge number,

$$N=\exp(C\,\kappa)\ , \tag{8.25}$$

where κ is the number of atoms and C a number of order one or larger, so indeed, N is astronomically large. As explained in the previous sections, we expect these environment states in general to be highly entangled, but in the first step of this argument, we consider simple, "pure" environment states in their energy eigenstates. These states will be referred to as $|\,ES,\,t\rangle$.

Now, we can consider the above process more carefully. The initial state at $t=0$ is

$$|\psi,0\rangle=|1\rangle\,|ES,0\rangle\ . \tag{8.26}$$

Then, at $t=t_1$, we may assume that the wave function is (apart from an overall phase rotation)

$$|\psi,t_1\rangle=\Big(\sqrt{0.6}\,|1\rangle+\sqrt{0.4}\,|2,\,\gamma\rangle\,e^{i\varphi(ES)}\Big)|ES,t_1\rangle\ . \tag{8.27}$$

Here, we take into account that the environment states may have caused a relative phase rotation $\varphi(ES)$. The point is that total energy is conserved, but a small (positive or negative) part of it may have been absorbed by the environment, a part that will be *different* when the atom has decayed. Since we are not closely watching the environment while doing the experiment, we do not have perfect control over this phase difference.

In Eq. (8.27), for simplicity, the ES states were considered to be in some eigenstate of the hamiltonian. But this is not the wave function that we are interested in. Both in terms of the SM degrees of freedom, and in terms of the ontological states of Section 7, the environment will be time dependent. Also, the ontological states $|O,k\rangle$ are expected to appear in the conventional quantum mechanical description as entirely entangled states, which we shall refer to as $|EES,k\rangle$:

$$|EES,k\rangle=\sum_{i=1}^{N}\alpha_i^{(k)}|ES_i\rangle\ . \tag{8.28}$$

[2] Historically, Schrödinger thought of an unstable atom, but a Stern-Gerlach experiment might be even more illustrative. For our present argument any quantum system can be used.

Let us reexpress the result in terms of a density matrix:

$$|EES,k\rangle\,\langle EES,k| = \sum_{i,j=1}^{N} \alpha_i^{(k)}\alpha^{*(k)}_j \begin{pmatrix} 0.6 & \sqrt{0.24}\,e^{-i\varphi_j} \\ \sqrt{0.24}\,e^{i\varphi_i} & 0.4\,e^{i(\varphi_i-\varphi_j)} \end{pmatrix} |ES_i,t_1\rangle\langle ES_j,t_1|\ , \quad (8.29)$$

where φ_i stands for $\varphi(ES_i)$. Furthermore, N is the number (8.25) of environment states, and the index i labels them.

If all states $|EES,k\rangle$ occur with (approximately) equal probability $1/N$, we can use orthogonality,

$$\sum_k \alpha_i^{(k)}\alpha^{*(k)}_j = \delta_{ij}\ , \quad (8.30)$$

and the density matrix becomes

$$\sum_k \frac{1}{N}|EES,k\rangle\,\langle EES,k| = \frac{1}{N}\begin{pmatrix} 0.6\,\mathbb{I} & \sqrt{0.24}\,X \\ \sqrt{0.24}\,X^* & 0.4\,\mathbb{I} \end{pmatrix}\ , \quad (8.31)$$

where X stands for

$$X = \sum_i e^{-i\varphi_i}|ES_i\rangle\,\langle ES_i|\ . \quad (8.32)$$

The density matrix is that of a *microcanonical ensemble* where the total energy is fixed, *apart from small variations that allow our states to depend slowly on time*, so that all states that obey the restrictions dictated by the macroscopic description of the environment (including the total energy) have (approximately) equal probabilities. It is important to realize that this assumes that the EES states that we use all describe this subset of all macroscopic states. Indeed, this is what our ontological theory supposes, so there is no contradiction here.

If it weren't for the phases φ_i, the environment would merely contribute the identity as its density matrix. But now let us consider the phases in the off-diagonal part. Suppose *decoherence* takes place[29]. This means that the phases φ_i take all values, practically randomly, depending on the energy of the environment states. These energy eigenstates are delocalized. Therefore, if any matrix element is considered describing localized operators (using our *templates*), many of the ES states contribute, and their contributions are equal apart from the phases. Therefore, one expects the phases to cancel out. In short, *if used only in combination of localized observables or operators*, the off diagonal terms in our density matrix, the matrices X, cancel out to zero. The density matrix is therefore

$$\varrho = \frac{1}{N}\begin{pmatrix} 0.6 & 0 \\ 0 & 0.4 \end{pmatrix}\mathbb{I}\ . \quad (8.33)$$

Note that this density matrix was arrived at by performing probabilistic averages, not by demanding a wave function to collapse. It therefore completely agrees with the Schrödinger equation. But now it is of interest to see what it means in terms of the ontological states $|O,k\rangle$. We stated that the probability for starting out with any of these states was equal: $P_k = 1/N$. Suppose now that, at the start, we had one single pure state $|O,k_0\rangle$. The suspicion that we explained in Section 7 is that this ontological state will *either* lead to a dead cat *or* a live one, but never to a superposition. This agrees with our density matrix (8.33) if indeed the *probability* that state $|1\rangle$ was realized was 60% and the probability for $|2,\gamma\rangle$ was 40%. The probability for a superposition to arise is zero.[3]

Throughout the process, the probabilities for any of the ontological states to be realized were conserved. Therefore, we must conclude that, from the very start, the ontological states consisted for 60% of states that would later evolve into a live cat and 40% of states that would evolve into a dead one. If indeed we had picked a state at random, the 60/40 distribution would be that of the probabilities.

This, we now claim, is the origin of Born's rule. The ontological states only evolve *either* into pure states only describing a live cat, *or* into pure states only containing a dead cat, and never a superposition. The probabilities are simply in the number of ontological states with these properties. If one starts out picking one at random, then the probabilities will always be given by Born's rule. Notice that this argument identifies the Born rule probabilities with the relative abundances of the initial states that could have been picked "at random".

In deterministic physics, such as in the classical Van der Waals gas, the origin of probabilistic distributions can only be in the arbitrariness of the initial state (assuming infinitely precise equations of motion). If one assumes any kind of continuous distribution of positions and momenta of molecules at $t = 0$, then this determines the fate of the system, again in probabilistic terms. According to our ontological theory of quantum mechanics, the probabilities generated by Born's rule, are to be interpreted exactly in the same terms. If we do not know the initial state with infinite accuracy then we won't be able to predict the final state any better than that. The probabilistic distribution at $t = 0$ determines the probabilistic distribution at all later times.

It is only if we restrict ourselves to the usual templates of states containing only limited numbers of localized particles, that we are fooled into believing that the wave function suddenly collapsed when the cat's door was opened, since we are confronted with the density matrix (8.33); in reality the ontic states of the underlying automaton correspond to templates that are highly entangled with the environment, so that the density matrix (8.33) emerges naturally, fully in accordance with the Schrödinger equation.

9. Discussion

The number N of Eq. (8.25) stands for the *dimensionality* of the Hilbert space of environment states, and for the *total number* of allowed ontological states of the cellular automaton. In the latter terminology, superpositions are forbidden, so if states $|A\rangle$ and $|B\rangle$ are ontological states, then $|\psi\rangle = \lambda|A\rangle + \mu|B\rangle$, with $\lambda \neq 0$ and $\mu \neq 0$, is not such a state. This is how the states $|\text{live cat}\rangle$ and $|\text{dead cat}\rangle$ could emerge as ontological states, but not the state $\lambda|\text{live cat}\rangle + \mu|\text{dead cat}\rangle$. Thus, the initial state automatically collapses with the appropriate probabilities.

In our deterministic theory for QM, quantum superposition must be looked at as a property of the statistical approach to handling the extremely complex local equations of motion. Quantum wave functions were introduced for the convenience of the computation; linearity came as a handy tool for making calculations, but it so happens that quantum superpositions of ontological states themselves do not describe any real world, and this, as it turns out now, explains why we do not see quantum superpositions occurring in the macro world. By using the linearity of the

[3] Note, that the important assumption mentioned in Section 7 was used here. It was assumed that, in the deterministic theory, the question whether the cat is dead or alive can be settled by careful statistical analysis of the state of the sub-microscopic degrees of freedom of the system. If the wave function is delta-peaked on one of the sub-microscopic states, it will be delta-peaked as either a live cat or a dead cat.

Schrödinger equation, we automatically adopt the Born interpretation of the squared norms as probabilities, because only this way the linear evolution equation for the density matrix (8.31) can assure probability conservation.

One may even conclude that the absence of superimposed states in the macroscopic world, which is usually mistaken to imply a collapsing wave function, is actually an important argument *in favor* of microscopic hidden variables.

References

[1] B. Kaufman, *Phys. Rev.* **76** (1949) 1232; B. Kaufman and L. Onsager, *Phys. Rev.* **76** (1949) 1244.

[2] G. 't Hooft, '*Quantum Mechanics and determinism*, in Proceedings of the Eighth Int. Conf. on "Particles, Strings and Cosmology, Univ. of North Carolina, Chapel Hill, Apr. 10-15, 2001, P. Frampton and J. Ng, Eds., Rinton Press, Princeton, pp. 275 - 285; ITP-UU/01/18, SPIN-2001/11, arXiv:hep-th/0105105; id., *Determinism beneath Quantum Mechanics*, presented at "Quo vadis Quantum Mechanics?", Temple University, Philadelphia, September 25, 2002., ITP-UU-02/69, SPIN-2002/45, arXiv:quant-ph/0212095

[3] G. 't Hooft, *Classical cellular Automata and Quantum Field Theory*, in Proceedings of the Conference in Honour of Murray Gell-Mann's 80th Birthday "Quantum Mechanics, Elementary Particles, Quantum Cosmology and Complexity", Singapore, February 2010, H.Fritzsch and K.K.Phua, eds., World Scientific, pp 397 - 408, repr. in: Int. J. Mod. Phys. A25, no 23 (2010) 4385-4396.

[4] G. 't Hooft, *Entangled quantum states in a local deterministic theory*, 2nd Vienna Symposium on the Foundations of Modern Physics (June 2009), ITP-UU-09/77, SPIN-09/30; arXiv:0908.3408v1 [quant-ph].

[5] A.P. Balachandran and L. Chandar, *Nucl. Phys.***B 428** (1994) 435;

[6] D.B. Miller and E. Fredkin, *Two-state, Reversible, Universal cellular Automata in Three Dimensions*, Proc. 2nd Conf. on Computing Frontiers, Ischia, Italy: ACM 45, doi: 10.1145/1062271, arXiv:nlin/0501022.

[7] A.A. Sagle and R.E. Walde, *Introduction to Lie groups and Lie Algebras*, Academic Press, New York, 1973. ISBN 0-12-614550-4.

[8] A. Einstein, B. Podolsky and N. Rosen, *Can Quantum-Mechanical Description of Physical Reality Be Considered Complete?*, *Phys. Rev.* **47** (1935) 777.

[9] J. S. Bell, *Speakable and unspeakable in quantum mechanics* (Cambridge Univ. Press, Cambridge, 1987).

[10] M. Seevinck, *Parts and Wholes. An Inquiry into Quantum and Classical Correlations*, thesis presented at Utrecht University on October 27, 2008, arXiv:0811.1027[quant-ph].

[11] G. 't Hooft, *How a wave function can collapse without violating Schrödinger's equation, and how to understand Born's rule*, ITP-UU-11/43, SPIN-11/34, arXiv: 1112.1811[quant-ph]

[12] A. Aspect, P. Grangier, and G. Roger, *Experimental Realization of Einstein-Podolsky-Rosen-Bohm Gedankenexperiment: A New Violation of Bell's Inequalities*, Phys. Rev. Lett., **49** (1982) 91.

[13] J. Conway and S. Kochen, *The Strong Free Will Theorem*, arXiv:0807.3286 [quant-ph].

[14] J. Conway, in *New Scientist*, 6 may 2006, p. 8: *Free will - you only think you have it.*

[15] G. 't Hooft, *The Free-Will Postulate in Quantum Mechanics*, ITP-UU-07/4, SPIN-07/4, arXiv:quant-ph/0701097.

[16] A. Bassi and G.C. Ghirardi, Phys. Rep. **379** (2003) 257.

[17] P. Pearle, Phys. Rev. D **13** (1976) 857; id., Found. Phys. **12** (1982) 249.

[18] G.C. Ghirardi, A. Rimini and T. Weber, Phys. Rev. D **34** (1986) 470.

[19] M. Schlosshauer, *Decoherence, the measurement problem, and interpretations of quantum mechanics*, arXiv:quant-ph/0312059.

[20] S.L. Adler, *Quantum Theory as an Emergent Phenomenon*, Cambridge University Press, 2004.

[21] G. 't Hooft, *Determinism in Free Bosons*, *Int. J. Theor. Phys.* **42** (2003) 355, arXiv:hep-th/0104080.

[22] M. Blasone, P. Jizba and H. Kleinert, *Annals of Physics* **320** (2005) 468, arXiv: quant-ph/0504200; id., *Braz. J. Phys.* **35** (2005) 497, arXiv: quant-ph/0504047.

[23] M. Blasone, P. Jizba and G. Vitiello, *Dissipation and Quantization*, arXiv: hep-th/0007138.

[24] H. Th. Elze, *Deterministic models of quantum fields*, *J. Phys.*: Conf. Ser. **33** (2006) 399, arXiv:gr-qc/0512016v1.

[25] G. 't Hooft, *Emergent quantum mechanics and emergent symmetries*, presented at PASCOS 13, Imperial College, London, July 6, 2007; ITP-UU-07/39, SPIN-07/27; arXiv:hep-th/0707.4568.

[26] G. 't Hooft, *Quantum Gravity as a Dissipative Deterministic System*, *Class. Quant. Grav.* **16** (1999) 3263, arXiv:gr-qc/9903084.

[27] G. 't Hooft, *The mathematical basis for deterministic quantum mechanics*, in *Beyond the Quantum*, World Scientific, Th. M. Nieuwenhuizen et al, ed., pp.3-19, `arXiv: quant-ph/0604008`.

[28] E. Schrödinger, *Die gegenwärtige Situation in der Quantenmechanik*, Die Naturwissenschaften, **23** (1935) 807-812; 823-828; 844-849 [A 207d]. Transl.: *The present situation in quantum mechanics*, Translator: John D. Trimmer, Proc. of the Ame. Philosophical Society, **124**, 323-38; Section I.11 of Part I of *Quantum Theory and Measurement* (J.A. Wheeler and W.H. Zurek, eds., Princeton university Press, New Jersey 1983.

[29] E. Joos, *Decoherence Website*, http://www.decoherence.de/ .

EmerQuM 11: Emergent Quantum Mechanics 2011 IOP Publishing
Journal of Physics: Conference Series **361** (2012) 012025 doi:10.1088/1742-6596/361/1/012025

Noncommutative spectral geometry, dissipation and the origin of quantization

M Sakellariadou[1], A Stabile[2] and G Vitiello[2]

[1] Department of Physics, King's College, University of London, Strand WC2R 2LS, London, U.K.

[2] Facoltà di Scienze and Gruppo Collegato I.N.F.N., Università di Salerno, I-84100 Salerno, Italy

E-mail: mairi.sakellariadou@kcl.ac.uk, anstabile@gmail.com, vitiello@sa.infn.it

Abstract. We present a physical interpretation of the doubling of the algebra, which is the basic ingredient of the noncommutative spectral geometry, developed by Connes and collaborators as an approach to unification. We discuss its connection to dissipation and to the gauge structure of the theory. We then argue, following 't Hooft's conjecture, that noncommutative spectral geometry classical construction carries implicit in its feature of the doubling of the algebra the seeds of quantization.

1. Noncommutative Spectral Geometry

Unification of all forces, including gravity, remains one of the open issues in theoretical physics. As one approaches Planckian energy scales, the assumption that physics can be described by the sum of the Einstein-Hilbert and Standard Model (SM) actions breaks down and one must consider quantum gravity effects. In an attempt to provide a basis for describing the quantum nature of space-time, which may lead to the unification of all forces, Connes and collaborators developed Noncommutative Spectral Geometry (NCSG) which combines notions of noncommutative geometries [1, 2] with spectral triples. Within NCSG, the SM of electroweak and strong interactions is seen [3] as a phenomenological model, which specifies the geometry of space-time so that the Maxwell-Dirac action functional leads to the SM action.

This unification model lives by construction at high energy scales, offering an appropriate framework to address early universe cosmology [4] - [11]. This is however beyond the scope of this presentation. In what follows we attempt instead, to shed some light on how some criticisms raised against NCSG approach, and in particular its application in early universe cosmology, can be answered. More precisely, we will discuss the physical meaning of the choice of the *almost* commutative geometry and its relation to quantization [12]. In our discussion we will also consider the relation of the NCSG formalism with the gauge structure of the theory and with dissipation.

Let us first however review the main elements of NCSG. It is based on a two-sheeted space, made from the product of a four-dimensional smooth compact Riemannian manifold $\mathcal{M}$ (a continuous geometry for space-time) with a fixed spin structure, by a discrete noncommutative space $\mathcal{F}$ (an internal geometry for the SM) composed by only two points. The noncommutative nature of the discrete space $\mathcal{F}$ is given by a spectral triple $(\mathcal{A}, \mathcal{H}, \mathcal{D})$, where $\mathcal{A}$ is an involution of operators on the finite-dimensional Hilbert space $\mathcal{H}$ of Euclidean fermions, and $\mathcal{D}$ is a self-adjoint unbounded operator in $\mathcal{H}$. The space $\mathcal{H}$ is the Hilbert space $L^2(\mathcal{M}, S)$ of square integrable spinors S on $\mathcal{M}$ and the algebra $\mathcal{A}$ is the algebra $\mathcal{A} = C^\infty(\mathcal{M})$ of smooth functions on $\mathcal{M}$ and acts in $\mathcal{H}$ by multiplication operators. The operator D is the Dirac operator $\partial\!\!\!/_{\mathcal{M}} = \sqrt{-1}\gamma^\mu\nabla^s_\mu$ on the spin Riemannian manifold $\mathcal{M}$.

Within NCSG all information about space is encoded in the algebra of coordinates $\mathcal{A}$. Assuming the algebra $\mathcal{A}$ constructed in the geometry $\mathcal{M} \times \mathcal{F}$ is symplectic-unitary, it must be of the form [13]

$$\mathcal{A} = M_a(\mathbb{H}) \oplus M_k(\mathbb{C}) \ , \tag{1}$$

with $k = 2a$ and $\mathbb{H}$ being the algebra of quaternions. The field of quaternions $\mathbb{H}$ plays an important rôle in this construction and its choice remains to be explained. To obtain the SM one assumes quaternion linearity. The first possible value for the even number k is 2, corresponding to a Hilbert space of four fermions, but this choice is ruled out from the existence of quarks. The next possible value is $k = 4$ leading to the correct number of $k^2 = 16$ fermions in each of the three generations.

The noncommutative spectral geometry model is based upon the spectral action principle stating that, within the context of a product noncommutative geometry, the bare bosonic Euclidean action is given by the trace of the heat kernel associated with the square of the noncommutative Dirac operator and is of the form

$$\mathrm{Tr}(f(D/\Lambda)) \ ; \tag{2}$$

f is a cut-off function, Λ fixes the energy scale, D and Λ have physical dimensions of a mass. This action can be seen *à la* Wilson as the bare action at the mass scale Λ. The fermionic term can be included in the action functional by adding $(1/2)\langle J\psi, D\psi\rangle$, where J is the real structure on the spectral triple and ψ is a spinor in the Hilbert space $\mathcal{H}$ of the quarks and leptons.

For the four-dimensional Riemannian geometry, the trace $\mathrm{Tr}(f(D/\Lambda))$ is expressed perturbatively in terms of the geometrical Seeley-deWitt coefficients a_n, which are known for any second order elliptic differential operator, as [14, 15, 16, 17]

$$\begin{aligned}\mathrm{Tr}(f(D/\Lambda)) &\sim 2\Lambda^4 f_4 a_0 + 2\Lambda^2 f_2 a_2 + f_0 a_4 + \cdots \\ &\quad + \Lambda^{-2k} f_{-2k} a_{4+2k} + \cdots \ ,\end{aligned} \tag{3}$$

where the smooth even cut-off function f, which decays fast at infinity, appears through its momenta f_k given by:

$$\begin{aligned} f_0 &\equiv f(0) \\ f_k &\equiv \int_0^\infty f(u) u^{k-1} \mathrm{d}u \ , \quad \text{for} \ \ k > 0 \ , \\ \mathrm{f}_{-2k} &= (-1)^k \frac{k!}{(2k)!} f^{(2k)}(0) \ . \end{aligned}$$

Since its Taylor expansion at zero vanishes, the asymptotic expansion Eq. (3) reduces to

$$\mathrm{Tr}(f(D/\Lambda)) \sim 2\Lambda^4 f_4 a_0 + 2\Lambda^2 f_2 a_2 + f_0 a_4 \ . \tag{4}$$

The cut-off function f plays a rôle only through its three momenta f_0, f_2, f_4, which are three real parameters, related to the coupling constants at unification, the gravitational constant, and the cosmological constant, respectively. The term in Λ^4 gives a cosmological term, the term in Λ^2 gives the Einstein-Hilbert action functional, and the Λ-independent term yields the Yang-Mills action for the gauge fields corresponding to the internal degrees of freedom of the metric.

In this purely geometric approach to the SM, the fermions provide the Hilbert space of a spectral triple for the algebra, while the bosons are obtained through inner fluctuations of the Dirac operator of the product geometry. The computation of the asymptotic expression for the spectral action functional results to the full Lagrangian for the Standard Model minimally coupled to gravity, with neutrino mixing and Majorana mass terms.

Finally, let us clarify in which sense we talk of dissipation in what follows. This is necessary because the Standard Model, as is well known, is Quantum Field Theory (QFT) model describing a closed (nondissipative) system. Dissipation enters our discussion of the implications of the algebra doubling in the specific sense one observes that in electrodynamics neither the energy-momentum tensor of the matter field, nor that of the gauge field, are conserved. However, $\partial_\mu T^{\mu\nu}_{\rm matter} = eF^{\mu\nu} j_\mu = -\partial_\mu T^{\mu\nu}_{\rm gauge~field}$. Thus, the *total* $T^{\mu\nu}_{\rm total} = T^{\mu\nu}_{\rm matter} + T^{\mu\nu}_{\rm gauge~field}$, which is the energy-momentum tensor of the *closed* system {matter field, electromagnetic field} is conserved: each element of the couple is *open* (dissipating) on the other one, although the *closeness* of the total system is ensured. In this sense, dissipation considerations in our discussion below do not spoil the closeness of the SM.

2. The algebra doubling and the gauge structure

In this Section we study the relation between the two-sheeted space in the NCSG construction and the gauge structure of the theory. One central ingredient in NCSG is indeed the "doubling" of the algebra $\mathcal{A} \to \mathcal{A}_1 \otimes \mathcal{A}_2$ acting on the "doubled" space $\mathcal{H} = \mathcal{H}_1 \otimes \mathcal{H}_2$. Such a doubling is the formal realization of the NCSG two-sheeted space. Let us first observe that the doubling of the algebra is also present in the standard Quantum Mechanics (QM) formalism of the density matrix and Wigner function. Indeed, the expression of the Wigner function is [18]

$$W(p,x,t) = \frac{1}{2\pi\hbar}\int \psi^*\left(x - \frac{1}{2}y, t\right)\psi\left(x + \frac{1}{2}y, t\right) e^{-i\frac{py}{\hbar}} dy \ .$$

By putting $x_\pm = x \pm \frac{1}{2}y$, the associated density matrix is

$$W(x_+, x_-, t) \equiv \langle x_+|\rho(t)|x_-\rangle = \psi^*(x_-, t)\psi(x_+, t) \ , \tag{5}$$

The coordinate $x(t)$ of a quantum particle is thus split into two coordinates $x_+(t)$ (going forward in time) and $x_-(t)$ (going backward in time). The forward and the backward in time evolution of the density matrix $W(x_+, x_-, t)$ is then described by "two copies" of the Schrödinger equation:

$$i\hbar \frac{\partial \langle x_+|\rho(t)|x_-\rangle}{\partial t} = H\langle x_+|\rho(t)|x_-\rangle \ , \tag{6}$$

where H is given in terms of the two Hamiltonian operators $H_\pm$ as $H = H_+ - H_-$. The connection with Alain Connes' discussion of spectroscopic experiments and the algebra doubling is thus evident: the density matrix and the Wigner function *require* the introduction of a "doubled" set of coordinates $(x_\pm, p_\pm)$ and of their respective algebras. Use of Eq. (6) shows immediately that the eigenvalues of H are directly the Bohr transition frequencies $h\nu_{nm} = E_n - E_m$, which was the first hint towards an explanation of spectroscopic structure.

The need to double the degrees of freedom is implicit even in the classical theory when considering the Brownian motion and it is related to dissipation. In the classical Brownian theory one has the equation of motion [19]

$$m\ddot{x}(t) + \gamma\dot{x}(t) = f(t) \ , \tag{7}$$

where $f(t)$ is a random (Gaussian distributed) force: $< f(t)f(t') >_{\rm noise} = 2\gamma k_B T\ \delta(t - t')$. By averaging over the fluctuating force f, one obtains [19]

$$< \delta[m\ddot{x} + \gamma\dot{x} - f] >_{\rm noise} = \int \mathcal{D}y < \exp[\frac{i}{\hbar}\int dt\ L_f(\dot{x}, \dot{y}, x, y)] >_{\rm noise} \ , \tag{8}$$

where

$$L_f(\dot{x}, \dot{y}, x, y) = m\dot{x}\dot{y} + \frac{\gamma}{2}(x\dot{y} - y\dot{x}) + fy \; . \tag{9}$$

Note that $\hbar$ is introduced solely for dimensional reasons. We thus see that the constraint condition at the classical level introduced a new coordinate y, and the system equations are obtained:

$$m\ddot{x} + \gamma\dot{x} = f \; , \quad m\ddot{y} - \gamma\dot{y} = 0 \; . \tag{10}$$

The x-system is an *open* system. In order to set up the canonical formalism it is required to *close* the system; this is the rôle of the y-system, which is the time-reversed copy of the x-system. The $\{x - y\}$ system is thus a closed system. We also remark that the exact expression for the imaginary part of the action reads [20, 21]

$$\mathrm{Im}\mathcal{S}[x, y] = \frac{1}{2\hbar}\int_{t_i}^{t_f}\int_{t_i}^{t_f} dt \; ds \, N(t-s) \; y(t) \; y(s) \; , \tag{11}$$

where $N(t-s)$ denotes the quantum noise in the fluctuating random force given by the Nyquist theorem [20]. From Eq. (11) we see that nonzero y yields an "unlikely process" in the classical limit "$\hbar \to 0$", in view of the large imaginary part of the action. At quantum level, instead, nonzero y may allow quantum noise effects arising from the imaginary part of the action [20]. We thus see that the second sheet cannot be neglected: in the perturbative approach one may drop higher order terms in the action functional expansion, since they correspond to unlikely processes at the classical level. However, these terms may be responsible for quantum noise corrections and thus, in order to not preclude the quantization effects, one should keep them.

Let us unveil now the relation between the two-sheeted space in the NCSG construction and the gauge structure of the theory. Consider the equation of the classical one-dimensional damped harmonic oscillator

$$m\ddot{x} + \gamma\dot{x} + kx = 0 \; , \tag{12}$$

with time independent m, γ and k, which is a simple prototype of open systems. In the canonical formalism for open systems, the doubling of the degrees of freedom is required in such a way as to complement the given open system with its time-reversed image, playing the rôle of the "bath", thus obtaining a globally closed system for which the Lagrangian formalism is well defined. Thus we consider the oscillator in the *doubled* y coordinate

$$m\ddot{y} - \gamma\dot{y} + ky = 0 \; . \tag{13}$$

The system of the oscillators Eq. (12) and Eq. (13) is then a closed system described by the Lagrangian density Eq. (9) where we put $f = kx$. The canonically conjugate momenta p_x and p_y can now be introduced as customary. Let us use the coordinates $x_1(t)$ and $x_2(t)$: $x_1(t) = (x(t) + y(t))/\sqrt{2}$ and $x_2(t) = (x(t) - y(t))/\sqrt{2}$. The motion equations are

$$m\ddot{x}_1 + \gamma\dot{x}_2 + kx_1 = 0 \; , \qquad m\ddot{x}_2 + \gamma\dot{x}_1 + kx_2 = 0 \; , \tag{14}$$

and $p_1 = m\dot{x}_1 + (1/2)\gamma x_2$; $p_2 = -m\dot{x}_2 - (1/2)\gamma x_1$. Following Refs. [22, 23, 24, 25] we can now put $B \equiv \gamma c/e$, $\epsilon_{ii} = 0$, $\epsilon_{12} = -\epsilon_{21} = 1$ and introduce the vector potential as

$$A_i = \frac{B}{2}\epsilon_{ij}x_j \quad (i, j = 1, 2) \; . \tag{15}$$

The Lagrangian can be written then in the familiar form

$$L \;=\; \frac{1}{2m}(m\dot{x}_1 + \frac{e_1}{c}A_1)^2 - \frac{1}{2m}(m\dot{x}_2 + \frac{e_2}{c}A_2)^2 - \frac{e^2}{2mc^2}({A_1}^2 + {A_2}^2) - e\Phi \; , \tag{16}$$

which describes two particles with opposite charges $e_1 = -e_2 = e$ in the (oscillator) potential $\Phi \equiv (k/2e)({x_1}^2 - {x_2}^2) \equiv \Phi_1 - \Phi_2$ with $\Phi_i \equiv (k/2/e){x_i}^2$ and in the constant magnetic field $\boldsymbol{B}$ defined as $\boldsymbol{B} = \boldsymbol{\nabla} \times \boldsymbol{A} = -B\hat{\mathbf{3}}$. Remarkably, we have the Lorentzian-like (pseudoeuclidean) metric in Eq. (16). The "minus" sign, implied by the doubling of the degrees of freedom, is crucial in our derivation (and in the NCSG construction).

In conclusion, the doubled coordinate, e.g., x_2 acts as the gauge field component A_1 to which the x_1 coordinate is coupled, and *vice versa*. The energy dissipated by one of the two systems is gained by the other one and vice versa, in analogy to what happens in standard electrodynamics as observed at the end of Section I. The interpretation is recovered of the gauge field as the bath or reservoir in which the system is embedded [24, 25]. The gauge structure thus appears intrinsic to the doubling procedure.

Such a conclusion can be also reached in the case of a fermion field. For brevity we do not report here the discussion for the fermion case, which can be found in [24, 25, 12]. We only observe that, considering as an example the Lagrangian of the massless free Dirac field $L = -\overline{\psi}\gamma^\mu\partial_\mu\psi$, the field algebra is doubled by introducing the fermion tilde-field $\tilde{\psi}(x)$ and the Lagrangian is rewritten as

$$\hat{L} = L - \tilde{L} = -\overline{\psi}\gamma^\mu\partial_\mu\psi + \overline{\tilde{\psi}}\gamma^\mu\partial_\mu\tilde{\psi}\ . \tag{17}$$

The tilde-system is a "copy" (with the same spectrum and couplings) of the ψ-system. The Hamiltonian for the system is of the form $\hat{H} = H - \tilde{H}$. The key point is that the matrix elements of the Lagrangian Eq. (17) in a conveniently introduced space of states $\mathcal{H}_\theta \equiv \{|0(\theta)\rangle\}$, where $|0(\theta)\rangle$ denotes the ground state (see [24, 25, 21, 12]), as well as of a more general Lagrangian than the simple one presently considered, are invariant under the simultaneous local gauge transformations of ψ and $\tilde{\psi}$. The label θ in $\mathcal{H}_\theta$ denotes the angle of a Bogoliubov transformation (see below). The tilde term $\overline{\tilde{\psi}}\gamma^\mu\partial_\mu\tilde{\psi}$ transforms in such a way to compensate the local gauge transformation of the ψ kinematic term, i.e. $\overline{\tilde{\psi}}(x)\gamma^\mu\partial_\mu\tilde{\psi}(x) \to \overline{\tilde{\psi}}(x)\gamma^\mu\partial_\mu\tilde{\psi}(x) + g\partial^\mu\alpha(x)\tilde{j}_\mu(x)$. This suggests to introduce the field A'_μ by

$$g j^{\bar{\mu}}(x) A'_{\bar{\mu}}(x) \cong \overline{\tilde{\psi}}(x)\gamma^{\bar{\mu}}\partial_{\bar{\mu}}\tilde{\psi}(x)\ , \quad \bar{\mu} = 0, 1, 2, 3\ . \tag{18}$$

where the bar over μ means no summation over repeated indices. The symbol $\cong$ denotes equality among matrix elements in $\mathcal{H}_\theta$. Thus we find that A'_μ transforms as $A'_\mu(x) \to A'_\mu(x) + \partial_\mu\alpha(x)$, and one may identify, in $\mathcal{H}_\theta$, A'_μ with the conventional U(1) gauge vector field.

One can also show that the variations of the gauge field tensor $F'_{\mu\nu}$ have their source in the current $\tilde{j}_\mu$, which suggests that the tilde field plays the rôle of the "bath" or "reservoir". Such an interpretation in terms of a reservoir, may thus be extended also to the gauge field A'_μ, which indeed acts in a way to "compensate" the changes in the matter field configurations due to the local gauge freedom.

Finally, it can be shown that in the formalism of the algebra doubling a relevant rôle is played by the noncommutative q-deformed Hopf algebra [26], pointing to a deep physical meaning of the noncommutativity in this construction. Indeed, the map $\mathcal{A} \to \mathcal{A}_1 \otimes \mathcal{A}_2$ is just the Hopf coproduct map $\mathcal{A} \to \mathcal{A} \otimes \mathbb{1} + \mathbb{1} \otimes \mathcal{A} \equiv \mathcal{A}_1 \otimes \mathcal{A}_2$ which duplicates the algebra. The Bogoliubov transformation of "angle" θ relating the fields $\psi(\theta; x)$ and $\tilde{\psi}(\theta; x)$ to $\psi(x)$ and $\tilde{\psi}(x)$, is known to be obtained by convenient combinations of the *deformed* coproduct operation of the form $\Delta a_q^\dagger = a_q^\dagger \otimes q^{1/2} + q^{-1/2} \otimes a_q^\dagger$, where $q \equiv q(\theta)$ is the deformation parameters and $a_q^\dagger$ are the creation operators in the q-deformed Hopf algebra [26]. These deformed coproduct maps are noncommutative and the deformation parameter is related to the condensate content of $|0(\theta)\rangle$. It is interesting to observe that the q-derivative is a finite difference derivative [26], which has to be compared with the fact that in the NCSG construction the derivative in the discrete direction is a finite difference quotient.

A relevant point is that the deformation parameter *labels* the θ-representations $\{|0(\theta)\rangle\}$ and, for $\theta \neq \theta'$, $\{|0(\theta)\rangle\}$ and $\{|0(\theta')\rangle\}$ are unitarily inequivalent representations of the canonical (anti-)commutation rules. This is a characteristic feature of quantum field theory [21, 27]. Its physical meaning is that an order parameter exists, which assumes different θ-dependent values in each of the representations. In other words, the deformed Hopf algebra structure induces the *foliation* of the whole Hilbert space into physically inequivalent subspaces.

3. Dissipation and quantization

By discussing classical, deterministic models, 't Hooft has conjectured that, provided some specific energy conditions are met and some constraints are imposed, loss of information might lead to a quantum evolution [28, 29, 30]. In agreement with 't Hooft's conjecture, on the basis of the discussion in the previous Sections and following Refs. [31, 32], we propose [12] that the NCSG classical construction carries *implicit* in its feature of the doubling of the algebra the seeds of quantization.

We consider the classical damped harmonic x-oscillator described by Eq. (12) and its time–reversed image Eq. (13). The Casimir operator C and the (second) $SU(1,1)$ generator J_2 are [23]

$$\mathcal{C} = \frac{1}{4\Omega m}\left[\left(p_1^2 - p_2^2\right) + m^2\Omega^2\left(x_1^2 - x_2^2\right)\right] , \qquad J_2 = \frac{m}{2}\left[\left(\dot{x}_1 x_2 - \dot{x}_2 x_1\right) - \Gamma r^2\right] , \tag{19}$$

where $\mathcal{C}$ is taken to be positive and

$$\Gamma = \frac{\gamma}{2m} , \; \Omega = \sqrt{\frac{1}{m}(k - \frac{\gamma^2}{4m})} , \text{ with } k > \frac{\gamma^2}{4m} .$$

The system's Hamiltonian can be written as [31, 32]

$$H = \sum_{i=1}^{2} p_i \, f_i(q) , \tag{20}$$

with $p_1 = \mathcal{C}$, $p_2 = J_2$, $f_1(q) = 2\Omega$, $f_2(q) = -2\Gamma$, $\{q_i, p_i\} = 1$ and the other Poisson brackets are vanishing. The Hamiltonian Eq. (20) belongs to the class of Hamiltonians considered by 't Hooft. There, the $f_i(q)$ are nonsingular functions of the canonical coordinates q_i and the equations for the q's, namely $\dot{q}_i = \{q_i, H\} = f_i(q))$, are decoupled from the conjugate momenta p_i. A complete set of observables, called *beables*, then exists, which Poisson commute at all times. The meaning of this is that the system admits a deterministic description even when expressed in terms of operators acting on some functional space of states $|\psi\rangle$, such as the Hilbert space [29]. Such a description in terms of operators and Hilbert space, does not imply *per se* quantization of the system. Quantization is achieved only as a consequence of dissipation. The Hamiltonian is written as $H = H_{\rm I} - H_{\rm II}$, with

$$H_{\rm I} = \frac{1}{2\Omega\mathcal{C}}(2\Omega\mathcal{C} - \Gamma J_2)^2 \; , \; H_{\rm II} = \frac{\Gamma^2}{2\Omega\mathcal{C}} J_2^2 \tag{21}$$

and we impose the constraint $J_2|\psi\rangle = 0$, which defines physical states and guaranties that H is bounded from below. We can then write

$$H|\psi\rangle = H_{\rm I}|\psi\rangle = 2\Omega\mathcal{C}|\psi\rangle = \left(\frac{1}{2m}p_r^2 + \frac{K}{2}r^2\right)|\psi\rangle , \tag{22}$$

with $K \equiv m\Omega^2$. We thus realize that $H_{\rm I}$ reduces to the Hamiltonian for the two-dimensional "isotropic" (or "radial") harmonic oscillator $\ddot{r} + \Omega^2 r = 0$.

The physical states are invariant under time-reversal ($|\psi(t)\rangle = |\psi(-t)\rangle$) and periodical with period $\tau = 2\pi/\Omega$. Note that $H_{\rm I} = 2\Omega\mathcal{C}$ has the spectrum $\mathcal{H}_{\rm I}^n = \hbar\Omega n$, $n = 0, \pm1, \pm2, ...$; since our choice has been that $\mathcal{C}$ is positive, only positive values of n will be considered.

By exploiting the periodicity of the physical states $|\psi\rangle$ and following Ref. [33], one obtains

$$\frac{\langle\psi(\tau)|H|\psi(\tau)\rangle}{\hbar}\tau - \phi = 2\pi n \quad , \quad n = 0, 1, 2, \ldots \, .$$

Using $\tau = 2\pi/\Omega$ and $\phi = \alpha\pi$, where α is a real constant, leads to

$$\mathcal{H}_{\rm I,eff}^n \equiv \langle\psi_n(\tau)|H|\psi_n(\tau)\rangle = \hbar\Omega\left(n + \frac{\alpha}{2}\right) \, . \tag{23}$$

The index n signals the n dependence of the state and the corresponding energy. $\mathcal{H}_{\rm I,eff}^n$ gives the effective nth energy level of the system, namely the energy given by $\mathcal{H}_{\rm I}^n$ corrected by its interaction with the environment. We conclude that the dissipation term J_2 of the Hamiltonian is responsible for the zero point ($n = 0$) energy: $E_0 = (\hbar/2)\Omega\alpha$. In QM the zero point energy is formally due to the nonzero commutator of the canonically conjugate q and p operators: the zero point energy *is* the "signature" of quantization. Our discussion thus shows that dissipation manifests itself as "quantization". In other words, the (zero point) "quantum contribution" E_0 to the spectrum of physical states signals the underlying dissipative dynamics.

Consider the defining relation for temperature in thermodynamics (with $k_{\rm B} = 1$): $\partial S/\partial U = 1/T$. Using $S \equiv (2J_2/\hbar)$ and $U \equiv 2\Omega\mathcal{C}$, Eq. (20) gives $T = \hbar\Gamma$. Provided S is identified with the entropy, $\hbar\Gamma$ can be regarded as the temperature. Thus, the "full Hamiltonian" Eq. (20) plays the rôle of the free energy $\mathcal{F}$, and $2\Gamma J_2$ represents the heat contribution in H (or $\mathcal{F}$). The fact that $2J_2/\hbar$ behaves as the entropy is not surprising since it controls the dissipative (thus irreversible loss of information) part of the dynamics.

The thermodynamical picture outlined above is also consistent with the results on the canonical quantization of open systems in quantum field theory [34].

4. The dissipative interference phase

Dissipation implies the appearance of a "dissipative interference phase" [35]. This provides a relation between dissipation and noncommutative geometry in the plane of the doubled coordinates and thus with the NCSG construction.

In the (x_+, x_-) plane (cf. Section 2), the components of forward and backward in time velocity $v_\pm = \dot{x}_\pm$ are given by

$$v_\pm = \frac{\partial H}{\partial p_\pm} = \pm\frac{1}{m}\left(p_\pm \mp \frac{\gamma}{2}x_\mp\right) , \tag{24}$$

and they do not commute

$$[v_+, v_-] = i\hbar\,\frac{\gamma}{m^2} \, . \tag{25}$$

It is thus impossible to fix these velocities v_+ and v_- as being identical [35]. By putting $mv_\pm = \hbar K_\pm$, a canonical set of conjugate position coordinates (ξ_+, ξ_-) may be defined by $\xi_\pm = \mp L^2 K_\mp$ so that

$$[\xi_+, \xi_-] = iL^2. \tag{26}$$

Equation (26) characterizes the noncommutative geometry in the plane (x_+, x_-). In full generality, one can show [35] that an Aharonov–Bohm-type phase interference can always be associated with the noncommutative (X, Y) plane where

$$[X, Y] = iL^2 \; ; \tag{27}$$

L denotes the geometric length scale in the plane. Consider a particle moving in the plane along two paths, $\mathcal{P}_1$ and $\mathcal{P}_2$, starting and finishing at the same point, in a forward and in a backward direction, respectively. Let $\mathcal{A}$ denote the resulting area enclosed by the paths. The phase interference ϑ may be be shown [35] to be given by

$$\vartheta = \frac{\mathcal{A}}{L^2} \ . \tag{28}$$

and Eq. (27) in the noncommutative plane can be written as

$$[X, P_X] = i\hbar \quad \text{where} \quad P_X = \left(\frac{\hbar Y}{L^2}\right) \ . \tag{29}$$

The quantum phase interference between two alternative paths in the plane is thus determined by the noncommutative length scale L and the enclosed area $\mathcal{A}$.

Notice that the existence of a phase interference is connected to the zero point fluctuations in the coordinates; indeed Eq. (27) implies a zero point uncertainty relation $(\Delta X)(\Delta Y) \geq L^2/2$.

In the dissipative case, $L^2 = \hbar/\gamma$, we conclude that the quantum dissipative phase interference $\vartheta = \mathcal{A}/L^2 = \mathcal{A}\gamma/\hbar$ is associated with the two paths $\mathcal{P}_1$ and $\mathcal{P}_2$ in the noncommutative plane, provided $x_+ \neq x_-$.

5. Conclusions

We have shown that the central ingredient in the NCSG, namely the doubling of the algebra $\mathcal{A} = \mathcal{A}_1 \otimes \mathcal{A}_2$ acting on the space $\mathcal{H} = \mathcal{H}_1 \otimes \mathcal{H}_2$ is related to dissipation and to the gauge field structure. Thus the two-sheeted geometry must be considered to be the construction leading to gauge fields, required to explain the Standard Model. By exploiting 't Hooft's conjecture, according which loss of information within the framework of completely deterministic dynamics, might lead to a quantum evolution, we have argued that dissipation, implied by the algebra doubling, may lead to quantum features. We thus suggest that the NCSG classical construction carries *implicit* in the doubling of the algebra the seeds of quantization.

We have shown that in Alain Connes' two-sheeted construction, the doubled degree of freedom is associated with "unlikely processes" in the classical limit, and it may thus be dropped in higher order terms in the perturbative expansion. However, since these higher order terms are the ones responsible for quantum corrections, the second sheet cannot be neglected if one does not want to preclude quantization effects. In other words, the second sheet cannot be neglected once the universe entered the radiation dominated era. However, at the Grand Unified Theories scale, when inflation took place, the effect of gauge fields is fairly shielded.

References

[1] Connes A 1994 *Noncommutative Geometry* (New York: Academic Press)
[2] Connes A and Marcolli M 2008 *Noncommutative Geometry, Quantum Fields and Motives* (India: Hindustan Book Agency)
[3] Chamseddine A H, Connes A and Marcolli M 2007 *Adv. Theor. Math. Phys.* **11** 991
[4] Nelson W and Sakellariadou M 2010 *Phys. Rev.* D **81** 085038
[5] Nelson W and Sakellariadou M 2009 *Phys. Lett.* B **680** 263
[6] Marcolli M and Pierpaoli E *Preprint* hep-th/0908.3683
[7] Buck M, Fairbairn M and Sakellariadou M 2010 *Phys. Rev.* D **82** 043509
[8] Nelson W, Ochoa J and Sakellariadou M 2010 *Phys. Rev. Lett.* **105** 101602
[9] Nelson W, Ochoa J and Sakellariadou M 2010 *Phys. Rev.* D **82** 085021
[10] Sakellariadou M 2011 *Int. J. Mod. Phys.* D **20** 785
[11] Sakellariadou M 2010 PoS CNCFG **2010** 028 [arXiv:1101.2174 [hep-th]]
[12] Sakellariadou M, Stabile A and Vitiello G 2011 *Phys. Rev.* D **84** 045026
[13] Chamseddine A H and Connes A 2007 *Phys. Rev. Lett.* **99** 191601

[14] Chamseddine A H and Connes A 2006 *J. Math. Phys.* **47** 063504
[15] Chamseddine A H and Connes A 1996 *Phys. Rev. Lett.* **77** 4868
[16] Chamseddine A H and Connes A 1977 *Comm. Math. Phys.* **186** 731
[17] Chamseddine A H and Connes A 2010 *Comm. Math. Phys.* **293** 867
[18] Feynman R P 1972 *Statistical Mechanics: A set of Lectures* (Reading, MA: W.A. Benjamin)
[19] Blasone M, Srivastava Y N, Vitiello G and Widom A 1998 *Annals Phys.* **267** 61
[20] Srivastava Y N, Vitiello G and Widom A 1995 *Annals Phys.* **238** 200 (1995)
[21] Blasone M, Jizba P and Vitiello 2011 *Quantum Field Theory and its macroscopic manifestations* (London: Imperial College Press)
[22] Tsue Y, Kuriyama A and Yamamura M 1994 *Prog. Theor. Phys.* **91** 469
[23] Blasone M, Graziano E, Pashaev O K and Vitiello G 1996 *Annals Phys.* **252** 115
[24] Celeghini E, Graziano E, Nakamura K and Vitiello G 1992 *Phys. Lett.* **B285** 98
[25] Celeghini E, Graziano E, Nakamura K and Vitiello G 1993 *Phys. Lett.* **B304**, 121
[26] Celeghini E, De Martino S, De Siena S, Iorio A, Rasetti M and Vitiello G 1998 *Phys. Lett.* **A244** 455
[27] Umezawa H, Matsumoto H and Tachiki M 1982 *Thermo Field Dynamics and condensed states* (Amsterdam, The Netherlands: North-Holland)
[28] 't Hooft G 1999 *Class. Quant. Grav.* **16** 3263
[29] 't Hooft G 1999 *Basics and Highlights of Fundamental Physics (Erice)*
[30] 't Hooft G 2007 *J. Phys.: Conf. Series* **67** 012015
[31] Blasone M, Jizba P and Vitiello 2001 *Phys. Lett.* A **287** 205
[32] Blasone M, Celeghini E, Jizba P and Vitiello G 2003 *Phys. Lett.* A **310** 393
[33] Berry M V 1984 *Proc. Roy. Soc. Lond.* **A392** 45
[34] Celeghini E, Rasetti M and Vitiello G 1992 *Annals Phys.* **215** 156
[35] Sivasubramanian S, Srivastava Y, Vitiello G and Widom A 2003 *Phys. Lett.* **A311** 97

Journal of Physics: Conference Series **361** (2012) 012026 doi:10.1088/1742-6596/361/1/012026

Quantum mechanics and local Lorentz symmetry violation

Petr Jizba[a] **and Fabio Scardigli**[b,c]

[a] FNSPE, Czech Technical University in Prague, Břehová 7, 115 19 Praha 1, Czech Republic
[b] Institute of Physics, Academia Sinica, Taipei 115, Taiwan
[c] Leung Center for Cosmology and Particle Astrophysics (LeCosPA), Department of Physics, National Taiwan University, Taipei 106, Taiwan

E-mail: p.jizba@fjfi.cvut.cz

E-mail: fabio@phys.ntu.edu.tw

Abstract. We show how a Brownian motion on a short scale can originate a relativistic motion on scales larger than the particle's Compton wavelength. Thus, Lorentz symmetry appears to be not a primitive concept, but rather it statistically emerges when a coarse graining average over distances of order, or longer than the Compton wavelength, is taken. We also present the generalizations needed to accommodate in our scheme the doubly special relativistic dynamics. In this way, a previously unsuspected, common stochastic origin of the two frameworks is revealed for the first time. Issues such as generalized commutation relations are also discussed.

1. Introduction

Lorentz symmetry (LS) is one of the cornerstones of contemporary physics. Up to now LS has been confirmed to unprecedented precision, and during the last century it has powerfully constrained theories in a way that has proved instrumental in discovering new laws of physics. Moreover, the mathematical structure of the Lorentz group is compellingly simple and elegant. It thus seems natural to assume that Lorentz invariance is an exact symmetry of nature valid for an arbitrary boost. Yet, in the last years several research lines have pursued the aim to reduce Lorentz invariance from a fundamental symmetry down to an emergent phenomenon. From a purely conceptual standpoint, the most cogent reason is that an infinite volume of the Lorentz group is experimentally untestable (since, unlike the rotation group, the Lorentz group is non-compact). Why should one then assume that the exact LS holds when this hypothesis cannot be tested, not even in principle? Recent developments in quantum gravity point to a breakdown of Lorentz symmetry at some critical energy or boost. Seed of this has been hinted already in quantum field theory where both the ultraviolet divergences and Landau poles are direct artefacts of the assumption that the spectrum of field degrees of freedom is boost invariant. However, in quantum gravity the profound difficulties associated with the problem of time [1, 2] indicate that an underlying preferred time may be necessary in order to reconcile gravitational and quantum physics. In particular, general arguments imply that a radical departure from standard space-time symmetries at the Planck scale is necessary [3, 4, 5, 6, 7, 8, 9, 10, 11, 12, 13, 14, 15]. Though, none of the existent reasonings amount to a convincing argument that the LS breaking is an inevitable aspect of quantum gravity but taken together they do motivate serious attempts to address possible observable consequences of a violation of the LS, and to strengthen observational bounds. What should be perhaps emphasized is that the idea of LS violation is not new and it has been considered by a number of authors over the last forty years or so (see, e.g. Refs. [16, 17, 18] and citations therein).

In this paper we will illustrate what quantum mechanics may have to say about the emergent LS in a very simple (but quite generic) setup. In particular, we show that a relativistic quantum mechanics, as formulated through path integrals (PI), bears in itself a seed of understanding how the LS can be broken at short spatio-temporal scales and yet emerge as an apparently exact symmetry at large scales. Our argument is based upon a recent observation [19, 20, 21] that PI for both fermionic and bosonic relativistic particles may be interpreted (when analytically continued to imaginary times) as describing a doubly-stochastic process that operates on two vastly different spatio-temporal scales. The short spatial scale, which is much smaller than the Compton length, describes a non-relativistic Wiener process with a fluctuating Newtonian mass. This might be visualized as if the particle would be randomly propagating (in the sense of Brownian motion) through a granular or "polycrystalline" medium. The large spatial scale corresponds, on the other hand, to distances that are much larger than particle's Compton length. At such a scale the particle evolves according to a genuine relativistic motion, with a sharp value of the mass coinciding with the Einstein rest mass. Particularly striking is the fact that when we average the particle's velocity over the correlation distance (i.e., over particle's Compton wavelength) we obtain the velocity of light c. So the picture that emerges from this analysis is that the particle (with a non-zero mass!) propagates over the correlation distance $1/mc$ (hereafter $\hbar = 1$) with an average velocity c, while at larger distance scales (i.e., when a more coarse grained view is taken) the particle propagates as a relativistic particle with a sharp mass and an average velocity that is smaller than c. This bears a strong resemblance with Feynman's chessboard PI for a relativistic Dirac fermion in $1+1$ dimensions [22, 23]. In passing, we may stress that the outlined superposition of two stochastic processes with widely separated times scales fits the conceptual framework which is often referred to as a superstatistics [24].

The outlined scenario can be also conveniently applied in various doubly special relativistic (DSR) models. In those models a further invariant scale ℓ, besides the speed of light c, is introduced, and ℓ is assumed typically to be of the order of the Planck length. In the present framework, the scale ℓ can be naturally identified with the minimal grain size of the polycrystalline medium. By following the same strategy as in the special relativistic context, i.e., analysing the structure of paths which enter the Feynman summation, one can again identify correlation lengths, canonical commutation relations and the respective Hausdorff dimensions.

The purpose of this paper is to call attention to such a peculiar behaviour of relativistic PI at short space-temporal scales — a fact already recognized by Feynmann — and bring the ensuing implications to the attention of our particle-physics and cosmology colleagues.

2. Path integrals and superstatistics paradigm

We will start with the well known fact that when a conditional probability density function (PDF's) is formulated through a PI, then it satisfies the Chapman–Kolmogorov equation (CKE) for continuous Markovian processes, namely

$$P(x_b, t_b|x_a, t_a) \;=\; \int_{-\infty}^{\infty} \mathrm{d}x\, P(x_b, t_b|x, t)P(x, t|x_a, t_a)\,. \tag{1}$$

Conversely, any probability satisfying CKE possesses a PI representation [22, 25].

In physics one often encounters probabilities formulated as a superposition of PI's, e.g.

$$\bar{P}(x_b, t_b|x_a, t_a) \;=\; \int_0^{\infty} \mathrm{d}v\; \omega(v, t_{ba}) \int_{x(t_a)=x_a}^{x(t_b)=x_b} \mathcal{D}x\mathcal{D}p\; e^{\int_{t_a}^{t_b} \mathrm{d}\tau (ip\dot{x} - vH(p,x))}\,. \tag{2}$$

Here $\omega(v, t_{ba})$ with $t_{ba} = t_b - t_a$ is a normalized PDF defined on $\mathbf{R}^+ \times \mathbf{R}^+$. The random variable v is in practice typically related to the inverse temperature, coupling constant, friction constant or volatility.

Since the Markovian behaviour is a fundamental component in many statistical models, it is important to know whether also $\bar{P}(\boldsymbol{x}_b, t_b|\boldsymbol{x}_a, t_a)$ can satisfy the CKE (1). The answer is affirmative provided $\omega(v, t)$ fulfils a certain simple functional equation. Following Ref. [19] we define a re-scaled weight function

$$w(v,t) \; \equiv \; \omega(v/t,t)/t\,, \tag{3}$$

and calculate its (one-sided) Laplace transform

$$\tilde{w}(p_v,t) \; \equiv \; \int_0^\infty \mathrm{d}v\, e^{-p_v v} w(v,t)\,. \tag{4}$$

Then $\bar{P}(\boldsymbol{x}_b, t_b|\boldsymbol{x}_a, t_a)$ satisfies CKE only if

$$\tilde{w}(p_v, t_1 + t_2) \; = \; \tilde{w}(p_v, t_2)\tilde{w}(p_v, t_1)\,. \tag{5}$$

Assuming continuity in t, $w(p_v, t)$ is unique and can be explicitly written as (see [19]):

$$\tilde{w}(p_v, t) \; = \; [G(p_v)]^t \; = \; e^{-tF(p_v)}\,. \tag{6}$$

A function $F(p_v)$ must increase monotonically in order to allow for inverse Laplace transform, and satisfy the condition $F(0) = 0$ to ensure that ω is normalized to one. Finally the Laplace inverse of $\tilde{w}(p_v, t)$ yields $\omega(v, t)$.

Once the above conditions are satisfied, then $\bar{P}(\boldsymbol{x}_b, t_b|\boldsymbol{x}_a, t_a)$ possesses a path integral representation on its own. The new Hamiltonian is given by the relation $\bar{H}(\boldsymbol{p}, \boldsymbol{x}) \; = \; F(H(\boldsymbol{p}, \boldsymbol{x}))$. Here one must worry about the notorious operator-ordering problem, not knowing in which temporal order $\boldsymbol{p}$ and $\boldsymbol{x}$ must be taken in F. At this stage it suffices to observe that when H is $\boldsymbol{x}$ independent, the former relation is exact. The issue of general H's and the ensuing operator ordering was discussed in detail in Ref. [19].

3. Emergent Special Relativity

As an application of the fundamental relation (2), the relativistic path integral for a scalar particle can be rewritten in the form [see also Ref. [19, 20, 21]]

$$\begin{aligned}
&\int_{\boldsymbol{x}(0)=\boldsymbol{x}'}^{\boldsymbol{x}(t)=\boldsymbol{x}} \mathcal{D}\boldsymbol{x} \frac{\mathcal{D}\boldsymbol{p}}{(2\pi)^D} \exp\left\{ \int_0^t \mathrm{d}\tau \left[\mathrm{i}\boldsymbol{p}\cdot\dot{\boldsymbol{x}} \; - \; c\sqrt{\boldsymbol{p}^2 + m^2c^2} \right] \right\} \\
&= \int_0^\infty \mathrm{d}\tilde{m} \sqrt{\frac{c^2 t}{2\pi\tilde{m}}} \; \mathrm{e}^{-tc^2(\tilde{m}-m)^2/2\tilde{m}} \int_{\boldsymbol{x}(0)=\boldsymbol{x}'}^{\boldsymbol{x}(t)=\boldsymbol{x}} \mathcal{D}\boldsymbol{x} \frac{\mathcal{D}\boldsymbol{p}}{(2\pi)^D} \exp\left\{ \int_0^t \mathrm{d}\tau \left[\mathrm{i}\boldsymbol{p}\cdot\dot{\boldsymbol{x}} \; - \; \frac{\boldsymbol{p}^2}{2\tilde{m}} \; - \; mc^2 \right] \right\} \\
&= \int_0^\infty \mathrm{d}\tilde{m}\, f_{\frac{1}{2}}\left(\tilde{m}, tc^2, tc^2m^2\right) \int_{\boldsymbol{x}(0)=\boldsymbol{x}'}^{\boldsymbol{x}(t)=\boldsymbol{x}} \mathcal{D}\boldsymbol{x} \frac{\mathcal{D}\boldsymbol{p}}{(2\pi)^D} \exp\left\{ \int_0^t \mathrm{d}\tau \left[\mathrm{i}\boldsymbol{p}\cdot\dot{\boldsymbol{x}} \; - \; \frac{\boldsymbol{p}^2}{2\tilde{m}} \; - \; mc^2 \right] \right\},
\end{aligned} \tag{7}$$

where $t_b - t_a \equiv t - 0 = t$, and

$$f_p(z,a,b) \; = \; \frac{(a/b)^{p/2}}{2K_p(\sqrt{ab})} \, z^{p-1} \, \mathrm{e}^{-(az+b/z)/2}\,, \tag{8}$$

is the generalized inverse Gaussian distribution [26] (K_p is the modified Bessel function of the second kind with index p). The structure of (7) implies that $\tilde{m}$ can be interpreted as a Newtonian mass which takes on continuous values distributed according to $f_{\frac{1}{2}}(\tilde{m}, tc^2, tc^2m^2)$ with $\langle\tilde{m}\rangle = m + 1/tc^2$ and $\text{var}(\tilde{m}) = m/tc^2 + 2/t^2c^4$. As a result one may view a single-particle relativistic theory as a single-particle non-relativistic theory where the particle's Newtonian mass $\tilde{m}$ represents a fluctuating parameter which approaches on average the Einstein rest mass m in the large t limit. We stress that the time t in question should be understood as a time after which the observation (i.e., the position measurement) is made. In particular, during the period t the system remains unperturbed. In this respect the smearing distribution $f_{\frac{1}{2}}(\tilde{m}, tc^2, tc^2m^2)$ represents a temporal coarse-grained distribution for a Newtonian mass — the longer the time between measurements, the poorer the resolution of mass fluctuations. One can thus justly expect that in the long run all mass fluctuations will be washed out and only a sharp time-independent effective mass will be perceived. The form of $\langle\tilde{m}\rangle$ identifies the time scale at which this happens with $t \sim 1/mc^2$. The latter is the time for light to cross the particle's Compton wavelength — i.e., the Compton time t_C. The expression for $\langle\tilde{m}\rangle$ suggests however also another interesting physical implication. As we have seen, when $t \gg 1/mc^2$ then $\langle\tilde{m}\rangle$ rapidly converges to the relativistic value m, signalling that the motion becomes genuinely relativistic at large times. Note that for t large enough we surely have $m > 1/tc^2$ which we can read as $mc^2t > 1$. The latter means that, for large t, the relativistic Heisenberg inequality for the energy/time variables is satisfied, $\Delta E \Delta t \geq 1$. On the other hand, for $t \ll 1/mc^2$, the fluctuations of the Newtonian mass $\tilde{m}$ around the average m are huge. The motion takes place inside a specific space-grain, and in each space-grain the motion is a classical, i.e. non relativistic, Brownian motion controlled by the Hamiltonian $\boldsymbol{p}^2/2\tilde{m}$. There the relativistic Heisenberg uncertainty relation is clearly violated, in fact $mc^2t < 1$ (remind that m is the Einstein rest mass). However, if we compute the non-relativistic Heisenberg relation, using the Newtonian mass $\tilde{m}$ and the non-relativistic kinetic energy $E_{kin} \sim \tilde{m}v^2$, we find

$$\Delta t \Delta E_{kin} \;=\; \left\langle \tilde{m}\frac{(\Delta x)^2}{\Delta t} \right\rangle \;\sim\; 1\,. \tag{9}$$

So the non-relativistic Heisenberg relation is not violated, confirming that the motion on short scales is genuinely classical (i.e. non-relativistic, see also [21]).

Until now, we have considered the conventional PI for a relativistic particle in Polyakov's gauge [27]. This has fixed the distribution describing the fluctuations of the Newtonian mass $\tilde{m}$ to $f_{\frac{1}{2}}(\tilde{m}, tc^2, tc^2m^2)$. Reciprocally, the form of the smearing function $f_{\frac{1}{2}}$ fixes the gauge, which in this case corresponds to the Polyakov gauge. So the use of smearing functions bypasses the conventional Dirac–Bergman methodology for quantization of constrained systems. In fact, in [21] we arrive at the correct special-relativistic CCR

$$[\hat{\boldsymbol{x}}_j, \hat{\boldsymbol{p}}_i] \;=\; \mathrm{i}\left(\delta_{ij} + \frac{\hat{\boldsymbol{p}}_i\hat{\boldsymbol{p}}_j}{m^2c^2}\right), \tag{10}$$

without using the machinery of Dirac brackets.

Fluctuations of the Newtonian mass can be portrayed as originating from particle's evolution in an "inhomogeneous" or a "polycrystalline" medium. Granularity, as known from solid state systems, typically leads to corrections in the local dispersion relation [28] and hence to alterations in the local effective mass. The following picture thus emerges: on the short-distance scale, a non-relativistic particle can be envisaged as propagating through a single grain with a local mass $\tilde{m}$, in a classical Brownian motion. This fast-time process has a time scale $\sim 1/\tilde{m}c^2$. An averaged value of the time scale can be computed with the help of the smearing distribution $f_{\frac{1}{2}}(\tilde{m}, tc^2, tc^2m^2)$, which gives a transient temporal scale $\langle 1/\tilde{m}c^2\rangle = 1/mc^2$. The latter coincides with particle's Compton time t_C. At time scales much longer than t_C (large-distance scale), the probability that the particle encounters a grain which endows it with a mass $\tilde{m}$ is $f_{\frac{1}{2}}(\tilde{m}, tc^2, tc^2m^2)$.

We may also observe that the averaged (or coarse-grained) velocity over the correlation time $t = 1/mc^2$ equals the speed of light c. In fact

$$\langle|\boldsymbol{v}|\rangle_{t=1/mc^2} = \left.\frac{\langle|\boldsymbol{p}|\rangle}{\langle\tilde{m}\rangle}\right|_{t=1/mc^2} = \frac{1}{2m}\int_0^\infty d\tilde{m}\, f_{\frac{1}{2}}(\tilde{m}, 1/m, m)\sqrt{\left(\frac{8m\tilde{m}c^2}{\pi}\right)} = c. \qquad (11)$$

So on a short-time scale of order λ_C the Klein–Gordon particle propagates with an averaged velocity which is the speed of light c. But if one checks the particle's position at widely separated intervals (much larger than λ_C), then many directional reversals along a typical PI trajectory will take place, and the particle's net velocity will be then less than c — as it should be for a massive particle (see Fig. 1). In addition, the time-compounded smearing distribution tends for large times rapidly to the delta-function distribution $\delta(\tilde{m} - m)$ thanks to the central limit theorem. This means that the particle acquires a sharp mass equal to Einstein's (i.e., Lorentz invariant) mass, and the process (not being hindered by fluctuating masses) turns out to be purely Brownian.

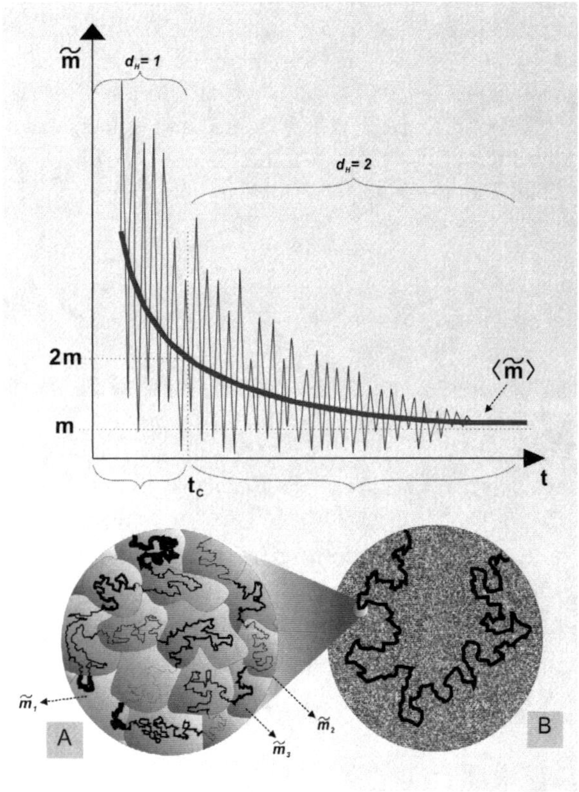

Figure 1. The roughness of the representative trajectories in the relativistic path integral (7) depends on a spatial/temporal scale. On a fine scale (A), where $t \ll t_C$ (or $\ell \ll \lambda_C$) a particle can be considered as propagating with a sharp Newtonian mass $\tilde{m}$ into a single spatial grain with a Brownian motion controlled by the Hamiltonian $\boldsymbol{p}^2/2\tilde{m}$. On the intermediate scale of order λ_C the particle propagates with an average velocity equal to the speed of light c. On a coarser scale (B) the particle appears to follow a Brownian process with a sharp Lorentz invariant mass m, and the particle's net velocity is then less than c — as it should be for a massive relativistic particle governed by the relativistic Hamiltonian $c(\boldsymbol{p}^2 + m^2c^2)^{1/2}$.

On a speculative vein, one can fit the above observation into the currently much debated emergent relativity theory, i.e. the approach that tries to view either special or general theory of relativity not as primitive concepts but rather as theories that statistically emerge from a deeper (essentially non-relativistic) level of dynamics [29, 30, 31, 18, 32, 33, 34, 15, 35, 36]. All these remarks extend directly also to certain interacting systems (for the Dirac's Hamiltonian see, for instance, Ref. [20])

At first sight it may seem rather surprising that a LS process may emerge from a superposition of two non-relativistic stochastic processes. However, in many doubly stochastic systems the statistically emergent behaviour has a structure vastly different from those of the respective defining processes. Hydrodynamic turbulence provides an example, where the emergent *velocity increments* and their ensuing Kolmogorov scaling can be understood as originating from two stochastic processes (energy dissipation and chaotic force) operating on two vastly different time scales, despite the fact that none of the processes exhibits any particular scaling structure [37]. Analogous situations are also known from financial markets, e.g., credit risk models or stochastic volatility models.

In what follows we propose an unifying approach for Special and Doubly Special Relativity, based on the existing laws of quantum mechanics as formulated through PI.

4. Emergent Doubly Special Relativity

Doubly Special Relativity (DSR) stands among the prominent ideas introduced in physics during the last decade, and also among the most controversial ones. Many foundational issues about this theory are still being debated, in particular for example the multi-particle sector of the theory (the so called Soccer Ball Problem) [38]. An important area of investigation has been that of the relations between DSR and other theoretical construction of modern physics. In a nutshell, DSR is a theory which coherently tries to implement a second invariant, besides the speed of light, into the transformations among inertial reference frames. This new invariant comes directly from the research in quantum gravity, and it is usually assumed to be an observer-independent length-scale — the Planck length ℓ_p, or its inverse, i.e., the Planck energy $E_p = c\ell_p^{-1}$. Thus, it is not so surprising that the relations mainly studied are those between DSR and various quantum gravity models [39, 40, 41]. In a particularly suggestive approach [42], DSR has been presented as the low energy limit of Quantum Gravity. Connections between DSR and other theories (non commutative geometry, AdS space-time, etc.) have also been recently investigated [43].

Clearly, the most important connection is the one between DSR and Special Relativity (SR) itself. However in literature does not exist any conceptually deeper elaboration of this connection, apart form the obvious statement that for energy scales much smaller than E_p, DSR should reduce to conventional SR, with leading corrections of first or higher order in the ratio of energy scales to E_p. In this respect, the findings of the present paper seem to open up new vistas. In fact, when the microstructure of space-time is considered, then Special Relativity or DSR seem to emerge from particular choices of such microstructure itself, and from a non-relativistic Hamiltonian mechanics.

To extend our reasonings to DSR, we start by considering the modified invariant, or deformed dispersion relation,

$$\frac{\eta^{ab} p_a p_b}{(1-\ell_p p_0)^2} \; = \; m^2 c^2 \,, \tag{12}$$

proposed by Magueijo and Smolin [44, 45], in which m plays the role of the DSR invariant mass. Assuming a metric signature $(+,-,-,-)$, we can solve (12) in respect to p_0, which essentially coincides with the physical Hamiltonian $\bar{H} = cp_0$, the generator of the temporal translations in respect to the coordinate time t. Our starting Hamiltonian is therefore

$$\bar{H} \; = \; c\, \frac{-m^2c^2\ell \mp \sqrt{\boldsymbol{p}^2(1-m^2c^2\ell^2)+m^2c^2}}{1-m^2c^2\ell^2} \,, \tag{13}$$

which we assume as the transformed Hamiltonian $\bar{H}(p,x) = F(H(p,x))$ entering the proper path integral representation of $\bar{P}(x_b,t_b|x_a,t_a)$, see (6) and the comments below [see also Ref. [19]].

Now, in close analogy with (7), it is possible to show the superstatistics identity (see [21])

$$\int_{\boldsymbol{x}(0)\,=\,\boldsymbol{x}'}^{\boldsymbol{x}(t)\,=\,\boldsymbol{x}} \mathcal{D}\boldsymbol{x}\frac{\mathcal{D}\boldsymbol{p}}{(2\pi)^D}\exp\left\{\int_0^t \mathrm{d}\tau\left[\mathrm{i}\boldsymbol{p}\cdot\dot{\boldsymbol{x}} + c\frac{\left(m^2c^2\ell - \sqrt{\boldsymbol{p}^2(1-m^2c^2\ell^2)+m^2c^2}\right)}{(1-m^2c^2\ell^2)}\right]\right\}$$

$$= \int_0^\infty \mathrm{d}\tilde{m}\, f_{\frac{1}{2}}\left(\tilde{m}, tc^2\lambda, tc^2m^2\lambda\right)\int_{\boldsymbol{x}(0)\,=\,\boldsymbol{x}'}^{\boldsymbol{x}(t)\,=\,\boldsymbol{x}} \mathcal{D}\boldsymbol{x}\frac{\mathcal{D}\boldsymbol{p}}{(2\pi)^D}\exp\left\{\int_0^t \mathrm{d}\tau\left[\mathrm{i}\boldsymbol{p}\cdot\dot{\boldsymbol{x}} - \frac{\boldsymbol{p}^2}{2\tilde{m}} - E_0\right]\right\}, \quad (14)$$

where $E_0 = mc^2/(1+mc\ell)$ is the particle's rest energy [see, e.g., [45]] and $\lambda = 1/(1-m^2c^2\ell^2)$ is the deformation parameter. From (8), it is easy to see that $\langle\tilde{m}\rangle = m + 1/(tc^2\lambda)$ and $\mathrm{var}(\tilde{m}) = m/tc^2\lambda + 2/t^2c^4\lambda^2$.

From the structure of $\langle\tilde{m}\rangle$ we can obtain further useful insights. Similarly as in the SR framework, the fluctuating Newtonian mass $\tilde{m}$ converges rapidly, at long times t, to the SR rest mass m. But this time the rate of convergence is controlled also by the parameter λ. Reminding that $E_p = c/\ell_p$, we see that $\lambda = 1/(1-E^2/E_p^2)$. So, $\langle\tilde{m}\rangle$ can converge rapidly to the Einstein value m, even at short times, provided that the particle's energy E be close to the Planck energy E_p. The correlation distance is now given by $\sim 1/(mc\lambda)$, and since $\lambda > 1$, then $1/(mc\lambda) < 1/(mc)$ always.

It is not hard to show([cf. [21]) that the PI identity (14) implies the commutators

$$[\hat{\boldsymbol{x}}_i, \hat{\boldsymbol{p}}_j]_{\mathrm{DSR}} = \mathrm{i}\left(\delta_{ij} + \frac{\kappa^2 - m^2c^2}{\kappa^2m^2c^2}\,\hat{\boldsymbol{p}}_i\hat{\boldsymbol{p}}_j\right). \quad (15)$$

Here $\kappa = 1/\ell$. The CCR (15) resembles the Snyder version of the deformed CCR associated to the dispersion relation (12) [cf. Refs. [46, 48, 47, 49]]. To be precise, the Snyder fundamental commutation relation [see Ref. [46]] in the notation of the present paper (for $\hbar = 1$, the Snyder fundamental length $a \equiv \ell = 1/\kappa$) would read

$$[\hat{\boldsymbol{x}}_i, \hat{\boldsymbol{p}}_j]_{\mathrm{Snyder}} = \mathrm{i}\left(\delta_{ij} + \frac{\hat{\boldsymbol{p}}_i\hat{\boldsymbol{p}}_j}{\kappa^2}\right). \quad (16)$$

So the prefactor of the deforming term in the Snyder commutator is a *constant* related to the fundamental length, while in the DSR commutator (15) the prefactor varies with the Einstein mass of the particle considered. The minimal length interval ℓ is typically set to be the Planck length ℓ_p, or more generally to be the Compton length λ_C (which reduces to the Planck length for a Planck mass). For definiteness we will in the following identify ℓ with ℓ_p. In this connection, note that when $mc \to \kappa$, i.e., when m coincides with the Planck mass, then the CCR (15) becomes non-relativistic. This can also be directly seen from (14), where for $m \to M_p$ then $\lambda \to \infty$, and the smearing distribution $f_{\frac{1}{2}}(\tilde{m}, tc^2\lambda, tc^2m^2\lambda) \to \delta(m-\tilde{m})$, which yields the usual PI for a Wiener process.

Should we have used instead of (12) a different DSR dispersion relation, for example

$$\frac{p_0^2 - \mathbf{p}^2}{1-(\ell_p p_0)^2} = m^2c^2, \quad (17)$$

(which is discussed in Ref. [50]), we would have obtained results analogous to (14), although less significant from the physical point of view (see again [21]).

For sake of completeness, we should also note that commutator (15) does not coincide with the corresponding commutator of the Magueijo–Smolin paper [45], although it comes from the same dispersion relation (12), proposed in [45].

Moreover, the commutator (15) does not enjoy an important property of the Magueijo-Smolin's commutator. In fact, when the energy of the boosted system approaches the Planck energy, then the Magueijo-Smolin's commutator reads

$$[\hat{\boldsymbol{x}}_j, \hat{\boldsymbol{p}}_i]_{\mathrm{MS}} \;=\; \mathrm{i}\,\delta_{ij}\left(1 - \frac{E}{E_p}\right), \tag{18}$$

so it predicts the appearance of a classical (non-quantal) world at the Planck scale. On the contrary, when the energy of the particle under boost approaches the Planck value, our commutator (15) becomes

$$[\hat{\boldsymbol{x}}_j, \hat{\boldsymbol{p}}_i]_{\mathrm{DSR}} \;=\; \mathrm{i}\,\delta_{ij}\,. \tag{19}$$

This qualitative difference in behaviour can be traced back to the fact that commutators in (doubly-)special relativity depend on two things; First, the fundamental commutators are essentially the Dirac brackets of the canonical variables. The explicit definition of the Dirac brackets for a specific physical system depends on the choice of a gauge (gauge fixing condition), that is, for relativistic systems, on the choice of a specific physical time. So the commutation relations are generally gauge fixing dependent in both the SR and DSR systems. Second, the fundamental commutator $[\hat{\boldsymbol{x}}_j, \hat{\boldsymbol{p}}_i]$ depends (through the Jacobi identities) on the whole symplectic structure of the system (and therefore also on the commutator $[\hat{\boldsymbol{x}}_j, \hat{\boldsymbol{x}}_i]$, for example). These are not specified by a particular DSR model, but they have to be chosen aside. Of course, one obtains different theories for different choices of $[\hat{\boldsymbol{x}}_j, \hat{\boldsymbol{x}}_i]$.

In our specific path-integral approach, the gauge which is automatically incorporated in the path integral is the Polyakov gauge. We obtain the same fundamental commutator of Ghosh [48]. For Ghosh [48] and Mignemi [51] the $\hat{\boldsymbol{x}}_j$'s do not commute, $[\hat{\boldsymbol{x}}_j, \hat{\boldsymbol{x}}_i] \neq 0$, while on the contrary Magueijo–Smolin in [45] require the $\hat{\boldsymbol{x}}_j$'s to commute (see their formula (60)).

So, the DSR theory is not only defined by the dispersion relation, but also by the gauge fixing and the choice of the symplectic structure, which is essentially arbitrary. In principle, therefore, only the experiment can effectively discriminate among different models. It should now become clear why our model can produce commutators different from those of Ref. [45], although both models share the same deformed dispersion relation. As a small further note, we may add that from the Magueijo–Smolin paper [45] it is not clear if the proposed commutators satisfy the Jacobi identities or not (not enough commutators are, in fact, explicitly specified to enable the reader to verify the Jacobi identity).

5. Discussion and perspectives

In this paper, we have argued that the Lorentz invariance is, in fact, not present at the microscopic level but emerges only as a large distance/low-energy property. We have bolstered this viewpoint by showing that both the SR and DSR systems can be mathematically seen as statistically arising from an underlying non-relativistic Wiener process. Such an emergence results from the superposition of two stochastic processes. On a short spatial scale (much shorter than particle's Compton wavelength) the particle moves according to a Brownian, non-relativistic, motion. Its Newtonian mass fluctuates according to an inverse Gaussian distribution.

The time-compounded smearing distribution tends, however, rapidly to the delta-function distribution due to the central limit theorem. This happens at the time scale of the order of Compton time, at which the relative mass fluctuation is of order unity. On a time scale much larger than the Compton time, the particle then behaves as a relativistic particle with a sharp mass equal to Einstein's (i.e., Lorentz invariant) mass. In this case the particle moves with a net velocity which is less than c. Here the reader may notice a close analogy with the Feynman chessboard PI. In contrast to the chessboard PI, our approach is not confined to only $1+1$ dimensional Dirac fermions.

The presented concept of *statistical emergence*, which is shared both by SR and DSR, can offer a new valid insight into the Planck-scale structure of space-time. The existence of a discrete polycrystalline substrate might be welcomed in various quantum gravity constructions. In fact, it has been speculated for long time that quantum gravity may lead to a discrete structure of space and time which can cure classical singularities. This idea has been embodied, in particular, in Loop Quantum Cosmology [52, 53]. A similar proposal was put forward in Ref. [54] in connection with the space-time foam. It should be stressed that many condensed matter systems show that a discrete sub-structure might lead to a genuinely relativistic dynamics at low energies [15], without any internal inconsistency. A good prototypical example are wide single-layer carbon crystals (graphene), where an effective theory emerges in which conducting electrons behave, at low temperatures, as massless relativistic Dirac fermions with a "terminal velocity" equal to the Fermi velocity in the crystal.

On the basis of these results, two major issues appear worth of further research. First, the rôle of the smearing distribution. In the presented approach, the specific form of the smearing distribution is a mere by-product of the superstatistics PI paradigm. Our heuristic picture of a polycrystalline medium which we affiliate with the smearing distribution is clearly not the only possibility. Furthermore, a deeper understanding of a dynamical origin of our smearing would be highly desirable. In fact, the exact LS is due to a very special form of the Newtonian-mass distribution. The exact LS of a spacetime has no fundamental significance in the superstatistics PI, but it is only an accidental symmetry of the spatially coarse-grained theory. This "fortuitousness" is controlled by a specific form of the grain distribution. In fact, we have seen that a small departure from its shape brings a departure from the LS and leads, e.g., to the DSR. In addition, in Ref. [21], we have shown that the DSR systems are stable under a small change in the form of the grain distribution.

The second issue concerns the extension of our schema to curved spacetimes. This could be an important step in addressing the issue of quantum gravity. In this connection we may notice a conceptual similarity with the Hořava–Lifshitz gravity with anisotropic scaling. There, similarly as in our case, space and time are not equivalent at the fundamental level, and therefore the theory is intrinsically non-relativistic. The relativistic concept of time together with its Lorentz invariance emerges at distances much larger than Compton wavelength.

Acknowledgments

A particular thank goes to S. Mignemi for his enlightening emails on the DSR commutators and DSR symplectic structure. We are grateful also for comments from H. Kleinert, C. Schubert, F. Bastianelli, and L.S. Schulman. The figure is due to the art of M. Nespoli.

References

[1] J. Butterfield and C.J. Isham, *The arguments of time*, in: J. Butterfield (Ed.), (Oxford University Press, London 2006).
[2] K.V. Kuchař, *Time and Interpretations of Quantum Gravity*, in: G. Kunstatter, D.E. Vincent and J.G. Williams (Eds.), Proceedings of the Fourth Canadian Conference on General Relativity and Relativistic Astrophysics, 1618 May 1991, University of Winnipeg, (World Scientific, Singapore, 1992), p. 211.
[3] G. 't Hooft, Class. Quant. Grav. **13**, 1023 (1996).
[4] S. Hawking, D. Page, D.N.and C.N. Pope, Nucl. Phys. **170**, 283 (1980); A. Ashtekar, C. Rovelli, L. Smolin and A. Weaving, Phys. Rev. Lett. **69**, (1992) 237.
[5] T. Damour and A.M. Polyakov, Nucl. Phys. B **423**, 532 (1994.
[6] V.A. Kostelecky and S. Samuel, Phys. Rev. D **39**, 683 (1989).
[7] R. Gambini, J. Pullin, Phys. Rev. D **59**, 124021 (1999).
[8] J. Alfaro, H.A. Morales-Tecotl and L.F. Urrutia, Phys. Rev. Lett. **84**, 2318 (2000).

[9] M. Hayakawa, Phys. Lett. B **478**, 394 (2000).
[10] S.M. Carroll, J.A. Harvey, V.A. Kostelecky, C.D. Lane and T. Okamoto, Phys. Rev. Lett. **87**, 141601 (2001).
[11] A. Anisimov, T. Banks, M. Dine and M. Graesser, Phys. Rev. D **65**, 085032 (2002).
[12] H. Kleinert, *Multivalued Fields in in Condensed Matter, Electrodynamics, and Gravitation*, (World Scientific, Singapore, 2008).
[13] P. Jizba, H. Kleinert and F. Scardigli, Phys. Rev. D **81**, 084030 (2010).
[14] G. 't Hooft, J. Stat. Phys. **53**, 323 (1988); in 37th International School of Subnuclear Physics: Basics and Highlights of Fundamental Physics, Erice, Italy, 1999, edited by A. Zichichi (World Scientific, London, 2001).
[15] G.E. Volovik, *The Universe in a Helium Droplet*, (Clarendon Press, Oxford, 2003).
[16] V. Fock, *The Theory of Space-time and Gravitation*, (Pergamon Press, New York, 1964); P. Kosinski et al., Mod. Phys. Lett. A **10**, 2599 (1995); J. Lukierski et al., Ann. Phys. N.Y. **243**, 90 (1995); J. Lukierski and A. Nowicki, hep-th/0203065; hep-th/0207022.
[17] G. Amelino-Camelia, Int. J. Mod. Phys. D **11** (2002) 35; J. Magueijo and L. Smolin, Phys. Rev. Lett. **88** (2002) 190403; G. Amelino-Camelia, Nature **418** (2002) 34.
[18] T. Jacobson and A.C. Wall, Found. of Phys. **40**, 1076 (2010); [arXiv:0804.2720]. T. Jacobson, S. Liberati and D. Mattingly, Annals Phys. **321**, 150 (2006). T. Jacobson, S. Liberati and D. Mattingly, Nature **424**, 1019 (2003).
[19] P. Jizba and H. Kleinert, Phys. Rev. E **78**, 031122 (2008).
[20] P. Jizba and H. Kleinert, Phys. Rev. D **82**, 085016 (2010).
[21] P. Jizba and F. Scardigli, arXiv:1105.3930.
[22] R.P. Feynman and A.R. Hibbs, *Quantum Mechanics and Path Integrals* (McGraw-Hill, New York, 1965).
[23] B. Gaveau, T. Jacobson, M. Kac and L.S. Schulman, Phys. Rev. Lett. **53**, 419 (1984); T. Jacobson and L.S. Schulman, J. Phys. A **17**, 375 (1984).
[24] C. Beck, Phys. Rev. Lett. **87**, 180601 (2001).
[25] M. Kac, *Probability and related Topics in Physical Sciences* (Interscience, New York, 1959).
[26] W. Feller, *An Introduction to Probability Theory and its Applications, Vol. II* (John Wiley, London, 1966).
[27] A.M. Polyakov, *Gauge Fields and Strings* (Harwood, New York, 1987).
[28] see, e.g., E.A. Johnson and A. MacKinnon, J. Phys.: Condens. Matter **5** (1993) 5859.
[29] D. Bohm, *The Special Theory of Relativity*, (Routledge, New York, 1996).
[30] D. Bohm and B.J. Hiley, *The Undivided Universe*, (Routledge, New York, 1993).
[31] N.R. Douglas and N.A. Nekrasov Rev. Mod. Phys. **73**, 977 (2001).
[32] R.B. Laughlin, Int. J. Mod. Phys. A **18**, 831 (2003).
[33] C.D. Froggatt and H.B. Nielsen, *Origin of Symmetries*, (World Scientific, Singapore, 1991).
[34] J.D. Bjorken, *Emergent Gauge Bosons*, hep-th/0111196.
[35] G. Calcagni, Phys. Rev. Lett. **104**, 251301 (2010).
[36] O. Lauscher, M. Reuter, JHEP **10**, 050 (2005).
[37] C. Beck, Physica D **193**, 195 (2004).
[38] J. Kowalski-Glikman, *Introduction to Doubly Special Relativity*, Lecture Notes in Physics. 669. Springer. pp. 131 - 159 (2005); arXiv:hep-th/0405273.
[39] L. Smolin, arXiv:1007.0718; arXiv:0808.3765.
[40] F. Girelli, E.R. Livine, D. Oriti, Phys. Rev. D **81**, 024015 (2010).
[41] C. Rovelli, S. Speziale, Phys. Rev. D **67**, 064019 (2003).
[42] F. Girelli, E.R. Livine, D. Oriti, Nucl. Phys. B **708**, 411 (2005).
[43] F. Girelli, E.R. Livine, arXiv:0708.3813.
[44] J. Magueijo and L. Smolin, Phys. Rev. Lett. **88**, 190403 (2002); S. Mignemi, Phys. Lett. A **316**, 173 (2003).
[45] J. Magueijo and L. Smolin, Phys. Rev. D **67**, 044017 (2003).
[46] H.S. Snyder, Phys. Rev. **71**, 38 (1947).
[47] S. Ghosh, Phys. Rev. D **74**, 084019 (2006).
[48] S. Ghosh, Phys. Lett. B **648**, 262 (2007).
[49] R. Banerjee, S. Kulkarni and S. Samanta, JHEP **05**, 077 (2006).
[50] D. Kimberly, J. Magueijo, and J. Medeiros, Phys. Rev. D **70**, 084007 (2004).
[51] S. Mignemi, Phys. Rev. D **68**, 065029 (2003).
[52] A. Ashtekar, Gen. Rel. Grav. **41**, 707 (2009).
[53] M. Bojowald, Class. Quant. Grav. **26**, 075020 (2009).
[54] A. Vilenkin, Nuc. Phys. B **252**, 141 (1985).

A holographic map of action onto entropy

D Acosta[1,a], P Fernández de Córdoba[2,b], J M Isidro[2,c], J L G Santander[3,d]

[1]Departamento de Matemáticas, Universidad de Pinar del Río, Pinar del Río, Cuba
[2]Instituto Universitario de Matemática Pura y Aplicada, Universidad Politécnica de Valencia, Valencia 46022, Spain
[3]Cátedra Energesis de Tecnología Interdisciplinar, Universidad Católica de Valencia, Guillem de Castro 94, Valencia 46003, Spain

E-mail: [a]dago@mat.upr.edu.cu

E-mail: [b]pfernandez@mat.upv.es

E-mail: [c]joissan@mat.upv.es

E-mail: [d]martinez.gonzalez@ucv.es

Abstract. We propose a holographic correspondence between the action integral I describing the mechanics of a finite number of degrees of freedom in the bulk, and the entropy S of the boundary (a holographic screen) enclosing that same volume. The action integral must be measured in units of (i times) Planck's constant, while the entropy must be measured in units of Boltzmann's constant. In this way we are led to an intriguing relation between the second law of thermodynamics and the uncertainty principle of quantum mechanics.

1. Introduction

There are compelling reasons to believe that quantum mechanics must be an emergent phenomenon [2, 10, 11, 14, 15, 18, 19, 20, 21, 23, 24]. Actually not just quantum mechanics, but also gravity and spacetime appear to be emergent phenomenona as well (for a comprehensive review see [31] and refs. therein). The guiding principle in all emergent theories is the fact that they provide a coarse–grained description of some underlying theory [9]. Due to our ignorance of a full microscopic description, emergent phenomena are in principle amenable to a thermodynamical description.

It is the purpose of this contribution to develop an approach to emergent quantum mechanics from the *entropic* point of view presented in ref. [1]. We take the view that, apart from other important reasons [3, 16, 25, 26, 28, 29, 32], quantum theory must be an emergent phenomenon *also* because the spacetime it is defined on is an emergent concept. There exist in the literature a number of different approaches to account for the emergent nature of spacetime, too numerous to quote here in detail. Here we will follow the holographic [22, 34] proposal presented in ref. [35]. Thus gravity and quantum mechanics share the common feature of being effective, thermodynamical descriptions of their respective underlying theories.

We should point out that quantum mechanics can be recast in thermodynamical terms [30], although without making use of the properties of emergence and holography used here. On the contrary, our approach hinges crucially on the notions of emergence and holography.

Altogether, our approach will provide us with *a holographic, entropic picture of emergent quantum mechanics*.

2. The correspondence

2.1. The main result

Our main result can be summarised in the holographic correspondence

$$\text{(bulk)}\ \frac{\mathrm{i}I}{\hbar} \leftrightarrow \frac{S}{k_B}\ \text{(boundary)}. \tag{1}$$

This correspondence can be explained as follows. Let a finite 3–dimensional volume $\mathcal{V}$ be given, such that it is bounded by a closed 2–dimensional surface $\mathcal{S}$ (a holographic screen, see [35]). Let a finite number of quantum–mechanical degrees of freedom be defined within $\mathcal{V}$, described by the action integral I. The screen $\mathcal{S}$ is assumed to carry N information bits. These bits encode the holographic projection, onto $\mathcal{S}$, of the degrees of freedom within $\mathcal{V}$. Since we do not know the specific mechanism whereby the holographic principle projects the mechanics within $\mathcal{V}$ onto its boundary $\mathcal{S} = \partial\mathcal{V}$, we assign the screen an entropy S, which measures our ignorance about the specific nature of the degrees of freedom on the surface. Thus I describes a mechanical system in the bulk, while S describes its corresponding thermodynamics on the boundary. In this setup, space merely plays the role of a *storage device for information*; space has already emerged within $\mathcal{S}$, while it does not yet exist outside $\mathcal{S}$ [35]. It will be observed that each side of the dimensionless correspondence (1) is measured in units of the corresponding quantum—the quantum of action (Planck's constant $\hbar$) on the mechanical side, the quantum of entropy (Boltzmann's constant k_B) on the thermodynamical side. Finally, there is a relative factor of i, whose origin will be elucidated presently. For the moment we note that the semiclassical limit $\hbar \to 0$ in the bulk corresponds to letting $k_B \to 0$ on the boundary. Last but not least, the two quantities I and S separately obey a corresponding extremum principle. Eqn. (1) differs from an ana! logous correspondence, presented in [1], by a factor of 2, to be explained later.

2.2. A quantum of entropy

The starting point in ref. [35] is a classical point particle of mass M approaching a holographic screen $\mathcal{S}$, from that side of the latter on which spacetime has already emerged. At a distance from $\mathcal{S}$ equal to 1 Compton length, the particle causes the entropy S of the screen to increase by the amount

$$\Delta S = 2\pi k_B, \tag{2}$$

where k_B is Boltzmann's constant. The above can also be understood as meaning that $2\pi k_B$ is the *quantum* by which the entropy of the screen increases, whenever a particle crosses $\mathcal{S}$. The factor 2π on the right–hand side is conventional. Relevant is only the fact that the entropy increase of the screen appears quantised in units of k_B.

2.3. Two thermodynamical languages

Thermodynamics can be conveniently expressed in either of two equivalent languages, respectively called the energy representation and the entropy representation [8]. Any given thermodynamical system can be completely described if one knows its *fundamental equation*. The latter contains all the thermodynamical information one can obtain about the system. The fundamental equation can be expressed in either of two equivalent ways, respectively called the *energy representation* and the *entropy representation*. In the energy representation one has a fundamental equation $E = E(S, \ldots)$, where the energy E is a function of the entropy S, plus of whatever additional variables may be required. In the entropy representation one solves for the entropy in terms of the energy to obtain a fundamental equation $S = S(E, \ldots)$.

Here we will argue that quantum mechanics as we know it (*i.e.*, on spacetime) corresponds to the energy representation, while quantum mechanics on a holographic screen (*i.e.*, in the absence of spacetime) will correspond to the entropy representation. Our goal is to describe the laws of *entropic quantum mechanics*, that is, the thermodynamical laws on the boundary $\mathcal{S}$ that correspond to the quantum mechanics within $\mathcal{S} = \partial\mathcal{V}$.

One must bear in mind, however, that standard thermodynamical systems admit both representations (energy and entropy) simultaneously, which representation one uses being just a matter of choice. In our case this choice is dictated, for each fixed observer, by that side of the screen on which the observer wants to study quantum mechanics. For example there is no energy variable beyond the screen, as there is no time variable, but an observer can assign the screen an entropy, measuring the observer's ignorance of what happens beyond the screen. This notwithstanding, the analogy with thermodynamical systems we have just sketched can be quite useful.

2.4. A (classical) holographic dictionary

Assume that we are give a foliation of 3–space by 2–dimensional holographic screens $\mathcal{S}_j$: $R^3 = \cup_{j\in\mathcal{J}}\mathcal{S}_j$, where the index j runs over some (continuous) set $\mathcal{J}$. For reasons to be explained presently we will restrict our attention to potentials such that the $\mathcal{S}_j$ are all closed surfaces; we denote the finite volume they enclose by $\mathcal{V}_j$, so $\partial\mathcal{V}_j = \mathcal{S}_j$.

One can formulate a *holographic dictionary* between gravitation, on the one hand, and thermodynamics, on the other [35]. Let V_G denote the gravitational potential created by a total mass $M = \int_{\mathcal{V}} \mathrm{d}^3V \, \rho_M$ within the volume $\mathcal{V}$. Then the following two statements are equivalent:
i) there exists a gravitational potential V_G satisfying Poisson's equation $\nabla^2 V_G = 4\pi G\rho_M$, such that a test mass m in the background field created by the mass distribution ρ_M experiences a force $\mathbf{F} = -m\nabla V_G$;
ii) given a foliation of 3–space by holographic screens, $R^3 = \cup_{j\in\mathcal{J}}\mathcal{S}_j$, there are two scalar quantities, called entropy S and temperature T, such that the force acting on a test mass m is given by $F\delta x = \int_{\mathcal{S}} T\delta \mathrm{d}S$. The latter integral is taken over a screen that does not enclose m.
Moreover, the thermodynamical equivalent of the gravitational theory includes the following *dictionary entries* [35]:

$$\frac{1}{k_B}S(x) = \frac{-1}{4\hbar c L_P^2} V_G(x) A(V_G(x)), \tag{3}$$

$$2\pi k_B T(x) = \frac{\mathrm{d}V_G}{\mathrm{d}n}, \tag{4}$$

$$\frac{k_B}{2}\int_{\mathcal{S}} \mathrm{d}^2 a \, T = L_P^2 M c^2. \tag{5}$$

In (3), (4) and (5) we have placed all thermodynamical quantities on the left, while their mechanical analogues are on the right. As in ref. [35], the area element d^2a on $\mathcal{S}$ is related to the infinitesimal number of bits $\mathrm{d}N$ on it through $\mathrm{d}^2a = L_P^2 \mathrm{d}N$. We denote the area of the equipotential surface passing through the point x by $A(V_G(x))$, while $\mathrm{d}V_G/\mathrm{d}n$ denotes the derivative of V_G along the normal direction to the same equipotential. The above expressions tell us how, given a gravitational potential $V_G(x)$ and its normal derivative $\mathrm{d}V_G/\mathrm{d}n$, the entropy S and the temperature T can be defined *as functions of space*.

Specifically, eqn. (3) expresses the proportionality between the area A of the screen $\mathcal{S}$ and the entropy S it contains. This porportionality implies that gravitational equipotential surfaces get translated, by the holographic dictionary, as *isoentropic surfaces*, above called holographic screens $\mathcal{S}$.

Equation (4) expresses the Unruh effect: an accelerated observer experiences the vacuum of an inertial observer as a thermal bath at a temperature T that is proportional to the observer's acceleration $\mathrm{d}V_G/\mathrm{d}n$.

Finally, eqn. (5) expresses the first law of thermodynamics and the equipartition theorem. The right–hand side of (5) equals the total rest energy of the mass enclosed by the volume $\mathcal{V}$, while the left–hand side expresses the same energy content as spread over the bits of the screen $\mathcal{S} = \partial\mathcal{V}$, each one of them carrying an energy $k_BT/2$. It is worthwhile noting that equipartition need not be postulated. Starting from (4) one can in fact prove the following form of the equipartition theorem [1]:

$$\frac{k_B}{2}\int_{\mathcal{S}}\mathrm{d}^2a\,T = \frac{A(\mathcal{S})}{4\pi}U(\mathcal{S}), \qquad A(\mathcal{S}) = \int_{\mathcal{S}}\mathrm{d}^2a. \tag{6}$$

Above, U can be an arbitrary potential energy, subject only to the requirement that its equipotential surfaces are closed. We will henceforth mean eqn. (6) when referring to the first law and the equipartition theorem. In all the above we are treating the area as a continuous variable, but in fact it is quantised [35]. If $N(\mathcal{S})$ denotes the number of bits of the screen $\mathcal{S}$, then

$$A(\mathcal{S}) = N(\mathcal{S})L_P^2. \tag{7}$$

However, in the limit $N \to \infty$, when $\Delta N/N << 1$, this approximation of the area by a continuous variable is accurate enough. We will see later on that letting $N \to \infty$ is equivalent to the semiclassical limit in quantum mechanics.

3. The emergence of quantum mechanics

We intend to write a holographic dictionary between quantum mechanics, on the one hand, and thermodynamics, on the other. This implies that we will need to generalise eqns. (3), (4) and (6) so as to adapt them to our quantum–mechanical setup. Thus we will replace the classical particle of [35] with a quantum particle, subject to some potential energy U of nongravitational origin, but we will take (2) to hold for a quantum particle as well. We will assume U to be such that its equipotential surfaces $\mathcal{S}_j$ are closed, in agreement with our assumptions about the foliation. Let $H = K + U$ be the classical Hamiltonian function on R^3 whose quantisation leads to the quantum Hamiltonian operator $\hat{H} = \hat{K} + \hat{U}$ that governs our quantum particle.

3.1. A (quantum) holographic dictionary

Inside the screen, spacetime has already emerged. This gives us the energy representation of quantum mechanics—the one we are used to: a time variable with a conserved Noether charge, the energy, and wavefunctions depending on the spacetime coordinates. We have the uncertainty relation

$$\Delta\hat{Q}\,\Delta\hat{P} \geq \frac{\hbar}{2}. \tag{8}$$

Expectation values are computed as functional integrals, with a density function d_I given by

$$d_I = \exp\left(\frac{\mathrm{i}}{\hbar}I\right). \tag{9}$$

Above, $I = \int \mathrm{d}tL$ is the action integral satisfying the Hamilton–Jacobi equation.

We can now posit the quantum–mechanical analogues of eqns. (3), (4) and (6). In the energy representation these analogues read, respectively,

$$\frac{1}{k_B}\hat{S}(x) = \frac{1}{4\hbar c L_P}A(U(x))|\hat{U}(x)|, \tag{10}$$

$$2\pi k_B \hat{T}(x) = L_P \frac{\mathrm{d}\hat{U}}{\mathrm{d}n}, \tag{11}$$

$$\frac{k_B}{2} \int_{\mathcal{S}} \mathrm{d}^2 a \, \hat{T} = \frac{A(\mathcal{S})}{4\pi} \hat{U}(\mathcal{S}). \tag{12}$$

3.2. Emergence of the holographic correspondence

It is well known, in the theory of thermodynamical fluctuations, that the probability density function d_S required to compute expectation values of thermodynamical quantities is given by the exponential of the entropy [8]:

$$d_S = \exp\left(\frac{S}{k_B}\right). \tag{13}$$

Comparing (13) with (9) we arrive at the holographic correspondence (1)

$$\frac{\mathrm{i}I}{\hbar} \leftrightarrow \frac{S}{k_B} \tag{14}$$

between the energy representation and the entropy representation.

We would like to point out that an analogous correspondence has been given in [1], the only difference being a factor of 2 in the denominator on the right–hand side. This factor of 2 is easy to account for. In [1], one compares the semiclassical wavefunction in the energy representation, given by $\psi = \exp(\mathrm{i}I/\hbar)$, with the square root of the probability density function d_S in the entropy representation, given by $\sqrt{d_S} = \exp(S/2k_B)$. Instead, here we are equating the probability densities d_I and d_S rather than the wavefunctions. See refs. [4, 5] for specific instances of the correspondence (14).

3.3. Emergence of the wavefunction

The equation $U(x^1, x^2, x^3) = U_0$, where U_0 is a constant, defines an equipotential surface in R^3. As U_0 runs over all its possible values, we obtain a foliation of R^3 by equipotential surfaces. Following [35], we will identify equipotential surfaces with holographic screens. Hence forces will arise as entropy gradients.

Assume that ψ is nonvanishing at a certain point in space. Consider an infinitesimal cylinder around this point, with height L_P and base area equal to the area element $\mathrm{d}^2 a$. Motivated by the proportionality between area and entropy, already mentioned, we postulate that there is an infinitesimal entropy flow $\mathrm{d}S$ *from the particle to the area element* $\mathrm{d}^2 a$:

$$\mathrm{d}S = C \, 2\pi k_B L_P |\psi|^2 \mathrm{d}^2 a. \tag{15}$$

Here C is a dimensionless numerical constant, to be determined presently. A closed surface Σ receives an entropy flux $S(\Sigma)$:

$$S(\Sigma) = C(\Sigma) 2\pi k_B L_P \int_\Sigma \mathrm{d}^2 a \, |\psi|^2. \tag{16}$$

The constant $C(\Sigma)$ will in general depend on the particular surface chosen; the latter may, but need not, be a holographic screen. The key notion here is that the integral of the scalar field $|\psi|^2$ over any surface carries an entropy flow associated. When the surface Σ actually coincides with a holographic screen $\mathcal{S}$, and when the latter is not a nodal surface of ψ, the constant $C(\mathcal{S})$ may be determined by the requirement that the entropy flux from the particle to the screen equal the quantum of entropy (2). Thus

$$\frac{1}{C(\mathcal{S})} = L_P \int_{\mathcal{S}} \mathrm{d}^2 a \, |\psi|^2. \tag{17}$$

Let us now read eqn. (17) in reverse, under the assumption that one knows the proportionality constants $C(\mathcal{S}_j)$ for a given foliation $R^3 = \cup_{j\in\mathcal{J}}\mathcal{S}_j$. This amounts to a knowledge of the integrands, *i.e.*, of the probability density $|\psi|^2$ within the surface integral (17) on each and every $\mathcal{S}_j$. From these tomographic sections of all probability densities *there emerges the complete wavefunction ψ on all of R^3*, at least up to a (possibly point–dependent) phase $\mathrm{e}^{\mathrm{i}\alpha}$.

Thus the integrand of (17) gives the surface density of entropy flow into the holographic screen $\mathcal{S}_j$, and the wavefunction ψ becomes (proportional to) the square root of this flow. The collection of all these tomographic sections of ψ along all possible screens amounts to a knowledge of the complete wavefunction. Hence *a knowledge of the different surface densities of entropy flux across all possible screens is equivalent to a knowledge of the quantum–mechanical wavefunction ψ*. This is how the quantum–mechanical wavefunction ψ emerges from the holographic screens.

4. The Unruh equation of state

In this section we will rewrite the dictionary entries (10), (11) and (12), postulated to hold in the energy representation of quantum mechanics, in the entropy representation. For this purpose we first need to solve the eigenvalue equation $\hat{S}\phi = S\phi$ on the screen, so the latter will be kept fixed. That is, we will not consider a variable surface $\mathcal{S}_j$ of the foliation, but rather a specific surface corresponding to a fixed value of the index j. Observe also a difference in notation: ϕ instead of ψ. This is to stress the fact that, by (10), entropy eigenstates ϕ cannot be eigenstates of the complete Hamiltonian $\hat{H}$, but only of the potential energy $\hat{U}$. Once we have solved the eigenvalue equation

$$\hat{U}\phi = U\phi, \tag{18}$$

then the same ϕ diagonalise $\hat{S}$:

$$\hat{S}\phi = S\phi, \qquad S = \frac{k_B}{4\hbar c L_P} A(\mathcal{S})|U(\mathcal{S})|. \tag{19}$$

Thermodynamical quantities will now arise as expectation values of operators in the entropic eigenstates $\phi(\mathcal{S})$.

We first deal with (10). Clearly its reexpression in the entropy representation will be the thermodynamical fundamental equation $S = S(A)$, since the extensive parameter corresponding to the holographic screen is the area A. Then we have

$$\langle\hat{S}\rangle = \frac{k_B}{4\hbar c L_P} A(\mathcal{S})|U(\mathcal{S})|. \tag{20}$$

Availing ourselves of the freedom to pick the origin of potentials at will, let us set $|U(\mathcal{S})| = \hbar c/L_P$. Thus

$$\langle\hat{S}\rangle = \frac{k_B}{4L_P^2} A, \tag{21}$$

which is the celebrated Bekenstein–Hawking law. It arises as a thermodynamical fundamental equation in the entropy representation.

Our holographic screen is treated thermodynamically as a stretched membrane, so the generalised force conjugate to the extensive parameter A is the surface tension σ. Then the equation of state corresponding to (21) is

$$\sigma = \frac{k_B\langle\hat{T}\rangle}{4L_P^2}. \tag{22}$$

Rewrite the above as $2\pi k_B\langle\hat{T}\rangle = 8\pi L_P^2\sigma$ and recall that σ is the normal component of force per unit length on the screen. Since force is proportional to acceleration, the above equation of state turns out to be equivalent to the Unruh law.

Finally we turn to the first law of thermodynamics and the equipartition theorem. Taking the expectation value, in the entropic eigenstates ϕ, of the operator equation (12), produces the thermodynamical expression for the equipartition theorem:

$$\frac{k_B}{2}\int_{\mathcal{S}} \mathrm{d}^2 a\, \langle \hat{T} \rangle = \frac{A(\mathcal{S})}{4\pi} \langle \hat{U}(\mathcal{S}) \rangle. \tag{23}$$

5. The second law of thermodynamics, revisited

The second law of thermodynamics,

$$\Delta S \geq 0, \tag{24}$$

has been related to the Heisenberg uncertainty principle in ref. [30]. In ref. [13] it has been argued that the second law of thermodynamics has a quantum–mechanical reexpression in the Bell inequalities. In ref. [1] we have established a link between (24) and the Hilbert space of entropic quantum mechanics. Here we would like to propose yet another quantum–mechanical interpretation of the second law, one that combines the uncertainty principle with the notion of emergence.

From eqn. (2) one derives the obvious inequality

$$\Delta S \geq \pi k_B \tag{25}$$

which looks like some refinement of the second law (24)—the latter would be recovered in the semiclassical limit $k_B \to 0$. Therefore let us, for the sake of the argument, consider eqn. (25) as a more precise statement of the second law than (24). As such (25) is reminiscent the uncertainty principle (8) of quantum mechanics. However the left–hand side of (25) contains just one uncertainty, instead of a product of two uncertainties as usual. This reflects the fact that the variable on the left, S, is *selfconjugate*—its dimension equals that of the quantum k_B on the right–hand side[1]. We can include a dimensionless formal parameter! τ in the left–hand side that will make (25) resemble the uncertainty principle in its standard form. This can be done as follows.

Let N denote the total number of bits on $\mathcal{S}$. Whenever a quantum particle hits the screen we have $\Delta N = 1$, and the ratio $\Delta N/N$ will be small if N is large enough. In this limit we can treat N as a continuous variable, that we re-denote by τ in order to interpret it as a continuous, dimensionless parameter:

$$\tau := N, \qquad \text{when} \qquad \frac{\Delta N}{N} << 1. \tag{26}$$

This is the limit $N \to \infty$ referred to in (7). Compatibility with all the above requires this limit to correspond to $k_B \to 0$ or, equivalently, to $\hbar \to 0$. In other words, the large area limit for a holographic screen corresponds to the semiclassical approximation in quantum mechanics.

We have $\Delta\tau \geq 1$, the inequality allowing for the possibility of more than just one particle hitting $\mathcal{S}$. Thus multiplying the two inequalities $\Delta\tau \geq 1$ and $\Delta S \geq \pi k_B$ together we arrive at the following uncertainty principle *on the holographic screen*:

$$\Delta S \Delta\tau \geq \pi k_B. \tag{27}$$

The fact that k_B, though small, is nonvanishing, leads to the impossibility of having strictly reversible processes; reversibility is possible only in the limiting case of a vanishing value for

[1] Compare this situation with (q,p) and (H,t), which are conjugate pairs: the product of the two components of each pair has the dimension of $\hbar$. Angular momentum L is selfconjugate, in the sense that it carries the dimension of $\hbar$, but one writes the corresponding uncertainty principle as $\Delta L \Delta\varphi \geq \hbar/2$, where the dimensionless variable φ is an angle.

the quantum k_B. We conclude that quantisation appears as dissipative mechanism. The notion that information loss leads to a quantum behaviour lies at the heart of the notion of emergence [6, 7, 14, 15, 18, 19, 20, 21, 23, 24, 33].

We have derived the uncertainty principle (27) starting from the second law of thermodynamics (24). Let us now prove that the reverse path is also possible: from the uncertainty principle to the second law of thermodynamics. We start from (8) in the bulk rewritten as $\Delta I/\hbar \geq 1$, where $I = \int p dq$ is the action. On the boundary, the correspondence (1) allows to reexpress the above inequality as in (25). Along the way we have dropped irrelevant numerical factors.

Altogether, we have an equivalence between the uncertainty principle of quantum mechanics (either in the bulk (8) or on the boundary (27)), and a refined version of the second law of thermodynamics, one that includes a small but nonvanishing value of the corresponding quantum ($\hbar$ or k_B) on the right–hand side. This is in agreement with the results of [30]—now with the added bonus that our equivalence has the properties of emergence and holography.

6. Discussion

The entropy representation of quantum mechanics, as presented here, is a holographic projection of the energy representation of the same theory, as defined on spacetime. Our central claim, summarised by eqn. (1), expresses this holographic property.

There is, however, one additional property of quantum mechanics that is deeply encoded in eqn. (1); as such it is not immediately recognised. Namely, quantum mechanics is an emergent phenomenon *also* because quantum mechanics is defined on spacetime, and spacetime itself is an emergent phenomenon. Let us analyse this latter point in more detail.

Any model of emergent gravity must ultimately account for the laws governing the motion of material bodies. Thus, *e.g.*, the proposal made in [35] allows for a (somewhat heuristic) derivation of Newton's law of motion, $F = ma$, and of the relativistic generalisations thereof, as emergent, thermodynamical laws. Moreover, the intriguing presence of Planck's constant $\hbar$ [12] in the purely classical setup of ref. [35] makes one suspect that quantum mechanics also has a role to play in that setup. On the other hand, it is well known that Newton's law $F = ma$ can be recovered in the semiclassical limit of quantum mechanics, as being satisfied by the expectation values of certain operators (Ehrenfest's theorem). Last but not least, thermodynamics is the paradigm of emergent phenomena.

All these different pieces of evidence point toward one and the same conclusion—*viz.*, that if classical mechanics follows from the emergence property of spacetime, then the same should be true of quantum mechanics. Here and in ref. [1] we have exploited this point of view. We would like to stress that this conclusion is ultimately independent of the precise mechanism whereby spacetime emerges. Thus, although the holographic dictionary presented in previous sections hinges crucially on the emergence mechanism being precisely that of ref. [35], the holographic correspondence (1) is independent of that mechanism. As such, the holographic correspondence (1) should hold just as well in any other specific model for the emergence of spacetime (say, loop quantum gravity or any alternative thereto such as [17, 27]).

Acknowledgments

J.M.I. heartily thanks the organisers of the *5th International Heinz von Foerster Congress: Emergent Quantum Mechanics*, Vienna, Nov. 11-13, 2011, for the opportunity to present this talk, and for providing a congenial environment for scientific discussions. This work has been supported by Universidad Politécnica de Valencia under grant PAID-06-09.

Handle stets so, daß weitere Möglichkeiten entstehen.
—*Heinz von Foerster.*

References

[1] Acosta D, Fernández de Córdoba P, Isidro J M and Santander J L G 2011 An entropic picture of emergent quantum mechanics *Preprint* 1107.1898 [hep-th]

[2] Adler S 2004 *Quantum theory as an emergent phenomenon* (Cambridge: Cambridge University Press)

[3] Allahverdyan A, Balian R and Nieuwenhuizen T 2011 Understanding quantum measurement from the solution of dynamical models *Preprint* 1107.2138 [quant-ph]

[4] Banerjee R and Majhi B 2010 Statistical origin of gravity *Phys. Rev.* **D 81** 124006 (*Preprint* 1003.2312 [gr-qc])

[5] Banerjee R 2010 From black holes to emergent gravity *Int. J. Mod. Phys.* **D 19** 2365 (*Preprint* 1005.4286 [gr-qc])

[6] Blasone M, Jizba P and Vitiello G 2001 Dissipation and quantization *Phys. Lett.* **A 287** 205 (*Preprint* hep-th/0007138)

[7] Blasone M, Jizba P and Vitiello G 2004 *Dissipation, Emergent Quantization, and Quantum Fluctuations* in *Decoherence and Entropy in Complex Systems, Selected Lectures from DICE 2002*, ed Elze H T, Lecture Notes in Physics **633** (Berlin: Springer)

[8] Callen H 1960 *Thermodynamics* (New York: John Wiley)

[9] Carroll R 2010 *On the Emergence Theme of Physics* (Singapore: World Scientific)

[10] Carroll R 2011 Remarks on osmosis, quantum mechanics and gravity *Preprint* 1104.0383 [gr-qc]

[11] Carroll R 2011 Thermodynamics and scale relativity *Preprint* 1110.3059 [math-ph]

[12] Chen P and Wang C H 2011 Where is $\hbar$ hiding in entropic gravity? *Preprint* 1112.3078 [gr-qc]

[13] Durham I 2011 The necessity of entanglement and the equivalence of Bell's theorem with the second law of thermodynamics *Preprint* 1111.1746 [quant-ph]

[14] Elze H T 2009 The attractor and the quantum states, *Int. J. Qu. Info.* **7** 83 (*Preprint* 0806.3408 [quant-ph])

[15] Elze H T 2011 Linear dynamics of quantum–classical hybrids *Preprint* 1111.2276 [quant-ph]

[16] Faraggi A and Matone M 2009 The equivalence postulate of quantum mechanics: main theorems *Preprint* 0912.1225 [hep-th]

[17] Finster F and Grotz A 2011 A Lorentzian quantum geometry *Preprint* 1107.2026 [math-ph]

[18] Groessing G 2008 On the thermodynamic origin of the quantum potential *Physica* **A 388** 811 (*Preprint* 0808.3539 [quant-ph])

[19] Groessing G, Mesa Pascasio J and Schwabl H 2011 A classical explanation of quantization *Found. Phys.* **41** 1437 (*Preprint* 0812.3561 [quant-ph])

[20] Groessing G 2010 Sub–quantum thermodynamics as a basis of emergent quantum mechanics *Entropy* **12** 1975

[21] Groessing G, Fussy S, Mesa Pascasio J and Schwabl H 2011 An explanation of interference effects in the double slit experiment: classical trajectories plus ballistic diffusion caused by zero–point fluctuations *Ann. Phys.* **327** (2012) 421. *Preprint* 1106.5994 [quant-ph]

[22] 't Hooft G 1993 Dimensional reduction in quantum gravity, *Preprint* gr-qc/9310026

[23] 't Hooft G 2007 Emergent quantum mechanics and emergent symmetries *Preprint* 0707.4568 [hep-th]

[24] 't Hooft G 2011 How a wavefunction can collapse without violating Schroedinger's equation, and how to understand Born's rule *Preprint* 1112.1811 [quant-ph]

[25] Kauffman L 2011 Eigenforms and quantum physics *Preprint* 1109.1892 [math-ph]

[26] Khrennikov A 2010 An analog of Heisenberg uncertainty relation in prequantum classical field theory *Physica Scripta* **81** 065001 (*Preprint* 0912.1565 [quant-ph])

[27] Mäkelä J 2010 Notes concerning "On the origin of gravity and the laws of Newton" by E. Verlinde *Preprint* 1001.3808 [gr-qc]

[28] Munkhammar J 2010 Is holographic entropy and gravity the result of quantum mechanics? *Preprint* 1003.1262 [hep-th]

[29] Munkhammar J 2011 Canonical relational quantum mechanics from information theory *Preprint* 1101.1417 [physics.gen-ph]

[30] Olah N 2011 *Einsteins Trojanisches Pferd: eine Thermodynamische Deutung der Quantentheorie* (Wien: Springer)

[31] Padmanabhan T 2011 Lessons from classical gravity about the quantum structure of spacetime *J. Phys. Conf. Ser.* **306** 012001 (*Preprint* 1012.4476 [gr-qc])

[32] de la Peña L, Cetto A, Valdés–Hernández A and França H 2011 Origin and meaning of quantum nonlocality *Preprint* 1110.4641 [quant-ph]

[33] Sakellariadou M, Stabile A and Vitiello G 2011 Noncommutative spectral geometry, algebra doubling and the seeds of quantization *Phys. Rev.* **D 84** 045026 (*Preprint* 1106.4164 [math-ph])

[34] Susskind L 1994 The world as a hologram *Preprint* hep-th/9409089

[35] Verlinde E 2011 On the origin of gravity and the laws of Newton *JHEP* **1104** 029 (1001.0785 [hep-th])

Journal of Physics: Conference Series **361** (2012) 012028 doi:10.1088/1742-6596/361/1/012028

Classical-Quantum Coexistence: a 'Free Will' Test

Lajos Diósi

Wigner Research Center for Physics, H-1525 Budapest 114, P.O.Box 49, Hungary

E-mail: diosi.lajos@wigner.mta.hu

Abstract. Von Neumann's statistical theory of quantum measurement interprets the instantaneous quantum state and derives instantaneous classical variables. In reality, quantum states and classical variables coexist and can influence each other in a time-continuous way. This has been motivating investigations since a long time in quite different fields from quantum cosmology to optics as well as in foundations. Different theories (mean-field, Bohm, decoherence, dynamical collapse, continuous measurement, hybrid dynamics, e.t.c.) emerged for what I call 'coexistence of classical continuum with quantum'. I apply to these theories a sort of 'free will' test to distinguish 'tangible' classical variables useful for causal control from useless ones.

1. Introduction

It is now widely agreed that the fundamental description of physical systems is the quantum theory. The state of the given closed system is described by the density matrix $\hat\rho$, the Hamiltonian $\hat H$ governs its dynamics by the von Neumann equation of motion:

$$\frac{d\hat\rho}{dt}=\frac{-i}{\hbar}[\hat H,\hat\rho]\,. \tag{1}$$

This yields the unitary evolution

$$\hat\rho\longrightarrow\hat U(t)\hat\rho\hat U^\dagger(t) \tag{2}$$

where $\hat U(t)$ solves $i\hbar d\hat U/dt=\hat H\hat U$ with the initial condition $\hat U(0)=\hat 1$. The density matrix $\hat\rho$ is interpreted statistically by the concept of measurement [1]. The role of the latter is that it assigns classical variables z to the quantum density matrix $\hat\rho$:

$$\hat\rho\Longrightarrow z\,. \tag{3}$$

Without such assignment, the quantum theory would be incomplete because the matrix $\hat\rho$ itself cannot be identified as classical variables. A matrix element, say ρ_{11}, of $\hat\rho$ does not behave like true classical variables should. How should true classical variables behave? The answer is a new 'Free Will Test' to distinguish the formal classical variables from the true ones which we call *tangible*.

2. Free will

Suppose we have assigned a classical variable z to the quantum state $\hat\rho$ at a given time. We expect of a tangible classical z that we can use it at our free will to control the future dynamics of our system. We expect in particular that we can make the future unitary evolution depend on

z. If the resulting theory remains consistent then z is tangible classical variable and the given assignment $\hat\rho \Rightarrow z$ passes the Free Will Test.

FWT is satisfied within standard quantum theory. As we said, the assignment $\hat\rho \Rightarrow z$ is given by a von Neumann measurement. The mathematical ingredient is a complete orthogonal set of projectors $\hat P_z$ where the labels z, also called measurement outcomes, are the possible classical values that will be assigned to $\hat\rho$ at random with probability $p_z = \mathrm{tr}(\hat P_z\hat\rho)$, respectively. The emergence of the classical variable z is accompanied by the instantaneous 'collapse' of the state:

$$\hat\rho \longrightarrow \frac{1}{p_z}\hat P_z\hat\rho\hat P_z \equiv \hat\rho_z\ . \tag{4}$$

Now, according to FWT, we assume a z-dependent unitary dynamics $\hat U_z$, evolve each $\hat\rho_z$ and average them over the statistics p_z of the classical variable z, yielding:

$$\hat\rho \longrightarrow \hat\rho' = \sum_z p_z\hat U_z\hat\rho_z\hat U_z^\dagger = \sum_z \hat U_z\hat P_z\hat\rho\hat P_z\hat U_z^\dagger\ . \tag{5}$$

We find that the map $\hat\rho \to \hat\rho'$ is linear. This linearity is necessary [2] for the statistical interpretation [1] of $\hat\rho'$, it maintains consistency of the theory. Hence we see the measurement outcomes z are tangible classical variables, FWT has been passed by standard quantum mechanics.

FWT is powerful. In standard quantum mechanics it singles out quantum measurement theory as the only way to assign tangible classical variables to the quantum state. Measurement outcomes are the only tangible classical variables. This also implies that a deterministic assignment $\hat\rho \Rightarrow z$ will never pass FWT, will never yield a tangible z. Had we assigned z to the upper-left matrix element of $\hat\rho$ the resulting dynamics $\hat\rho \to \hat\rho'$ would be non-linear which is a typical inconsistency indicating that z is not tangible . A frequently used non-tangible classical variable is the quantum mean, e.g., of the coordinate $\hat q$:

$$z = \mathrm{tr}(\hat q\hat\rho)\ . \tag{6}$$

For this deterministic way of assignment the z-dependent dynamics

$$\hat\rho \longrightarrow \hat\rho' = \hat U_z\hat\rho\hat U_z^\dagger \tag{7}$$

is non-linear and invalidates the statistical interpretation of $\hat\rho'$. Tangible classical variable is the Husimi assignment of phase-space coordinates $z \equiv (q,p)$ of a quantum harmonic oscillator [3]. It can be underlied by the measurement of the overcomplete set of coherent state projectors $\hat P_z = |z\rangle\langle z|$. The probability of the assignment is $p_z = \mathrm{tr}(\hat P_z\hat\rho)$. The consistency (linearity) of the z-dependent dynamics follows from the same structure (4-5) as in case of standard projective measurements.

3. Classical-quantum coexistence

We experience quantum and classical phenomena in our surrounding. Classical ones are apparent, they emerge from the quantum ones according to quantum theory. This explanation can be extended for all phenomena except for quantum gravity where the emergence of classical continuum is still waiting for a consistent theory. For this failure, different things within standard theory can be blamed. Even the standard approach itself can be wrong if classical phenomena are the basic ones and quantum ones only emerge from it [4, 5, 6]. In any case, a minimum lesson exists. We should extend our investigations for all possible mathematical models of the mere coexistence of the quantum and classical phenomena. I avoid talking about their interaction because this term is reserved for something which is dynamical. Instead, I talk about their mutual influence. Quantum and classical phenomena coexist where coexistence includes their mutual dependence:

$$\text{Classical} \Longleftrightarrow \text{Quantum} \tag{8}$$

Mathematically this means that I allow for any coupling between the equation governing $\hat{\rho}(t)$ of the quantum sector and the equation governing $z(t)$ of the classical sector, respectively.

The mathematical form of the influence of the classical on the quantum looks trivial. The Hamiltonian will be made z-dependent:

$$\frac{d\hat{\rho}}{dt} = \frac{-\mathrm{i}}{\hbar}[\hat{H}(z), \hat{\rho}] \, . \tag{9}$$

This dependence may be time-local or time-non-local, it must respect causality anyway: $\hat{H}$ at time t should not depend on $z(t')$ at $t' > t$.

The influence of quantum on classical, called back-reaction, is the problem. In Sec. 1 we mentioned that the measurement theory can assign a certain z to the current state $\hat{\rho}$. It is also reassuring that the influence (4) of z on $\hat{\rho}$ is part of von Neumann theory. There is a major defect: the von Neumann assignment of z is instantaneous, it cannot assign the continuum $z(t)$ to $\hat{\rho}(t)$. The von Neumann theory of measurement has been generalized for time-continuous measurements so that in principle we possess a model of quantum-classical coexistence. The realistic non-Markovian continuous measurement theory is in its infancy, faces particular issues.

4. Five theories of coexistence

Parallel to the measurement approach, many different—though necessarily interrelated—proposals have targeted the quantum-classical coexistence and the implicit back-reaction. Some of them are very popular, some of them are less so. This time we ask the same question: do they pass FWT?

4.1. Mean-field

Robust quantum fields in quantum cosmology (also in nuclear, molecular, laser physics, e.t.c.) may look like classical fields. They can be replaced by classical fields in mean-field theory [7, 8]. This is a very successful method to identify the classical continuum z by the quantum mean value of the robust quantum field $\hat{q}$ in question:

$$z = \mathrm{tr}(\hat{q}\hat{\rho}) \, . \tag{10}$$

The mean-field $z(t)$ is smooth and causal. To apply FWT, we make $\hat{H}(t)$ depend on $z(t)$:

$$\frac{d\hat{\rho}}{dt} = \frac{-\mathrm{i}}{\hbar}[\hat{H}(z), \hat{\rho}] \, . \tag{11}$$

This means a non-linear evolution for $\hat{\rho}$, it denies the statistical interpretation of $\hat{\rho}(t)$. The mean-field z is not tangible. Hence the corresponding assignment $\hat{\rho} \Rightarrow z$ may be a good approximation but it is flawed fundamentally (as it has already been known independently of FWT).

4.2. De-Broglie—Bohm

This is the oldest theory [9, 10] of assigning classical continuum to the quantum state. The assignment of the classical variable $z(t)$ concerns the coordinate operator $\hat{q}$, the theory is restricted for pure state density matrices $\hat{\rho} = \hat{\rho}^2$. The initial assignment $\hat{\rho}(0) \to z(0)$ is random;

its statistics is the same as if $\hat{q}$ were measured in standard quantum measurement. This feature can be preserved by the following deterministic evolution of the de-Broglie—Bohm trajectories:

$$m\frac{d^2z}{dt^2} = -V'(z) - V'_{\hat{\rho}}(z)\ . \tag{12}$$

The classical continuum variable $z(t)$ senses the $\hat{\rho}$-dependent quantum potential $V_{\hat{\rho}}(z)$ in addition to the potential V already present in $\hat{H}$. The trajectory $z(t)$ is smooth and causal.

Does FWT pass? If we causally control $\hat{H}$ by z, e.g.: we add a deliberate potential $V(z)$, the theory remains consistent with quantum mechanics. If, however, we apply a more general casual control and add a potential $V(z(t');t)$ at time t, controlled by $z(t')$ at an earlier time $t' < t$, then the theory won't be consistent with standard quantum mechanics. The de-Broglie—Bohm trajectory z is useful, e.g., in quantum chemistry calculations [11, 12] but it may prove not to be tangible, not to be fundamental assignment of classical continuum to the quantum system unless we confirm or ignore FWT.

4.3. Decoherence

This is the key mechanism [13, 14, 15] of how quantum systems may, without measurement, express classical features. At certain specific circumstances, the assignment $\hat{\rho} \rightarrow z$ can be introduced even to closed, unitarily evolving, quantum systems. A rigorous mathematical model [16, 17, 18] can be based on the class (history-) operator

$$\hat{C}_z = \hat{P}^{(n)}_{z_n} \dots \hat{P}^{(2)}_{z_2}\hat{P}^{(1)}_{z_1}, \tag{13}$$

the time-ordered product of n Hermitian projectors chosen from n independent complete orthogonal sets. The candidate classical variable is the time-series $z = (z_1, z_2, \dots z_n)$ of the labels. The assignment is random, with the probability $p_z = \mathrm{tr}(\hat{C}^\dagger_z\hat{C}_z\hat{\rho})$. The consistency of the assignment requires the decoherence condition:

$$\mathrm{tr}(\hat{C}^\dagger_z\hat{C}_u\hat{\rho}) = 0 \quad \text{for } z \neq u. \tag{14}$$

Whether FWT passes or doesn't? We conjecture it does, at least under largely general circumstances. The correct answer needs future investigations, see the proofs [19, 20] of a related simple test of dynamic robustness.

If the classical variable z turns out to be tangible still it needs a refinement since the construction (13) is discrete, the time-series of classical variables $z_1, z_2, \dots, z_n$ is discrete, not a continuum yet. To yield the tangible classical continuum $z(t)$ one has to take a certain time-continuous limit of the theory. This limit is non-trivial, it may just coincide [21] with theories of time-continuous measurement.

4.4. Measurement

We learned in Sec. 2 that a quantum measurement assigns a tangible classical variable z to the quantum state $\hat{\rho}$. We can use sequential measurements to assign the time-series $z_1, z_2, \dots, z_n$. The problem is the same as with the decoherence theory. To assign a tangible classical continuum $z(t)$ one has to take the time-continuous limit.

The fundamental need of the continual assignment $\hat{\rho} \Rightarrow z$ led to the theory of time-continuous quantum measurement [22]; the theory was invented and fully developed independently in other important contexts [23, 24, 25]. Let's restrict ourselves for presenting the structure of assignment itself:

$$z = \mathrm{tr}(\hat{q}\hat{\rho}) + \text{white-noise}\ . \tag{15}$$

This variable is the time-continuous outcome of the time-continuous measurement. As such, it inherits the tangibility of single measurement outcomes: $z(t)$ can safely be used to control the Hamiltonian at any later time. In other words, the assignment passes FWT. It is remarkable that the white-noise contribution turns the non-tangible mean-field (10) into the tangible classical variable. The mathematics of time-continuous measurement theory re-appears in dynamical collapse theories [26, 27]. Despite vast research on their ontological status, the tangible variable z has traditionally been abandoned, with rare exceptions [26, 28].

Unfortunately, we face a novel issue with the assignment (15). The white-noise function is everywhere singular hence the classical variable $z(t)$ becomes singular everywhere. The white-noise is an artefact rooted in the simple memoryless (Markovian) model of time-continuous measurement. We could get smooth coloured noise and smooth tangible $z(t)$ in a non-Markovian theory [29]:

$$z = \mathrm{tr}(\hat{q}\hat{\rho}) + \text{colored-noise} \,. \tag{16}$$

This brings causality issues in. Recent discussions predicts that the tangible non-Markovian assignments $\hat{\rho} \Rightarrow z$ are restricted by time-delay/retrodiction [30, 31]. At the current time t the assignment (16) is tangible for earlier times $t' < t$ and $z(t')$ is not (yet) tangible at the current time t if t' gets closer to t than the non-Markovian memory of the measurement.

4.5. Hybrid dynamics

When a classical continuum z and a quantum system coexist we can assume dynamical coupling for them. We could just use an interaction Hamiltonian $\hat{H}_I(z)$ which is a Hermitian operator as well as a function of the classical dynamical variable. For a long time it has been known that such purely canonical hybrid dynamics does not work. Nonetheless, there are various phenomenological constructions of hybrid dynamics and they describe the coexistence between classical continuum and quantum variables, see Salcedo's summary [32] and references therein.

Suppose a quantum system described by the density matrix $\hat{\rho}$ coexists with a classical system described by the density $\rho_C(z)$ over variables z which may be the canonical coordinates $\{z\} = \{(q,p)\}$. If the quantum and classical systems have not yet influenced each other, i.e., they are independent, then their composite state can be described by the hybrid density $\hat{\rho}(z) = \hat{\rho}\rho_C(z)$. If we let them influence each other, the product structure is no longer valid. The evolution

$$\hat{\rho}\rho_C(z) \longrightarrow \hat{\rho}(z;t) \tag{17}$$

preserves positivity and normalization. The evolution must be linear for the sake of statistical interpretation of $\hat{\rho}(z;t)$ [2]. This is the formal criterion of the statistical consistency discussed by [32] in details. In proposals of hybrid dynamics of ensemble coupling [33] or coupling the mean-field to the classical sector, the hybrid state $\hat{\rho}(z)$ would not follow a linear evolution, in accordance with the analysis in [32]. This does not necessarily mean the failure of FWT.

FWT is a different and more powerful test w.r.t. the statistical test in [32]. Some hybrid theories intend to replace or modify the standard statistical interpretation of the density matrix. In all these cases FWT requests that the given non-standard theory remain self-consistent under the z-dependent causal control of the dynamics, no matter if $\hat{\rho}(z;t)$ evolves linearly or not, no matter if the hybrid dynamics is a sort of canonical one or something radically different. However, the theories with non-linear evolution of the hybrid density will easily loose their self-consistence under the z dependent causal control.

There are many versions of hybrid dynamics [34, 35, 36, 37, 38] which are still to be checked against FWT. Some are likely to pass (see, e.g. [39]), especially those which couple certain measurement outcomes—tangible variables—to the classical sector, see [35] coupling the Markovian outcome (15), [36] coupling the non-Markovian continuum of Husimi-variables.

5. Closing remarks

The whole diversity of foundational theories of quantum-classical relationship, from measurement theory to hybrid dynamics, distills to the mathematical issue of coexistence of classical continuum and quantum variables. This coexistence must respect certain basic constraints on one hand and might be radically different from standard quantum theory on the other. Only we have to formulate our basic constraints mathematically. This concept was raised in [40, 41]. and further explained in [42]: 'In our longstanding struggles with the problem of Classical vs. Quantum, the main issue to overcome has always been the painful lack of a consistent model that "couples" the coexisting classical and quantum entities. Aren't quantum dynamics and measurement too restrictive? Are there any other consistent mechanisms?'.

One of the non-trivial mathematical constraints is the Free Will Test. It has nothing direct to do with the Free Will Theorem [43]. A traditional definition of 'free will' in quantum theory is our freedom to choose the measurement setup, see e.g. in [44]. My definition is our free choice (control) of future dynamics (and measurement setups, if you like) in function of our knowledge obtained earlier. The deliberate z-dependent causal control of the dynamics will possess its mathematical representation within all plausible theory of classical-quantum coexistence. The condition that the theory remains self-consistent must be decidable in all theories. If not, it means that the theory should first be concretized. That FWT may not be satisfied by a number of popular theories of coexistence was also recognized in [40, 41]. (Tangible and non tangible classical variables were called just *true* and *not true* ones, respectively.)

I haven't considered the status of my proposal among current competing metaphysical concepts [45]. It may become part of them or contribute to them in some form. We should not overemphasize the metaphysical aspect at the expense of physics proper. The proposal is objective and mathematical, I could well replace the anthropic term FWT by FCT: Free Control Test. It calls for direct checks (i.e.: calculations) in a number of theories, it shows the mathematical barriers against tangible classical variables other than measurement outcomes in whatever sophisticated formalisms.

Acknowledgments

This work was supported by the Hungarian Scientific Research Fund under Grant No. 75129, the EU COST Action MP1006 'Fundamental Problems in Quantum Physics'. The author gratefully thanks the organizers of the International Conference on 'Emergent Quantum Mechanics'.

References

[1] von Neumann J 1932 *Mathematische Grundlagen der Quantenmechanik* (Berlin: Springer)
[2] Diósi 2011 L *Short Course in Quantum Information Theory* (Berlin: Springer)
[3] Husimi K 1940 *Proc. Phys. Math. Soc. Jpn.* **22** 264
[4] t' Hooft G 1999 *Class. Quant. Grav.* **16** 3263
[5] Adler S 2004 *Quantum Theory as an Emergent Phenomenon* (Cambridge: Cambridge University Press)
[6] Wetterich C 2011 *Preprint* arXiv:1201.6212 (this volume)
[7] Moller C 1962 in *Les Theories Relativistes de la Gravitation* ed A Lichnerowicz and M.A. Tonnelat (Paris: CNRS)
[8] Rosenfeld L 1963 *Nucl. Phys.* **40** 353
[9] Bohm D 1952 *Phys. Rev.* **85** 166; *Phys. Rev.* **85** 180
[10] Dürr D and Teufel S 2009 *Bohmian Mechanics* (Berlin: Springer)
[11] Prezhdo O V and Brooksby C 2001 *Phys. Rev. Lett.* **86** 3215
[12] Wyatt R E and Trahan C J 2005 *Quantum dynamics with trajectories* (Berlin: Springer)
[13] Zeh H D 1971 in *Foundations of Quantum Mechanics* ed B. d'Espagnat (New York: Academic Press) p 263
[14] Zurek W 1991 *Phys. Today* **44** 36
[15] Giulini D, Joos E, Kiefer C, Kupsch J, Stamatescu I O and Zeh H D 2003 *Decoherence and the Appearance of a Classical World in Quantum Theory* (Berlin: Springer)
[16] Griffiths R B 1984 *J. Stat. Phys.* **36** 219
[17] Omnès R 1992 *Rev. Mod. Phys.* **64** 339

[18] Gell-Mann M and Hartle J B 1993 *Phys. Rev.* **D 47** 3345
[19] Diósi L 2004 *Phys. Rev. Lett.* **92** 170401
[20] Halliwell J J 2009 *Quant. Inf. Proc.* **8** 479
[21] Diósi L, Gisin N, Halliwell J and Percival I C 1995 *Phys. Rev. Lett.* **74** 203
[22] Diósi L 1988 *Phys. Lett.* **129 A** 419
[23] Belavkin V P 1989 *Phys. Lett.* **140 A** 355
[24] Carmichael H 1993 *An Open Systems Approach to Quantum Optics* (Berlin: Springer)
[25] H.M. Wiseman H M and Milburn G J 2010 *Quantum Measurement and Control* (Cambridge: Cambridge University Press)
[26] Diósi L 1989 *Phys. Rev.* **A 40** 1165
[27] Bassi A and Ghirardi G C 2003 *Physics Reports* **379** 257
[28] Dowker F and Herbauts I 2005 *Found. Phys. Lett.* **18** 499
[29] Strunz W T, Diósi L and Gisin N 1999 *Phys. Rev. Lett.* **82** 1801
[30] Diósi L 2008 *Phys. Rev. Lett.* **100** 080401; **101** 149902(E)
[31] Wiseman H M and Gambetta J M 2008 *Phys. Rev. Lett.* **101** 140401
[32] Salcedo L L 2012 Statistical consistency of quantum-classical hybrids *Preprint* arXiv:1201.4237
[33] Reginatto M and Hall J W 2009 *J. Phys.: Conf. Ser.* **174** 012038
[34] Diósi L 1995 Quantum dynamics with two Planck constants and the semiclassical limit *Preprint* arXiv:quant-ph/9503023
[35] Diósi L and Halliwell J J 1998 *Phys. Rev. Lett.* **81** 2846
[36] Diósi L, Gisin N and Strunz W T 2000 *Phys. Rev.* **A 61** 22108
[37] Elze H T, Gambarotta G and Vallone F 2011 *J. Phys. Conf. Ser.* **306** 012010
[38] Milburn G J 2012 Decoherence and the conditions for the classical control of quantum systems *Preprint* arXiv:1201.5111
[39] Elze H T 2011 *Preprint* arXiv:1202.2472 (this volume)
[40] Diósi L 1999 *Talk at* 'Complexity, Computation and the Physics of Information' (Cambridge, 5-23 July 1999) www.rmki.kfki.hu/~diosi/slides/cambridge.pdf
[41] Diósi L 2004 *Talk at* 'Quantum Theory Without Observers II' (Bielefeld, 2-6 February 2004) www.rmki.kfki.hu/~diosi/slides/bielefeld.pdf
[42] Diósi L 2006 *AIP Conf. Proc.* **844** 133
[43] Conway J and Kochen S 2006 *Found. Phys.* **36** 1441
[44] t' Hooft G 2007 The Free-Will Postulate in Quantum Mechanics *Preprint* arXiv:quant-ph/0701097
[45] Callender C 2009 in *Compendium of Quantum Physics* ed Greenberger D M, Hentschel K and Weinert F (Berlin: Springer) p 384

EmerQuM 11: Emergent Quantum Mechanics 2011 IOP Publishing
Journal of Physics: Conference Series **361** (2012) 012029 doi:10.1088/1742-6596/361/1/012029

Wave-particle duality in classical mechanics

Alexander Y Davydov
AlgoTerra LLC, 249 Rollins Avenue # 202, Rockville, MD 20852, USA
E-mail: alex.davydov@algoterra.com

Abstract. Until recently, wave-particle duality has been thought of as quantum principle without a counterpart in classical physics. This belief was challenged after (i) finding that average dynamics of a classical particle in a strong inhomogeneous oscillating field resembles that of a quantum object and (ii) experimental discovery of "walkers" - macroscopic droplets that bounce on a vertically vibrating bath of the same fluid and can self-propel via interaction with the surface waves they generate. This paper exposes a new family of objects that can display both particle and wave features all together while strictly obeying laws of the Newtonian mechanics. In contrast to the previously known duality examples in classical physics, oscillating field or constant inflow of energy are not required for their existence. These objects behave deterministically provided that all their degrees of freedom are known to an observer. If, however, some degrees of freedom are unknown, an observer can describe such objects only *probabilistically* and they manifest *weird* features similar to that of quantum particles. We show new classical counterparts of such quantum phenomena as particle interference, tunneling, above-barrier reflection, trapping on top of a barrier, and spontaneous emission of radiation. In the light of these findings, we hypothesize that quantum mechanics may *emerge* as approximation from a more profound theory on a deeper level.

1. Introduction

Wave-particle duality is one of the fundamental principles of quantum mechanics which is directly linked to many of its mysteries. It postulates a dual nature of particles and radiation in microworld by enforcing the trade-off between the particle-like and the wave-like behaviour of a quantum object, depending on experimental conditions. Since its original formulation as a hypothesis by Louis-Victor de Broglie in 1923 [1], the duality principle has been supported by numerous experiments involving photons [2], electrons [3-6], neutrons [7], atoms and dimers [8], small van der Waals clusters [9] and, more recently, C_{60} fullerenes [10, 11]. Perhaps the most famous among these is the double-slit experiment, in which, let say, electrons are shot at a barrier with two slits, travel through the slits and are detected on a wall behind the barrier. The interference pattern formed by the electrons on the wall strongly suggests that they behave like waves in this setup. The interference pattern remains even when the electrons are shot separately, with interval between shots being large enough to ensure that only a single electron is present in the system at a time. If, however, an experimentator decides to determine through which slit each electron passes and sets up a detector close to one of the slits, mysteriously electrons stop creating an interference pattern, i.e., they start behaving like classical particles.

Until quite recently, it was widely believed that wave-particle duality has peculiarly quantum nature, an attribute of behaviour of matter at small length and energy scales that does not have a counterpart in classical physics. This belief remained unshaken by some known examples of 'classical tunneling' - crossings of potential barriers by objects of insufficient kinetic energy under condition that these objects are extended in space so that not all their parts are crossing the barrier region simultaneously (e.g., the Fosbury flop in high jump or a long train passing over the short hill) [12, 13]. The situation changed when in 2005 I. Dodin and N. Fisch showed that the average dynamics of a classical particle in a high-frequency oscillating field resembles motion of a quantum particle in a conservative field, with particle average displacement during the oscillation period playing the role of an effective de Broglie wavelength λ. In a quasiclassical field, with a spatial scale large compared to λ, the guiding-center motion is adiabatic. Otherwise, a particle exhibits quantized eigenstates in ponderomotive potential wells, tunnels through "classically forbidden" regions, and experiences stochastic reflection from attractive potentials [14-16]. In another development at approximately the same time, Y. Couder *et. al* discovered that macroscopic oil droplets can bounce on the vertically oscillating liquid bath and interact with the surface waves they generate to form stable self-propelled symbiotic structures called "walkers" [17, 18]. Walkers were shown experimentally to exhibit interference in double-slit experiment [19] along with other quantum-like phenomena such as unpredictable tunneling [20] and orbit quantization in a rotating frame [21].

The goal of this paper is to further extend the list of classical examples of wave-particle duality by presenting a rather general mechanism by which such examples can be produced in the framework of purely classical physics. Herein we present new classical counterparts of such quantum phenomena as particle interference, tunneling, above-barrier reflection, trapping on top of a potential barrier, and spontaneous emission of radiation. This list is likely to be continued increasing the number of links between classical and quantum theories. These links deserve serious attention and would be good topics for future research. Nevertheless, based on what is known so far including the findings presented in this paper, one cannot help but wonder whether the relationship between quantum and classical physics is much closer than is currently acknowledged. Especially intriguing is the question of whether quantum mechanics is an *incomplete* theory that emerges from the underlying classical sub-dynamics. This question has been the subject of long and on-going debates among physicists starting from the birth of quantum theory and has lead to many interesting theoretical developments including the stochastic mechanics introduced by E. Nelson [22, 23], statistical mechanics of matrix models of S. Adler [24], and emergent quantum mechanics based on nonequilibrium thermodynamics on a sub-quantum level pioneered by G. Grössing [25].

2. Classical mechanics of soft bodies

Let us define a *soft body* (SB) as a collection of identical point particles (*monads*) of mass m coupling pairwise with attractive central force. One can think of SB as a cloud consisting of interacting monads. Monads are confined to the cloud as a whole since parting the cloud into distant pieces would require a significant amount of energy. However, the cloud's overall shape can easily vary with time, thus distinguishing a soft body from a rigid one. The dynamics of monads inside SB is determined by coupling between them as well as by external forces that monads can also 'feel'. For simplicity sake, we assume that monad-monad interaction is given by an attractive harmonic potential, i.e., the force exerted upon a monad at location $\mathbf{r}_i$ by another monad at location $\mathbf{r}_j$ is proportional to a distance $\mid \mathbf{r}_i - \mathbf{r}_j \mid$. It follows that monads can penetrate through each other effortlessly (cf. *asymptotic freedom*) and, therefore, internal dynamics of monads is collisionless and can be treated analytically. Another assumption is concerned with how an external force acts on a monad. Let $U(\mathbf{r}, t)$ denote the external potential and N be a total number of monads in SB. We assume that a monad located at $\mathbf{r}_i$ effectively 'experiences' the external potential $U(\mathbf{r}_i, t)/N$. This calibration condition ensures that, when all monads happen to be co-located at the same point $\mathbf{r}$, the total external potential of SB equals

$U(\mathbf{r}, t)$. Thus, total potential energy of SB can be written as

$$U_{tot}(\mathbf{r}_1, \mathbf{r}_2, \ldots, \mathbf{r}_N) = \frac{1}{4} m\omega_0^2 \sum_{all\ pairs}^{\frac{1}{2}N(N-1)} \mid \mathbf{r}_i - \mathbf{r}_j \mid^2 + \frac{1}{N} \sum_{i=1}^{N} U(\mathbf{r}_i, t), \tag{1}$$

where ω_0 denotes the characteristic angular frequency of oscillations of SB. In what follows, we will be concerned with dynamics of the *center of mass* of SB moving in a smooth external potential in the non-relativistic limit. The main question is: How can we describe this motion effectively using as few degrees of freedom as possible?

2.1. The simplest case: SB comprised from two monads in one dimension

Consider the simplest possible instance of SB - two monads in one dimension. The total mass of SB reads $M = 2m$. Let x_1 and x_2 denote the coordinates of two monads. Assume that the external potential is time-independent: $U(x, t) \equiv U(x)$. Also assume that, at any time instance, a linear size of SB is much smaller than the characteristic length at which $U(x)$ changes substantially. It is convenient to introduce new variables as follows

$$X = \frac{1}{2}(x_1 + x_2),\ \xi = \frac{1}{2}(x_1 - x_2) \tag{2}$$

where X is a position of the center of mass (CM) and ξ is an internal degree of freedom of SB. Expanding the potential $U(x)$ in power series around CM, we obtain for the total potential energy U_{tot} (up to the higher order terms)

$$U_{tot}(x_1, x_2) = \frac{1}{2} U(x_1) + \frac{1}{2} U(x_2) + \frac{1}{4} m\omega_0^2 (x_1 - x_2)^2 \cong U(X) + \frac{1}{2} U''(X)\xi^2 + \frac{1}{2} M\omega_0^2 \xi^2, \tag{3}$$

where prime denotes differentiation with respect to X. The total kinetic energy T_{tot} of SB expressed via new variables reads

$$T_{tot} = \frac{1}{2} M\dot{X}^2 + \frac{1}{2} M\dot{\xi}^2, \tag{4}$$

where the overdot means differentiation with respect to time. From the action integral

$$S = \int_{t_1}^{t_2} (T_{tot} - U_{tot}) dt \cong \int_{t_1}^{t_2} [\frac{1}{2} M\dot{X}^2 + \frac{1}{2} M\dot{\xi}^2 - U(X) - \frac{1}{2} M(\omega_0^2 + \frac{1}{M} U''(X))\xi^2] dt \tag{5}$$

we can then write down the Euler-Lagrange equations [26] as conditions for the action S to be stationary, by the Hamilton's principle, as follows

$$M\ddot{X} = -U'(X) - \frac{1}{2} U'''(X)\xi^2 \tag{6}$$

$$\ddot{\xi} + [\omega_0^2 + \frac{1}{M} U''(X)]\xi = 0 \tag{7}$$

Notice rather remarkable relationship between the Eqs. (6) and (7). Equation (6) expresses the Newton's second law for the motion of CM, with the last term on the r.h.s. describing influence of the internal degree of freedom on this motion. Equation (7) can be easily identified as the time-dependent harmonic oscillator equation (TDHOE) for the internal degree of freedom ξ, with time dependence entering the picture via the term $U''[X(t)]$. Thus, we see that CM *parametrically* affects oscillations of the internal degree of freedom, which in its turn, can directly influence the motion of CM (in the presence of position-dependent external potential). Also notice the mathematical equivalence of TDHOE to the time-independent Schrödinger equation, the root of interesting links between quantum and classical physics [27].

2.2. Conservation of energy

When an external potential of SB is (explicitly) time-independent, the total energy is a conserved quantity. In this case, by combining (3) and (4), we obtain

$$E = \frac{1}{2}\,M\dot{X}^2 \,+\, U(X) + \frac{1}{2}\,M\dot{\xi}^2 \,+\, \frac{1}{2}\,M\omega_0^2\xi^2 \,+\, \frac{1}{2}\,U''(X)\xi^2 = constant \tag{8}$$

The first two terms in (8) give contribution that can be associated with the kinetic and potential energies of CM, the third and fourth terms add up to the 'internal energy', while the fifth term describes the coupling between (in this case) two degrees of freedom of SB and can result in energy exchange between them. Let us emphasize that, unless $\xi \equiv 0$ or the external potential U is a constant or a linear function of X, the sum $\frac{1}{2}\,M\dot{X}^2 \,+\, U(X)$ does not conserve during motion. This is the main cause of quantum-like behaviour of the CM of SB under various circumstances, as will be clear later.

2.3. Time-dependent harmonic oscillator

It is convenient to rewrite Eq.(7) as follows

$$\ddot{\xi} + \Omega^2(t)\xi = 0 \tag{9}$$

with $\Omega^2(t) \equiv \omega_0^2 + \frac{1}{M}\,U''[X(t)]$. The time-dependent parameter $\Omega(t)$ is allowed to take not only real but pure imaginary values as well, so that $\Omega^2(t)$ may become negative for some time intervals without any contradiction [27]. Two linearly independent solutions of Eq.(9) can be written in the polar form

$$\xi_1 = \rho(t)cos[\theta(t)] \;;\; \xi_2 = \rho(t)sin[\theta(t)] \tag{10}$$

and combined into one complex-valued function φ as follows

$$\varphi(t) = \xi_1(t) + i\xi_2(t) = \rho(t)e^{i\theta(t)} \tag{11}$$

After substituting (11) into Eq.(10) and separately equating real and imaginary terms on both sides, we obtain two equations

$$\frac{1}{\rho}\,\frac{d}{dt}(\rho^2\dot{\theta}) = 0 \tag{12}$$

and

$$\ddot{\rho} + \Omega^2(t)\rho - \rho\,\dot{\theta}^2 = 0 \tag{13}$$

Equation (12) can be thought of as the angular momentum conservation law (for two-dimensional motion of a point particle of mass m in the plane (ξ_1,ξ_2)) and recast as

$$\frac{1}{2}M\rho^2\dot{\theta} = J = constant, \tag{14}$$

where constant J has a dimension of *action* (or angular momentum) and plays an essential role in the theory to be developed below. Expressing $\dot{\theta}$ via ρ from (14) and substituting it into (13), we obtain the celebrated *Ermakov-Pinney* equation [28, 29]

$$\ddot{\rho} + \Omega^2(t)\rho - \frac{4J^2}{M^2\rho^3} = 0 \tag{15}$$

Alternatively, one can assume that $\dot{\theta} > 0$ for all times and, from (14), find $\rho = \sqrt{2J/(M\dot{\theta})}$ and consequently $\ddot{\rho} = -\sqrt{\frac{J}{2M}}\,(\dddot{\theta}\cdot\dot{\theta} - \frac{3}{2}\ddot{\theta}^2)/\dot{\theta}^{5/2}$. After substituting these results into (13), one finds that the phase function $\theta(t)$ must satisfy the equation

$$\Omega^2(t) = \dot{\theta}^2 + \frac{1}{2}\,\{\theta;t\}, \tag{16}$$

where $\{\theta;t\}$ denotes the *Schwarzian derivative* [30] of $\theta(t)$ defined as $\{\theta;t\} \equiv \dddot{\theta}/\dot{\theta} - \frac{3}{2}(\ddot{\theta}/\dot{\theta})^2$. Hence the (non-conserved) sum of kinetic and potential energies of the time-dependent harmonic oscillator, which coordinate is given by $\xi_1(t)$, can be written in the form

$$\epsilon_1(t) = \frac{1}{2}\,M\dot{\xi}_1^2 + \frac{1}{2}\,M\Omega^2(t)\xi_1^2 = J\omega_{\texttt{eff}}(t), \tag{17}$$

where the *effective* angular frequency $\omega_{\texttt{eff}}(t)$ is defined as

$$\omega_{\texttt{eff}}(t) \equiv \dot{\theta} + \frac{1}{2}\,cos^2(\theta)\cdot(\dddot{\theta}\cdot\dot{\theta} - \ddot{\theta}^2)/\dot{\theta}^3 + \frac{1}{2}sin(2\,\theta)\cdot\ddot{\theta}/\dot{\theta}\,, \tag{18}$$

with $\theta(t)$ being the solution of Eq.(16) for a given $\Omega^2(t)$.

The formulae (17) and (18) are general and therefore must be applicable in the particular case when $\Omega^2(t) \equiv \omega_0^2 = constant$. In this case, the general solution of Eq.(16) is given by

$$\theta(t) = \arctan\left[\kappa\cdot\tan\left(\omega_0 t + \alpha_0\right)\right] \tag{19}$$

where κ is an arbitrary positive number, and α_0 denotes the phase constant. The total energy of the oscillator is, of course, time-invariant in this case and reads

$$\epsilon_1 = J\omega_0/\kappa \tag{20}$$

In the light of (17) and (18), we can rewrite the energy conservation law for SB (8) as follows

$$E = \frac{1}{2}\,M\dot{X}^2 + U(X) + J\omega_{\texttt{eff}}(t) = constant \tag{21}$$

2.4. Equation of motion for center of mass

Assuming that $\omega_0^2 \gg \left|U''(X)/M\right|$ for any X, we can employ the Wentzel-Kramers-Brillouin (WKB) method [31] to obtain the approximate solution of Eq.(7):

$$\xi(t) \approx \frac{\sqrt{2J/M}}{\sqrt[4]{\omega_0^2 + U''[X(t)]/M}}\cdot\cos\left(\int_{t_0}^{t}\sqrt{\omega_0^2 + U''[X(\tau)]/M}\;d\tau + \alpha\right) \tag{22}$$

where α is a phase constant, and we have used the condition that, in absence of the external potential (for free SB), $\xi(t) = \sqrt{2J/(M\omega_0)}\cos\left(\omega_0 t + \tilde{\alpha}\right)$ with $\tilde{\alpha} \equiv \alpha - \omega_0 t_0$. Notice the dependence of $\xi(t)$ not only upon the value of X at moment t but also upon the whole path history $X(\tau)$ at earlier times $\tau < t$ via the integral in the argument of the cosine function in (22). Interestingly, this feature is reminiscent of "path memory" of walkers, which is believed to be responsible for quantization of their orbits in a rotating bath [21]. Substituting (22) into (6), we finally obtain the integro-differential equation of motion for a CM of SB:

$$M\ddot{X} = -U^{'}(X) - \frac{J\,U^{'''}(X)}{M\sqrt{\omega_0^2 + U''(X)/M}} \cdot \cos^2\left(\int_{t_0}^{t} \sqrt{\omega_0^2 + U''[X(\tau)]/M}\, d\tau + \alpha\right) \qquad (23)$$

In case $J = 0$, i.e., when the internal degree of freedom of SB is not excited, Eq.(23) coincides with the second Newton's law for motion of a point particle of mass M in the external potential $U(X)$. However, if $J \neq 0$, the behaviour of CM can be significantly altered in comparison with that of a point particle of the same mass, presenting a great surprise to an observer who is unaware of that difference and treats SB as a point particle. It is worth emphasizing that Eq.(23) is valid even for a wider class of potentials that explicitly depend on time $U(X,t)$, while the energy conservation law (expressed by (8) or (21)) does not hold in this case.

To further simplify our analysis, henceforth we accept a crude approximation of (23) given by

$$M\ddot{X} = -U^{'}_{\mathtt{eff}}(X,t) \qquad (24)$$

$$U_{\mathtt{eff}}(X,t) \equiv U(X) + \frac{J \cdot U^{''}(X)}{M\omega_0} \cdot \cos^2(\omega_0 t + \alpha) \qquad (25)$$

where the phase constant α has been redefined to absorb $-\omega_0 t_0$.

3. Classical examples of wave-particle duality and analogues of quantum mechanical phenomena

Here we present new examples of wave-particle duality in the realm of classical mechanics based on the theory developed in section 2.

3.1. Classical analogue of tunneling

Consider a potential barrier of a Gaussian shape $U(X) = A\exp\left(\frac{(X-X_0)^2}{2\sigma^2}\right)$. Assume that a SB of mass M approaches this barrier from the left with CM moving at constant velocity V such that the corresponding kinetic energy $KE = \frac{1}{2}MV^2$ (far from the barrier) is smaller than the barrier height A, as shown in Fig.1. Analysis of Eqs. (24) and (25) reveals that such SB may either reflect back or go through the barrier, depending on the value of the phase $\theta = \omega_0 t + \alpha$ at moment when the CM is near the barrier top. Figure 2 shows the CM trajectories for different values of phase constant α chosen at random from the continuous uniform distribution on the interval $[0,2\pi[$. Thus, α plays the role of a *hidden variable* that completely determines the result of the interaction of SB with the potential barrier. For an observer who treats this SB as a point particle with the equivalent mass, its behaviour seems unpredictable with apparent violation of the energy conservation law for some short time intervals and, therefore, can be classified as *tunneling*.

3.2. Classical analogue of above-barrier reflection

Consider now the case when $KE > A$ as depicted in Fig.3. With appropriate choice of parameters, it is possible to obtain above-barrier reflection of a SB as shown in Fig.4.

3.3. Classical analogue of particle trapping on top of a barrier and spontaneous emission

Another example of paradoxical behaviour of SB can be obtained if we select $KE > A$ but $KE - A \ll A$, see Fig.5. In this case, apart from the above-barrier reflection, we observe a new phenomenon: the CM can be confined for considerable interval oscillating around the

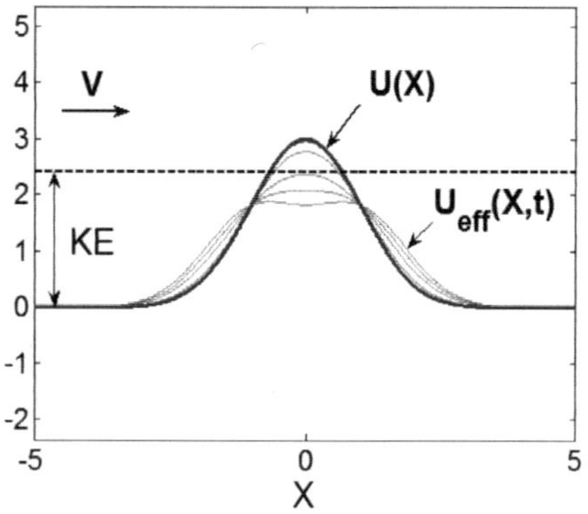

Figure 1. Setup for classical analogue of quantum tunneling. See text for details.

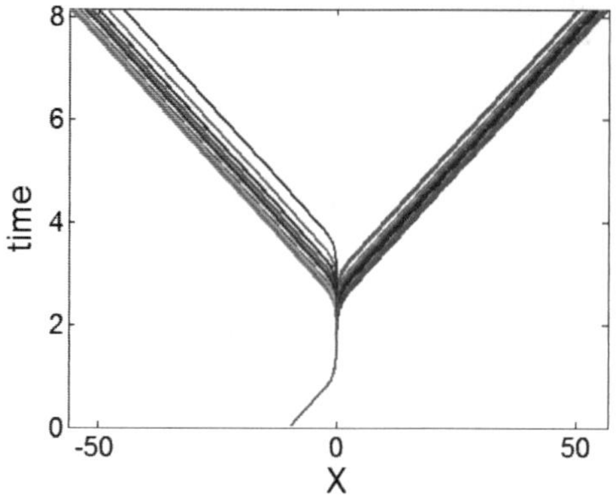

Figure 2. Trajectories for 100 different values of phase constant α.

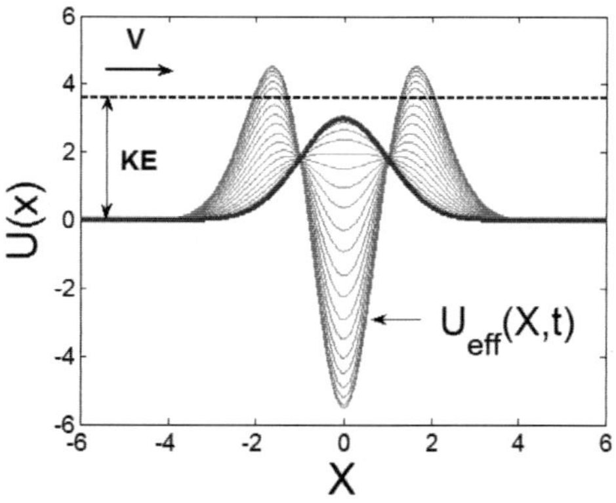

Figure 3. Setup for classical analogue of above-barrier reflection. See text for details.

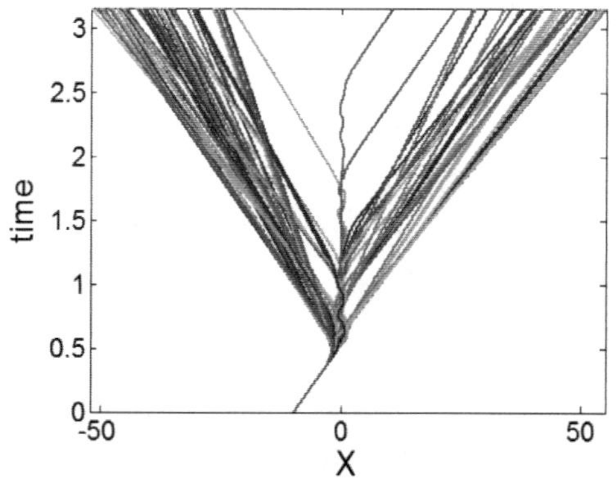

Figure 4. Trajectories for 100 different values of phase constant α illustrating classical above-barrier reflection.

top of the potential barrier and then be released in either direction with equal probability and approximately the same velocity as it had before the confinement. This behaviour is similar to that of a quantum particle confined in a potential plateau region, as has been discussed in detail in [32]. Numerical simulations show that direction of the *emission* of a trapped SB is highly sensitive to the phase θ of the internal degree of freedom, i.e., small change in the value of θ can lead to emission in the opposite direction.

3.4. Interference in 1D

Interference phenomena are ubiquitous in Nature and are most easily understood using the theory of waves. For instance, colourful patterns produced by soap bubbles can be explained by constructive and destructive interference of light waves. It is much more demanding to explain thin-film interference using the representation of light as a collection of photons, requiring concepts of quantum theory, such as probability amplitude, to be invoked. Interestingly, Sir Isaac Newton - who considered light as being composed of tiny particles which he called 'corpuscles' - made some ingenious arguments in attempt to interpret the results of an experiment measuring the relationship between the thickness of a sheet of glass and partial reflection of light off it.

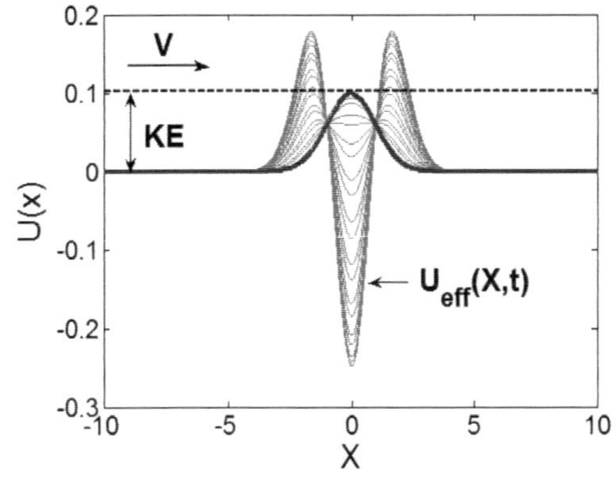

Figure 5. Setup for classical analogues of paradoxical confinement of a particle on top of a potential barrier and spontaneous emission. See text for details.

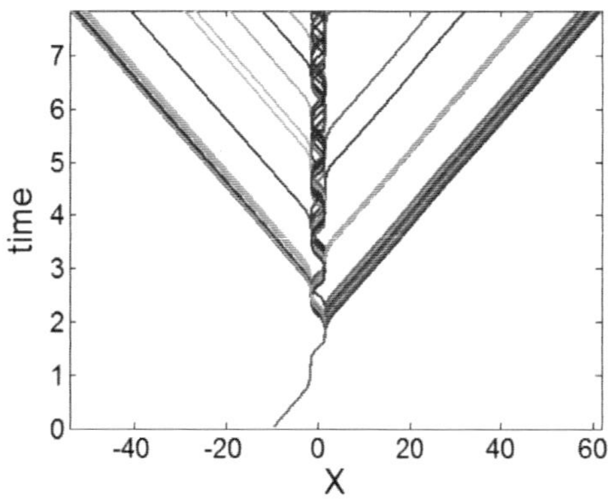

Figure 6. Trajectories for 100 different values of phase constant α illustrating classical confinement on top of a barrier and subsequent spontaneous emission.

In the words of R. Feynman [33], "...to account for the fact that the thickness of the glass determines the amount of partial reflection, Newton proposed this idea: Light striking the first surface sets off a kind of wave or field that travels along with the light and predisposes it to reflect or not reflect off the second surface. He called this process 'fits of easy reflection or easy transmission' that occur in cycles, depending on the thickness of the glass".

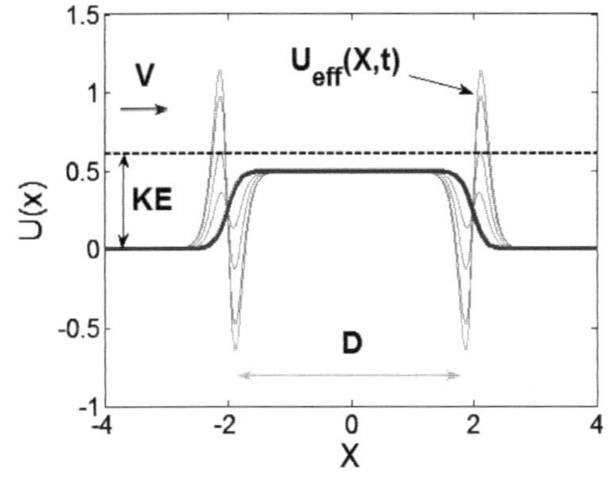

Figure 7. Scattering of a SB off a 'soft rectangular' potential barrier. See text for details.

Under specific circumstances, a SB can exhibit behaviour alike interference of waves in thin films, although it can be revealed only *statistically* as a build-up process of an interference-like pattern from single events and, besides, has a temporal character. Consider an ensemble of soft bodies of the same type prepared such that all members have fixed kinetic energy KE of their CM but different phase constants α_i drawn from the continuous uniform distribution on the interval $[0, 2\pi[$. Each SB moves from the left toward the potential barrier of 'soft rectangular' shape $U(X) = \frac{A}{2}\left(\tanh\left(\frac{X+d}{L}\right) - \tanh\left(\frac{X-d}{L}\right)\right)$ shown in Fig. 7. The width of the barrier is given by $D \equiv 2d$. The kinetic energy of CM exceeds the barrier height: $KE > A$. We define the fraction of soft bodies in the ensemble trapped inside the barrier at some time instance $t = \tau_m$ as the *trapping coefficient*. The *transmission coefficient* is defined along the similar lines. Results of numerical simulations demonstrate that both trapping and transmission coefficients,

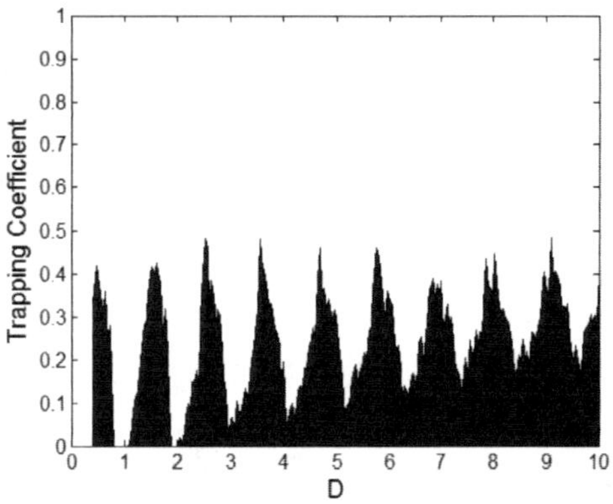

Figure 8. Trapping coefficient as a function of barrier width D estimated numerically after fixed time interval since the first collision of a SB with the barrier.

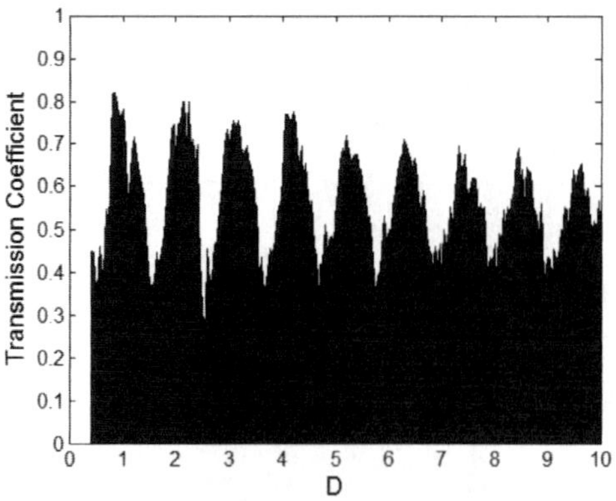

Figure 9. Transmission coefficient as a function of barrier width D estimated numerically after fixed time interval since the first collision of a SB with the barrier.

estimated at some fixed τ_m, depend quasi-periodically upon the width of barrier D, as shown in Figures 8 and 9. This oscillatory dependence fades away with increasing τ_m, as more and more soft bodies of the ensemble which are temporally confined inside the barrier start to leak out in either back or forward direction until none are left.

4. Conclusions

We have presented a new family of classical objects (soft bodies) obeying laws of the Newtonian mechanics that can exhibit particle and wave properties simultaneously. Such systems extend the list of known examples of the wave-particle duality in classical mechanics. They do not require an oscillating external field or constant inflow of energy to manifest their dual properties and behave deterministically provided that all degrees of freedom are known to an observer. If some degrees of freedom are unknown, soft bodies can be described only *probabilistically* and manifest *weird* features similar to that of quantum particles including interference, tunneling, above-barrier reflection, trapping on top of a barrier, and spontaneous emission of radiation. Further developments are under way. In the light of these findings, it is tempting to hypothesize that quantum mechanics may *emerge* as approximation from a more profound theory on a deeper level. If true, the road to achieving complete understanding of quantum theory may lie through an attempt to construct an *average dynamics* of classical objects like soft bodies, in which unobserved degrees of freedom have been effectively integrated out. This average dynamics will necessarily have a probabilistic nature. One can speculate that a concept of *wave function* may emerge as a mathematical tool to cope with lack of information about all degrees of freedom of a soft body, and the Schrödinger equation may even be *deduced* from the first principles. Such program is in line with the vision of A. Einstein who predicted: "Assuming the success of efforts to accomplish a complete physical description, the statistical quantum theory would, within the framework of future physics, take an approximately analogous position to the statistical mechanics within the framework of classical mechanics. I am rather firmly convinced that the development of theoretical physics will be of this type, but the path will be lengthy and difficult." [34]. The present paper advocates making new steps along this path.

Acknowledgements
I thank Andrej Ahmeteli for interesting discussions and bringing the Ref. [13] to my attention. Also I am grateful to Ilya Dodin for providing me with references to his work on quantum-like behaviour of classical particles in oscillating fields.

References

[1] de Broglie L 1923 Waves and quanta *Nature* **112** 540
[2] Taylor G I 1909 Interference fringes with feeble light *Proc. Camb. Philos. Soc.* **15** 114
[3] Davisson C J and Germer L H 1927 *Nature* **119** 558-60
[4] Merkli P G, Missiroli G F and Pozzi G 1976 *Am. J. Phys.* **44** 306
[5] Missiroli G F, Pozzi G and Valdré U 1981 *J. Phys.* E **14** 649
[6] Tonomura A, Endo J, Matsuda T, Kawasaki T and Ezawa H 1989 *Am. J. Phys.* **57** 117
[7] Halban H v Jr and Preiswerk P 1936 *C.R. Acad. Sci.* **203** 73-75
[8] Estermann I and Stern O 1930 *Z. Phys.* **61** 95-125
[9] Schöllkopf W and Toennies J P 1994 Nondestructive mass selection of small van der Waals clusters *Science* **266** 1345-48
[10] Arndt M, Nairz O, Vos-Andreae J, Keller C, van der Zouw G and Zeilinger A 1999 Wave-particle duality of C_{60} molecules *Nature* **401**(6754) 680
[11] Nairz O, Arndt M and Zeilinger A 2003 Quantum interference experiments with large molecules *Am. J. Phys.* **71** (no. 4) 319-25
[12] Vlasov A D and Lupandin O S 1979 *From Ptolemy's Epicycles to Magic Nuclei and Plankeons: Model Concepts in Physics* (Moscow, USSR: Atomizdat) p 103
[13] Cohn A and Rabinowitz M 1990 Classical tunneling *Int. J. Theor. Phys.* **29** 215
[14] Dodin I Y and Fisch N J 2005 Quantumlike dynamics of classical particles in ponderomotive potentials *Phys. Rev. Lett.* **95** 115001
[15] Dodin I Y and Fisch N J 2006 Nonadiabatic tunneling in ponderomotive barriers *Phys. Rev.* E **74** 056404
[16] Dodin I Y and Fisch N J 2007 Particle manipulation with nonadiabatic ponderomotive forces *Phys. Plasmas* **14** 055901
[17] Couder Y, Protiere S., Fort E. and Boudaoud A 2005 *Nature* **437** 208
[18] Protiere S., Boudaoud A and Couder Y 2006 *J. Fluid Mech.* **554** 85
[19] Couder Y and Fort E 2006 Single-particle diffraction and interference at a macroscopic scale *Phys. Rev. Lett.* **97** 154101
[20] Eddi A, Fort E, Moisy F and Couder Y 2009 Unpredictable tunneling of a classical wave-particle association *Phys. Rev. Lett.* **102** 240401
[21] Fort E, Eddi A, Boudaoud A, Moukhtar and Couder Y 2010 Path-memory induced quantization of classical orbits *PNAS* **107** (no. 41) 17515
[22] Nelson E 1966 Derivation of the Schrödinger equation from Newtonian mechanics *Phys. Rev.* **150** 1079-85
[23] Nelson E 1985 *Quantum Fluctuations* (Princeton, NJ: Princeton University Press)
[24] Adler S L 2004 *Quantum Theory as an Emergent Phenomenon: The Statistical Mechanics of Matrix Models as the Precursor of Quantum Field Theory* (Cambridge, UK: Cambridge University Press)
[25] Grössing G 2010 Sub-quantum thermodynamics as a basis of emergent quantum mechanics *Entropy* **12** 1975-2044
[26] Arnold V I 1978 *Mathematical Methods of Classical Mechanics* (New York: Springer-Verlag)
[27] Davydov A 2011 On close relationship between classical time-dependent harmonic oscillator and non-relativistic quantum mechanics in one dimension *Int. J. Theor. Phys.* **50** 1451-67
[28] Ermakov V P 1880 *Univ. Izv. Kiev* **20** 1
[29] Pinney E 1950 *Proc. Am. Math. Soc.* **1** 681
[30] Ovsienko V and Tabachnikov S 2005 *Projective Differential Geometry Old and New* (Cambridge: Cambridge University Press)
[31] Berry M V and Mount K 1972 Semiclassical approximations in wave mechanics *Rep. Prog. Phys.* **35** 315
[32] Garrido P L, Goldstein S, Lukkarinen J and Tumulka R 2011 Paradoxical reflection in quantum mechanics *Am. J. Phys.* **79** (no. 12) 1218 31
[33] Feynman R P 1988 *QED: The Strange Theory of Light and Matter* (Princeton, NJ: Princeton University Press)
[34] Einstein A 1970 *Albert Einstein: Philosopher-Scientist*, ed Schilpp P A (Peru, IL: Open Court) p 672

Classical signal model reproducing quantum probabilities for single and coincidence detections

Andrei Khrennikov, Börje Nilsson, and Sven Nordebo
International Center for Mathematical Modelling
in Physics and Cognitive Sciences
Linnaeus University, Växjö, S-35195, Sweden

E-mail: Andrei.Khrennikov@lnu.se

Abstract. We present a simple classical (random) signal model reproducing Born's rule. The crucial point of our approach is that the presence of detector's threshold and calibration procedure have to be treated not as simply experimental technicalities, but as the basic counterparts of the theoretical model. We call this approach threshold signal detection model (TSD). The experiment on coincidence detection which was done by Grangier in 1986 [22] played a crucial role in rejection of (semi-)classical field models in favour of quantum mechanics (QM): impossibility to resolve the wave-particle duality in favour of a purely wave model. QM predicts that the relative probability of coincidence detection, the coefficient $g^{(2)}(0)$, is zero (for one photon states), but in (semi-)classical models $g^{(2)}(0) \geq 1$. In TSD the coefficient $g^{(2)}(0)$ decreases as $1/\mathcal{E}_d^2$, where $\mathcal{E}_d > 0$ is the detection threshold. Hence, by increasing this threshold an experimenter can make the coefficient $g^{(2)}(0)$ essentially less than 1. The TSD-prediction can be tested experimentally in new Grangier type experiments presenting a detailed monitoring of dependence of the coefficient $g^{(2)}(0)$ on the detection threshold. Structurally our model has some similarity with the prequantum model of Grössing et al. Subquantum stochasticity is composed of the two counterparts: a stationary process in the space of internal degrees of freedom and the random walk type motion describing the temporal dynamics.

1. Introduction

We study the old problem of a possibility to construct a classical field model reproducing probabilistic predictions of quantum mechanics. The common opinion, see Bell [1], is that this is impossible to do, but see [2]–[16] for numerous attempts to proceed with classical wave (oscillatory) prequantum models. The main argument is that for composite quantum systems (e.g., a pair of entangled photons) the correlation predicted by QM (and confirmed by experiment [17]–[19]) cannot be reproduced by models with (local) hidden variables. This is also the common opinion that "Bell's theorem" [1] is formulated in so general abstract framework that it rejects all (local) models with hidden variables, including variables of classical field (e.g., electromagnetic) type. Although nowadays Bell's inequality and entangled systems are the hot topics [20], [21], the problem of a possibility of the classical field description of a single quantum system is also of large importance for quantum theory, including quantum information theory. One of the most important tests of a possibility to represent a photon simply as a pulse of classical electromagnetic field is the experiment on coincidence detection in two output channels of the polarization beam splitter (PBS) [22], [23]. QM predicts that if the source can be considered as one-photon source, then the probability of coincidence detection is equal to zero: one photon cannot be split between two channels, it is either in one or another channel. Any pulse of the classical electromagnetic field is split between two output channels of PBS. Hence, the probability of coincidence detection is nonzero. Of course, the real experimental situation is more complicated: for any "one-photon source" the probability of emission of e.g. two photons is nonzero (although it can be made very small), there is also the contribution of noise, including so called duck counts (i.e., counts in detectors in the absence of the source). Therefore, instead of the absolute probability of coincidence detection P_{12}, experimenters use the relative probability [22], [23]:

$$g^{(2)}(0) = \frac{P_{12}}{P_1 P_2}, \tag{1}$$

where P_{12} is the probability of coincidence detection in channels $i = 1, 2$ and P_i are probabilities of detection in corresponding channels. Known (semi-) classical models of QM predict that $g^{(2)}(0) \geq 1$. If photon is not a classical pulse, one can expect that $g^{(2)}(0) < 1$ (even by taking into account noise and emission of double photons). The first experiment of this type was performed by [22], see also [23] and [24] for review. It was shown that the number of double clicks is relatively small. This experiment played an important role in quantum foundations. As is commonly accepted, this experiment justified rejection of (semi-)classical field theories as "prequantum theories"; in particular, photons cannot be interpreted as pulses of classical electromagnetic field.

The aim of this paper is to show that the main reason of dis-matching of predictions of (semi-)classical field models with QM and experimental data is that such models do not take into account the impact of *detectors* to creation of the quantum statistics.

We present a classical field model which in combination with the procedure of detection of random signals by *threshold type detectors* (which are properly calibrated) reproduces quantum probabilities and, in particular, the coefficient $g^{(2)}(0) << 1$ for a sufficiently high detection threshold $\mathcal{E}_d$. Hence, the prediction of our model differs crucially from the prediction of the known classical field models which do not take into account the evident fact that measurement is not simply detection (of the monitoring-type) of continuous classical signals, but creation of discrete counts with the aid of threshold detectors. Detectors also have to be properly calibrated.

Our model, *threshold signal detection model* (TSD), predicts that the quantum prediction, $g^{(2)}(0) << 1$, matches better measurements for high value of the detection threshold. The basic prediction of TSD is that the coefficient

$$g^{(2)}(0) \leq \frac{K}{\mathcal{E}_d^2},$$

where $K > 0$ is a constant depending on the signal. This K depends on the brightness of the source: higher brightness implies larger K. It also depends on elements of the density matrix $\rho = (\rho_{ij})$ corresponding to the prequantum signal. (In TSD ρ corresponds to the normalization (by the trace) of the covariance operator B of the prequantum random signal, i.e., $\rho = B/\mathrm{TrB}$.) Larger ρ_{12} implies larger K.

It is interesting to compare this prediction with the real experiment. Unfortunately, it seems that Grangier's type experiments with detailed monitoring of dependence of the coincidence probability on the value of the threshold have never been done; more specifically, that is: In Grangier's experiment, did the calibration of detectors play a crucial role to eliminate coincidences?

This is clearly a crucial question. It is hard to give a definite answer as to how exactly it was influencing the result at a fundamental level because Grangier et al [22], [23] didn't study the influence of threshold on the $g^{(2)}(0)$ parameter he was measuring. It is nevertheless clear that with a lower threshold, he would have gotten a less good result. Let me translate the part of

Journal of Physics: Conference Series **361** (2012) 012030 doi:10.1088/1742-6596/361/1/012030

Grangier's thesis [23] where he explains for the first time the role of the threshold and how its level was chosen:

"[...] In this configuration, the threshold has a double role of acquisition of timing information and of selection of the pulses (the too weak pulses are not taken into account). The problems connected to the choice of the discriminator threshold and of the high voltage of the photomultipliers are discussed in detail in reference [17]. We have in the present experiment chosen a **rather high threshold**, which amount to give the priority of the stability of the counting rates and the reproducibility of the results, rather than to the global detection efficiencies." (I stressed with bold the important fact that Grangier proceeded with rather high threshold.)

TSD provides a strong motivation to perform Grangier's type experiment with monitoring of dependence of the coincidence probability on the detection threshold and source's brightness.

TSD can be considered as measurement theory for recently developed *prequantum classical statistical field theory*, PCSFT, [32],[33]. The latter reproduced all quantum averages and correlations including correlations for entangled quantum states. In particular, PCSFT correlations violated Bell's inequality. The main problem for matching of PCSFT and conventional QM was that PCSFT (nor other classical field models) was not able to describe probabilities of discrete clicks of detectors. In particular, PCSFT is theory of correlations of *continuous signals*. "Prequantum observables" are given by quadratic forms of signals. These forms are unbounded and this is not surprising that correlations of such observables can violate Bell's type inequalities, see [34] for discussion and an elementary example. The condition of coincidence of ranges of values of quantum observables and corresponding "prequantum variables" plays a crucial role in Bell's argument. TSD solved the measurement problem of PCSFT. In the same way as in Bell's consideration, TSD operates with discrete observables. In particular, in the case of photon polarization (its projection to a fixed axis) TSD operates with dichotomous variables taking values ± 1.

TSD/PCSFT for composite quantum systems was presented in [35]. However, random signals considered in [35] have a complex structure of temporal correlations, even in the case of a single system. In the present paper, we use simply a combination of the Wiener process (as the career of temporal correlations) and a Gaussian stationary process valued in the space of internal degrees of freedom of a quantum system (e.g., its polarization). We do not consider spatial degrees of freedom.

To escape misunderstanding, we stress that the presented detection model is purely classical; in particular, detectors are classical detectors which are sensitive to the energy density of a signal in the domain of detection V. Such a detector clicks at the first instance of time $\tau = \tau_{\rm d}$ when the total energy of signal in V approaches the detection threshold $\mathcal{E}_d$. Since we consider random signals, this instant of detection is a random variable $\tau = \tau(\omega)$. Our main aim is to find its average $\bar{\tau}$. The quantity $1/\bar{\tau}$ determines the probability of detection. At the moment we proceed with such an operational description of the classical detectors. We plan to consider a more detailed scheme of the classical threshold detection in another paper in which we shall study classical signals with spatial degrees of freedom, i.e., signals in physical space-time. (In the present paper we restrict the model to "internal degrees of freedom" such as polarization. The model with spatial degrees of freedom is essentially more complicated, since it involves processes with infinite-dimensional state space.)

2. Threshold detection

2.1. The class of random signals

We consider a special class of classical random signals. The model is phenomenological: we cannot present physical motivations for selection of this class of signals, besides the fact that detection of such signals with the aid of threshold type detectors reproduces the correct quantum probabilities. Another class of classical random signals serving for the same aim was introduced in [35]. Signals considered in the present paper have essentially simpler temporal structure, simply the Wiener process.

Structurally our model has some similarity with the prequantum model of Grössing et al [15]. Subquantum stochasticity is combined of the two counterparts: a stationary process in the space of internal degrees of freedom and the random walk type motion describing the temporal dynamics.

We start with stationary signals. We proceed in the finite dimensional state space corresponding to internal degrees of freedom such as polarization. Generalization to spatial degrees of freedom is evident, but it has essentially more complicated mathematical structure.

Let H be the m dimensional complex Hilbert space. Let $\phi \equiv \phi(\omega)$ be the H-valued Gaussian random variable with zero average and the covariance operator B. This operator is Hermitian and positively defined. Take in H an orthonormal basis $\{e_j\}$ and consider corresponding signal's components $\phi_j(\omega) = \langle\phi(\omega), e_j\rangle$. The mathematical expansion of the random vector $\phi(\omega)$ with respect to the basis $\phi(\omega) = \sum_j \phi_j(\omega)e_j$ physically corresponds to splitting of the signal into disjoint channels. We remark that correlations of signal's components are given by

$$E\phi_i\overline{\phi_j} = \langle Be_i, e_j\rangle = b_{ij}.$$

Now we introduce the temporal stochastics by simply using the one dimensional Wiener process $w(t)$ which is independent from the stationary process $\phi(\omega)$. Consider the random (non-stationary) signal $\phi(t,\omega) = w(t)\phi(\omega)$ and its components corresponding to internal degrees of freedom: $\phi_j(t,\omega) = w(t)\phi_j(\omega)$. We remark that correlations of signal's components are given by

$$E\phi_i(t)\overline{\phi_j(s)} = \min(t,s)b_{ij}.$$

The energy of the ith component of the complex signal $\phi(t)$ (at the instance of time t) is given by the square of its absolute value

$$\mathcal{E}_i(t,\omega) = |\phi_i(t,\omega)|^2.$$

The total energy of the signal is

$$\mathcal{E}_i(t,\omega) = \sum_i |\phi_i(t,\omega)|^2 = \|\phi(t,\omega)\|^2.$$

2.2. The scheme of threshold measurement

We consider the measurement scheme in which each channel, $i = 1, 2, \ldots m$, goes to a threshold type detector. We assume that all detectors have the same threshold $\mathcal{E}_d > 0$. The detection procedure under consideration is reduced to the condition of the energy level approaching the detection threshold. The instant of time τ corresponding to the signal's detection ("click") by jth detector is determined by the condition:

$$\mathcal{E}_j(\tau,\omega) = \mathcal{E}_d. \tag{2}$$

We remark that the instant of the signal detection is a random variable:

$$\tau = \tau(\omega).$$

Mathematically our aim is to find average of the instance of detection, $\bar{\tau} = E\tau$. The quantity $1/\bar{\tau}$ will be used to find the probability of detection, "how often the detector produces clicks," see section 3.

We apply the mathematical expectation (average) operator to both sides of the detection condition (2) and we obtain

$$E\mathcal{E}_j(\tau(\omega),\omega) = \mathcal{E}_d, \tag{3}$$

or

$$Ew^2(\tau(\omega),\omega)E|\phi_j(\omega)|^2 = \mathcal{E}_d. \tag{4}$$

To find the first average, we use the formula of total probability:

$$Ew^2(\tau(\omega),\omega) = \int_0^\infty d\tau P(\tau(\omega)=\tau)Ew^2(\tau,\omega).$$

We know that, for the fixed τ, $Ew^2(\tau,\omega)=\tau$. Hence,

$$Ew^2(\tau(\omega),\omega) = \int_0^\infty d\tau P(\tau(\omega)=\tau)\tau = E\tau \equiv \bar{\tau}.$$

Thus, the detection condition (4) has the form:

$$\bar{\tau}E|\phi_j(\omega)|^2 = \mathcal{E}_d, \tag{5}$$

or

$$\bar{\tau}b_{ii} = \mathcal{E}_d. \tag{6}$$

We remark that $\tau = \tau_i, i = 1,2,...,m$. Thus

$$\frac{1}{\bar{\tau}_i} = \frac{b_{ii}}{\mathcal{E}_d}. \tag{7}$$

3. Probabilities of clicks in detection channels

Hence, during a long period of time T such a detector clicks N_{click}-times, where

$$N_i \approx \frac{T}{\bar{\tau}_i} = \frac{b_{ii}T}{\mathcal{E}_d}. \tag{8}$$

To find the probability of detection and match the real detection scheme which is used in quantum experiments we have to use a proper normalization of N_i, This is an important point of our considerations. (The normalization problem is typically ignored in standard books on quantum foundations, cf., however, [34].) In QM-experiments probabilities are obtained through normalization corresponding to the sum of clicks in all detectors involved in the experiment, e.g., spin up and spin down detectors.

Hence, the total number of clicks:

$$N = \sum_i N_i = \frac{T\sum_i b_{ii}}{\mathcal{E}_d} = \frac{T\mathrm{TrB}}{\mathcal{E}_d}. \tag{9}$$

We remark that the total number of clicks does not depend on the split of the signal into disjoint channels, i.e., on the selection of an orthonormal basis $\{e_j\}$ in H. In fact,

$$E\|\phi(\omega)\|^2 = \mathrm{TrB}.$$

The probability of detection for the jth detector is given by

$$P_i = N_i/N = \frac{b_{ii}}{\mathrm{TrB}}. \tag{10}$$

In fact, this is the Born's rule of QM. Consider the operator

$$\rho = B/\mathrm{TrB}. \tag{11}$$

This is the Hermitian positive trace one operator; so it has all properties of the *density operator* used in QM to describe the state of a quantum system.

Set $\widehat{C}_i = |e_i\rangle\langle e_i|$, the orthogonal projection onto the vector e_i. Then the equality for the probability of detection (10) can be written as

$$P_i = \mathrm{Tr}\rho\widehat{\mathrm{C}}_\mathrm{i}. \tag{12}$$

This is the QM-rule for calculation of probabilities of detection.

4. Coincidence detection

Coincidence of clicks corresponds to matching of two conditions of threshold approaching corresponds to two constraints:

$$\mathcal{E}_1(\tau_1(\omega), \omega) = \mathcal{E}_d, \mathcal{E}_2(\tau_2(\omega), \omega) = \mathcal{E}_d = \mathcal{E}_d, \tag{13}$$

where matching has the form

$$\tau_1(\omega) = \tau_2(\omega) = \tau(\omega). \tag{14}$$

Our aim is to estimate the probability of coincidence P_{12}. To shorter notatin, we set $\mathcal{E}_i(\omega) \equiv \mathcal{E}_i(\tau(\omega); \omega)$ or even simply $\Gamma_i, i = 1, 2$.

We consider the set of random parameters corresponding to coincidence detection: $A_{12} = \{\omega : \mathcal{E}_i = \mathcal{E}_d, i = 1, 2\}$. ($A_{12}$ is the event of coincidence detection). We have to estimate its probability, $P_{12} = P(A_{12}))$. We shall get a rather rough estimate which, nevertheless, will be sufficient for our purpose. However, we shall see that better estimates of this probability will clarify essentially inter-relation between our "prequantum classical field theory", QM, and experiment. In principle, one may hope to derive an approximative expression for P_{12} as we did for probabilities $P_j, j = 1, 2$. However, this is a complicated probabilistic problem.

We remark that A_{12} is a subset of the set

$$A_{1\times 2} = \{\omega : \mathcal{E}_1\mathcal{E}_2 = \mathcal{E}_d^2\}.$$

Hence, $P(A_{12}) \leq P(A_{1\times 2})$. And the set $A_{1\times 2}$ is a subset of the set

$$A_{1\times 2\geq \mathcal{E}_d^2} = \{\omega : \mathcal{E}_1\mathcal{E}_2 \geq \mathcal{E}_d^2\}$$

and hence $P(A_{12}) \leq P(A_{1\times 2\geq \mathcal{E}_d^2})$. The latter probability we can (roughly) estimate by using Chebyshov inequality (which usage is standard for such estimates, cf. [36]). In the simplest form, for a random variable $u = u(\omega)$ and a constant $k > 0$, this inequality has the form: $P(\omega : u \geq k) \leq \frac{E|u(\omega)|}{k}$. In our case $u = \mathcal{E}_1\mathcal{E}_2, k = \mathcal{E}_d^2$. We have

$$P(A_{12}) \leq \frac{E\mathcal{E}_1\mathcal{E}_2}{\mathcal{E}_d^2}. \tag{15}$$

We find this average by using the formula of total probability [36]:

$$E\mathcal{E}_1(\omega)\mathcal{E}_2(\omega) = \int_0^\infty E\mathcal{E}_1(\tau, \omega)\mathcal{E}_2(\tau, \omega)P(\tau(\omega) = \tau)d\tau. \tag{16}$$

Thus our main problem is to find the correlation of two energies for each instant of time τ. We have

$$E\mathcal{E}_1(\tau,\omega)\mathcal{E}_2(\tau,\omega) = Ew^4(\tau)E|\phi_1(\omega)|^2|\phi_2(\omega)|^2.$$

The first factor is known, $Ew^4(\tau) = 3\tau^2$.

To find the second factor, we shall use general theory of Gaussian integrals on complex Hilbert space [37], see appendix. Consider in H (m-dimensional complex space) projection operators $\widehat{A}_k = |e_k\rangle\langle e_k|, k = 1, ..., m$. Set $f_{A_k}(\phi) = \langle \widehat{A}_k\phi, \phi\rangle, \phi \in H$, the quadratic form corresponding to the operator $\widehat{A_k}$. By (27), appendix, we obtain

$$Ef_{A_1}f_{A_2} = \text{Tr}B\widehat{A}_1\text{Tr}B\widehat{A}_2 + \text{Tr}B\widehat{A}_2B\widehat{A}_1, \tag{17}$$

where B is the covariance operator. We remark that

$$b_{ij} = \rho_{ij}\text{TrB}, \tag{18}$$

where $\rho = (\rho_{ij})$ is a density operator. So, we consider the prequantum random signal corresponding to the quantum state ρ and we study the problem of detection coincidence for such a signal $\phi(t) \equiv \phi_\rho(t,\omega)$. We have

$$\text{Tr}B\widehat{A}_i = b_{ii}, i = 1, 2.$$

In the same way

$$\text{Tr}B\widehat{A}_2B\widehat{A}_1 = |\langle e_1|B|e_2\rangle|^2 = |b_{12}|^2.$$

Finally, we obtain (for the fixed instant of time τ)

$$E\mathcal{E}_1(\tau,\omega)\mathcal{E}_2(\tau,\omega) = 3\tau^2(b_{11}b_{22} + |b_{12}|^2).$$

The formula of total probability (16) implies

$$E\mathcal{E}_1(\tau(\omega);\omega)\mathcal{E}_2(\tau(\omega);\omega) = 3\overline{\tau^2}(b_{11}b_{22} + |b_{12}|^2),$$

where $\overline{\tau^2} = E\tau^2$. By (11) we can rewrite this answer in terms of quantum mechanical density matrix

$$E\mathcal{E}_1(\tau(\omega);\omega)\mathcal{E}_2(\tau(\omega);\omega) = 3\overline{\tau^2}(\text{TrB})^2(\rho_{\text{ii}}\rho_{\text{jj}} + |\rho_{\text{ij}}|^2),$$

Thus by the Chebyshov inequality

$$P(A_{12}) \leq \frac{3(\text{TrB})^2\overline{\tau^2}}{\mathcal{E}_d^2}(\rho_{ii}\rho_{jj} + |\rho_{ij}|^2). \tag{19}$$

The quantity

$$\Delta \equiv \sqrt{\overline{\tau^2}} \tag{20}$$

can be interpreted as the time parameter scaling the time intervals between coincidence clicks. Therefore $\overline{\mathcal{E}}_\Delta = \Delta\text{TrB}$ is average of signal's energy distributed between clicks. Suppose that

$$\epsilon = \frac{\overline{\mathcal{E}}_\Delta}{\mathcal{E}_d} << 1. \tag{21}$$

In this case

$$P(A_{12}) \leq 3\epsilon^2(\rho_{ii}\rho_{jj} + |\rho_{ij}|^2). \tag{22}$$

As usual [], set $g^{(2)}(0) = \frac{P_{12}}{P_1 P_2}$, where P_1 and P_2 are probabilities of detection in channels i and j, respectively (it is assumed that the later probabilities are positive). We proved, see section 3, that in TSD model, these probabilities are equal to quantum probabilities: $P_1 = \rho_{11}$ and $P_2 = \rho_{22}$. Hence,

$$g^{(2)}(0) = \frac{P_{12}}{P_1 P_2} \leq 3\epsilon^2 \Big(1 + \frac{|\rho_{ij}|^2}{\rho_{ii}\rho_{jj}}\Big). \tag{23}$$

Since $\epsilon \sim \frac{1}{\mathcal{E}_d}$, by increasing the parameter $\mathcal{E}_d$ the coefficient $g_{12}(0)$ can be done less than 1. Hence, our prequantum classical field model can violate (under the selection of a proper threshold) the inequality $g^{(2)}(0) \geq 1$ which is valid for "standard" classical signal theory – its standard version which does not take into account the evident fact that measurement is not simply detection (of the monitoring-type) of continuous classical signals, but creation of discrete counts with the aid of threshold detectors. Detectors also have to be properly calibrated.

5. Appendix: Gaussian integrals

Details of theory of integration with respect to Gaussian measures on complex Hilbert spaces can be found in [37].

Let W be a real Hilbert space. Consider a σ-additive Gaussian measure p on the σ-field of Borel subsets of W. This measure is determined by its covariance operator $B : W \to W$ and mean value $m \in W$. For example, B and m determine the Fourier transform of p :

$$\tilde{p}(y) = \int_W e^{i(y,\phi)} dp(\phi) = e^{\frac{1}{2}(By,y)+i(m,y)}, y \in W.$$

(In probability theory it is called the characteristic functional of the probability distribution p.) In what follows we restrict our considerations to *Gaussian measures with zero mean value*: $(m, y) = \int_W (y, \psi) dp(\psi) = 0$ for any $y \in W$. Sometimes there will be used the symbol p_B to denote the Gaussian measure with the covariance operator B and $m = 0$. We recall that the covariance operator B is defined by its bilinear form $(By_1, y_2) = \int (y_1, \phi)(y_2, \phi) dp(\phi), y_1, y_2 \in W$

Let Q and P be two copies of a real Hilbert space. Let us consider their Cartesian product $H = Q \times P$, "phase space," endowed with the symplectic operator $J = \left(\begin{array}{cc} 0 & 1 \\ -1 & 0 \end{array} \right)$. Consider the class of Gaussian measures (with zero mean value) which are invariant with respect to the action of the operator J; denote this class $\mathcal{S}(H)$. It is easy to show that $p \in \mathcal{S}(H)$ if and only if its covariance operator commutes with the symplectic operator, [32].

As always, we consider complexification of H (which will be denoted by the same symbol), $H = Q \oplus iP$. The complex scalar product is denoted by the symbol $\langle \cdot, \cdot \rangle$. The space of bounded Hermitian operators acting in H is denoted by the symbol $\mathcal{L}_s(H)$.

We introduce the complex covariance operator of a measure p on the complex Hilbert space $H : \langle Dy_1, y_2 \rangle = \int_H \langle y_1, \phi \rangle \langle \phi, y_2 \rangle dp(\phi)$. Let p be a measure on the Cartesian product $H_1 \times H_2$ of two complex Hilbert spaces. Then its covariance operator has the block structure

$$D = \left(\begin{array}{cc} D_{11} & D_{12} \\ D_{21} & D_{22} \end{array} \right), \tag{24}$$

where $D_{ii} : H_i \to H_i$ and $D_{ij} : H_j \to H_i$. The operator is Hermitian. Hence $D_{ii}^* = D_{ii}$, and $D_{12}^* = D_{21}$.

Let H be a complex Hilbert space and let $\widehat{A} \in \mathcal{L}_s(H)$. We consider its quadratic form (which will play an important role in our further considerations) $\phi \to f_A(\phi) = \langle \widehat{A}\phi, \phi \rangle$. We make a trivial, but ideologically important remark: $f_A : H \to \mathbf{R}$, is a "usual function" which is defined point wise. We use the equality, see, e.g., [32]:

$$\int_H f_A(\phi) dp_D(\phi) = \mathrm{Tr}\, D\widehat{A} \tag{25}$$

Journal of Physics: Conference Series **361** (2012) 012030 doi:10.1088/1742-6596/361/1/012030

Let p be a Gaussian measure of the class $\mathcal{S}(H_1 \times H_2)$ with the (complex) covariance operator D and let operators $\widehat{A}_i$ belong to the class $\mathcal{L}_s(H_i), i = 1, 2$. Then

$$\int_{H_1 \times H_2} f_{A_1}(\phi_1) f_{A_2}(\phi_2) dp(\phi) = \mathrm{Tr} D_{11} \widehat{A}_1 \; \mathrm{Tr} D_{22} \widehat{A}_2 + \mathrm{Tr} D_{12} \widehat{A}_2 D_{21} \widehat{A}_1 \qquad (26)$$

This equality is a consequence of the following general result [32]:
Let $p \in \mathcal{S}(H)$ with the (complex) covariance operator D and let $\widehat{A}_i \in \mathcal{L}_s(H)$. Then

$$\int_H f_{A_1}(\phi) f_{A_2}(\phi) dp(\phi), = \mathrm{Tr} D \widehat{A}_1 \mathrm{Tr} D \widehat{A}_2 + \mathrm{Tr} D \widehat{A}_2 D \widehat{A}_1. \qquad (27)$$

Summary on coincidence probability: *Prequantum classical field model with threshold and properly calibrated detectors violates predictions of standard classical and semiclassical models, cf. [24], and gives the prediction compatible with known experimental data. More detailed experiments of Grangier's type with monitoring of the dependence of* $g^{(2)}(0)$ *on the detection threshold and source's brightness are demanded.*

References

[1] Bell J 1987 *Speakable and Unspeakable in Quantum Mechanics* (Cambridge: Cambridge Univ. Press)
[2] De Broglie L 1924 *Recherches sur la Thorie des Quanta* (Paris)
[3] Schrödinger E 1935 *Naturwissenschaften* **23** 807
[4] Schrödinger E 1935 *Naturwissenschaften* **23** 823
[5] Schrödinger E 1935 *Naturwissenschaften* **23** 844
[6] Einstein A and Infeld L 1961 *Evolution of Physics: The Growth of Ideas from Early Concepts to Relativity and Quanta* (New-York: Simon and Schuster)
[7] Lamb W E 2001 *The Interpretation of Quantum Mechanics* (Princeton: Rinton Press, Inc.)
[8] Lande A 1965 *New Foundations of Quantum Mechanics* (Cambridge: Cambridge Univ. Press)
[9] Louisell H H 1973 *Quantum Statistical Properties of Radiation* (New York: J. Wiley)
[10] Mandel L and Wolf E *Optical Coherence and Quantum Optics* (Cambridge: Cambridge University Press)
[11] De la Pena L and Cetto A 1996 *The Quantum Dice: An Introduction to Stochastic Electrodynamics* (Dordrecht: Kluwer)
[12] Boyer T 1980 *Foundations of Radiation Theory and Quantum Electrodynamics* (New York: Plenum Press) p 141
[13] Cole D C, Rueda A, and Danley K 2001 *Phys. Rev.* A **63** 054101
[14] Grössing G 2009 *Physica A* **388** 811
[15] Grössing G, Mesa Pascasio J, Schwabl M 2011 *Found. Phys.* **41** 1437
[16] Nieuwenhuizen Th 2006 *Proc. Int. Conf. Quantum Theory: Reconsideration of Foundations-3* (Växjö) vol 810 (Melville, NY: American Institute of Physics) p 198
[17] Aspect A 1983 *Three Experimental Tests of Bell Inequalities by the Measurement of Polarization Correlations between Photons* (Orsay)
[18] Zeilinger A 2010 *Dance of the Photons: From Einstein to Quantum Teleportation* (New-York: Farrar, Straus and Giroux)
[19] Weihs G 1999 *An Experiment to Test Bell's Inequality under Einstein Locality* (Innsbruck: University of Innsbruck Press)
[20] Accardi L, Adenier G, Fuchs C A, Jaeger G, Khrennikov A, Larsson J-A, Stenholm S 2009 *Proc. Int. Conf. Foundations of Probability and Physics-5* vol 1101 (Melville, NY: American Institute of Physics)
[21] Adenier G, Khrennikov A, Lahti P, Manko , and Nieuwenhuizen Th 2008 *Proc. Int. Conf. Quantum theory: reconsideration of foundations-4* (Växjö) vol 962 (Melville, NY: American Institute of Physics)
[22] Grangier P 1986 *Etude Expérimentale de Propriétés Non-classiques de la Lumière: Interférence à un seul Photon* (Centre D'Orsay: Université de Paris-Sud)
[23] Grangier P, Roger G, and Aspect A 1986 *EPL* **1** 173
[24] Beck M 2007 *J. Opt. Soc. Am.* B **24** 2972
[25] Khrennikov A 2005 *J. Phys. A: Math. Gen.* **38** 9051
[26] Khrennikov A 2006 *Found. Phys. Letters* **18** 637
[27] Khrennikov A 2006 *Physics Letters A* **357** 171

[28] Khrennikov A 2006 *Found. Phys. Lett.* **19** 299
[29] Khrennikov A 2006 *Nuovo Cimento* B **121** 505
[30] Khrennikov A 2009 *EPL* **88** 40005.1
[31] Khrennikov A 2010 *EPL.* **90** art-number 40004
[32] Khrennikov A 2010 *J. of Russian Laser Research* **31** 191
[33] Khrennikov A, Ohya M and Watanabe N 2010 *J. Russian Laser Research* **31** 462
[34] Khrennikov A 2009 *Contextual approach to quantum formalism* (Berlin-Heidelberg-New York: Springer)
[35] Khrennikov A 2011 Quantum probabilities and violation of CHSH-inequality from classical random signals and threshold type properly calibrated detectors *Preprint* arXiv:1111.1907v2 [quant-ph]
[36] Ritov S M 1966 *Introduction to Statistical Radiophysics* (Moscow: Nauka, Fizmatlit)
[37] Khrennikov A 2007 *J. Math. Phys.* **48** Art. No. 013512

EmerQuM 11: Emergent Quantum Mechanics 2011 IOP Publishing
Journal of Physics: Conference Series **361** (2012) 012031 doi:10.1088/1742-6596/361/1/012031

Quantum fermions and quantum field theory from classical statistics

Christof Wetterich
Institut für Theoretische Physik
Universität Heidelberg
Philosophenweg 16, D-69120 Heidelberg
E-mail: c.wetterich@thphys.uni-heidelberg.de

Abstract. An Ising-type classical statistical ensemble can describe the quantum physics of fermions if one chooses a particular law for the time evolution of the probability distribution. It accounts for the time evolution of a quantum field theory for Dirac particles in an external electromagnetic field. This yields in the non-relativistic one-particle limit the Schrödinger equation for a quantum particle in a potential. Interference or tunneling arise from classical probabilities.

1. Introduction

It is often believed that quantum physics and classical statistical physics are fundamentally different on a conceptual level. For both branches of physics probabilities play a central role. Classical statistical physics involves only the concept of a positive definite probability distribution $\{p_\tau(t)\}$, rules for the computation of expectation values of observables and correlation functions in terms of $\{p_\tau(t)\}$, and a dynamical equation describing the time evolution of the probabilities. In quantum physics the complex wave function plays a central role. It is often said that quantum physics is the physics of phases in addition to the probabilities. Observables and correlations are associated to non-commuting operators. These properties lead to characteristic features such as interference, entanglement, tunneling, fermions and bosons, that are all not familiar in macroscopic classical statistical physics.

We present here a microscopic classical statistical model that accounts for all characteristic features of quantum mechanics. We consider a generalized Ising-type model for discrete Ising spins on a lattice. The discrete variables can be associated with simple yes/no alternatives or bits. The states τ of the classical statistical ensemble are configurations of Ising spins or lattice-ordered sequences of bits. The probabilities p_τ are positive and normalized, $p_\tau \geq 0$, $\Sigma_\tau p_\tau = 1$. We will propose a particular evolution law for the time dependence of the probability distribution $\{p_\tau(t)\}$ that respects these properties. We will show that for this law the Ising-type classical statistical model describes a quantum field theory for Dirac-fermions in an external electromagnetic field. Specializing to probability distributions that describe one-particle states we recover the Dirac equation in an external electromagnetic field. In the limit of a static electric potential and for a non-relativistic particle this yields the Schrödinger equation for the quantum wave function in a potential. The characteristic features of quantum mechanics such as interference in a double slit experiment are therefore accounted for by a suitable evolution law for classical probabilities. Quantum mechanics, quantum fermions and quantum field theory can be obtained from classical statistics.

This result has far reaching conceptual, and in the future perhaps also practical, consequences. Quantum systems turn out to be a particular class of classical statistical systems, characterized by particular properties of the evolution law. They are now embedded in a wider class of classical statistical systems with different evolution laws. This opens the possibility to describe phenomena as decoherence [1] or syncoherence [2] not only by embedding a quantum subsystem into a larger quantum system that includes the environment. Many characteristic features of decoherence and syncoherence may find a description in terms of probabilities for the subsystem alone, but with an effective evolution law different from the unitary quantum evolution. Precision tests of quantum mechanics can be described in a coherent conceptual framework for "zwitter"-particles whose properties interpolate continuously between a quantum particle and a classical particle [3]. On the other hand, a unified conceptual framework can also help to understand the observed quantum features of interference and tunneling in experiments with classical objects, as droplets of a liquid [4]. Finally, one may hope that the equivalent classical statistical description of quantum systems may permit to devise new calculational tools.

It may be worthwhile to specify clearly what we mean by a classical statistical system. The first ingredient is the classical probability distribution $\{p_\tau(t)\}$. The second one is the rule for the computation of expectation values of observables. Classical observables A have a fixed value A_τ in every state τ. These are the only possible outcomes of a measurement. The expectation value obeys $\langle A\rangle = \sum_\tau A_\tau p_\tau$. Third, one needs to specify a law for the time evolution of $\{p_\tau(t)\}$. In our model the Ising spins are the fundamental entities. We do not assume the existence of some underlying law that determines the time evolution of a given spin configuration. Our concept is "probabilistic realism", where the description of nature is genuinely probabilistic. The basic dynamical law should therefore be formulated as an evolution law for the probabilities. Conceptually, this is similar to a formulation of classical mechanics with the Liouville equation as basic dynamical equation. (Newton's laws for the motion of planets can, of course, be recovered from the Liouville equation for the particular ensemble which describes isolated bodies.) In classical mechanics we can also start from Newton's equation and derive the Liouville equation, interpreting the probabilities as a "lack of knowledge" of a deterministic system. This possibility is not available in our setting since we have no evolution law for individual spin configurations. (It is not formally excluded that an evolution law for individual spin configurations exists, similar to cellular automata [5]. It will, however, be difficult to find it and we do not need it.) Probabilities are not the lack of knowledge of an otherwise more complete system, but simply the basic concept for formulating laws of nature. On a fundamental level we only require that the evolution law preserves positivity and normalization of the probabilities and that it is causal in the sense that the distribution $\{p_\tau(t+\epsilon)\}$ at time $t+\epsilon$ can be computed from the distribution at time t. This typically leads to a differential evolution equation. In practice, the proposed evolution law will also preserve locality in space. (No signals travel faster than light.)

Our derivation of the Schrödinger equation in terms of an Ising type classical statistical ensemble and the description of the outcome of a double slit experiment only involves these three basic ingredients of a classical statistical system. In particular, we consider "interval observables", which take the value one if a particle is present in a certain region of space, and zero if it is absent. These observables are directly formulated in terms of the discrete classical observables for the Ising spins. The "discreteness" of individual particles is therefore a natural outcome. The continuous aspects of the wave function will be related to the description of probabilities by continuous functions. Particle-wave duality finds a natural explanation in our probabilistic setting. We implement the classical statistical rule that only values A_τ in the spectrum of an observable are found in measurements. This distinguishes our approach from Bohm's quantum mechanics [6] or other approaches where the wave function is a classical field, which has a continuous spectrum and is, in principle, observable.

The fourth ingredient for a classical statistical system is a rule for the computation of a correlation function that describes a sequence of ideal measurements. It should be emphasized that such a rule is an independent postulate for any probabilistic description. As every experimentalist knows there are many ways to perform a sequence of two measurements - "ideal" ones and "less ideal" ones. Correlations describe the outcome of a sequence of two measurements. They depend on the precise way in which measurements are performed. "Less ideal" measurements will yield correlations which differ from "ideal" measurements. The dependence of the correlation function $\langle BA\rangle$ on the precise way of performing a sequence of measurements for A and B is reflected mathematically in the many different possibilities to define a product structure between observables. A product structure assigns to two observables A and B a third observable C, namely $(B, A) \to C = B \circ A$. Translated to experiments, C corresponds to the "combined observable" whose expectation value is the correlation function. In general, the correlation function $\langle BA\rangle$ depends on the conditional probabilities $w(B_\rho|A_\tau)$ to find for B the value B_ρ if a measurement of A has found the value A_τ. These conditional probabilities depend on the precise way how the sequence of measurements is performed.

Since there are many ways to perform sequences of measurements, and many associated correlation functions, one wants to define the notion of "ideal measurements". Correlations differing from the ones for ideal measurements can then be attributed to "imperfect measurements". It is at this point that some thought is needed. For macroscopic classical statistical systems an idealized sequence of measurements is generally associated to the classical correlation function $\langle BA\rangle = \sum_\tau p_\tau B_\tau A_\tau$. While well adapted to large macroscopic systems, as two temperature measurements at different positions in a liquid, it often fails for isolated microscopic subsystems as an isolated atom. For an ideal measurement of properties of a subsystem the correlation function has to be computable from the information available for the description of the subsystem. This requirement often conflicts with the classical correlation function [2]. Classical correlation functions often involve properties of the subsystem as well as its environment (apparatus). The appropriate correlation functions for ideal measurements differ from the classical correlation function in these cases.

We emphasize that the issue of correlations is not relevant for the description of the time evolution of expectation values of observables. Thus our description of interference, tunneling or entanglement in terms of the time evolution of a "classical" probability distribution does not involve any choice of the correlation function. This choice matters, however, if we want to describe the outcome of sequences of ideal measurements. It is important for an understanding why Bell's inequalities [7, 8] are violated for the correlations in quantum systems [2].

The present work is based on a series of papers by the author [9], [10], [3], [2] and has substantial overlap with ref. [9].

2. Generalized Ising model

We will consider an Ising-model type system for discrete Ising-spins, $s_\gamma(\vec{x}) = \pm 1$, with $\vec{x}$ points on a suitable three-dimensional lattice and γ denoting different "species" of Ising-spins, $\gamma = 1 \dots N_s$. For $N_s = 4$ our system will be equivalent to a quantum field theory for Majorana or Weyl spinors, while for $N_s = 8$ we will describe Dirac spinors. The states τ are sequences or configurations of Ising-spins, $\tau = \{s_\gamma(\vec{x})\}$. Instead of Ising-spins, we will actually use occupation numbers or bits $n_\gamma(\vec{x}) = \big(s_\gamma(\vec{x}) + 1\big)/2$, such that the states τ describe bit sequences of numbers $n_\gamma(\vec{x}) = 0, 1$, $\tau = \big\{n_\gamma(\vec{x})\big\}$. We will consider a setting with $L^3/8$ lattice points (L even) such that the number of states is $2^{(N_s L^3/8)}$. The continuum limit $L \to \infty$ is taken at the end.

The sequences or configurations of occupation numbers $\big\{n_\gamma(\vec{x})\big\}$ show already a strong analogy to the basis states of a quantum theory for an arbitrary number of fermions, formulated in the occupation number basis for positions. We will exploit this analogy in order to formulate a fundamental "evolution law" for the time evolution of the probability distribution $p_\tau(t)$, such that our system describes a relativistic quantum field theory for fermions.

2.1. Classical wave function

In order to define our model we need to specify an evolution law for the time dependence of the probability distribution $\{p_\tau(t)\}$. A useful concept for this purpose is the classical wave function $\{q_\tau(t)\}$ [3], defined by

$$q_\tau(t) = s_\tau(t)\sqrt{p_\tau(t)}\ ,\ p_\tau(t) = q_\tau^2(t)\ ,\ s_\tau(t) = \pm 1. \tag{1}$$

This is a real function which is given by the square roots of the probabilities up to signs $s_\tau(t)$. In terms of the wave function the expectation values of classical observables obey the "quantum rule"

$$\langle A(t)\rangle = \langle q(t)\hat{A}q(t)\rangle\ ,\ \langle A(t)\rangle = \sum_\tau p_\tau(t)A_\tau, \tag{2}$$

with $\hat{A}$ a diagonal operator $(\hat{A})_{\tau\rho} = A_\tau\delta_{\tau\rho}$. The signs s_τ do not appear in eq. (2). At this stage the use of a wave function seems redundant. Nevertheless, it permits the use of the quantum formalism for a description of a classical statistical system.

Consistent evolution laws correspond to rotations of the vector $q_\tau(t)$,

$$q_\tau(t) = \sum_\rho R_{\tau\rho}(t,t')q_\rho(t')\ ,\ \sum_\rho R_{\tau\rho}(t,t')R_{\sigma\rho}(t,t') = \delta_{\tau\sigma}. \tag{3}$$

The positivity of $p_\tau = q_\tau^2$ is automatic, and the normalization of the distribution remains preserved provided it was normalized for some initial time t_{in}, since the length of a vector or $\Sigma_\tau q_\tau^2 = \Sigma_\tau p_\tau$ is preserved by rotations. We will specify a linear evolution for the wave function. In other words, we will consider rotation matrices $R_{\tau\rho}$ that are independent of the wave function. The classical wave function specifies the probability distribution uniquely. The specification of an evolution law for the wave function therefore defines the dynamics of the classical statistical system completely.

Quantum wave functions are usually defined in a complex Hilbert space. Indeed, many characteristic quantum features are closely associated to the "physics of phases". In our setting we will define a complex quantum wave function by introducing a complex structure in the real space spanned by $\{q_\tau\}$. The $2^{(N_sL^3/8)}$ real components of the vector $\{q_\tau\}$ can then be associated to a complex vector with dimension $2^{(N_sL^3/8)-1}$. There is no conceptual difference between the classical and quantum wave function in our setting. However, more of the information contained in a complex wave function $\psi_\alpha = \psi_{R,\alpha} + i\psi_{I,\alpha}$ is now directly associated to classical probabilities. In quantum mechanics one usually distinguishes between the probabilities $p_\alpha = |\psi_\alpha|^2$ and the phases, whereas in our approach both positive numbers $\psi_{R,\alpha}^2$ and $\psi_{I,\alpha}^2$ will be associated to classical probabilities p_τ.

The only information contained in the classical wave function which goes beyond the set of probabilities $\{p_\tau\}$ concerns the signs $\{s_\tau\}$. The choice of $\{s_\tau\}$ does not affect the expectation values of classical observables and may therefore be considered as a type of gauge choice. A linear evolution law (3) involves, however, directly these signs and one may wonder if the evolution of the classical wave function involves physics beyond the classical statistical setting, thereby "smuggling in" quantum mechanics in some hidden way. This is actually not the case. For computing $q(t+\epsilon)$ from a given wave function at time t according to eq. (3), we only have for every τ the choice between the two values $s_\tau(t) = \pm 1$. Only one of them is consistent with the normalization of the probabilities. This fixes for all t the sign distribution $\{s_\tau(t)\}$ in terms of some initial distribution" $\{s_\tau(t_0)\}$ for $t = t_0$. For a differential evolution equation it is easy to see [3] that s_τ is constant as long as $p_\tau > 0$. Possible sign jumps of s_τ can occur for times where $p_\tau(t) = 0$, and are decided by the normalization of the probability distribution.

One may visualize this issue by a simple periodic time evolution for a two state system with $\tau = 0, 1$ [11]. A periodic change between the probabilities p_0 and $p_1 = 1 - p_0$ can be described by the second order evolution equation $\partial_t^2 p_0 = 2\omega^2(1 - 2p_0)$, with solution $p_0 = \cos^2(\omega t + \alpha)$. In terms of the wave function this becomes a first order equation

$$\partial_t \begin{pmatrix} q_0 \\ q_1 \end{pmatrix} = \omega \begin{pmatrix} 0, 1 \\ -1, 0 \end{pmatrix} \begin{pmatrix} q_0 \\ q_1 \end{pmatrix}. \tag{4}$$

The probability distribution is fixed for all t by the initial values $p_0(t_0)$ and $\partial_t p_0(t_0)$ for the second order evolution equation and equivalently by $p_0(t_0), s_0(t_0)s_1(t_0)$ for eq. (4). Only the relative sign $s_0 s_1$ matters, such that the needed "initial data" are the same for both descriptions. (It is actually possible to compute the signs $s_\tau(t)$ from "all time probabilities" [11].) For an expression of the relevant information contained in the wave function it is important that the latter is real. This contrasts with the discussion of classical mechanics in a Hilbert space by Koopman and von Neumann [12], where phases of a complex wave function appear as new degrees of freedom beyond the probability distribution.

2.2. Grassmann wave function

A second useful formal tool for the description of our Ising type model is a map to an equivalent description in terms of Grassmann variables. This will make the close connection to fermions most transparent [10]. The Grassmann formulation relies on the isomorphism between states τ and the basis elements g_τ of a real Grassmann algebra that can be constructed from the Grassmann variables $\psi_\gamma(x)$. Each basis element g_τ is a product of Grassmann variables $g_\tau = \psi_{\gamma 1}(x_1)\psi_{\gamma 2}(x_2)\ldots$ which is ordered in some convenient way. If a Grassmann element g_τ contains a given variable $\psi_\gamma(x)$ we put the number $n_\gamma(x)$ in the sequence τ to 0, while we take $n_\gamma(x) = 1$ if the variable $\psi_\gamma(x)$ does not appear in the product.

An arbitrary element g of the Grassmann algebra can be expanded in terms of the basis elements

$$g = \sum_\tau q_\tau g_\tau. \tag{5}$$

A time dependent wave function $\{q_\tau(t)\}$ can therefore be associated to a time dependent element of the Grassmann algebra $g(t)$, provided the coefficients $q_\tau(t)$ are real and normalized according to $\sum_\tau q_\tau^2(t) = 1$. We will formulate the fundamental evolution law as an evolution law for the "Grassmann wave function" $g(t)$.

For this purpose we will formulate in the next section a Grassmann functional integral for a quantum field theory of massless Majorana fermions. It will involve $N_s L^3/8$ Grassmann variables $\psi_\gamma(t, x)$ for every discrete time point t. The Grassmann wave function $g(t)$ and the fundamental evolution law can be extracted from this functional integral. This will establish a map between a quantum field theory for fermions and Ising type classical statistical ensembles with dynamics specified by an appropriate evolution law. From the point of view of the Ising type classical statistical ensemble the functional integral is a pure technical device for the specification of an evolution law for the classical wave function $\{q_\tau(t)\}$ and probability distribution $\{p_\tau(t)\}$. Conceptually, we could start equivalently by specifying the matrix $R_{\tau\rho}(t, t - \epsilon)$ in eq. (3). In practice, the functional integral formulation is much simpler since it provides for an easy realization of appropriate symmetries of the evolution law, as Lorentz symmetry in our particular model.

3. Evolution law, action and functional integral

3.1. Action and functional integral

We start with the action S and Lagrangian $L(t)$ for the regularized quantum field theory of massless Majorana spinors ($N_s = 4$)

$$S = \sum_{t=t_{in}}^{t_f-\epsilon} L(t) \ , \ L(t) = \sum_x \psi_\gamma(t,x) B_\gamma(t+\epsilon,x). \tag{6}$$

Here we define

$$B_\gamma(t+\epsilon,x) = \prod_j \Big(\sum_{v_j=\pm1} \Big) Y_{\gamma\delta}(\{v\}) \psi_\delta(t+\epsilon, x_k + v_k\Delta), \tag{7}$$

with

$$Y_{\gamma\delta}(\{v\}) = \frac{1}{8}\Big[1 - \sum_k (v_k + w_k \tilde{I}) T_k - v_1 v_2 v_3 \tilde{I}\Big]_{\gamma\delta}. \tag{8}$$

Thus $B_\gamma(t+\epsilon,x)$ involves a linear combination of Grassmann variables at lattice sites which are diagonal neighbors of x, corresponding to corners of a cube with basis length 2Δ, with x being at its center. The sums extend over $j,k = 1\ldots3$ and each corner corresponds to a particular combination of the three signs $v_j = \pm1$. We define w_k by $w_1 = -v_2 v_3, w_2 = v_1 v_3, w_3 = -v_1 v_2$. Eqs. (6)-(8) and the following imply a summation over repeated indices $\gamma, \delta = 1\ldots4$.

We have also introduced the three real symmetric 4×4 matrices T_k as

$$T_1 = \begin{pmatrix} 0, 1 \\ 1, 0 \end{pmatrix} , \ T_2 = \begin{pmatrix} 0, c \\ -c, 0 \end{pmatrix} , \ T_3 = \begin{pmatrix} 1, & 0 \\ 0, & -1 \end{pmatrix} , \ T_k^T = T_k, \tag{9}$$

where $c = i\tau_2$, τ_k are Pauli matrices and 1 stands for the unit 2×2 matrix. The product of these matrices yields the real antisymmetric 4×4 matrix

$$\tilde{I} = -\begin{pmatrix} c, 0 \\ 0, c \end{pmatrix} = T_1 T_2 T_3 \ , \ \tilde{I}^T = -\tilde{I}. \tag{10}$$

The sum in eq. (6) extends over discrete time points t_n, with $\int_t = \epsilon \sum_t = \epsilon \sum_n$, $t_{n+1} - t_n = \epsilon, n \in \mathbb{Z}$, $t_{in} \le t_n \le t_f$. The time-continuum limit is taken as $\epsilon \to 0$ for fixed t_{in}, t_f. Similarly, we sum in eq. (6) over points x of a lattice. For this purpose we consider for every given t a cubic lattice with lattice distance 2Δ and $\int_x = 8\Delta^3 \sum_x$. We may take $\epsilon = \Delta$ such that the space-time lattice points belong to a hypercubic bcc lattice with distance 2Δ between nearest neighbors, which we call the "fundamental lattice". For even $t = 2n\epsilon$ the space lattice points are even, $x_k = 2m_k'\Delta$, with $n, m_k' \in \mathbb{Z}$. This will be called the even sublattice. The odd sublattice consists of the odd time points $t = (2n+1)\epsilon$ for which the space points are also odd, $x_k = (2m_k'+1)\Delta$. The action (6) involves indeed only Grassmann variables for points on the fundamental lattice. For even t eq. (6) involves Grassmann variables $\psi_\gamma(t,x)$ living on the even sublattice, while the combination $B_\gamma(t+\epsilon,x)$ lives on the odd sublattice. For odd t the role of the sublattices is exchanged, now with ψ on the odd and B on the even sublattice. This construction eliminates "lattice doublers" - a more detailed discussion of the lattice implementation can be found in ref. [9].

The action (6) is an element of a real Grassmann algebra - all coefficients multiplying the Grassmann variables $\psi_\gamma(t,x)$ are real. Within the Grassmann algebra the operation of transposition amounts to a total reordering of all Grassmann variables. The action (6) is antisymmetric under this operation, $S^T = -S$. If we define formally the Minkowski action $S_M = iS$ the latter is hermitean, $S_M = S_M^\dagger$ since $S_M^* = -S_M$.

The action (6) is Lorentz-symmetric in the continuum limit $\epsilon = \Delta, \Delta \to 0$. In the continuum limit we write the action as

$$S = \int_{t,x} \{\psi_\gamma \partial_t \psi_\gamma - \psi_\gamma (T_k)_{\gamma\delta} \partial_k \psi_\delta\}. \tag{11}$$

It involves now four Grassmann functions $\psi_\gamma(t,x), \gamma = 1 \dots 4, x = (x_1, x_2, x_3)$. The integral extends over three dimensional space and time, with $\partial_t = \partial/\partial t$ and $\partial_k = \partial/\partial x_k$.

The Lorentz invariance of the action (11) is most easily established by employing the real matrices

$$\gamma^0 = \begin{pmatrix} 0 & , & \tau_1 \\ -\tau_1 & , & 0 \end{pmatrix} \; , \; \gamma^k = -\gamma^0 T_k, \tag{12}$$

such that

$$S = -\int_{t,x} \bar{\psi} \gamma^\mu \partial_\mu \psi \; , \; \bar{\psi} = \psi^T \gamma^0, \tag{13}$$

where $\mu = (0,k)$ and $\partial_0 = \partial_t$. The real 4×4 Dirac matrices γ^μ obey the Clifford algebra, $\{\gamma^\mu, \gamma^\nu\} = 2\eta^{\mu\nu}$, with signature of the metric given by $\eta_{\mu\nu} = diag(-1,1,1,1)$.

The functional integral is defined by the partition function

$$Z = \int \mathcal{D}\psi \bar{g}_f[\psi(t_f)] e^{-S} g_{in}[\psi(t_{in})] \; , \; \int \mathcal{D}\psi = \prod_{t,x} \int (d\psi_4(t,x) \dots d\psi_1(t,x)). \tag{14}$$

The boundary terms g_{in} and $\bar{g}_f$ only depend on the Grassmann variables $\psi(t_{in})$ and $\psi(t_f)$, respectively. The boundary terms $\bar{g}_f$ and g_{in} are related to each other [9], such that the functional integral (14) is fully specified by the choice of g_{in}. In turn, g_{in} is given by the initial values for the classical wave function, cf. eq. (5).

3.2. Grassmann wave function from functional integral

In this section we compute for the functional integral (14) a Grassmann wave function $g(t)$, which is an element of the Grassmann algebra constructed from the Grassmann variables $\psi_\gamma(t,x)$ at given t. The functional integral (14) involves variables for arbitrary time points t_n. In order to construct a wave function $g(t)$ which only refers to a particular time t we have to "integrate out" the information referring to other time points $t' \neq t$ [10, 13]. This can be done by decomposing the action (6)

$$S = S_< + S_> \; , \; S_< = \sum_{t'<t} L(t') \; , \; S_> = \sum_{t' \geq t} L(t'). \tag{15}$$

The wave function $g(t)$ obtains now by integrating out all Grassmann variables for $t' < t$

$$g(t) = \int \mathcal{D}\psi(t' < t) e^{-S_<} g_{in}. \tag{16}$$

As a consequence of its definition (16) the Grassmann wave function obeys the time evolution

$$g(t+\epsilon) = \int \mathcal{D}\psi(t) e^{-L(t)} g(t) = \int \mathcal{D}\psi(t) \exp\big\{-\sum_x \sum_\gamma \psi_\gamma(t,x) B_\gamma(t+\epsilon,x)\big\} g(t). \tag{17}$$

This determines $\{q_\tau(t+\epsilon)\}$ in terms of $\{q_\tau(t)\}$. Thus the action (6)-(8) specifies the dynamics how the probability distribution $\{p_\tau(t)\}$ evolves in time. The particular dynamics of a given model is determined by the form of B_γ in eq. (8). In a similar way we can relate $g(t+2\epsilon)$ to

$g(t)$. Performing the necessary Grassmann integrals over $\psi(t)$ and $\psi(t+\epsilon)$ one obtains [9] for $\epsilon \to 0$ the differential evolution equation

$$\partial_t g(t) = \frac{1}{2\epsilon}\left[g(t+2\epsilon) - g(t)\right] = \mathcal{K}g(t) = \sum_\tau q_\tau(t)\mathcal{K}g_\tau = \sum_{\tau,\rho} q_\tau(t) g_\rho K_{\rho\tau} = \sum_\tau \partial_t q_\tau(t) g_\tau, \quad (18)$$

with Grassmann evolution generator $\mathcal{K}$

$$\mathcal{K} = \sum_x \frac{\partial}{\partial\psi_\gamma(x)}(T_k)_{\gamma\delta}\partial_k\psi_\delta(x). \quad (19)$$

From eq. (18) we can directly extract the evolution equation for the classical wave function (3) in a continuous version,

$$\partial_t q_\tau(t) = \sum_\rho K_{\tau\rho} q_\rho(t) \ , \ K_{\rho\tau} = \int \mathcal{D}\psi \tilde{g}_\rho \mathcal{K} g_\tau. \quad (20)$$

We identify this evolution equation with the generalized Schrödinger equation for a quantum wave function for the special case of a real wave function and purely imaginary and hermitean Hamiltonian $H = i\hbar K$. It specifies the evolution of the probabilities for the Ising-type classical statistical ensemble.

4. Majorana fermions

4.1. Conserved quantities

For a physical interpretation of the evolution equation (18), (19) it is useful to look for conserved quantities. They correspond to time independent expectation values of classical statistical observables. For example, we may consider the observable measuring the occupation number $N_\gamma(x)$ of the bit γ located at x. The spectrum of possible outcomes of measurements consists of values 1 or 0, depending if a given state $\tau = [n_\gamma(x)]$ has this particular bit occupied or empty.

For the Grassmann basis element g_τ associated to τ one finds $N_\gamma(x) = 0$ if g_τ contains a factor $\psi_\gamma(x)$, and $N_\gamma(x) = 1$ otherwise. We can associate to this observable a Grassmann operator $\mathcal{N}_\gamma(x)$ obeying (no summation here)

$$\mathcal{N}_\gamma(x) g_\tau = \big(N_\gamma(x)\big)_\tau g_\tau \ , \ \mathcal{N}_\gamma(x) = \frac{\partial}{\partial\psi_\gamma(x)}\psi_\gamma(x). \quad (21)$$

Two occupation number operators $\mathcal{N}_{\gamma_1}(x_1)$ and $\mathcal{N}_{\gamma_2}(x_2)$ commute.

In general, we may associate to each classical observable A a diagonal quantum operator $\hat{A}$ acting on the wave function, defined by $(\hat{A}q)_\tau = A_\tau q_\tau$ with $\hat{A}$ a diagonal operator $\hat{A}_{\tau\rho} = A_\tau \delta_{\tau\rho}$. In the Grassmann formulation one uses the associated Grassmann operator $\mathcal{A}$ obeying

$$\mathcal{A} g_\tau = A_\tau g_\tau \ , \ \langle A \rangle = \int \mathcal{D}\psi \tilde{g} \mathcal{A} g. \quad (22)$$

Here $\tilde{g}$ is conjugate to g, i.e. for $g = \sum_\tau q_\tau g_\tau$ one has $\tilde{g} = \sum_\tau q_\tau \tilde{g}_\tau$. The conjugate basis elements of the Grassmann algebra $\tilde{g}_\tau$ are defined [10] by the relation

$$\tilde{g}_\tau g_\tau = |0\rangle = \prod_x \prod_\gamma \psi_\gamma(x) \ , \ \int \mathcal{D}\psi |0\rangle = 1, \quad (23)$$

(no sum over τ) and the requirement that no variable $\psi_\gamma(x)$ appears both in $\tilde{g}_\tau$ and g_τ.

Using $\partial_t q = Kq$ (20) and $\partial_t g = \mathcal{K}g$ (18) the time evolution of the expectation value is expressed in terms of the commutators

$$\partial_t \langle A \rangle = \langle q[\hat{A}, K]q \rangle = \int \mathcal{D}\psi \tilde{g}[\mathcal{A}, \mathcal{K}]g. \quad (24)$$

Conserved quantities are represented by Grassmann operators that commute with $\mathcal{K}$, $[\mathcal{A}, \mathcal{K}] = 0$.

4.2. Particle states

Our system admits a conserved particle number, corresponding to the sum over all occupation numbers $N_\gamma(x)$ in the lattice, or to the Grassmann operator $\mathcal{N}$,

$$\mathcal{N} = \int_y \frac{\partial}{\partial\psi_\gamma(y)}\psi_\gamma(y) \ , \ [\mathcal{N}, \mathcal{K}] = 0. \tag{25}$$

We can decompose an arbitrary Grassmann element into eigenstates of $\mathcal{N}$, $g = \Sigma_m A_m g_m$, $\mathcal{N} g_m = m g_m$. For an eigenstate g_m the probabilities in the Ising-type classical statistical ensemble differ from zero only if the total number of "occupied bits" equals precisely m. The time evolution does not mix sectors with different particle number m, such that the coefficients A_m are time independent, $\partial_t g = \Sigma_m A_m \partial_t g_m$, $\partial_t g_m = \mathcal{K} g_m$. We can restrict our discussion to eigenstates of $\mathcal{N}$.

A static vacuum state g_0 with a fixed particle number m_0 obeys

$$\mathcal{K} g_0 = 0 \ , \ \mathcal{N} g_0 = m_0 g_0 \ , \ \int \mathcal{D}\psi \tilde{g}_0 g_0 = 1. \tag{26}$$

An example for a possible vacuum state is the totally empty state $g_0 = |0\rangle$ (eq. (23)). Another example is the totally occupied state with $g_0 = 1$, $m_0 = B = N_s L^3/8$. There are many possible vacuum states, for example a half filled state with $m_0 = B/2$.

We next define creation and and annihilation operators $a^\dagger_\gamma(x)$, $a_\gamma(x)$ as

$$a^\dagger_\gamma(x) g = \frac{\partial}{\partial\psi_\gamma(x)} g \ , \ a_\gamma(x) g = \psi_\gamma(x) g \ , \ \{a^\dagger_\gamma(x), \, a_\epsilon(y)\} = \delta_{\gamma\epsilon}\delta(x-y) \ , \ \mathcal{N} = \int_x a^\dagger_\gamma(x) a_\gamma(x). \tag{27}$$

Acting with the creation operator on the vacuum produces one-particle states

$$g_1(t) = \int_x q_\gamma(t,x) a^\dagger_\gamma(x) g_0 = \mathcal{G}_1 g_0 \ , \ (\mathcal{N} - m_0) g_1 = g_1. \tag{28}$$

The time evolution of the one-particle wave function q_γ is given by

$$\partial_t g_1 = \mathcal{K} g_1 = \int_x (\partial_t q \frac{\partial}{\partial\psi}) g_0 = \int_x q \left[\mathcal{K}, \frac{\partial}{\partial\psi}\right] g_0. \tag{29}$$

Inserting eq. (19) yields the Dirac equations for a real wave function

$$\gamma^\mu \partial_\mu q = 0. \tag{30}$$

We emphasize that the Dirac equation follows for arbitrary static states g_0 which obey $\mathcal{K} g_0 = 0$. Since there is no mass term and the Dirac-matrices as well as the wave function are real, eq. (30) describes the time evolution of the wave function for a single massless Majorana fermion.

States with of n fermions can be constructed by applying n creation operators $a^\dagger_\gamma(x)$ on the vacuum. For example, the two-fermion state obeys

$$g_2(t) = \frac{1}{\sqrt{2}} \int_{x,y} q_{\gamma\epsilon}(t,x,y) a^\dagger_\gamma(x) a^\dagger_\epsilon(y) g_0. \tag{31}$$

Since the creation operators anticommute the two-particle wave function is antisymmetric, as appropriate for fermions. From there we recover the standard interference rules for identical fermions. The linearity of the evolution equation in q is crucial in this respect. It guarantees the superposition principle for the wave function, and therefore interference effects for the probabilities.

5. Dirac fermions in electromagnetic fields

Dirac spinors can be composed of two different Majorana spinors with equal mass ($N_s = 8, B = L^3$). We extend our model by introducing in the action a mass term $\sim m$ and a coupling $\sim e$ to an external electromagnetic potential $A_\mu(x)$, (with suppressed spinor indices)

$$\begin{aligned} S &= \int_{t,x} \{\psi_1(\partial_t - T_k\partial_k + m\gamma^0\tilde{I})\psi_1 + \psi_2(\partial_t - T_k\partial_k + m\gamma^0\tilde{I})\psi_2\} \\ &\quad -e\int_{t,x} \{\psi_1(A_0 - A_kT_k)\psi_2 - \psi_2(A_0 - A_kT_k)\psi_1\}. \end{aligned} \tag{32}$$

We introduce a complex structure by defining the four-component complex Grassmann variables $\psi_D = \psi_1 + i\psi_2$, $\psi_D^* = \psi_1 - i\psi_2$. In terms of ψ_D the action (32) takes the familiar form

$$S = -\int_{t,x} \bar{\psi}_D(\gamma^\mu\partial_\mu - m\tilde{I})\psi_D - ie\int_{t,x} \bar{\psi}_D\gamma^\mu A_\mu\psi_D \ , \ \bar{\psi}_D = \psi_D^\dagger\gamma^0. \tag{33}$$

5.1. Evolution equation for Dirac fermions

For the action (32) the Grassmann evolution equation $\partial_t g = \mathcal{K}g$ involves $\mathcal{K} = \mathcal{K}_0 + \mathcal{K}_m + \mathcal{K}_A$

$$\mathcal{K}_0 + \mathcal{K}_m = \int_x \sum_{a=1,2} \frac{\partial}{\partial\psi_a(x)}(T_k\partial_k - m\gamma^0\tilde{I})\psi_a(x), \tag{34}$$

$$\mathcal{K}_A = e\int_x \left[\frac{\partial}{\partial\psi_1(x)}(A_0(x) - A_k(x)T_k)\psi_2(x) - \frac{\partial}{\partial\psi_2(x)}(A_0(x) - A_k(x)T_k)\psi_1(x)\right].$$

It describes the dynamics for an arbitrary number of charged relativistic fermions (and their antiparticles) in external electromagnetic fields.

The evolution equation for the classical wave function (20), and therefore for the probability distribution, obtains again from eq. (18). In summary, we have formulated a time evolution equation for a classical wave function $q_\tau(t)$ which describes an arbitrary number of charged electrons in external electric and magnetic fields. Its restriction to one-particle states yields the relativistic Dirac equation. As usual, a non-relativistic approximation will yield the familiar Schrödinger equation for a particle in a potential or moving in external magnetic fields.

5.2. Classical observables

Classical observables can be constructed from linear combinations of products of occupation numbers. We combine the index $a = 1, 2$ for the two Majorana spinors ψ_a with the index γ into a common index $\epsilon = (\gamma, a), \epsilon = 1 \ldots 8$. The Grassmann operators corresponding to the occupation numbers $N_\epsilon(x)$ obey eq. (21). A local particle number can be defined as

$$\mathcal{N}(x) = \sum_\epsilon \mathcal{N}_\epsilon(x) \ , \ [\mathcal{N}(x), \mathcal{K}] = (T_k)_{\eta\alpha}\partial_k\mathcal{M}_{\alpha\eta} \ , \ \mathcal{M}_{\epsilon\eta}(x) = \frac{\partial}{\partial\psi_\epsilon(x)}\psi_\eta(x). \tag{35}$$

According to eq. (24) the time evolution of the classical mean local particle number obeys

$$\partial_t\langle\mathcal{N}(x)\rangle = (T_k)_{\eta\alpha}\partial_k\langle\mathcal{M}_{\alpha\eta}\rangle. \tag{36}$$

This involves the expectation values of "off-diagonal" operators $\mathcal{M}_{\alpha\eta}$. Products of operators $\mathcal{M}_{\alpha\eta}$ can be defined by the non-commutative quantum product and we see here a first appearance of non-commutative structures. The total particle number $\mathcal{N} = \int_x \mathcal{N}(x)$, $[\mathcal{N}, \mathcal{K}] = 0$, is conserved.

6. Quantum mechanics for particle in a potential

6.1. Dirac equation

We next concentrate on one-particle states, described by Grassmann elements

$$g_1(t) = \int_x \left(q_{1,\gamma}(t,x) \frac{\partial}{\partial \psi_{1,\gamma}(x)} + q_{2,\gamma}(t,x) \frac{\partial}{\partial \psi_{2,\gamma}(x)} \right) g_0. \tag{37}$$

Here g_0 is some arbitrary static "vacuum state". For $\mathcal{K} g_0 = 0$ we find for the one-particle wave function the Dirac equation

$$\gamma^\mu (\partial_\mu + ieA_\mu)\varphi_D = im\bar{\gamma}\varphi_D \ , \ \varphi_D = q_1 + iq_2. \tag{38}$$

This equation holds for an arbitrary representation of the Dirac matrices γ^μ. The matrix $\bar{\gamma}$ (often called γ^5) obeys $\bar{\gamma} = -i\gamma^0\gamma^1\gamma^2\gamma^3$. In our conventions [14] the Dirac equation obeys parity, charge conjugation and time reversal symmetry.

The derivation of the Dirac equation for a one-particle state (37) has only used the general evolution equation for Grassmann elements $g(t)$ and the condition $\mathcal{K} g_0 = 0$. It therefore describes the dynamics for a very extended family of classical probability distributions or classical wave functions. Indeed, it is sufficient that g_0 is an arbitrary static state (not necessarily a priori with a fixed particle number). This reflects the physical property that isolated one-particle states can occur under a wide variety of circumstances, and that for sufficient isolation the properties of the environment do not matter for the dynamics of the isolated particle. In this context we emphasize, however, that our model only describes external electromagnetic fields while we do not account for the fields generated by the particles that may be present in the environment. In this sense our setting describes "real physics" only in a situation where the electromagnetic fields generated by all present particles can be approximated by a "mean field" that is independent of individual particle positions. This is precisely the setting of standard one-particle quantum mechanics.

6.2. Particle observables

Let us consider the expectation value of the local occupation number $\mathcal{N}(x)$ defined by eq. (35). For $g_0 = |0\rangle$ its expectation value obeys the simple relation

$$\langle \mathcal{N}(x) \rangle = \sum_\epsilon q_\epsilon^2(t,x) \ , \ \int_x \sum_\epsilon \left(q_\epsilon(x)\right)^2 = \int_x \varphi_D^\dagger(x)\varphi_D(x) = 1. \tag{39}$$

The normalization (39) is preserved by the unitary time evolution of the one-particle wave function. Indeed, we can write the Dirac equation (38) as a Schrödinger-type equation $i\hbar\partial_t\varphi_D = H\varphi_D$, with hermitean Hamiltonian $H = i\hbar T_k \partial_k + \hbar m \gamma^0 \bar{\gamma} + \hbar e(A_0 - T_k A_k)$.

For a pure one-particle state the expectation value $\langle N(x) \rangle$ can be interpreted in a natural way as the probability density to find the particle at the position x. This is precisely the standard interpretation of $\psi_D^\dagger(x)\psi_D(x)$ in one-particle quantum mechanics,

$$w(x) = \langle N(x) \rangle = \varphi_D^\dagger(x)\varphi_D(x) \ , \ \int_x w(x) = 1. \tag{40}$$

Since $N(x)$ is a classical observable, we can define the position of the particle as a classical observable, and similarly for functions of X,

$$X = \int_x x N(x) \ , \ f(X) = \int_x f(x) N(x). \tag{41}$$

The expectation value in classical statistics coincides with the standard quantum mechanics rule

$$\langle X\rangle = \langle\int_x xN(x)\rangle = \int_x xw(x) = \int_x \varphi_D^\dagger(x)x\varphi_D(x)\ ,\ \langle f(X)\rangle = \int_x \varphi_D^\dagger(x)f(x)\varphi_D(x). \tag{42}$$

In particular, one obtains the same formula for the dispersion $\langle X_kX_k\rangle - \langle X_k\rangle\langle X_k\rangle$ as in quantum mechanics. We conclude that measurements of the position of a particle, or more generally the distribution of positions in an ensemble of one-particle states, can be described equivalently in an Ising type a classical statistical ensemble with classical observables, or in quantum mechanics. The complete time evolution of the distribution of positions is identical in both descriptions. This covers, in particular, the characteristic quantum interference in a double slit experiment. The time evolution (20), (38) for the classical wave function and associated classical probability distribution $\{p_\tau\}$ produces for one-particle states exactly the quantum mechanical interference pattern.

6.3. Schrödinger equation

Standard quantum mechanics for an electron in a potential is recovered from the non-relativistic approximation to the Dirac equation. This is well known, and we sketch here for completeness only the case $A_k = 0$. The nonrelativistic approximation becomes valid if eA_0 and $iT_k\partial_k$ are small compared to m. Since $(\gamma^0\bar{\gamma})^2 = 1$, it is convenient to choose a new basis where $\gamma^0\bar{\gamma} = diag(1,1,-1,-1)$. With $V(x) = \hbar eA_0(x)$ and $M = \hbar m$ the Hamiltonian takes the form

$$H = \begin{pmatrix} M & , & \sigma_k p_k \\ \sigma_k^\dagger p_k & , & -M \end{pmatrix} + V(x)\ ,\ p_k = -i\hbar\partial_k, \tag{43}$$

with matrices $\sigma_1 = -1\ ,\ \sigma_2 = i\tau_2\ ,\ \sigma_3 = i\tau_3$. The appearance of $\hbar$ only fixes the units for p_k, M and $V(x)$.

We next decompose φ_D into two-component wave functions, $\varphi_D^T = (\chi^T, \rho^T)$. For the non-relativistic electron we consider the approximate solution $\rho = A\chi$, where $A = \sigma_k^\dagger p_k/(2M)$ is determined by requiring in leading order $\partial_t\rho = A\partial_t\chi$. Insertion into eq. (38) yields for $\psi = \exp(iMt/\hbar)\chi$ the standard Schrödinger equation for a particle in a potential V,

$$i\hbar\partial_t\psi = \left(\frac{p_kp_k}{2M} + V\right)\psi. \tag{44}$$

All quantum mechanical phenomena extracted from solutions of the Schrödinger equation are described by our time evolution equation for a classical statistical ensemble of Ising-spins on a lattice. For a potential realizing a double-slit situation the standard interference pattern will appear behind the slits. This holds provided that the initial state at some time t_0 corresponds to a one-particle state describing a particle moving towards the slits. Interference is realized by classical probabilities.

For the choice $g_0 = |0\rangle$ the evolution law for the classical probabilities associated to a one particle state is rather simple. The only non-vanishing probabilities $p_\epsilon(x) = q_\epsilon^2(x)$ occur if precisely one lattice site is occupied with a given species ϵ. The evolution of $q_\epsilon(t,x)$ obeys eq. (38). We recall, however, that the same physical situation arises for arbitrary static g_0. The time evolution of probabilities can then get rather complicated in the Ising-spin picture whereas it remains simple in the fermionic picture.

6.4. Particle-wave duality

The discreteness of measurement values in quantum mechanics can be traced back to the discrete occupation numbers of the Ising-type model. In quantum mechanics we may define an "interval observable" $J_{\mathcal{R}}$ by a function

$$J_{\mathcal{R}}(x) = \left\{ \begin{array}{ll} 1 & \text{if } x \in \mathcal{R} \\ 0 & \text{otherwise} \end{array} \right) \; , \; J_{\mathcal{R}}^2 = J_{\mathcal{R}}, \tag{45}$$

such that its spectrum consists of the discrete values 0 and 1. According to the rules of quantum mechanics, the possible outcomes of a measurement of $J_{\mathcal{R}}$ are 0 or 1. The interpretation in quantum mechanics is simple: either the particle is within the region (interval) $J_{\mathcal{R}}$, in which case the measurement value $J_{\mathcal{R}} = 1$ will be found, or it is outside this region, and $J_{\mathcal{R}} = 0$ will be found. Particles are discrete objects - they are either inside or outside an interval.

In our classical statistical Ising-type setting $J_{\mathcal{R}}$ is a classical observable, given by

$$J_{\mathcal{R}} = \int_{\mathcal{R}} N(x) = \sum_{\mathcal{R}} N_L(x). \tag{46}$$

The sum $\sum_{\mathcal{R}}$ extends over all lattice points within the region $\mathcal{R}$. The classical observable $N_L(x)$ corresponds to a normalization of occupation numbers for the discrete lattice where $\left(N_{L,\epsilon}\right)_\tau = 0, 1$. The sum over species $N_L(x) = \sum\limits_{\epsilon} N_{L,\epsilon}(x)$, cf. eq. (35), can therefore take in any classical state τ only the discrete values $\left(N_L(x)\right)_\tau = (0, 1 \ldots, 8)$, according to the total number of eight species. In consequence, for any state τ of the classical statistical ensemble, $(J_{\mathcal{R}})_\tau$ is a positive integer or zero. According to the standard rule of classical statistics these integers describe the possible outcomes of measurements.

For a one particle state the total particle number equals one,

$$1 = \int\limits_V N(x) = \sum_V N_L(x), \tag{47}$$

where $\sum\limits_V$ extends now over all lattice points in the total volume. Since $\mathcal{R}$ must be contained in V the maximal allowed value for $J_{\mathcal{R}}$ in a one-particle state is one, such that $(J_{\mathcal{R}})_\tau = 0, 1$ are the only possible values of the classical observable for such a state. This is precisely the quantum mechanical rule. No new postulate is necessary for this measurement in quantum mechanics - the quantum rule is inferred from the standard rule of classical statistics. Of course, the expectation value of $J_{\mathcal{R}}$ for a one particle state is the same in the quantum mechanical and the classical statistical description

$$\langle J_{\mathcal{R}} \rangle = \int_x \varphi_D^\dagger(x) J_{\mathcal{R}}(x) \varphi_D(x) = \int_{\mathcal{R}} \varphi_D^\dagger(x) \varphi_D(x). \tag{48}$$

While our model connects the discrete particle aspects in quantum mechanics directly to the discrete classical Ising-spins, the continuous wave aspects also arise in a natural way. The quantum wave function is continuous because probability distributions and associated classical wave functions are continuous (at least piecewise). As we have mentioned already, the characteristic interference effects for waves and the superposition arise from the linearity of the fundamental evolution equation in the classical wave function.

7. Conclusions

We have derived the Schrödinger equation for a quantum particle in a potential from a classical statistical ensemble for Ising-spins. The dynamics of the classical statistical system has to be specified by an appropriate evolution equation for the probability distribution. For this purpose we employ the classical wave function which is defined as the positive or negative root of the probability distribution. The proposed evolution equation is a linear differential equation for the classical wave function. The wave function at time t obtains from the wave function at t' by a rotation - this guarantees the preservation of the normalization of the probability distribution.

We have exploited a map between the classical wave function and a Grassmann wave function which is an element of a real Grassmann algebra. In turn, the time evolution of the Grassmann wave function can be associated to a Grassmann functional integral. This allows us to formulate the evolution equation for the classical wave function in terms of the action of a functional integral. The symmetries of the model, as Lorentz symmetry and electromagnetic gauge symmetry for our model of Dirac spinors in an external electromagnetic field, can be easily implemented in this way. Since the classical wave function is real we have to formulate the model in terms of a real Grassmann algebra.

Our model is regularized on a lattice of space points. On the one hand, this guarantees that mathematical expressions are well defined for a finite number of lattice points, with continuum limit of an infinite number of points taken at the end. On the other hand, the concept of classical Ising-spins or associated occupation numbers at every point x is well defined. The discreteness of the particle aspects of quantum mechanics can be traced back to the discrete occupation numbers that can only take the values zero or one. We also have used discrete time steps, and we have formulated the lattice action such that the Grassmann wave function $g(t+2\epsilon)$ obtains from $g(t)$ by a rotation. The time evolution is unitary not only in the limit $\epsilon \to 0$, but also for finite ϵ. A unitary time evolution for infinitesimal time steps is easily achieved by any antisymmetric evolution generator $K_{\tau\rho}$ for the wave function, $q_\tau(t)$, cf. eq. (20). Generating a unitary evolution also for finite smallest time steps imposes restrictions on the form of the lattice action. This, together with the requirement of a real Grassmann action, requires some care for the construction of the model and explains the specific form of the action (6)-(8) in comparison with other possible lattice actions.

The proposed evolution equation for the classical statistical ensemble of Ising-spins does not only lead to the Schrödinger equation for non-relativistic one-particle states. It entails the full dynamical equations for a quantum field theory of Dirac fermions in an external electromagnetic field. The dynamics of states with an arbitrary number of fermions, including the characteristic interference patterns for indistinguishable fermions in quantum mechanics, is correctly described.

At this point it seems worthwhile to ask some questions about the origin of characteristic features of quantum mechanics in our classical statistical setting. Particle-wave duality is realized by the discreteness of Ising-spins on one side, and the continuous probability distribution or classical wave function on the other side. Interference arises from the formulation of the basic evolution law in terms of the classical wave function. While the classical wave function is real, it can nevertheless take positive and negative values which can add to zero locally if two wave functions are added. The superposition principle or linearity of the quantum evolution finds a direct origin in the formulation of a dynamical law for the classical statistical ensemble that is linear in the classical wave function. The characteristic physics of phases in quantum mechanics is connected to the presence of a complex structure within the real Grassmann algebra. Planck's constant $\hbar$ appears purely as a conversion factor of units. The uncertainty relations can be obtained directly from the possible solutions of the Schrödinger equation.

Finally, one of the most characteristic elements of the mathematical formulation of quantum mechanics is the presence of a non-commutative product for operators and associated observables. These structures are obviously present in our formulation and clearly very useful for a discussion of solutions of the Schrödinger equation or Dirac equation. The essence of the emergence of non-commutative structures is the coarse graining of information [15], [3], [2].

This issue has not been addressed in the present note. It seems reasonable to expect, however, that a system that is governed by the Dirac equation on microphysical scales - say lattice distances shorter than the Planck length - will also show similar properties at "macroscopic scales" associated to coarse graining. (Such macroscopic scales can still be much smaller than all characteristic scales of atom physics or elementary particle physics.) The basic reason is that the form of the Dirac equation for one-particle states is essentially fixed by the symmetries. It will not be altered if the coarse graining respects the symmetries. While it remains an interesting task to perform this coarse graining explicitly, the main message of this note is already very clear at the present stage: Quantum field theory and quantum particles can be obtained from a classical statistical ensemble.

References

[1] H. D. Zeh, Found. Phys. **1** (1970) 69; W. Zurek, Rev. Mod. Phys. **75** (2003) 715
[2] C. Wetterich, Annals of Phys. **325** (2010) 852; Ann. Phys. (Berlin) **522** (2010) 467; Journal of Phys. **174** (2009) 012008
[3] C. Wetterich, Phys. Lett. **A376** (2012) 706; Annals of Phys. **325** (2010) 1359; Ann. Phys. (Berlin) 522 (2010) 807; arXiv: 1003.0772;
[4] Y. Couder, E. Fort, Phys. Rev. Lett. **97** (2006) 154101; A. Eddi, E. Fort, F. Moisy, Y. Couder, Phys. Rev. Lett. **102** (2009) 240401
[5] G. t'Hooft, Int. J. Mod. Phys. **A25** (2010) 4384
[6] D. Bohm, Phys. Rev. **85** (1952) 166
[7] J. S. Bell, Physica 1 (1964) 195
[8] J. Clauser, M. Horne, A. Shimony, R. Holt, Phys. Rev. Lett. **23** (1969) 880; J. Bell, "Foundations of Quantum Mechanics", ed. B. d'Espagnat (New York: Academic, 1971) p. 171; J. Clauser, M. Horne, Phys. Rev. **D10** (1974) 526; J. Clauser, A. Shimony, Rep. Prog. Phys. **41** (1978) 1881
[9] C. Wetterich, arXiv: 1111.4115
[10] C. Wetterich, Annals of Phys. **325** (2010) 2750; Annals of Phys. **326** (2011) 2243
[11] C. Wetterich, arXiv: 1002.2593
[12] B. Koopman, Proc. Nat. Acad. Sci. **17** (1931) 315; J. von Neumann, Ann. Math. **33** (1932) 587; **33** (1932) 789; D. Mauro, Phys. Lett. **A315** (2003) 28
[13] C. Wetterich, in "Decoherence and Entropy in Complex Systems", ed. T. Elze, p. 180, Springer Verlag 2004, arXiv: quant-ph/0212031
[14] C. Wetterich, Nucl. Phys. **B852** [FS] (2011) 174
[15] C. Wetterich, arXiv:1005.3972

Gravitational strings beyond quantum theory: Electron as a closed heterotic string

Alexander Burinskii

Lab. of Theor. Phys. , NSI, Russian Academy of Sciences, B. Tulskaya 52 Moscow 115191 Russia

E-mail: bur@ibrae.ac.ru

Abstract. The observable parameters of the electron indicate unambiguously that its gravitational background should be the Kerr-Newman solution without horizons. This background is not flat and has a non-trivial topology created by the Kerr singular ring. This ring was identified with a closed gravitational string. We discuss the relation of this string to the closed heterotic string of the low energy string theory and show that travelling waves along the KN string give rise to the Dirac theory of electron. Gravitational strings form a bridge between gravity and quantum theory, indicating a new way to consistent Quantum Gravity. We explain the pointlike experimental exhibition of the electron and argue that the predicted closed string may be observed by the novel experimental method of the "nonforward" Compton scattering.

1. Introduction

Modern physics is based on Quantum theory and Gravity. Both theories are confirmed experimentally with great precision. Nevertheless, they are conflicting and cannot be unified in a whole theory. One of the principal contradictions between Quantum theory and Gravity is the question on the shape and size of an electron. Quantum theory states that electron is pointlike and structureless [1, 2], and it is experimentally supported by the high energy scattering. In the same time, gravity based on the Kerr-Newman (KN) solution indicates unambiguously that electron should form a closed string of the Compton size. Contrary to the Schwarzschild solution, which displays the 'range' of a gravitational field (radius of the horizon $r_g = 2m$) proportional to the mass of the source [1], the KN geometry of the rotating bodies presents a new dimensional parameter $a = J/m$, which grows with angular momentum J and has the reverse mass-dependence. As a result, the zone of gravitational interaction is determined by parameter a, which increases for the large angular momentum and small masses, and therefore, it turns out to be essential for elementary particles. The reason for that is a specific structure of the KN gravitational field, which concentrates near the Kerr singular ring, forming a closed gravitational waveguide – a type of the closed gravitational string [3].

In 1968 Carter obtained that the KN solution for the charged and rotating black holes has $g = 2$ as that of the Dirac electron, [4, 5], which allowed one to consider KN solution as consistent with gravity electron model, [3, 4, 5, 6, 7, 8, 9, 10, 11, 12, 13, 14, 15, 16, 17, 18]. Mass of the electron in the considered units is $m \approx 10^{-22}$, while $a = J/m \approx 10^{22}$. Therefore, $a >> m$, and the black hole horizons disappear, opening a two-sheeted spacetime with a nontrivial topological

[1] We use the units $\hbar = G = c = 1$

defect in the form of the naked Kerr singular ring. The Kerr ring takes the Compton radius, corresponding to the size of a "dressed" electron in QED and to the limit of localization of the electron in the Dirac theory.

There appear two questions: 1) How does the KN gravity know about one of the principal parameters of Quantum theory? and 2) Why does Quantum theory work successfully on the flat spacetime, ignoring the stringlike peculiarity of the background gravitational field? A small and slowly varying gravitational field could be ignored; however, the stringlike KN singularity forms a branch line of the spacetime and creates a two-sheeted topology, ignorance of which cannot be justified. A simple answer to these questions is to assume that there is a general underlying theory providing the consistency of quantum theory and gravity. In this paper we suggest an approach which allows one to resolve this puzzle from a rather unexpected suggestion, that the underlying theory is the Einstein-Maxwell gravity as a fundamental part of the theory of superstrings. In this case, quantum theory should follow from the theory of superstring, and we make here first steps in this direction. Starting from description of the structure of the KN spacetime in Section II, we show in Section III the relationships of the Kerr singular ring to a heterotic string of the low energy string theory, and then, in Section III we show 'emergence' of the Dirac equation and the corresponding wave functions from the physical model of the lightlike travelling waves, propagating along the KN circular ring. In Section IV we describe a model for regularization of the KN solution by the Higgs fields, which allows us to regularize the KN background, retaining its asymptotical KN form. It justifies the use of flat background in the Dirac theory and QED, answering the second above-mentioned question.

So far as gravity predicts the existence of the closed Kerr string on the boundary of the Compton area, such a string, if it really exists, should be experimentally observable, and there appears the question while it was not obtained earlier by the high energy scattering. In the Conclusion, we give an explanation for this fact and argue that the KN string may apparently be detected by the novel experimental method based on the theory Generalized Parton Distributions (GPD) [19, 20] which represents a new regime for probing the transverse shape of the particles by the "nonforward Compton scattering" [21].

2. The Kerr-Newman background

The Kerr-Newman solution in the Kerr-Schild form has the metric

$$g_{\mu\nu} = \eta_{\mu\nu} + 2Hk_\mu k_\nu, \tag{1}$$

where

$$H = \frac{mr - e^2/2}{r^2 + a^2\cos^2\theta}, \tag{2}$$

and $\eta_{\mu\nu}$ is metric of auxiliary Minkowski space in the Cartesian coordinates (t, x, y, z), and the Kerr coordinates r and θ are the spheroidal oblate coordinates, which are related to the Cartesian coordinates as follows

$$\begin{aligned} x + iy &= (r + ia)e^{i\phi}\sin\theta \\ z &= r\cos\theta. \end{aligned} \tag{3}$$

The function H is singular at $r = 0, \ \cos\theta = 0$, corresponding to the Kerr singular ring. The KN electromagnetic potential has the form

$$\alpha^\mu_{KN} = Re\frac{e}{r + ia\cos\theta}k^\mu. \tag{4}$$

The KN potential as well as the KN metric are aligned with the null direction k^μ, which forms the Kerr principal null congruence (PNC) determined by the differential form [5]

$$k_\mu dx^\mu = dt + \frac{z}{r}dz + \frac{r(xdx + ydy)}{r^2 + a^2} - \frac{a(xdy - ydx)}{r^2 + a^2}. \tag{5}$$

The potential and metric are singular at the Kerr ring, which forms a branch line of the Kerr spacetime in two sheets, corresponding to $r > 0$ and $r < 0$ in the Kerr oblate coordinate system. Vector field k^μ forms Principal Null Congruence (PNC) of KN space, which is determined by the Kerr theorem in twistor terms, [12, 13]. The Kerr PNC is smoothly propagated via the Kerr disk $r = 0$ from the 'negative' sheet ($r < 0$) of spacetime to the 'positive' one ($r > 0$) (see Fig.1), and therefore, it covers the KN space twice: $k^{\mu(+)}$ for $r > 0$ and $k^{\mu(-)}$ for $r < 0$, leading to different metrics and different electromagnetic field on the 'positive' and 'negative' sheets [22]. Therefore, the Kerr ring creates a two-sheeted background topology.

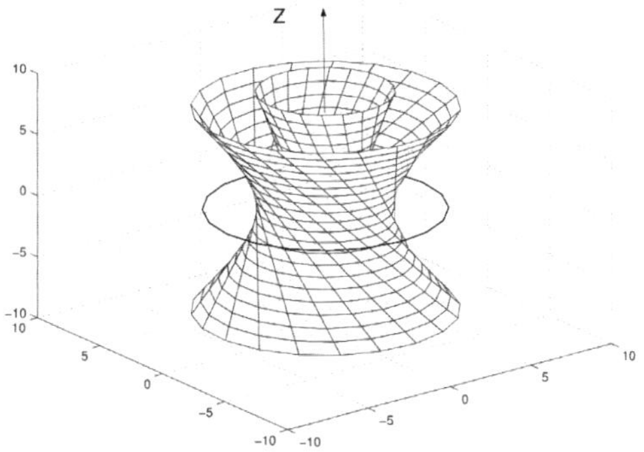

Figure 1. Vortex of the Kerr congruence. Twistor null lines are focused on the Kerr singular ring, forming a circular gravitational waveguide, or string with lightlike excitations.

This two-sheetedness forms a principal puzzle of the Kerr geometry over a period of four decades. In in 1967 Keres [23] and then Israel [6] truncated negative KN sheet, $r < 0$,, replacing it by the *rotating disklike source* at $r = 0$, spanned by the Kerr singular ring of the Compton radius $a = \hbar/2m$. Then Hamity assumed in [24] that the disk is to be rigidly rotating, which led to a reasonable interpretation of the matter of the source as an exotic stuff with the zero energy density and negative pressure. The matter distribution was singular at the disk boundary, forming an additional closed string source, and López suggested in [9] to regularize this source, covering the Kerr singular ring by a disk-like ellipsoidal surface. As a result the KN source was turned into a rotating and charged oblate bubble with a flat interior, and further, the bubble source was realized as a regular soliton-like model [17], formed by a domain wall interpolating between the external KN solution and a flat pseudo-vacuum state inside the bubble.

Alternatively, from the very beginning, there were considered the string-like models of the KN source, which retained Kerr's two-sheeted topology, forming a closed 'Alice' string [3, 13, 11, 25]. The Kerr singular ring was considered as a waveguide for electromagnetic travelling waves generating the spin and mass of the KN solution in accordance with the old Wheeler's "geon" model of 'mass without mass' [7, 8, 26, 27].

3. The Kerr singular ring as a closed string

Exact *non-stationary* solutions for electromagnetic excitations on the Kerr-Schild background, [11, 22, 44], showed that there are no smooth harmonic solutions. The typical exact electromagnetic solutions on the KN background take the form of singular beams propagating along the rays of PNC, contrary to smooth angular dependence of the wave solutions used in perturbative approach!

Position of the horizon for the excited KS black holes solutions is determined by function H which has for the exact KS solutions the form, [5],

$$H = \frac{mr - |\psi|^2/2}{r^2 + a^2 \cos^2 \theta}, \tag{6}$$

where $\psi(x)$ is related to the vector potential of the electromagnetic field

$$\alpha = \alpha_\mu dx^\mu = -\frac{1}{2} Re \; [(\frac{\psi}{r + ia\cos\theta})e^3 + \chi d\bar{Y}], \; , \tag{7}$$

where $\chi = 2\int(1 + Y\bar{Y})^{-2}\psi dY$, and the vector field α satisfies the alignment condition

$$\alpha_\mu k^\mu = 0. \tag{8}$$

The equations (1)and (6) display compliance and elasticity of the horizon with respect to the electromagnetic field.

The Kerr-Newman solution corresponds to $\psi = q = const.$. However, any nonconstant holomorphic function $\psi(Y)$ yields also an exact KS solution, [5]. On the other hand, any nonconstant holomorphic functions on sphere acquire at least one pole. A single pole at $Y = Y_i$

$$\psi_i(Y) = q_i/(Y - Y_i) \tag{9}$$

produces the beam in angular directions

$$Y_i = e^{i\phi_i} \tan \frac{\theta_i}{2}. \tag{10}$$

The function $\psi(Y)$ acts immediately on the function H which determines the metric and the position of the horizon. The analysis showed, [22], that electromagnetic beams have very strong back reaction to metric and deform topologically the horizon, forming the holes which allows matter to escape interior (see fig.3).

The exact KS solutions may have arbitrary number of beams in different angular directions $Y_i = e^{i\phi_i} \tan \frac{\theta_i}{2}$. The corresponding function

$$\psi(Y) = \sum_i \frac{q_i}{Y - Y_i}, \tag{11}$$

leads to the horizon with many holes. For the parameters of an electron, the BH horizons disappear, however the radiating singular beams appear unavoidable, and in the far zone the beams tend to the known exact singular pp-wave solutions. The considered in [28] multi-center KS solutions showed that the beams are extended up to the other matter sources positioned at infinity.

The stationary KS beam-like solutions may be generalized to the time-dependent wave pulses, [11].

Due to the factor $\frac{1}{r+ia\cos\theta}$ in the vector potential α, any electromagnetic excitation of the KN geometry generates the travelling waves along the Kerr singular ring (at $r = \cos\theta = 0$), and simultaneously, the pole in Y creates an 'axial' singular beam, which is topologically coupled with the Kerr ring, see Fig.2. Therefore, the EM excitations on the KN background have a paired character, creating simultaneously a 'circular' travelling wave and the coupled with it a propagating outward 'axial' travelling wave.[2] The vector field k^μ is constant along the 'axial' beams, and asymptotically (by $r \to \infty$) the beams tend to the well known pp-wave (plane fronted) solutions, for which k_μ is a covariantly constant Killing direction. Adapting the z-axis of the coordinate system along the asymptotic pp-wave direction, and using the light-cone Cartesian coordinates $u = (t+z)/\sqrt{2}, \quad v = (z-t)/\sqrt{2}$ and $\zeta = (x+iy)/\sqrt{2}, \quad \bar{\zeta} = (x-iy)/\sqrt{2}$, one writes the outgoing pp-wave metric in the form

$$ds^2 = 2dudv + 2d\zeta d\bar{\zeta} + 2H(v,\zeta,\bar{\zeta})dv^2. \quad (12)$$

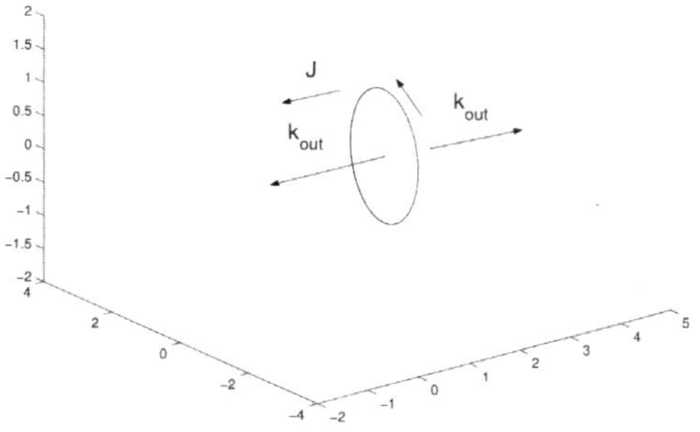

Figure 2. Skeleton of the Kerr geometry [11] formed by the topologically coupled 'circular' and 'axial' strings.

The pp-waves take important role in superstring theory. In the nonperturbative approach based on analogues between the strings and solitons, the pp-wave solutions are considered as fundamental strings forming fundamental classical solutions to the low-energy string theory [30, 31, 32, 33, 34]. The pp-waves may carry traveling electromagnetic and gravitational waves which represent propagating modes of the fundamental string [35]. In particular, the generalized pp-waves represent the singular strings with travelling electromagnetic waves [11, 37].

The string solutions are compactified to four dimensions, and the singular pp-waves are regarded as the massless fields around a lightlike singular source of the fundamental heterotic string. Indeed, metric of the fundamental string solution for travelling waves in z+ direction takes the form [34],

$$ds^2_{str} = e^{2\phi}(2dudv + 2H(v,\zeta,\bar{\zeta})dv^2) + 2d\zeta d\bar{\zeta}, \quad (13)$$

which differs from (12) only by factor $e^{2\phi}$ determined by the extra dilaton field ϕ, which deforms the stringy metric in the longitudinal null direction (note, that the used in string theory 'stringy' metric ds^2_{str} differs from the used in gravity 'Einstein' metric ds^2_E by the conformal rescaling $ds^2_{str} = e^{2\phi}ds^2_E$). In particular, for $\phi = 0$ the both metrics are equivalent. The "solution-generating transforms" [38] allowed Sen to get corresponding charged string with the lightlike moving current and oscillations, which was interpreted as a charged (and superconducting) heterotic string [34]. There appears also an extra axion field and the resulting dilaton field turns out to be singular at the string core.

[2] Similar spinor solutions were obtained on the KN background in the frame of the coupled Einstein-Maxwell-Dirac equations, [36]. The resulting massless spinor solutions turn out to be singular at the Kerr ring and the fermionic wave excitations generate simultaneously the travelling waves along the Kerr ring and the coupled 'axial' singular pp-waves.

It is suspected that the singular source of the string will be smoothed out in the full string theory which should take into account all orders in α'. Meanwhile, it was shown that the pp-waves have remarkable property that all the α'- corrections in the string equation of motion are automatically zero [31].

It has been noticed that the field structure of the Kerr singular ring is lightlike [8], similarly to the closed fundamental or the heterotic string [39]. The twisted Kerr congruence (Fig. 1.) represents a "hedgehog" defocused by the rotation. The null lines of the Kerr congruence are focused only in equatorial plane ($z = \cos\theta = 0$). The Eq. (5) shows that PNC takes near the ring the form

$$\tilde{k} = k|_{r=\cos\theta=0} = dt - (xdy - ydx)/a = dt - ad\phi, \tag{14}$$

and therefore, the lightlike vector field k_μ near the ring is tangent to the Kerr string world sheet, and the Kerr ring is sliding along itself with the speed of the light. As a consequence, the vector potential (4) and the metric (1) near the Kerr ring are aligned with the local direction of the Kerr string reproducing the structure of the closed pp-string. Therefore, the electromagnetic excitations of the KN background create the circular travelling waves along the Kerr ring coupled with the 'axial' outgoing lightlike pp-beams.

This similarity of the KN ring with the closed string is not incidental, since many solutions to the Einstein-Maxwell theory turn out to be particular solutions to the low energy string theory with a zero (or constant) axion and dilaton fields. Indeed, after compactification the bosonic part of the action for the four-dimensional low-energy string theory takes the form, [40],

$$S = \int d^4x\sqrt{-g}(R - 2(\partial\phi)^2 - e^{-2\phi}F^2 - \frac{1}{2}e^{4\phi}(\partial a)^2 - aF\tilde{F}), \tag{15}$$

and contains the usual Einstein term

$$S_g = \int d^4x\sqrt{-g}R \tag{16}$$

completed by the kinetic term for dilaton field $-2(\partial\phi)^2$ and by the scaled by $e^{-2\phi}$ electromagnetic part $e^{-2\phi}F^2$. The last two terms are related with axion field a and represent its nonlinear coupling with dilaton field

$$-\frac{1}{2}e^{4\phi}(\partial a)^2 \tag{17}$$

and interaction of the axion with the dual electromagnetic field

$$\tilde{F}_{\mu\nu} = \epsilon^{\lambda\rho}_{\mu\nu}F_{\lambda\rho}. \tag{18}$$

It follows immediately that *any solution of the Einstein gravity, and in particular the Kerr solution, is to be exact solution of the effective low energy string theory* with a zero (or constant) axion and dilaton fields.[3] The relationship between the classical cosmic superconducting strings and the heterotic superstrings was first mentioned by Witten in [30]. The KN string is closed, charged and has the lightlike current, what are the characteristic features of the heterotic strings.[4]

The stringy analog to the Kerr-Newman solution with nontrivial axion and dilaton fields was obtained by Sen [33], and it was shown in [10], that the field around the singular string in the

[3] Situation turns out to be more intricate for the Einstein-Maxwell solutions since the electromagnetic invariant F^2 plays the role of the source of dilaton field, while the term $F\tilde{F}$ turns out to be the source of the axion field.

[4] The characteristic peculiarities of the heterotic strings were identified by Witten with fundamental cosmic strings in [30]. There was also mentioned their relation with axion field and positioning at the boundary of a domain wall.

'axidilatonic' Kerr-Sen solution is very similar to the field around the heterotic string. The usual KN metric written in the Kerr-Schild tetrad form is

$$ds_E^2 = 2e^3e^4 + 2e^1e^2, \tag{19}$$

where e^3 and e^4 are real, and e^3 is directed along the Kerr congruence, $e^{3\mu} \sim k^\mu$, while e^1 and e^2 are the two complex conjugate null vectors orthogonal to k^μ. The metric of the axidilatonic Kerr-Sen solution has the form (contrary to (13) we use here the usual Einstein metric)

$$ds_E^2 = 2e^3\tilde{e}^4 + 2e^1e^2e^{-2(\Phi-\Phi_0)}, \tag{20}$$

which shows that the KN metric is deformed by the dilaton field $e^{-2(\Phi-\Phi_0)}$ is the orthogonal to the Kerr congruence directions. Near the Kerr ring, the lightlike direction e^3 is tangent to the ring, and the Kerr-Sen metric is to be deformed by dilaton in the orthogonal to the ring direction. The explicit form of the dilaton

$$e^{-2(\Phi-\Phi_0)} = 1 + \frac{Q^2}{M(r^2 + a^2\cos^2\theta)} \tag{21}$$

shows that it is singular near the Kerr ring, where $\cos\theta = 0$, $r \to 0$, and therefore, metric of the closed heterotic string of the Kerr-Sen solution is singular.

This indicates that the Kerr singular ring forms really a heterotic string with has to acquire the lightlike traveling modes. The structure of the Lagrangian (15) shows that the axion field involves the dual magnetic field, and therefore, the complex axidilaton combination $\lambda = a + ie^{i\phi}$ may generate the duality rotation creating a twist of the electromagnetic travelling waves. Exact solutions of this type represent especial interest, but so far are unknown. Note that the complex axidilaton field λ was interpreted in F-theory [39] as a complex structure of a torus forming an extra two-dimensional compact space.

The low energy string theory is classical one. Assuming that the lightlike string forms a core of the electron structure, we have to obtain a bridge to quantum theory of electron. The lightlike traveling waves along the KN closed string generate the spin and mass of the KN particle. Physically, it resembles the original Wheeler's idea of a "geon", [27], in which 'mass without mass' emerges from the system of circulating massless particles. It hints that the adequate first quantized quantum theory should be constructed from the initially massless fields, and we show further, that starting from the massless fields one can derive the usual Dirac equation for the massive electron.

4. Travelling waves and 'emergence' of the Dirac equation

The Wheeler's model of "geon" represents a cloud of the lightlike particles, which are held by own gravitational field. A "microgeon" with the Kerr geometry [8] represents a degenerate case of a sole 'photon' (or other lightlike particle) circulating on the lightlike orbit of the Compton size. It may be considered as a corpuscular analogue to travelling waves along the KN gravitational string.

The observable parameters of the electron indicate unambiguously the Kerr-Newman background spacetime and the Compton radius of the corresponding Kerr string. On the other hand, the Compton radius plays also peculiar role in the Dirac theory, as a limit of localization of the wave packet. Localization beyond the Compton zone creates a "zitterbewegung" affecting "*...such paradoxes and dilemmas, which cannot be resolved in frame of the Dirac electron theory...*" (Bjorken and Drell, [41]). Dirac wrote in his Nobel Prize Lecture : *"The variables α (velocity operators, AB) also give rise to some rather unexpected phenomena concerning the motion of the electron. .. It is found that an electron which seems to us to be moving slowly,* *must actually have a very high frequency oscillatory motion of small amplitude superposed on the regular motion which appears to us. As a result of this oscillatory motion, the velocity of the electron at any time equals the velocity of light."*

These words by Dirac may be considered as a direct hint regarding the inner structure of the electron, the core of which should therefore be represented by a circulating lightlike particle, and consequently, its wave function should satisfy the *massless* wave equation, while the mass should appear as an averaged energy of the circular motion. It corresponds also to generation of the mass in stringy models from excitations of the traveling modes.

4.1. Wheeler's microgeon – mass without mass

The local 4-momentum of a massless particle circulating around z-axis has to be lightlike,

$$p_x^2 + p_y^2 + p_z^2 = E^2, \tag{22}$$

while the effective mass-energy has to be related with an averaged orbital motion,

$$< p_x^2 > + < p_y^2 >= \tilde{m}^2. \tag{23}$$

The lightlike particle circulating along the Kerr ring of radius a will generate the angular momentum

$$\tilde{J}_z = Ea \equiv \tilde{m}a, \tag{24}$$

which is in agreement with the Kerr relation $J_z = ma$ providing consistent source of the mass $\tilde{m} = m$ and angular momentum for the KN solution , $\tilde{J}_z = J_z$. The wavelength of the lightlike particle will be

$$\lambda = 2\pi\hbar/E \equiv 2\pi\hbar/m \equiv 2\pi a\hbar/J_z. \tag{25}$$

For the half-integer z-projection of the spin, $J_z = n\hbar/2,\ n = 1, 2, 3..$, the travelling waves along the KN ring have the half-integer wavelengths. The ratio

$$\frac{2\pi a}{\lambda} = J/\hbar \tag{26}$$

shows the number of the wave-length trapped by the KN ring. For $J_z = \hbar/2$, we obtain that travelling mode should have one-half of the wavelength corresponding to a two-valued wave function, or to be treated on a covering two-sheeted manifold.

Averaging (22) under the condition (23) yields the relation

$$< p_x^2 + p_y^2 + p_z^2 >= \tilde{m}^2 + p_z^2 = E^2. \tag{27}$$

The simplest quantum analogue of this model corresponds to the wave equations obtained from the operator version of these relations: $\vec{p} \to \hat{\vec{p}} = -i\hbar\nabla, \quad \hat{E} = i\hbar\partial_t$.

4.2. Massive wave solutions of the scalar massless equation

We start from the solutions of the scalar massless equation

$$\partial_\mu \partial^\mu \phi = 0, \tag{28}$$

which should satisfy the wave analogue of the constraint (23),

$$(\partial_x^2 + \partial_y^2)\phi = -(\tilde{m}/\hbar)^2\phi. \tag{29}$$

For the considered orientation of spin, the system of the equations (28) and (29) is equivalent to the system

$$(\partial_x^2 + \partial_y^2)\phi = -(\tilde{m}/\hbar)^2\phi = (\partial_t^2 - \partial_z^2)\phi, \tag{30}$$

which shows explicitly that the variables may be separated by the ansatz

$$\phi = \mathcal{M}(x, y)\Phi_0(z, t), \tag{31}$$

where Φ_0 is the wave function of a massive particle which satisfies $(\partial_t^2 - \partial_z^2)\Phi_0 = -(\tilde{m}/\hbar)^2\Phi_0$ or equivalently

$$(\partial_\mu\partial^\mu - (\tilde{m}/\hbar)^2)\Phi_0(x, y) = 0. \tag{32}$$

The corresponding plane wave solution

$$\Phi_0(z, t) = \exp\frac{i}{\hbar}(zp_z - Et), \tag{33}$$

has the usual de Broglie periodicity. In the same time, the l.h.s. of (30) yields the equation

$$(\partial_x^2 + \partial_y^2)\mathcal{M}(x, y) = -(\tilde{m}/\hbar)^2\mathcal{M}(x, y), \tag{34}$$

which determines a factor of "internal vortex structure" $\mathcal{M}(x, y)$. The corresponding cylindrical solutions may be obtained in the polar coordinates ρ, θ. From the relation $x + iy = \rho e^{i\theta}$ we have

$$\rho = \sqrt{(x + iy)(x - iy)}, \quad e^{i\theta} = \sqrt{\frac{x + iy}{x - iy}}, \tag{35}$$

and the solutions of (34) are expressed via the Hankel functions of order ν, $\mathcal{H}_\nu(\frac{\tilde{m}}{\hbar}\rho)$

$$\mathcal{M}_\nu = \mathcal{H}_\nu(\frac{\tilde{m}}{\hbar}\rho)\exp\{i\nu\theta\}, \tag{36}$$

which are eigenfunctions of the operator $\hat{J}_z = \frac{\hbar}{i}\partial_\theta$ with eigenvalues $J_z = \nu\hbar$. For electron we put $J_z = \pm\hbar/2, \quad \nu = \pm 1/2$, and the corresponding functions

$$\mathcal{M}_{\pm 1/2} = \rho^{-1/2}\exp\{i(\frac{\tilde{m}}{\hbar}\rho \pm \frac{1}{2}\theta)\} \tag{37}$$

turn out to be anti-periodic in θ. One sees that the wave functions acquire singular ray along the vortex axis z, which forms the branch line of spacetime. The half-integer spin appears here from topological reason.

We introduce operators

$$\hat{m}_\pm = -i\hbar\partial_\pm \equiv -i\hbar(\partial_x \pm \partial_y) \tag{38}$$

and, using the relations

$$\frac{1}{\rho}(\partial_x + i\partial_y)\rho = \frac{1}{x - iy}, \quad (\partial_x + i\partial_y)\theta = \frac{i}{x - iy}, \tag{39}$$

obtain that the equation (34) may be split into the pair

$$\begin{aligned} \hat{m}_+\mathcal{M}_{-1/2}(x, y) &= \tilde{m}\mathcal{M}_{1/2}(x, y) \\ \hat{m}_-\mathcal{M}_{1/2}(x, y) &= \tilde{m}\mathcal{M}_{-1/2}(x, y), \end{aligned} \tag{40}$$

which is similar to splitting of the Dirac equation.

The wave function ϕ, solution of the massless equation (28) is factorized into the plane wave solution $\Psi_0(z,t)$ of the massive Klein-Gordon equation, and the extra singular factor $\mathcal{M}_\nu$, which turns the wave function into a singular string playing the role of a carrier of the de Broglie plane wave $\Psi_0(z,t)$. The resulting solution reproduces the old de Broglie wave-pilot conjecture.[5]

The obtained solutions may be generalized in diverse directions. First of all, there may be obtained the corresponding wave solutions, which are eigenfunctions of the operator of the total angular momentum $\mathbf{J^2}$ and simultaneously of the spin projection operator $\mathbf{J_z}$. The treatment of the spin in terms of the Pauli-Lubanski pseudovector $W^\mu = \frac{1}{2}\epsilon_{\mu\nu\rho\sigma}J^{\nu\rho}p^\sigma$ and the Casimir invariant $W^2 = W_\mu W^\mu$ allows one to consider the above solutions in a Lorentz covariant form for arbitrarily positioned and oriented wave functions. And finally, the corresponding solutions of the massless spinor equation may be obtained.

4.3. Massive wave solutions of the Dirac massless equation

Solutions of the Dirac massless equation

$$\gamma^\mu \partial_\mu \psi = 0 \tag{41}$$

may be constructed from the obtained above scalar wave functions $\mathcal{M}_\nu$. It is well known, that if a function $\phi(x)$ satisfies the equation $\partial_\mu\partial^\mu\phi = 0$, the corresponding spinor solution

$$\psi = \begin{pmatrix} A \\ B \\ C \\ D \end{pmatrix} \phi(x)$$

with arbitrary coefficients A, B, C, D will satisfy the massless Dirac equation (41). In particular, using the relations 38), one can show that the known two basic plane-wave solutions of the Dirac equation[6]

$$\gamma^\mu \partial_\mu \psi^r_D = m\psi^r_D \tag{42}$$

corresponding to positive energy $E > 0$,

$$\psi^r_D(x) = w^r(p)\exp\{-\frac{i}{\hbar}p_\mu x^\mu\}, \tag{43}$$

where $p_\mu = (E, 0, 0, p_z)$,

$$w^1 = \begin{pmatrix} 1 \\ 0 \\ \frac{p_z}{E+m} \\ 0 \end{pmatrix}, \quad w^2 = \begin{pmatrix} 0 \\ 1 \\ 0 \\ -\frac{p_z}{E+m} \end{pmatrix}, \tag{44}$$

being modified by the singular function $\mathcal{M}_\nu(x,y)$, yield the wave functions

$$\begin{aligned} \psi^1(x) &= \mathcal{M}_{-1/2}(x,y)w^1(p)\exp\{-\frac{i}{\hbar}p_\mu x^\mu\}, \\ \psi^2(x) &= \mathcal{M}_{1/2}(x,y)w^2(p)\exp\{-\frac{i}{\hbar}p_\mu x^\mu\}, \end{aligned} \tag{45}$$

[5] Note that structure of this solution is absolutely different from the well known Bohm model of the double solution.

[6] We use here notations of the book [41].

satisfying the massless Dirac equation (41)[7]. Like the massless scalar solutions, the wave functions of the massless Dirac equation form a singular string, modulated by the Dirac plane wave solution.

The performed separation of the variables allowed us to obtain the ordinary Dirac theory with the massive Dirac equation and the ordinary Dirac plane wave solutions from the solutions of the corresponding massless underlying theory, which regards zitterbewegung as a corpuscular analogue of the travelling waves. Note also that the considered separation of the variables is exact analogue of the usual Kaluza-Klein compactification. However, there is a principal difference that the "compactification" is performed in 4d spacetime, in which the Kerr ring plays the role of an extra dimension, performing the model of a "compactification without compactification".

4.4. Impact of the KN gravity and topology

Like the usual Dirac theory, the above treatment was performed on the flat background, ignoring Kerr-Newman gravity. There appears important question, how these solutions could be deformed by the KN gravitational field. The result may be elucidated by the known method of complex shift (see [13, 36, 43]). This method, initiated by Appel in 1887 , was used by Newman at al. to obtain KN solution from the Kerr one, and it was shown in [8] that the KN electromagnetic field and the wave excitations of the KN solutions may be obtained by the complex shift from the corresponding Coulomb solutions and spherical harmonics. Application of the complex shift to the empty Minkowski space displays strong deformation of its causal structure. In particular, spinor structure of the light cone, which describes a spherically symmetric hedgehog of the light-like directions, is deformed into a two-sheeted structure of the Kerr Congruence which should be described by the Kerr theorem in twistor terms. It follows that the two deforming effects – influence of the gravitational field and impact of the two-sheeted topology – are indeed separated, and the KN background creates the Kerr string and two-sheeted topology even in the limit of the zero gravitational field (zero mass of the source). On the other hand, the role of the KN gravitational field turns out to be very essential too. The known exact solutions on the Kerr-Schild background show that the necessary condition for the consistency with gravity is alignment of the all fields to the Kerr Principal Null Congruence (PNC) [5, 22, 44]. It puts very hard restrictions on the structure of the'solutions. In fact, the usual plane waves turn out to be inconsistent with the Kerr-Schild gravity and should be replaced by the pp-wave analogues, i.e. by fundamental strings, and, as it was described above, the wave excitations take the form of the coupled excitations of the axial and circular strings. Even the very weak Kerr-Schild gravity has very strong impact on the structure of the wave solutions leading to their strong localization near the fundamental strings. As a result, in the KN gravity appears new dimensional parameter – the Compton wavelength $a = J/m$, and the impact of the very weak gravitational field turns out to be essential on the Compton scale, which is very far from the claimed usually Planck scale.

5. Regularization: Electron as gravitating KN soliton

'Emergence' of the Dirac equation from gravitational model of the travelling waves circulating along the closed Kerr string indicates a principally new point of view that gravitation and superstring theory may form some more fundamental theory lying beyond the Dirac theory of electron. On the level of the low energy string theory, the Dirac field losses its meaning of the wave function and should be considered as a source of the charge and currents for the Maxwell equations. The fundamental heterotic strings are charged. The charges are distributed freely

[7] One sees that the new spinor components acquire explicit dependence $e^{i\pm\frac{1}{2}\theta}$ from angle θ. It corresponds to the spinor representation of the Lorentz group for the case when the axis of rotation is given explicitly [42] (Ch.3, sec.23 and App.a).

along closed strings and localized singularly at a core of the strings. The mechanism of such localization is going beyond the low energy string theory and belongs to an (unknown so far) effective theory of a more high level. As we have seen, metric of the KN and Kerr-Sen solution turns out to be singular too, and the raised in introduction question: "why does Quantum theory works successfully on the flat spacetime, ignoring the stringlike peculiarity of the background gravitational field?" has left unanswered. To justify consistency of Quantum theory with the experimentally observable parameters of the electron, there should also be performed a "regularization of the KN or the Kerr-Sen solution" which should retain the asymptotic form of the KN solution. Regularization of the BH solutions represents a very old problem which is close related with the more general problem of the regularization of the black hole singularity [45] and with the old problem of the regular source of the KN solution [14, 46, 47].

It is often mentioned that the heterotic strings are to be superconducting [30, 38, 34], which assumes the presence of the mechanism of broken gauge symmetry realized by the Higgs field. A hint on the important role of the Higgs field is also coming from the standard model. However, the Higgs field is absent in the Dirac theory of electron, in QED and also in the usual version of the low energy string theory. Meanwhile, the $U(1)\times\tilde{U}(1)$ Witten field model for superconducting cosmic strings [48] suggests that the Higgs field may effectively be used for regularization of the singular core of the heterotic strings. A supersymmetric version of the $U(1)\times\tilde{U}(1)$ Witten field model was applied for regularization of the KN solution in the gravitating soliton model [17], which was mentioned in Section II. Detailed treatment of this model goes out of the frame of this work, and in the next Section we consider it briefly, concentrating on its *gravitational aspect* and on the *interaction of the KN electromagnetic (EM) field with the Higgs field.*

The model of gravitating soliton [17] represents a regular version of the Löpez bubble model [9]. The KN source forms a domain wall separating external KN solution from the flat interior of the bubble which replaces the former Kerr singular ring. Therefore, interior of the bubble is flat and the Kerr singularity is removed.

Gravitational aspect of this model is related with a smooth metric, which interpolates between the external KN solution and the internal flat spacetime. The bubble boundary is determined from the equation $H = 0$, (2), which yields

$$r = r_e = e^2/(2m), \tag{46}$$

where r is the Kerr ellipsoidal radial coordinate, (3). As a result, the bubble takes the form an oblate disk of the Compton radius $a = \hbar/(2m)$ with the thickness $r_e = e^2/(2m)$, corresponding to the known 'classical size' of the electron. The KN electromagnetic field is regularized since the maximal value of the vector potential is realized in the equatorial plane $\cos\theta = 0$, on the stringy boundary of the bubble

$$|\alpha_\mu^{(str)}| \le e/r_e = 2m/e. \tag{47}$$

Note, that radius of the regularized closed string, being shifted to the boundary of the bubble, turns out to be slightly increased. This position of the string confirms the known suggestions that the heterotic strings have to be formed on the boundary of a domain wall [30].

Chiral sector. The domain wall should provide a smooth phase transition from the external KN solution to a flat false vacuum state inside the bubble. This phase transition is formed by a supersymmetric system of the chiral fields Φ^i, [17] and by the potential $V(r)$ generated from the specially adapted superpotential W, which provides the smooth transfer from the external KN 'vacuum state', $V_{ext} = 0$, to a flat internal 'pseudovacuum' state, $V_{int} = 0$. Note, that the considered chiral field model is a supersymmetric version of the suggested by Witten $U(1)\times\tilde{U}(1)$ chiral field model for superconducting strings [48], see details in [17, 14]. Therefore, the assumptions that the heterotic strings are to be superconducting [30] are also confirmed in this model.

Interaction of the Higgs field with the KN EM field is controlled by the system of equations

$$\partial^\nu \partial_\nu \alpha_\mu = I_\mu = e|\Phi|^2(\chi,_\mu + e\alpha_\mu), \tag{48}$$

which are equivalent to the equations used by Nielsen and Olesen for the stringlike vortex in a superconductor, [49]. The Higgs field $\Phi = |\Phi| \exp\{i\chi\}$ forms a superconducting condensate inside the bubble, which regularizes the KN EM field by mechanism of broken symmetry. Inside the bubble, the KN EM potential α^μ is eaten by the Higgs field and the current I_μ should vanish there, $I_\mu = 0$. It gives

$$\partial^\nu \partial_\nu \alpha_\mu = 0, \quad \chi,_\mu + e\alpha_\mu = 0, \tag{49}$$

which shows that gradient of the Higgs phase $\chi,_\mu$ compensates the gauge field α_μ. The field strength and currents are expelled to the string-like boundary of the bubble, and there appear the following two conditions:

$$\omega = \chi,_0 = -e\alpha_0^{(str)}, \quad \chi,_\phi = -e\alpha_\phi^{(str)}. \tag{50}$$

We have seen earlier, that the spacelike component $\alpha^{(str)}$ of the KN potential $\alpha_\mu^{(str)} = (\alpha_0^{(str)}, \alpha^{(str)})$ is tangent to the Kerr string. As a consequence, it forms a regular flow in ϕ-direction, $\alpha_\phi^{(str)}$, and creates a closed Wilson line along the Kerr string. As a result of (50), the KN gravitating soliton exhibits two important peculiarities:

- the quantum Wilson loop $e \oint \alpha_\phi^{(str)} d\phi = -4\pi m a$ leads to quantization of the total angular momentum of the soliton, $J = ma = n/2, \ n = 1, 2, 3, ...,$
- the Higgs field inside the bubble turns out to be oscillating with the frequency $\omega = 2m$, and therefore, it forms a coherent vacuum state, typical for the 'oscillon' bubble models.

The negative sheet of the metric disappears and metric turns out to be regularized and practically flat. The model contains the lightlike heterotic string on the border of the bubble, however, it is axially symmetric and the travelling waves are absent. Therefore, some extra excitations are needed to create the travelling waves.[8] The Compton size of the bubble does not contradict to the "dressed" QED electron, however, there is an essential difference. The dynamics of the virtual particles in QED is chaotic and can be conventionally separated from the "bare"electron, while the vacuum state inside the KN soliton forms a *coherent oscillating state* joined with the closed Kerr string. Therefore, the bubble source of the KN soliton and the Kerr String represent an *integral coherent structure* which cannot be separated from the "bare" particle.

6. Conclusion

We have showed that gravity definitely indicates presence of a closed string of the Compton radius $a = \hbar/(2m)$ in the electron background geometry. This string has gravitational origin and is closely related with the fundamental heterotic string of the low energy string theory. Starting from the corpuscular aspect of the underlying model of the massless travelling modes along the Kerr string, we showed 'emergence' of the usual Dirac equation and the corresponding 'massive' solutions with de Broglie periodicity. The original Dirac theory is modified in this case, and the wave functions acquire the singular stringlike carriers, so that the Dirac plane wave solutions turn out to be propagating along 'axial' singular strings. Contrary to the widespread opinion that quantum theory predominates over all other theories, the considered structure indicates the opposite: gravity, as a basic part of the underlying superstring theory, is to be lying beyond

[8] This point has interesting consequences which can be interpreted in the terms of quarks, and we expect to considered it elsewhere.

quantum theory. As we have seen, in the KN gravity appears a new dimensional parameter – the Compton wavelength $a = J/m$, and *the impact of the very weak gravitational field turns out to be essential on the Compton scale, which is very far from the usually claimed Planck scale.* The Kerr-Newman solution, together with interpretation of its source as a closed heterotic string, forms a bridge between gravity, superstring theory and the Dirac quantum theory, displaying a new way towards unification of quantum theory and gravity.

The observable parameters of the electron show unambiguously that the background of the electron should be determined by the KN geometry and should contain the Kerr string of the Compton radius. This radius is very big with respect to the modern scale of the experimental resolution, and it seems that this string should be experimentally detected. However, the high-energy scattering detects the pointlike structureless electron down to $10^{-16}cm$. One of the reasons of this fact was considered earlier in [12], where it was argued that the pointlike character of the interactions may be caused by the interactions of the KN particles via the considered above 'axial' pp-strings – carriers of the wave function. There are also diverse alternative explanations, for example, related with a known complex interpretation of the KN solution [8, 10, 13, 36], which shows that the real KN solution is generated by a complex pointlike source shifted in the imaginary direction. Here we present a new explanation of this fact related with the lightlike character of the closed heterotic string. Observation of the lightlike, relativistically rotating closed string should be accompanied by the *Lorentz contraction* for each line element of the closed string, and as a result, to a shrinking of the full image of this string to a point in the ultrarelativistic limit.

Indeed, observation of the lightlike objects is a very nontrivial process which depends essentially on the method of the observation. In particular, it has been shown by Penrose [50] that a *momentary photo-image* of a relativistic sphere shouldn't display the Lorentz contraction at all. Similar effect for a meter stick was obtained still earlier by Terrel [51]. It doesn't mean that the Einstein relativity is wrong, rather, the effect of the Lorentz contraction depends essentially on the method of observation. We note that the relativistic scattering differs absolutely from the *momentary photo-image*, and visibility of the Lorentz contraction by the relativistic scattering on the closed relativistic heterotic string is a-priori unclear.

For an intermediate *real* photon, propagating between the separated points P_1 and P_1, the usual null condition for the lightlike interval $s_{12}^2 = 0$ is valid. Such a scattering corresponds to an 'action-at-a-distance' and should lead to a pointlike 'contact' interaction which doesn't allow one to determine the form of a string or its size. An alternative regime may be related with scattering of a virtual photon, for which the strong lightlike condition $s_{12}^2 = 0$ is broken. Specifically, to recognize the stringlike relativistic object of a big size, there are necessary two special conditions:

a) the scattering should be *deeply virtual*, which means very large $Q^2 = q_{12}^2$, and $p \cdot q_{12}$,

b) the momentum transfer should be *relatively low* to provide the wavelengths comparable with extension of the object.

Both these conditions are satisfied in the novel method of scattering – the "non-forward Compton scattering", or Deeply Virtual Compton Scattering (DVCS), suggested by Radyushkin and Ji [19, 20] and broadly discussed last decade. It is projected now as a new regime of the high energy scattering, which could realize tomography of the elementary particles [21].

Experimental observation of the predicted closed heterotic string in the electron structure could have an extraordinary meaning leading to essential progress in understanding of Quantum theory and development of Quantum Gravity.

6.1. Acknowledgments

Author is thankful to organizers, and especially to G. Grössing, for very kind invitation to attend this conference and its brilliant organization, and also to many colleagues for useful discussions during the series of the summer 2011 conferences, in particular to A. Efremov, O. Teryaev, D. Gal'tsov, A. Khrennikov, A.Kamenschik, T. Nieuwenhuizen, Yu. Obukhov, O.Selyugin, K. Stepaniants, and especially to G. 't Hooft and also to A. Radyushkin for discussions concerning the GPD application.

References

[1] Wilczek F 2008 *The Lightness Of Being* (Basic Books)
[2] Susskind L 2008 *The Black Hole War* (Hachette Book Group US)
[3] Ivanenko D D and Burinskii A Ya 1975 Gravitational strings in the models of elementary particles *Izv. Vuz. Fiz.* **5** 135
[4] Carter B 1968 Global structure of the Kerr family of gravitational fields *Phys. Rev.* **174** 1559
[5] Debney G C, Kerr R P and Schild A 1969 Solutions of the Einstein and Einstein-Maxwell equations *J. Math. Phys.* **10** 1842
[6] Israel W 1970 Source of the Kerr metric *Phys. Rev.* D **2** 641
[7] Burinskii A Ya 1972 Microgeon with Kerr metric *Abstracts of the III Soviet Gravitational Conference* (Yerevan, 1972, p.217, in Russian).
[8] Burinskii A Ya 1974 Microgeons with spins *Sov. Phys. JETP* **39** 193
[9] López C A 1984 An extended model of the electron in general relativity *Phys. Rev.* D **30** 313
[10] Burinskii A 1995 Some properties of the Kerr solution to low-energy string theory *Phys. Rev.* D **52**, 5826 [arXiv:hep-th/9504139]
[11] Burinskii A 2004 Axial stringy system of the Kerr spinning particle *Grav. Cosmol.* **10** 50 [arXiv:hep-th/0403212]
[12] Burinskii A 2004 Twistorial analyticity and three stringy systems of the Kerr spinning particle *Phys. Rev.* D **70** 086006 [arXiv:hep-th/0406063]
[13] Burinskii A The Dirac-Kerr-Newman electron 2008 *Grav. Cosmol.* **14** 109 [arXiv:hep-th/0507109]
[14] Burinskii A 2002 Supersymmetric Superconducting Bag as a Core of Kerr Spinning Particle *Grav. Cosmol.* **8** 261 [arXiv:hep-th/0008129]
[15] Dymnikova I 2006 Spinning superconducting electrovacuum soliton *Phys. Lett.* B **639** 368
[16] Nieuwenhuizen Th M 2007 The Electron and the Neutrino as Solitons in Classical Electrodynamics In: *Beyond the Quantum* (eds. Th.M. Nieuwenhuizen et al., World Scientific, Singapure) pp.332-342
[17] Burinskii A 2010 Regularized Kerr-Newman solution as a gravitating soliton, *J. Phys. A: Math. Theor.* **43** 392001 [arXiv: 1003.2928].
[18] Burinskii A 2012 Gravity vs. Quantum theory: Is electron really pointlike? *J. Phys.: Conf. Ser.* (to appear) [arXiv:1112.0225]
[19] Radyushkin A V 1997 Nonforward parton distributions *Phys. Rev.* **D 56** 5524 [hep-ph/9704207]
[20] Ji X 1977 Gauge-invariant decomposition of nucleon spin *Phys. Rev. Lett.* **78** , 610 [hep-ph/9603249].
[21] Hoyer P and Samu Kurki Samu 2010 Transverse shape of the electron *Phys. Rev.* **D 81** 013002 [arXiv:0911.3011].
[22] Burinskii A 2009 First Award Essay of GRF: Instability of black hole horizon with respect to electromagnetic excitations *Gen. Rel. Grav.* **41** 2281 [arXiv:0903.3162 [gr-qc]].
[23] Keres H 1967 To physical interpretation of the solutions to Einstein equations *Zh.Exp. i Teor.Fiz (ZhETP)* **52** 768 (in Russian)
[24] Hamity V 1976 An interior of the Kerr metric *Phys. Lett.* A **56** 77
[25] Israel W 1977 Line sources in general relativity, *Phys. Rev.* **D 15** 935
[26] Burinskii A Ya 1971 On the particlelike solutions of the massles wave equations. *Abstracts of the VIII All-Union Conference on Elementary Particle Theory* (Uzhgorod, January 1971 pp.96-98 in Russian)
[27] Wheeler J A 1962 *Geometrodynamics* (Academic Press, New York)
[28] Burinskii A 2005 Wonderful consequences of the Kerr theorem *Grav. Cosmology* **11** 301 [hep-th/0506006]
[29] Green M B, Schwarz J H and Witten E 1987 *Superstring Theory V. I* (Cambridge University Press)
[30] E. Witten, "Cosmic Superstrings", Phys. Lett. B **153**, 43 (1985).
[31] Horowitz G and Steif A 1990 Spacetime Singularities in String Theory *Phys. Rev. Lett.* **66** 260
[32] Dabholkar A, Gibbons G, Harvey J and Ruiz Ruiz F 1990 Superstrings and Solitons *Nucl. Phys.* **B 340** 33
[33] Sen A 1992 Rotating charged black hole solution in heterotic string theory *Phys. Rev. Lett.* **69** 1006
[34] A. Dabholkar, J. Gauntlett, J. Harvey and D. Waldram, "Strings as Solitons & Black Holes as Strings", Nucl.Phys. **B 474**, 85 (1996) [arXiv:hep-th/9511053].
[35] Garfinkle D 1992 Black string traveling waves *Phys. Rev.* **D 46** 4286
[36] A. Burinskii, "Super-Kerr-Newman solution to broken N = 2 supergravity", Class. Quant. Grav. **16** 3497 (1999), [arXiv:hep-th/9903032v2],

[37] Tseytlin A 1995 Exact solutions of closed string theory *Class. Quant. Grav.* **12** 2365 [arXiv:hep-th/9505052]
[38] Sen A 1992 Macroscopic Charged Heterotic String *Nucl.Phys.* **B 388** 457 [arXiv:hep-th/9206016].
[39] Backer K, Backer M and Schwarz John H 2007 *String Theory and M-Theory: A Modern Introduction* (Cambridge University Press, ISBN 0-521-86069-5)
[40] Schapere A, Trivedi S and Wilczek F 1991 Dual dilaton dion *Modern Phys. Letters* A. **6** 2677
[41] Bjorken J and Drell S 1964 Relativistic Quantum Mechanics (McGraw Hill Book)
[42] Berestetskii V B, Pitaevskii L P and Lifshitz E M 1971-74 *Relativistic quantum theory* v.2 (Oxford, New York, Pergamon Press)
[43] Burinskii A 2003 Complex Kerr geometry and nonstationary Kerr solutions. *Phys. Rev. D* **67**, 124024 .
[44] Burinskii A 2010 Fluctuating twistor-beam solutions and holographic pre-quantum Kerr-Schild geometry *J. Phys.: Conf. Ser.* **222** 012044 [arXiv:1001.0332]
[45] Dymnikova I 2006 Vacuum Nonsingular Black Hole *Gen.Rel.Grav.* **24** 235
[46] Burinskii A, Elizalde E, Hildebrandt S R and Magli G 2002 *Phys. Rev. D* **65**, 064039 [arXiv:gr-qc/0109085]
[47] Gürses M and Gürsey F 1975 *J. Math. Phys.* **16** 2385
[48] Witten E 1985 Superconducting strings *Nucl. Phys.* **B 249** 557
[49] Nielsen H B and Olesen P 1973 *Nucl. Phys.* **B61** 45
[50] R. Penrose, "The Apparent shape of a relativistically moving sphere", Proc. Cambridge Phil. Soc. **55** 1371 (1959).
[51] J. Terrel, "Invisibility of the Lorentz Contraction", Phys. Rev. **116** 1041 (1959).

Journal of Physics: Conference Series **361** (2012) 012033 doi:10.1088/1742-6596/361/1/012033

Quantization in relativistic classical mechanics: the Stückelberg equation, neutrino oscillation and large-scale structure of the Universe

V D Rusov and D S Vlasenko

Department of Theoretical and Experimental Nuclear Physics, Odessa National Polytechnic University, 65044 Odessa, Ukraine

E-mail: siiis@te.net.ua

Abstract. Based on the Chetaev theorem on stable dynamical trajectories in the presence of dissipative forces, we obtain the generalized condition for stability of relativistic classical Hamiltonian systems (with an invariant evolution parameter) in the form of the Stückelberg equation. As is known, this equation is the basis of a competing paradigm known as parametrized relativistic quantum mechanics (pRQM).

It is shown that the energy of dissipative forces, which generate the Chetaev generalized condition of stability, coincides exactly with Bohmian relativistic quantum potential. Within the framework of Bohmian RQM supplemented by the generalized Chetaev theorem and on the basis of the principle of least action for dissipative forces, we show that the squared amplitude of a wave function in the Stückelberg equation is equivalent to the probability density function for the number of particle trajectories, relative to which the velocity and the position of the particle are not hidden parameters.

The conditions for reasonableness of trajectory interpretation of pRQM are discussed. Based on analysis of a general formalism for vacuum-flavor mixing of neutrino within the context of the standard and pRQM models we show that the corresponding expressions for the probability of transition from one neutrino flavour to another differ appreciably, but they are experimentally testable: the estimations of absolute value for neutrino mass based on modern experimental data for solar and atmospheric neutrinos show that the pRQM results have a preference. It is noted that the selection criterion of mass solutions relies on proximity between the average size of condensed neutrino clouds, which is described by the Muraki formula (29th ICRC, 2005) and depends on the neutrino mass, and the average size of typical observed void structure (dark matter + hydrogen gas), which plays the role of characteristic dimension of large-scale structure of the Universe.

1. Introduction

In this paper we continue to consider the question, which is formulated in [1, 2] in the following rather strict and paradoxical form: "Are the so-called quantization conditions that are imposed on the corresponding spectrum of a dynamical system possible in principle in classical mechanics, analogously to what is taking place in quantum mechanics?" As is shown in [1, 2], the answer to this question is positive and has been given more than 70 years ago by the Russian mathematician and mechanician N.G. Chetaev in his article "On stable trajectories in dynamics" [3, 4, 5, 6]. What is more, based on the Chetaev theorem on stable dynamical trajectories in the presence of dissipative forces [3, 4, 5, 6], we obtained the generalized condition for stability of Hamilton systems in the form of the Schrödinger equation [1, 2].

At the same time, it is necessary to remind that the essential difference of the Chetaev quantum mechanics with trajectories from the known alternative quantum theories [7-16] is the unconditional identity of the probability density function of the number of particle trajectories (obtained by the principle of least action of perturbations (see (28) in [1, 2]) and the probability density function for a particle to be in a certain place of the configuration space (obtained by the Bohm-Madelung equation of continuity (see (25) in [1, 2])). This exclusively important fact emphasizes naturally the physical identity of the probabilistic and trajectory interpretations of nonrelativistic quantum mechanics.

In this paper we show that similar results can be obtained in the framework of relativistic classical mechanics. In other words, based on generalized stability conditions for stability of Hamilton relativistic systems (with invariant evolution parameter) we obtained the relativistic Schrödinger equation or, more exactly, the so-called Stückelberg equation, which as is known is the basis of a competing paradigm known as parametrized relativistic quantum mechanics (pRQM) [17].

On the other hand, it is obvious that the fact of appearance of an alternative quantum theory causes immediately a number of questions connected, in the first place, with comparison between results of alternative and standard quantum theories obtained in the same experiments. As is exactly mentioned by Adler and Bassi [18], "...but apart from this history, there is another important motivation for considering modifications of quantum theory. This is to give a quantitative meaning to experiments testing quantum theory, by having an alternative theory, making predictions that differ from those of standard quantum theory, to which these experiments can be compared".

Thus, the main purpose of the present work is, on the one hand, derivation of the Schrödinger relativistic equation or, more exactly, so-called the Stückelberg equation based on the generalized condition for stability of Hamilton relativistic systems (with invariant evolution parameter) and, on the other hand, quantitative comparison between results of alternative (pRQM) and standard quantum theories by analysis of general formalism for vacuum-flavour mixing of neutrino.

2. The emergent Stückelberg equation in relativistic classical mechanics

In what follows, we present a generalization of the Chetaev theorem on stable trajectories in dynamics to the case of relativistic Hamiltonian $K(s, q^i, p_i)$, where s is the evolution parameter of a system $i = 0, 1, 2, 3$. With this purpose, we consider a material system, in which q^i and p_i are the generalized 4-space coordinates and momenta of the holonomic system in the field of potential forces which admit a force function of the form $U(q^i)$.

In the general case where the action S is the explicit function of the evolution parameter s, the total integral of the Hamilton-Jacobi differential equation corresponding to a given system takes the form:

$$S = f(s, q^i; \alpha^i) + C, \tag{1}$$

where α^i and C, are arbitrary constants, and the general solution of the mechanical problem is given, according to the well-known Jacobi theorem, by the formulas:

$$\beta^i = \partial S/\partial\alpha_i, \quad p_i = \partial S/\partial q^i, \quad i = 0, ..., 3, \tag{2}$$

where β^i are the new constants of integration. The possible motions of the mechanical system are defined by various values of the constants α^i and β^i.

We call a motion of the material system, whose stability will be studied, the unperturbed motion. First, we consider the stability of such motion with respect to the variables q^i under a perturbation of only the initial values of these variables (i.e. of values of the constants α^i and β^i) without disturbing forces. Omitting the details of the derivation which are given in [3, 4, 6, 19], we present the necessary Chetaev stability condition in this case as follows:

$$\Lambda = \frac{1}{m}\partial_i\partial^i S = 0, \tag{3}$$

We now complicate the problem. Let a really moving material system undergo the action of both the forces with force function U, theoretically considered above, and the unknown perturbation (dissipative) forces which are assumed to be potential and admit the dissipative force function Q. Then the real motion of the material system occurs in the field of forces with general force function $U^* = U + Q$. In this case, the real motion of the system does not coincide obviously at all with the theoretical one (without perturbation).

If we conserve the statement of the problem on the stability of real unperturbed motions in the theoretical field of forces with the function U at a perturbation of only initial data as above, then the necessary requirement of stability in the first approximation, e.g., in form (3), will not be efficient in the general case, because the new function S is unknown (as well as Q). However, it turns out that we can determine such conditions of stability, which are implicitly independent of the form of the unknown action functions S and the potential Q. Thus, we are based on the requirement of stability of form (3), by assuming the conditions of its existence (correctness, etc.) for real motions to be satisfied. In relation (3), we now replace the function S by a new function A defined by the equality:

$$\psi = A\exp(ikS), \tag{4}$$

where k is a constant; A is a function of the generalized coordinates q and the evolution parameter s.

The introduction of a real wave function, like the de Broglie "pilot-wave" [20], is extremely necessary from the physical viewpoint because of the following non-trivial reason. Since the dynamics of a physical system must undoubtedly conserve the Hamilton form of the equations of motion, the main "task" of such a real wave consists in the exact compensation of the action of dissipative forces, which are generated by the perturbation energy Q. We will show below that, in this case, such a procedure makes it possible not only to conserve the Hamilton form of the dynamics of a physical system, but allows one to determine the character of an analytic dependence of the energy Q of disturbing forces on the wave function amplitude A. Then relation (4) yields:

$$\partial_j S = \frac{1}{ik}\left(\frac{1}{\psi}\partial_j\psi - \frac{1}{A}\partial_j A\right) \tag{5}$$

and, hence, relation (3) looks like

$$\frac{1}{m}\partial_i\left[\frac{1}{\psi}\partial^i\psi - \frac{1}{A}\partial^i A\right] = 0. \tag{6}$$

On the other hand, we can write the Hamilton-Jacobi equations for a perturbed motion in the general case where the Hamiltonian H depends explicitly on the evolution parameter s:

$$\frac{\partial S}{\partial s} + \frac{1}{2m}\partial_i S\partial^i S + U + Q = 0,$$

or with the help of (4-6)

$$\frac{1}{2mk^2}\left(\frac{1}{\psi}\partial_i\psi - \frac{1}{A}\partial_i A\right)\left(\frac{1}{\psi}\partial^i\psi - \frac{1}{A}\partial^i A\right) = \frac{\partial S}{\partial s} + U + Q, \tag{7}$$

where $\partial S/\partial s$ can be determined with the help of (4). Adding relations (6) and (7), we obtain the necessary condition of stability (in the first approximation) in the form:

$$\frac{1}{2mk^2\psi}\partial_i\partial^i\psi - \frac{1}{2k^2A}\partial_i\partial^i A - \frac{1}{mk^2A}\partial_i A\left(\frac{1}{\psi}\partial_i\psi - \frac{1}{A}\partial_i A\right) - \\ -\frac{1}{ikA\psi}\left[A\partial_s\psi - \psi\partial_s A\right] - U - Q = 0. \tag{8}$$

In this place, we need to use a procedure for the compensation of the action of dissipative forces, which are generated by the perturbation energy Q in order to conserve the Hamilton form of the dynamics of a physical system (8). It is obvious that equality (8) will not contain Q, if the amplitude A is determined from the equation:

$$\frac{1}{2k^2A}\partial_i\partial^i A + \frac{i}{kA}\partial_i A\partial^i S - \frac{1}{ikA}\partial_s A + Q = 0, \tag{9}$$

which splits into two equations

$$Q = -\frac{1}{2mk^2A}\partial_i\partial^i A, \tag{10}$$

$$\partial_s A = -\frac{1}{m}\partial_i A\partial^i S, \tag{11}$$

after the separation of the real and imaginary parts. Here Q is the dissipation energy. Thus, if the properties of disturbing forces satisfy conditions (10) and (11), then the necessary condition of stability (8) has the form of a differential equation of the "Stückelberg" type:

$$\frac{i}{k}\partial_s\psi = -\frac{1}{2mk^2}\partial_i\partial^i\psi + U\psi, \tag{12}$$

where $q^\mu(s)$ gives the position of the physical system, whose possible trajectories in the 4-dimensional configuration space $q = (q_0, , q_3)$ are a solution of the system of the so-called governing equations [21]:

$$v_i = \frac{1}{m}\partial_j S, \tag{13}$$

where S is the phase of the wave function.

In other words, we obtained the following result: Eq. (3), which corresponds to the Chetaev stability condition, is transformed into an equation of the "Stückelberg" type (12) with the use of transformation (4). It should be noted that in spite of similarity of the Stückelberg and Schrödinger equations they are different. Evolution parameter differs from the time [17, 22]. It is obvious that, in the class of equations of type (12), the single-valued, finite, and continuous solutions for the function ψ in the stationary case are admissible only for the eigenvalues of the total energy E. Hence, the given stability of real motions takes place only for these values of the total energy E.

We now present a short analysis of the obtained results. It is known that, according to one of the theorems of stability theory [6], only two types of forces - dissipative and gyroscopic do not break the stability (if it is present) of a non-disturbed motion of holonomic mechanical systems. Therefore, by introducing a dissipative perturbation Q into the Hamilton-Jacobi Eq. (7) and by taking simultaneously the stability condition (3) into account, we reasonably expect that, under condition of the conservation of the stability of a mechanical system, it is possible to get a real functional dependence of the dissipation energy Q on characteristics of the wave function ψ (4).

Indeed, having obtained the condition for the stability of trajectories of a dynamical system in the form of an equation of the "Stückelberg" type (12), we established not only the physical sense of the perturbation energy Q, but showed also that it is a function of the amplitude of the wave function ψ (4) and takes form (10). It is a very important result. The subsequent content of the article is, as will be clear in what follows, a direct consequence of this result.

It is easy to show that namely the conclusion about the dissipative nature together with the simultaneous determination of a functional dependence of the energy of disturbing forces on the form (but not on the magnitude) of the amplitude of the wave function ψ allow one to generalize an equation of the "Stückelberg" type (12) to the case where the condition of stability (3) is not fulfilled, i.e., $\Lambda \neq 0$. Let us show this.

Obvious analysis of Eqs. (3) and (10)-(13) shows that with allowance for dimensions the expression:

$$\varepsilon = \frac{1}{2k}\Lambda = \frac{1}{2km}\partial_i\partial^i S, \tag{14}$$

is, in general case, the variations of particle kinetic energy predetermined accordingly by variations of its momentum.

Now, if to add and simultaneously to subtract the complex expression ($i\varepsilon$) and to substitute also Eq. (4) in the left-hand side of the Hamilton-Jacobi equation (7), we get the generalized equation corresponding to the extended Eq. (8):

$$\begin{aligned}\frac{1}{2mk^2\psi}\partial_i\partial^i\psi - \frac{1}{2mk^2A}\partial_i\partial^i A - \frac{i}{2mk}\partial_i\partial^i S - \frac{1}{mk^2A}\partial_i A\left(\frac{1}{\psi}\partial^i\psi - \frac{1}{A}\partial^i A\right) - \\ - \frac{1}{ikA\psi}\left[A\partial_s\psi - \psi\partial_s A\right] - U - Q = 0.\end{aligned} \tag{15}$$

Repeating the ideology of derivation of Eq. (12) it is easy from Eq .(15) to obtain an equation of the "Stückelberg" type, which is formally identical to Eq. (12), but already under a more general condition imposed on the disturbing energy and the wave function:

$$Q = -\frac{1}{2mk^2A}\partial_i\partial^i A, \tag{16}$$

$$\partial_s A = -\frac{A}{2m}\partial_i\partial^i S - \partial_i A\partial^i S. \tag{17}$$

Physical sense of (17) for the wave function amplitude consists in the fact that with regard for the formulas for a classical velocity $v_i = \partial_i S/m$ and the probability density $P(q,t) = [A(q,t)]^2$ (the substantiation of this formula will be given below) in the configuration space it can be easily transformed into the equation of continuity, which represents the invariability of the total number of "particles" (phase points) or, in other words, the probability conservation law.

We now return to our problem of quantization on the basis of the simplest example. Let us consider a material point with mass m in the field of conservative forces with the force function U, which depends, in the general case, on the evolution parameter. The problem on the stability of motions of such a point will be posed in the Cartesian coordinate system q_0, q_1, q_2, q_3. Denoting the momenta along the axes by p_0, p_1, p_2, p_3, respectively, we obtain known expression for the Hamiltonian::

$$K = \frac{1}{2m}p_i p^i. \tag{18}$$

In this case, conditions (16) and (17) for the structure of disturbing and compensating forces admit the relations:

$$Q = -\frac{\hbar^2}{2m}\frac{\partial_i \partial^i A}{A} \quad , \quad k = 1/\hbar \ , \tag{19}$$

$$\partial_s A = -\frac{A}{2m}\sum \frac{\partial^2 S}{\partial q_i^2} - \sum \frac{\partial A}{\partial q_i}\frac{p^i}{m}, \tag{20}$$

and the differential equation (12) defining stable motions takes the form:

$$i\hbar\frac{\partial \psi}{\partial s} = -\frac{\hbar^2}{2m}\partial_i \psi \partial^i \psi + U\psi. \tag{21}$$

which coincides with the Stückelberg equation. In our case, this equation restricts the choice of the integration constants in the full Hamilton-Jacobi integral. In what follows, we call Eq. (21) the Stückelberg-Chetaev equation, by emphasizing the specific feature of its origin.

It is of interest to consider the case related to the inverse substitution of the wave function (4) in the Schrödinger equation (21), that generates an equivalent system of equations known as the Bohm-Madelung system of equations [7, 8, 9]:

$$\partial_s A = -\frac{1}{2m}\left[A\partial_i \partial^i S + 2\partial_i A \partial^i S\right], \tag{22}$$

$$\partial_s S = -\left[\frac{1}{2m}\partial_i S \partial^i S + U - \frac{\hbar^2}{2m}\frac{\partial_i \partial^i A}{A}\right]. \tag{23}$$

It is very important that the last term in Eq. (23), which is the "quantum" potential of the so-called Bohm ψ-field [7, 8, 9, 23] in the interpretation of Ref. [7], coincides exactly with the dissipation energy Q in (19). At the same time, Eq. (22) is identical to the condition for $\partial A/\partial s$ in (17) and (20).

If we make substitution of the type:

$$P(q,s) = \psi\psi^* = [A(q,s)]^2, \tag{24}$$

then Eqs. (22) and (23) can be rewritten in the form:

$$\frac{\partial P}{\partial s} = -\frac{1}{m}\partial_i (P \cdot \partial^i S), \tag{25}$$

$$\partial_s S + \frac{\partial_i S \partial^i S}{2m} + U - \frac{\hbar^2}{4m}\left[\frac{\partial_i \partial^i P}{P} - \frac{1}{2}\frac{\partial_i P \partial^i P}{P^2}\right] = 0. \tag{26}$$

Here, Eq. (25) has a clear physical sense: $P(q,s)$ is the probability density to find a particle in a certain place of the 4-space, and $\partial_i S/m$ is, according to (13), the 4-velocity of this particle. In other words, Eq. (25) is nothing but the equation of continuity which indicates how the probability density $P(q,s)$ moves according to the laws of classical mechanics with a classical velocity at every point.

On the other hand, we can show that $P(q,s)$ is also the probability density function for the number of particle trajectories, that is substantiated in the following way. We assume that the influence of disturbing forces generated by the potential Q on a wave packet at an arbitrary point of the configuration space is proportional to the density of particle trajectories ($\psi\psi^* = A^2$) at this point. This implies that the disturbing forces do not practically perturb the packet, if the relation

$$\int Q\psi\psi^* dV \Rightarrow \min, \quad \int \psi\psi^* dV = 1, \tag{27}$$

is satisfied, where dV stands for an element of the configuration space volume. This means, in its turn, that the disturbing forces admit the absolute stability on the whole set of motions in the configuration space only if condition (27) is satisfied or, in other words, if the following obvious condition of the equivalent variational problem (for Q) fulfils:

$$\delta \int Q\psi\psi^* dV = \delta Q = 0. \tag{28}$$

The variational principle (28) is, in essence, the principle of least action of a perturbation.Below, we call it the principle of least action of perturbations by Chetaev [1]. Using the previous notations, we can write the following equality for Q:

$$Q = -\partial_s S - U - T = -\frac{\partial S}{\partial t} - U - \frac{1}{2}\partial_i S \partial^i S. \tag{29}$$

On the other hand, if (4) holds true, it is easy to show that:

$$\frac{1}{2m}\partial_i S \partial^i S = -\frac{1}{2mk^2\psi^2}\partial_i \psi \partial^i \psi + \frac{1}{2mk^2A^2}\partial_i A \partial^i A + ik\frac{1}{2mk^2A^2}\partial_i A \partial^i S. \tag{30}$$

Then it is necessary to perform the following subsequent substitutions. First, we substitute relation (30) in (29) and then introduce the result in the equation corresponding to the variational principle (28).

As a result of the indicated procedure of substitutions, we obtain the relation which is exactly equal to Eq. (15), i.e., to the extended equation (with regard for (3), (14), (15)] of type (8). Hence, the structural expressions and the necessary condition of stability which follow from it coincide with (16), (17), and (12), respectively. This means that, on the basis of the Chetaev variational principle (28), we get the independent confirmation of the fact that the physical nature of P(q, t) reflects really not only the traditional notion of the probability density for a particle to be at a certain place of the space according to the Bohm-Madelung equation of continuity (25) but plays also the role adequate to that of the probability density of the number of particle trajectories.

3. Neutrino oscillation measurements as verifiability criterion of different concepts of temporal evolution

Here the natural question raises: "What of the two equations realizing ideology of standard or parametrized RQM is substantiated and valid within the framework of modern experimental relativistic dynamics or, in other words, what of the two concepts of temporal evolution reflects nature of physical reality of our world more sufficiently?" It is interesting, that one of variants of the theoretical formulation of such a problem is stated for the first time by Fanchi [17, 22]. Because we will use the results of this work below, let us briefly consider the formulation of this problem.

Following [22], we consider the temporal model of vacuum-flavour mixing neutrino within the framework of the standard and pRQM models. At first, this is caused by the fact that, according to [22], differences between the standard and pRQM models are highlighted by developing a general formalism for vacuum-flavour mixing and then applying the formalism within the context of each theory. Secondly, the resulting two expressions for the probability of transition from one neutrino flavour to another are experimentally testable that is the one of purposes of our research, which is given in next sections.

Considering the temporal model of vacuum-flavour mixing, we imply the case of two-state mixing, i.e., we confine ourself to consideration of the two mass states $\{|\,\nu_j\rangle\}$ and two neutrino flavour states $\{|\,\nu_\alpha\rangle\}$, which may be written as the two-component column vectors:

$$\{|\nu_j\rangle\} = \begin{bmatrix} |\nu_1\rangle \\ |\nu_2\rangle \end{bmatrix}, \quad \{|\nu_\alpha\rangle\} = \begin{bmatrix} |\nu_e\rangle \\ |\nu_\mu\rangle \end{bmatrix}. \tag{31}$$

The temporal evolution equation for state may be written in terms of the temporal evolution operator $\mathbf{T}$ as:

$$\mathbf{T}|\nu_j\rangle = i\hbar\frac{\partial}{\partial\tau}|\nu_j\rangle = \mathbf{T}_j\ _{|\nu_j\rangle} \tag{32}$$

with the formal solution

$$\begin{bmatrix} |\nu_1(\tau)\rangle \\ |\nu_2(\tau)\rangle \end{bmatrix} = \begin{bmatrix} \exp(-i\frac{T_1\tau}{\hbar}) & 0 \\ 0 & \exp(-i\frac{T_2\tau}{\hbar}) \end{bmatrix} \begin{bmatrix} |\nu_1(0)\rangle \\ |\nu_2(0)\rangle \end{bmatrix}. \tag{33}$$

It is obvious [22], that the eigenvalue $\mathbf{T}_j$ of the temporal evolution operator $\mathbf{T}$ depends on the context of theory. In the standard theory, $\mathbf{T}_j$ is energy E_j of state j, while in pRQM it is the eigenvalue $\mathbf{K}_j$ of the mass operator for mass state j. Explicit free-particle expressions for the eigenvalues are:

$$(\mathbf{T}_j)_{Standard} = E_j = \left[p^2c^2 + m_j^2c^4\right]^{1/2}, \tau_{Standard} = t, \tag{34}$$

and

$$\begin{aligned}(\mathbf{T}_j)_{pPQM} = K_j &= \frac{\hbar^2}{2m_j}k_j^\mu k_{j\mu} = \\ &= \frac{\hbar^2}{2m_j}\left[(\omega_j/c^2) - \vec{k}_j\cdot\vec{k}_j\right], \quad (\tau)_{pRQM} = s.\end{aligned} \tag{35}$$

Here $\left\{k_j^\mu\right\} = \left\{\omega_j, \vec{k}_j\right\}$ denotes the energy-momentum four-vector in laboratory frame and m_j is the mass of state for a metric with signature $\{+,-,-,-\}$. According [22], in pRQM, all four component of the energy-momentum four-vector are observable and m_j is function of statistical values of the energy-momentum four-vector. In this case the energy ω_j of state j is treated in pRQM as an independent variable and does not satisfy an equation like (34), except in a statistical sense. By contrast m_j is an input parameter in the standard model. In this case the standard picture of neutrino oscillation by flavour mixing in vacuum assumes that the three-momentum is constant for all states and has magnitude $p^2 = h^2\vec{k}\cdot\vec{k}$.

It is known that a neutrino oscillation process is descried by hypothesizing that the mass basis $|\ \nu_j\rangle$ is related to the flavour basis $|\ \nu_\alpha\rangle$ by a unitary transformation U such that

$$\{|\nu_\alpha\rangle\} = U\{|\nu_j\rangle\} \equiv \begin{bmatrix} \cos\theta & \sin\theta \\ -\sin\theta & \cos\theta \end{bmatrix} \{|\nu_j\rangle\}, \tag{36}$$

and θ is the mixing angle of mass states in vacuum. For two flavours of neutrinos and the unitary transformation U we have the τ-dependent state:

$$|\nu_e(\tau)\rangle = e^{-iT_1\tau/\hbar}\cos\theta|\nu_1\rangle + e^{-iT_2\tau/\hbar}\sin\theta|\nu_2\rangle. \tag{37}$$

The probability of transition from $\nu_e(\tau)$ to ν_μ is the square of the magnitude of the transition matrix element

$$P(\nu_e \to \nu_\mu) = |\langle\nu_\mu|\nu_e(\tau)\rangle|^2 = \sin^2 2\theta \sin^2\frac{(T_2 - T_1)\tau}{2\hbar}. \tag{38}$$

Thus, the survival probability $P(\nu_e \to \nu_e)$ has the form

$$P\,(\nu_e \to \nu_e) = 1 - P(\nu_e \to \nu_\mu) = 1 - \sin^2 2\theta \sin^2 \frac{(T_2 - T_1)\tau}{2\hbar}. \quad (39)$$

It is obvious that, that standard model and pRQM model according to (34-35) differ in the explicit forms of T_j and τ. Therefore, the expression for the survival probability $P(\nu_e \to \nu_e)$, obtained in the framework of standard model and pRQM model, will have the next form:

$$\begin{aligned} P_{S\tan dard}(\nu_e \to \nu_e) &= 1 - \sin^2 2\theta \sin^2 \left[\frac{(m_2^2 - m_1^2)^2 c^4}{4\hbar E_v}\frac{L}{c}\right] = \\ &= 1 - \sin^2 2\theta \sin^2 \alpha_{S\tan dard}, \end{aligned} \quad (40)$$

where the coordinate time t is replaced in standard model by the distance L from the neutrino source to the detector, so that $t = L/c$. Note that for ultra-relativistic neutrino it is assumed that that $p_1 \approx p_2 = p$, and that $p/c \gg m_2, m_1$, and $p \approx E_\nu/c$, where $E_\nu = m_1c^2/(1-\beta^2)^{1/2}$ is the energy of the incident neutrino ($\beta = v/c$):

$$\begin{aligned} P_{pRQM}(\nu_e \to \nu_e) &= 1 - \sin^2 2\theta \sin^2 \left[\frac{(m_2 - m_1)c^2}{4\hbar}\frac{L}{c}\frac{(1-\beta^2)^{1/2}}{\beta}\right] = \\ &= 1 - \sin^2 2\theta \sin^2 \left[\frac{(m_2^2 - m_1^2)c^4}{4\hbar E_\nu}\frac{L}{c}\frac{m_1}{m_1 + m_2}\frac{1}{\beta}\right] = 1 - \sin^2 2\theta \sin^2 \alpha_{pRQM}. \end{aligned} \quad (41)$$

Based on well-known experiments on solar and reactor neutrino detection [24] one can to conclude that existence of neutrino oscillations is indisputably proved and to assume without loss of generality that the mass difference between neutrino flavour states is very small, i.e., $m_1 \approx m_2$. Thus, it is easy to see that in the case of ultra-relativistic neutrino (i.e., $\beta \to 1$) the dynamical factors $\alpha_{Standard}$ and α_{pRQM} are connected by the following simple expression:

$$\alpha_{pRQM} \cong 2\alpha_{Standard}. \quad (42)$$

or

$$\left(\Delta m_{21}^2\right)_{pRQM} \cong 2\left(\Delta m_{21}^2\right)_{Standard}, \quad (43)$$

where $\Delta m_{21}^2 = m_2^2 - m_1^2$.

In other words, the ratio of dynamical factors does not equal to 1 because of the difference in operators K_j between standard paradigm and pRQM. It was pointed out in Fanchi [22] that the pRQM mass operator is the direct result of the probabilistic basis of the theory. Consequently according to [22], the search for massive neutrinos is a test not only of the flavour-mixing hypothesis, but also the validity of standard model and pRQM.

4. Neutrino mass matrix via see-saw mechanism

It is obvious, that to analyse a general formalism for vacuum-flavour mixing of neutrino within the context of the standard and pRQM models we need the estimations of absolute value for neutrino mass based on modern experimental data for solar and atmospheric neutrinos. With that end in view we consider below the problem of determining the neutrino masses by use of special method for construction of neutrino mass matrix, which was described for the first time in [25], and advantages of the mass-squared differences and mixing parameters obtained from neutrino oscillation experiments.

It is known that one of the interesting mechanisms which can generate a small neutrino mass is the see-saw mechanism, in which the right-handed neutrino ν_R has a large Majorana mass M_N and the left-handed neutrino ν_L obtain a mass through leakage of the order of (m/M_N) with m is the Dirac mass. According to see-saw mechanism [26], the neutrino mass matrix M_ν is given by:

$$M_\nu \approx -M_D M_N^{-1} M_D^T,$$

where M_D and M_N are the Dirac and Majorana mass matrix respectively.

In order to obtain neutrino mass matrix M_ν that can give correct predictions on mass-squared differences and mixing parameters, Damanik [25] modified the the neutrino mass matrix M_ν by introducing one parameter δ to perturb the diagonal elements of M_ν. In the way that the perturbed mass matrix satisfies the requirement $Tr\,(M_\nu) \equiv 3P$. At the same time, the neutrino mass matrix M_ν can be written in the following form [25]:

$$M_\nu = \begin{pmatrix} P+2\delta & Q & Q \\ Q & P-\delta & Q \\ Q & Q & P-\delta \end{pmatrix}. \tag{44}$$

The eigenvalues of the neutrino mass matrix M_ν in (44) are:

$$\beta_{1,2} = P + \frac{Q}{2} + \frac{\delta}{2} \mp \frac{\sqrt{9\delta^2 - 6Q\delta + 9Q^2}}{2}, \tag{45}$$

$$\beta_3 = P - Q - \delta. \tag{46}$$

If the neutrino mass matrices M_ν in (44) is diagonalized by mixing matrix V in the equation

$$\begin{pmatrix} \nu_e \\ \nu_\mu \\ \nu_\tau \end{pmatrix} = V \begin{pmatrix} \nu_1 \\ \nu_2 \\ \nu_3 \end{pmatrix}, \tag{47}$$

where

$$V = \begin{pmatrix} \cos\theta & -\sin\theta & 0 \\ \sin\theta/\sqrt{2} & \cos\theta/\sqrt{2} & -1/\sqrt{2} \\ \sin\theta/\sqrt{2} & \cos\theta/\sqrt{2} & 1/\sqrt{2} \end{pmatrix}, \tag{48}$$

then we obtain

$$\tan^2(2\theta) = \frac{8Q^2}{(Q-3\delta)^2}, \tag{49}$$

and neutrino mass as follow:

$$m_1 = P + \frac{Q}{2} + \frac{\delta}{2} - \frac{\sqrt{9\delta^2 - 6Q\delta + 9Q^2}}{2}, \tag{50}$$

$$m_2 = P + \frac{Q}{2} + \frac{\delta}{2} + \frac{\sqrt{9\delta^2 - 6Q\delta + 9Q^2}}{2}, \tag{51}$$

$$m_3 = P - Q - \delta. \tag{52}$$

One can see that the obtained neutrino masses in Damanik's scenario [25] is an inverted hierarchy with masses: $\mid m_3 \mid < \mid m_1 \mid < \mid m_2 \mid$.

To solve the system of the equations (49)-(52) with respect to the values of neutrino masses it is necessary to supplement this system with the known values of mass-squared differences and mixing parameters, which obtained from neutrino oscillation experiments. From the recent global analysis of the neutrino oscillation data we refer to the following best-fit values of solar neutrino mass-squared differences [27]:

$$\Delta m_{21}^2 = (7.5^{+0.19}_{-0.20}) \times 10^{-5} \ \ eV^2, \tag{53}$$

with

$$\tan^2\theta_{12} = 0.452^{+0.035}_{-0.033}, \tag{54}$$

and for the atmospheric neutrino mass-squared differences [28]:

$$\Delta m_{32}^2 = (2.32^{+0.12}_{-0.08}) \times 10^{-3} \ \ eV^2. \tag{55}$$

If θ is the θ_{12} in (54), we have from (49) we have $\delta = -0.050909 \ Q$. If we insert this δ value into (50-52) then we obtain the neutrino mass as follow:

$$m_1 = P - 1.052610 \ Q, \tag{56}$$

$$m_2 = P + 2.001700 \ Q, \tag{57}$$

$$m_3 = P - 0.949910 \ Q. \tag{58}$$

Using the advantages of experimental data of neutrino oscillations in Eqs. (53) and (55) from Eqs.(56,57), we have the neutrino masses Table 1) obtained within the framework of standard model and pRQM model.

Table 1. Neutrino mass $|m_i|$ as function of dynamical factors $\alpha_{Standard}$ and α_{pRQM}.

	$\alpha_{Standard}$	α_{pRQM}
$\Delta m_{21}^2 \times 10^5 \ \ eV^2$	7.50	15.00
$\Delta m_{32}^2 \times 10^3 \ \ eV^2$	2.32	4.64
$tg^2\theta$	0.452	0.452
$\|m_1\|$	0.130955	0.185056
$\|m_2\|$	0.131141	0.185461
$\|m_3\|$	0.121975	0.172499

Thus, we estimated the neutrino masses within the framework of both models. So long as these estimations satisfy the known restriction on the total neutrino mass:

$$\sum m_a \leq 0.6eV, \quad \nu_a = \nu_e, \nu_\mu, \nu_\tau, \tag{59}$$

it is impossible to answer question, what of the two representations realizing the ideology of standard or parametrized RQM reflects the nature of physical reality of our world more sufficiently. A valid answer to this question can be given only when the experimental values of the neutrino masses will be obtained.

Below we will attempt to solve this problem even if at qualitative level using modern cosmological observations.

5. The dimension of the neutrino cloud and large-scale structure of Universe

In this section we compare the theoretical value of void structure of Universe obtained by use of foregoing alternative values of neutrino masses with the experimental value of average size of large-scale structure of Universe obtained on the basis of analysis of so-called the matter power spectrum of density fluctuation in space-time. The main purpose of such a comparison is the reveal of dominant role one of the two physical concepts of temporal evolution, which are considered above.

So, according to Muraki [30, 31], the possible role of massive neutrinos in our Universe is as follows. Massive neutrinos are attracted by the gravitational force. However, because of the degeneracy force, they will not collapse to form a supper black hall. The Jeans mass of neutrinos grows and fluctuation of the baryonic matter will be erased. The fluctuation of the baryonic matter can grow only at the border between neutrino bubbles. Consequently, it is possible to assume that the observed void structure of the Universe was formed by the neutrino degenerate force and grows of the density fluctuation at the border between voids.

Using the analogy with neutron star, whose radius [1] can be obtained by the balance between the degenerate force of neutrons and the gravitational force, Muraki [30, 31] in bounds of his script(scenario) got the next expression for the radius of a neutrino cloud:

$$r_\nu = 1.2\left(\frac{\hbar^2}{G}\right)\left(\frac{1}{M_\nu}\right)^{1/3}\left(\frac{1}{m_\nu}\right)^{8/3}, \tag{60}$$

where the neutrino cloud mass is

$$M_\nu = (4/3)\,\pi r_\nu^3 \times 110 \times 10^6\left[m^3\right] \times m_\nu\left[eV\right] \times 1.7 \times 10^{-36}\left[kg\right]. \tag{61}$$

However, in this scenario it is necessary to take into account also the effect of baryonic matter and dark matter [30, 31]. According to the last analysis of the WMAP7 data, total matter fraction Ω_M is $\Omega_M = 26.6 \pm 2.9\%$ [32], while the energy density of all types of neutrinos is $\Omega_\nu \leq 0.014$ [33] [2]. That means that the mass of (dark matter + hydrogen gas) is 19 times heavier than the mass of neutrinos. Therefore if we replace $110m_\nu$ by $2090m_\nu$ and at the same time take into account the contribution to the gravitational mass from other neutrinos, we will obtain modern expression of Muraki relation for the neutrino cloud diameter:

$$d_\nu = 7.1\left[Mpc\right] \times m_\nu^{-3/2}\left[eV\right]. \tag{62}$$

Using this expression it is easy to estimate the average diameter of neutrino cloud for different neutrino masses predetermined by the corresponding dynamical factors $\alpha_{Standard}$ and α_{pRQM} (see Table 1):

$$m_\nu^{pRQM} = 0.185461eV \rightarrow d_\nu^{pRQM} \cong 90Mpc, \tag{63}$$

[1] The canonical radius of a neutron star [29] can be obtained by the balance between the degenerate force of neutrons and the gravitational force [30, 31]: $r_n = 1.2(\hbar^2/G)(1/M_n)1/3(1/m_n)^{8/3} \simeq 10$ km, where M_n corresponds to a 1.4 solar mass.

[2] In terms of neutrino mass this restriction ($\Omega_\nu \leq 0.014$) means that $\sum m_\nu \leq 0.6\;\; eV$ [24, 33], which is satisfied by the values of triplet neutrino masses listed in Table 1.

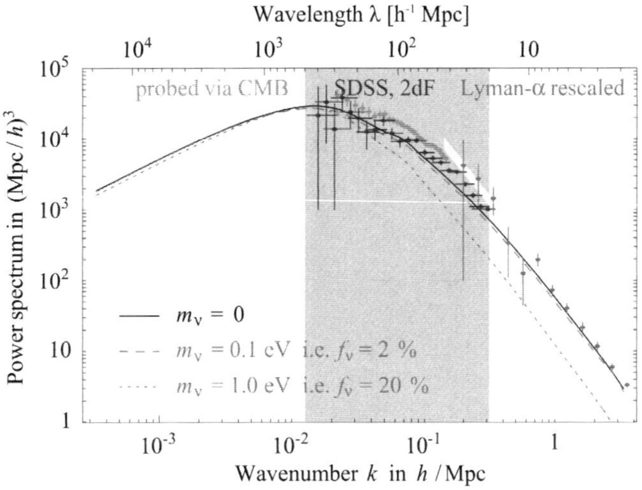

Figure 1. The matter power spectrum $P(k)$ predicted by the best-fit ΛCDM cosmological model (continuous curve) and how neutrino masses affect it (dashed curves). Measurements at different scales have been performed with different techniques that slightly overlap. The data points do not show the overall uncertainty that plagues galaxy surveys (SDSS, 2dF) at intermediate scales and especially Lyman-α data at smaller scales i.e. at larger k. Here $f_\nu \equiv \rho_\nu/\rho_{DM}$, where ρ_ν – neutrino density, ρ_{DM} – density of dark matter. The wide vertical stripe identifies an interval maximum best-fit ΛCDM cosmological model (continuous curve), corresponding to the range of wavelength values $\lambda = 20 \div 500$ $[h/Mpc]$. Adopted from [24].

$$m_\nu^{Standard} = 0.131141\ eV \to d_\nu^{Standard} \cong 150 Mpc. \qquad (64)$$

On the other hand, it is known, that the average size of typical observed void structure (dark matter + hydrogen gas), which plays the role of characteristic dimension of large-scale structure of the Universe, are predetermined by so-called of the matter power spectrum of density fluctuation in space-time (Fig. 1).

It is easy to show (see Figure 1 [24]), that the range of large-scale structure (LSS) set by the range of the wavelength $\lambda = 20 \div 500$ $[h^{-1}Mpc]$ $(h = 0.704$ [32]) is described by the average characteristic dimension of LSS of the Universe:

$$\langle \lambda_{LSS} \rangle \sim 90 \;\; Mpc. \qquad (65)$$

This is confirmed also by the known results of Tegmark *et al.* [34].

Comparative analysis of expressions (65) and (63,64) shows that the characteristic dimensions of LSS (65) and neutrino cloud (63) practically coincide:

$$\langle \lambda_{LSS} \rangle \sim d_\nu^{pRQM}. \qquad (66)$$

This result shows that the physical concept of temporal evolution described by Stückelberg equation is more adequate.

6. Conclusion

Certainly, it is impossible to consider obtained result as a proof, but it is a strong argument in favour of the Stückelberg concept of temporal evolution, which reflects the nature of physical reality of our world more sufficiently than the standard (Dirac) concept. It is clear that direct proof in favour of one or the other physical concept of temporal evolution will be possible only when the experimental value of neutrino mass will be determined.

In this sense, it is appropriate to note here that "... time and again a breakthrough in the field of new physics was related to neutrino. The first theory of β-decay proposed by Fermi was based on the Pauli hypothesis of existence of neutrino. Phenomenological (V-A) theory of weak interactions has arisen from the two component neutrino theory considered by Landau, Lee, Yang and Salam. Evidence in favour of the Standard Model of weak interaction of Glashow, Weinberg and Salam had been received on a neutrino beam (discovery of neutral currents)" [35]. At the present time, we observe a new phenomenon - neutrino oscillations, which result from temporal evolution of massive neutrinos and indicate new physics beyond the Standard Model. Undoubtedly, the discovery of neutrino mass in the near future (for example, by research of high-energy edge of the tritium β-spectrum) can become the basis for reconsideration not only of the key issues of quantum theory, but a holistic understanding of the nature of physical reality.

References

[1] Rusov V D, Vlasenko D S and Mavrodiev S Cht 2011 *Annals of Physics* **326** 1807-15
[2] Rusov V D, Vlasenko D S and Mavrodiev S Cht 2009 *Ukr. J. Phys.* **54** 1131-38
[3] Chetaev N G 1931 *Educational notes of Kazan University* 91 book **4** Mathematics N1 p 3
[4] Chetaev N G 1936 *Sci. Proc. Kazan Aircraft Inst.* **5** 3
[5] Chetaev N G 1962 *Stability of Motion. Works on Analytic Mechanics (in Russian)* (Moscow: Nauka)
[6] Chetaev N G 1961 *The Stability of Motion* (Pergamon Press)
[7] Bohm D 1952 *Phys. Rev.* **85** 166
Bohm D 1952 *Phys. Rev.* **85** 80
[8] Holland P R 1993 *Quantum Theory of Motion. An account of the de Broglie-Bohm Causal Interpretation of Quantum Mechanics* (Cambridge: Cambridge Univ. Press)
[9] Durr D and Teufel S 2009 *Bohmian Mechanics* (Berlin: Springer)
[10] Girardi G S, Rimini A and Weber T 1986 *Phys. Rev.* **34** 470
[11] Tumulka R 2006 *J. Stat. Phys.* **125** 821
[12] Nelson E 1966 *Phys. Rev.* **150** 1079
Nelson E 1985 *Quantum Fluctuations* (Princeton: Princeton Univ. Press)
[13] Nottale L 1993 *Fractal Space-Time and Microphysics: Towards a Theory of Scale Relativity* (Singapore: World Scientific)
[14] Santamato E 1984 *Phys. Rev.* **29** 216
Santamato E 1984 *J. Math. Phys.* **25** 2477
Santamato E 1985 *Phys. Rev.* **32** 2615
[15] Groessing G 2008 *Phys. Lett.* **372** 4556
Groessing G 2009 *Physica A* **388** 811
Groessing G 2010 *Entropy* **12** 4473
[16] Wetterich C 2010 *Annals of Physics* **325** 852
Wetterich C 2009 *J. Phys.: Conf. Ser* **174** 012008
[17] Fanchi J R 1993 *Parametrized Relativistic Quantum Theory* (Dordrecht: Kluwer Academic)
[18] Adler S L and Bassi A 2009 *Science* **325** 275 *arXiv*:0912.2211
[19] Chetaev N G 1958 *Sov Appl. Math. Mech.* **22** 487
[20] L. de Broglie 1928 *In Electrons et Photons: Rapports et Discussions du Cinquieme Conseil de Physique, edited by J. Bordet* (Paris: Gauthier-Villars) p 105
G. Bacciagaluppi and A. Valentini 2009 *Quantum theory at the Cross-Roads: Reconsidering the 1927 Solvay Conference* (Cambridge: Cambridge University Press)
[21] Durr D, Goldstein S and Zanghi N 1992 *J. Stat. Phys.* **67** 843
[22] Fanchi J R 1998 *Found. Phys.* **28** 1521
[23] Wyatt R E 2005 *Quantum Dynamics with Trajectories. Introduction to Quantum Hydrodynamics* (New York: Springer)
[24] Strumia A, Vissani F Neutrino masses and mixing and... *arXiv*:hep-ph/0606054
[25] Damanik A 2011 *Mod. Phys. Lett.* **26** 567
[26] Gell-Mann M, Ramond P and Slansky R 1979 *Suppergravity* eds P van Niewenhuzen and Freedman D Z (Amsterdam: North Holland)

[27] Gando A et al (The KamLAND Collaboration) Enhanced Constraints on θ_{13} from a Three-Flavor Oscillation Analysis of Reactor Antineutrinos at KamLAND *arXiv*:1009.4771
[28] Adamson P et al The MINOS Collaboration 2011 *Phys. Rev. Lett.* **106** 181801
[29] Potekhin A Yu 2010 *Physics-Yspekhi* **53** 1235-56
[30] Muraki Y 2005 *Neutrino mass and the Void structure of our Universe* 29th ICRC Pune 9 pp 227-30
[31] Muraki Y 1990 *The tau neutrino mass obtained from recent cosmic data* 21st ICRC 6-19 Jan. 1990 (Australia: Adelaide) vol 10 pp 54-57.
[32] Jarosik N et al. (WMAP Collaboration) Seven-Year Wilkinson Microwave Anisotropy Probe (WMAP) Observations: Sky Maps, Systematic Errors and Basic Results ApJS 192 (2011) 14 *arXiv*:1001.4744
[33] Rubakov V A 2011 *Physics-Yspekhi* **54** 633-41
[34] Tegmark M 2002 *Science* **296** 1427
[35] Bilenkii S M 2003 *Physics-Yspekhi* **46** 1137

EmerQuM 11: Emergent Quantum Mechanics 2011 IOP Publishing
Journal of Physics: Conference Series **361** (2012) 012034 doi:10.1088/1742-6596/361/1/012034

Eigenforms, Discrete Processes and Quantum Processes

Louis H. Kauffman

Department of Mathematics, 851 South Morgan Street, University of Illinois at Chicago, Chicago, Illinois 60607-7045

E-mail: kauffman@uic.edu

Abstract. This essay is a discussion of the concept of eigenform, due to Heinz von Foerster, and its relationship with discrete physics and quantum mechanics. We interpret the square root of minus one as a simple oscillatory process - a clock, and as an eigenform. By taking a generalization of this identification of i as a clock and eigenform, we show how quantum mechanics emerges from discrete physics.

1. Introduction

This paper is an attempt to bridge two worlds, the world of second order cybernetics and the world of foundations of physics. We will begin with elementary ideas of recursion in a cybernetic context and use these ideas to discuss discrete dynamical systems and show how quantum mathematical structures emerge from discrete dynamics and the desire to describe such dynamics to include temporal observations that keep track of the ticks of the clock.

Our essay begins with explication of the notion of eigenform as pioneered by Heinz von Foerster in his papers [4, 5, 6, 7] and explored in papers of the author [11, 12, 22, 23]. In [5] The familiar objects of our existence can be seen as tokens for the behaviours of the organism, creating apparently stable forms. Such an attitude toward objects makes it impossible to discriminate between the object as an element of a world and the object as a token or symbol that is simultaneously a process.

The notion of an eigenform is inextricably linked with second order cybernetics. One starts on the road to such a concept as soon as one begins to consider a pattern of patterns, the form of form or the cybernetics of cybernetics. Such concepts appear to loop around upon themselves, and at the same time they lead outward to new points of view. Such circularities suggest a possibility of transcending the boundaries of a system from within. When a circular concept is called into being, the boundaries turn inside out.

An object, in itself, is a symbolic entity, participating in a network of interactions, taking on its apparent solidity and stability from these interactions. We ourselves are such objects, we as human beings are "signs for ourselves", a concept originally due to the American philosopher C. S. Peirce [10]. Eigenforms are mathematical companions to Peirce's work.

In an observing system, what is observed is not distinct from the system itself, nor can one make a separation between the observer and the observed. The observer and the observed stand together in a coalescence of perception. From the stance of the observing system all objects are

Published under licence by IOP Publishing Ltd

non-local, depending upon the presence of the system as a whole. It is within that paradigm that these models begin to live, act and enter into conversation with us.

The central metaphor of this paper is the temporal nexus where time is implicit, and time is explicit and keeping time. In the nexus there is neither form nor sign, motion nor time. Time, the measurement of time and time's indication all emerge at once from the nexus in the form of action that is embodied in it. The metaphor suggests that it is no accident that deeper physical reality is revealed when mere numerical time t is replaced by the time of the nexus it. The time of the nexus is at once flowing, beyond motion, an eigenform, a geometric operator and a discrete dynamics counting below where counting cannot go.

Taking the point of view described above we rexamine the square root of minus one starting with the simple oscillatory processes that ensue from iterating the operator $Rx) = -1/x$ whose fixed points are i and $-i$. By looking at i in this way one finds the essential patterns of other discrete processes and we detail how this works in Section 7. Then by weaving the temporal process for i with the general discrete process we find, in a natural way, that discrete processes are described by the Heisenberg commutator

$$[p, q] = i\hbar.$$

This paper is a sketch of this point of view.

2. Objects as Tokens for Eigenbehaviours

In his paper "Objects as Tokens for Eigenbehaviours" [5] von Foerster suggests that we think seriously about the mathematical structure behind the constructivist doctrine that perceived worlds are worlds created by the observer. At first glance such a statement appears to be nothing more than solipsism. At second glance, the statement appears to be a tautology, for who else can create the rich subjectivity of the immediate impression of the senses? In that paper he suggests that the familiar objects of our experience are the fixed points of operators. These operators are the structure of our perception. To the extent that the operators are shared, there is no solipsism in this point of view. It is the beginning of a mathematics of second order cybernetics.

Consider the relationship between an observer O and an "object" A. "The object remains in constant form with respect to the observer". This constancy of form does not preclude motion or change of shape. Form is more malleable than the geometry of Euclid. In fact, ultimately the form of an "object" is the form of the distinction that "it" makes in the space of our perception. In any attempt to speak absolutely about the nature of form we take the form of distinction for the form. (paraphrasing Spencer-Brown [3]). It is the form of distinction that remains constant and produces an apparent object for the observer. How can you write an equation for this? We write

$$O(A) = A.$$

The object A is a fixed point for the observer O. The object is an eigenform. We must emphasize that this is a most schematic description of the condition of the observer in relation to an object A. We record only that the observer as an actor (operator) manages to leave the (form of) the object unchanged. This can be a recognition of symmetry, but it also can be a description of how the observer, searching for an object, makes that object up (like a good fairy tale) from the very ingredients that are the observer herself.

And what about this matter of the object as a token for eigenbehaviour? This is the crucial step. We forget about the object and focus on the observer. We attempt to "solve" the equation $O(A) = A$ with A as the unknown. Not only do we admit that the "inner" structure of the object is unknown, we adhere to whatever knowledge we have. We can start anew from the

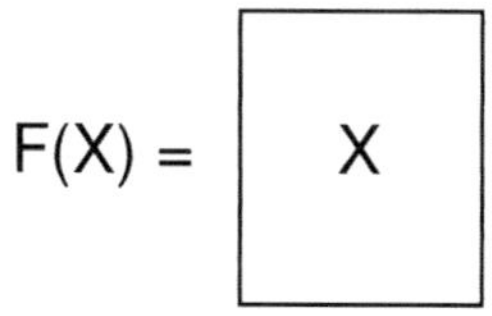

Figure 1.

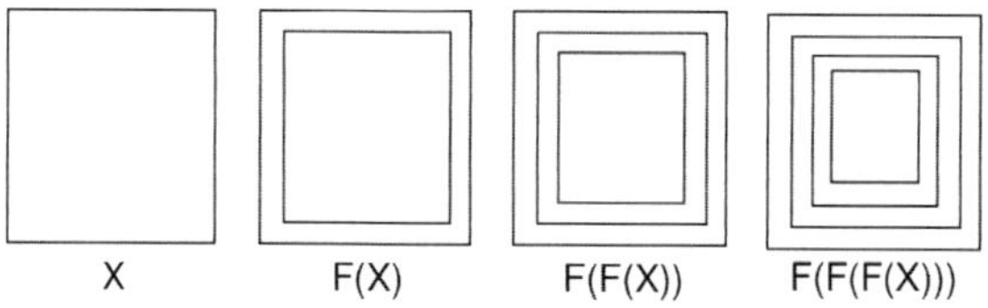

Figure 2.

dictum that the perceiver and the perceived arise together in the condition of observation. This is mutuality. Neither perceiver nor the perceived have priority over the other. A distinction has emerged and with it a world with an observer and an observed. The distinction is itself an eigenform.

3. The Eigenform Model

We have seen how the concept of an object has evolved. The notion of a fixed object has become the notion of a process that produces the apparent stability of an object. This process can be simplified in modelling to become a recursive process where a rule or rules are applied time and time again. The resulting object is the fixed point or eigenform of the process, and the process itself is the eigenbehaviour.

In this way we have a model for thinking about object as token for eigenbehaviour. This model examines the result of a simple recursive process carried to its limit. For example, suppose that we have a recursion as shown in Figure 1.

Each step in the process encloses the results of the previous step within a box. In Figure 2 we illustrate the first few steps of the process applied to an empty box X:

If we continue this process, then successive nests of boxes resemble one another, and in the limit of infinitely many boxes, we find that the infinite nest of boxes is invariant under the addition of one more surrounding box. Hence this infinite nest of boxes is a fixed point for the recursion. In other words, if J denotes the infinite nest of boxes, then $J = F(J)$ as shown in Figure 3. This equation is a description of a state of affairs. The form of an infinite nest of boxes is invariant under the operation of adding one more surrounding box.

In the process of observation, we interact with ourselves and with the world to produce stabilities that become the objects of our perception. These objects, like the infinite nest of boxes, often go beyond the specific properties of the world in which we operate. We make an imaginative leap to complete such objects to become tokens for eigenbehaviours. It is impossible to make an infinite nest of boxes. We do not make it. We imagine it. And in imagining that infinite nest of boxes, we arrive at the eigenform.

Sometimes one stylizes the structure by indicating where the eigenform X reenters its own indicational space by an arrow or other graphical device. See Figure 4 for this stylization for the case of the nested boxes.

Journal of Physics: Conference Series **361** (2012) 012034 doi:10.1088/1742-6596/361/1/012034

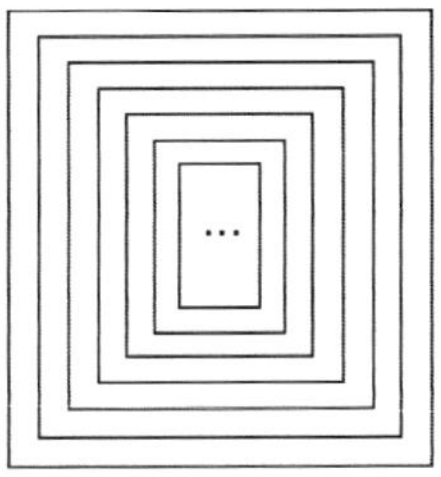

J = F(F(F(F(F(...)))))

J = F(J)

Figure 3.

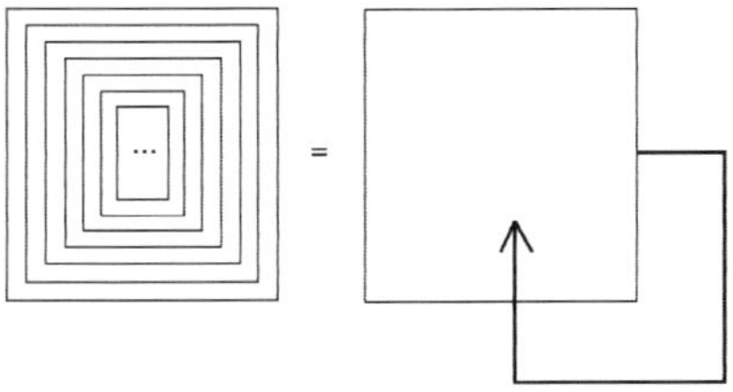

Figure 4.

An object is an amphibian between the symbolic and imaginary world of the mind and the complex world of personal experience. The object, when viewed as process, is a dialogue between these worlds. The object when seen as a sign for itself, or in and of itself, is imaginary. The perceiving mind is itself an eigenform of its own perception.

4. The Square Root of Minus One is a Clock

The purpose of this section is to place i, the square root of minus one, and its algebra in the context of eigenform and reflexivity. We begin by starting with a simple periodic process that is associated directly with the classical attempt to solve for i as a solution to a quadratic equation. We take the point of view that solving $x^2 = ax + b$ is the same (when $x \neq 0$) as solving

$$x = a + b/x,$$

and hence is a matter of finding a fixed point. In the case of i we have

$$x^2 = -1$$

and so desire a fixed point

$$x = -1/x.$$

There are no real numbers that are fixed points for this operator and so we consider the oscillatory process generated by

$$R(x) = -1/x.$$

The fixed point would satisfy

$$i = -1/i$$

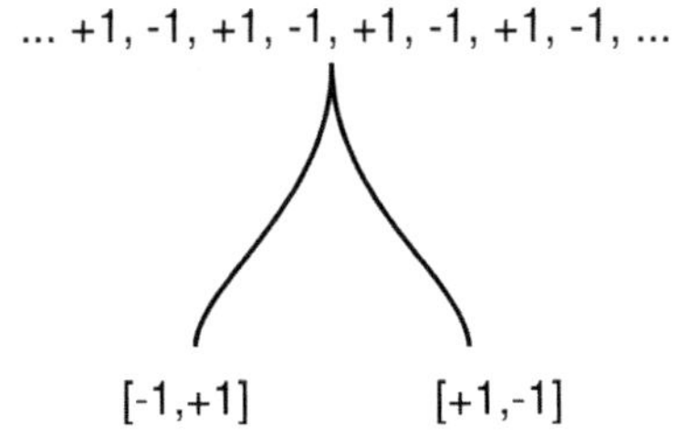

Figure 5.

and multiplying, we get that

$$ii = -1.$$

On the other hand the iteration of R yields

$$1, R(1) = -1, R(R(1)) = +1, R(R(R(1))) = -1, +1, -1, +1, -1, \cdots.$$

The square root of minus one is a perfect example of an eigenform that occurs in a new and wider domain than the original context in which its recursive process arose. The process has no fixed point in the original domain.

Looking at the oscillation between $+1$ and -1, we see that there are naturally two phase-shifted viewpoints. We denote these two views of the oscillation by $[+1, -1]$ and$[-1, +1]$. These viewpoints correspond to whether one regards the oscillation at time zero as starting with $+1$ or with -1. See Figure 5.

We shall let $I\{+1, -1\}$ stand for an undisclosed alternation or ambiguity between $+1$ and -1 and call $I\{+1, -1\}$ an iterant. There are two iterant views: $[+1, -1]$ and $[-1, +1]$.

Given an iterant $[a, b]$, we can think of $[b, a]$ as the same process with a shift of one time step. These two iterant views, seen as points of view of an alternating process, will become the square roots of negative unity, i and $-i$.

We introduce a temporal shift operator η such that

$$[a, b]\eta = \eta[b, a]$$

and

$$\eta\eta = 1$$

for any iterant $[a, b]$, so that concatenated observations can include a time step of one-half period of the process

$$\cdots abababab \cdots.$$

We combine iterant views term-by-term as in

$$[a, b][c, d] = [ac, bd].$$

We now define i by the equation

$$i = [1, -1]\eta.$$

This makes i both a value and an operator that takes into account a step in time.

We calculate

$$ii = [1,-1]\eta[1,-1]\eta = [1,-1][-1,1]\eta\eta = [-1,-1] = -1.$$

Thus we have constructed the square root of minus one by using an iterant viewpoint. In this view i represents a discrete oscillating temporal process and it is an eigenform for $R(x) = -1/x$, participating in the algebraic structure of the complex numbers. In fact the corresponding algebra structure of linear combinations $[a,b] + [c,d]\eta$ is isomorphic with 2×2 matrix algebra and iterants can be used to construct $n \times n$ matrix algebra. We treat this generalization elsewhere [20, 22, 23].

The Temporal Nexus. *We take as a matter of principle that the usual real variable t for time is better represented as it so that time is seen to be a process, an observation and a magnitude all at once.* This principle of "imaginary time" is justified by the eigenform approach to the structure of time and the structure of the square root of minus one.

As an example of the use of the Temporal Nexus, consider the expression $x^2 + y^2 + z^2 + t^2$, the square of the Euclidean distance of a point (x, y, z, t) from the origin in Euclidean four-dimensional space. Now replace t by it, and find

$$x^2 + y^2 + z^2 + (it)^2 = x^2 + y^2 + z^2 - t^2,$$

the squared distance in hyperbolic metric for special relativity. By replacing t by its process operator value it we make the transition to the physical mathematics of special relativity.

5. Quantum Physics, Eigenvalue and Eigenform

In quantum modelling [11], the state of a physical system is represented by a vector in a Hilbert space. As time goes on the vector undergoes a unitary evolution in the Hilbert space. Observable quantities correspond to Hermitian operators H and vectors v that have the property that the application of H to v results in a new vector that is a multiple of v by a real factor λ. Thus

$$Hv = \lambda v.$$

One says that v is an eigenvector for the operator H, and that λ is the eigenvalue.

The theory of eigenforms is a generalization that shifts eigenvectors to eigenforms. One can speculate that there might be a generalization of quantum mechanics where the eigenvectors are replaced by eigenforms and the Hilbert space involves superpositions of forms. In fact, we have formalized some of these ideas in our theory of quantum knots, but will not discuss that avenue in this paper. Here we are concerned with eigenforms in relation to discrete processes. The simplest case of this way of thinking is the treatment of i that we have used, making it into a discrete clock and making it possible to view i itself as an eigenform that represents this elementary discrete process. In the next sections we will analyse discrete processes in more generality and show how they partake of the patterns of elementary quantum physics.

6. The Wave Function in Quantum Mechanics and The Square Root of Minus One

One can regard a wave function such as $\psi(x,t) = exp(i(kx - wt))$ as containing a micro-oscillatory system with the special synchronizations of the iterant view $i = [+1,-1]\eta$. It is these synchronizations that make the big eigenform of the exponential work correctly with respect to differentiation, allowing it to create the appearance of rotational behaviour, wave behaviour and the semblance of the continuum. In other words, we are suggesting that once can take a temporal view of the well-known equation of Euler:

$$e^{i\theta} = cos(\theta) + isin(\theta)$$

by regarding the i in this equation as an iterant, as discrete oscillation between -1 and $+1$. One can blend the classical geometrical view of the complex numbers with the iterant view by thinking of a point that orbits the origin of the complex plane, intersecting the real axis periodically and producing, in the real axis, a periodic oscillation in relation to its orbital movement in the two dimensional space. The special synchronization is the algebra of the time shift embodied in

$$\eta\eta = 1$$

and

$$[a, b]\eta = \eta[b, a]$$

that makes the algebra of $i = [1, -1]\eta$ imply that $i^2 = -1$. This interpretation does not change the formalism of these complex-valued functions, but it does change one's point of view and in the next section we will show how the properties of i as a discrete dynamical system are found in any such system.

7. Time Series and Discrete Physics

We have just reformulated the complex numbers and expanded the context of matrix algebra to an interpretation of i as an oscillatory process and matrix elements as combined spatial and temporal oscillatory processes (in the sense that $[a, b]$ is not affected in its order by a time step, while $[a, b]\eta$ includes the time dynamic in its interactive capability, and 2×2 matrix algebra is the algebra of iterant views $[a, b] + [c, d]\eta$).

We now consider elementary discrete physics in one dimension. Consider a time series of positions

$$x(t) : t = 0, \Delta t, 2\Delta t, 3\Delta t, \cdots.$$

We can define the velocity $v(t)$ by the formula

$$v(t) = (x(t + \Delta t) - x(t))/\Delta t = Dx(t)$$

where D denotes this discrete derivative. In order to obtain $v(t)$ we need at least one tick Δt of the discrete clock. Just as in the iterant algebra, we need a time-shift operator to handle the fact that once we have observed $v(t)$, the time has moved up by one tick.

We adjust the discrete derivative. We shall add an operator J that in this context accomplishes the time shift:

$$x(t)J = Jx(t + \Delta t).$$

We then redefine the derivative to include this shift:

$$Dx(t) = J(x(t + \Delta t) - x(t))/\Delta t.$$

This readjustment of the derivative rewrites it so that the temporal properties of successive observations are handled automatically.

Discrete observations do not commute. Let A and B denote quantities that we wish to observe in the discrete system. Let AB denote the result of first observing B and then observing A. The result of this definition is that a successive observation of the form $x(Dx)$ is distinct from an observation of the form $(Dx)x$. In the first case, we first observe the velocity at time t, and then x is measured at $t + \Delta t$. In the second case, we measure x at t and then measure the velocity.

We measure the difference between these two results by taking a commutator

$$[A, B] = AB - BA$$

and we get the following computations where we write $\Delta x = x(t + \Delta t) - x(t)$.

$$x(Dx) = x(t)J(x(t + \Delta t) - x(t)) = Jx(t + \Delta t)(x(t + \Delta t) - x(t)).$$

$$(Dx)x = J(x(t + \Delta t) - x(t))x(t).$$

$$[x, Dx] = x(Dx) - (Dx)x = (J/\Delta t)(x(t + \Delta t) - x(t))^2 = J(\Delta x)^2/\Delta t$$

This final result is worth recording:

$$[x, Dx] = J(\Delta x)^2/\Delta t.$$

From this result we see that the commutator of x andDx will be constant if $(\Delta x)^2/\Delta t = K$ is a constant. For a given time-step, this means that

$$(\Delta x)^2 = K\Delta t$$

so that

$$\Delta x = \pm\sqrt{(K\Delta t)}$$

This is a Brownian process with diffusion constant equal to K.

Thus we arrive at the result that any discrete process viewed in this framework of discrete observation has the basic commutator

$$[x, Dx] = J(\Delta x)^2/\Delta t,$$

generalizing a Brownian process and containing the factor $(\Delta x)^2/\Delta t$ that corresponds to the classical diffusion constant. It is worth noting that the adjustment that we have made to the discrete derivative makes it into a commutator as follows:

$$Dx(t) = J(x(t + \Delta t) - x(t))/\Delta t = (x(t)J - Jx(t))\Delta t = [x(t), J]/\Delta t.$$

By replacing discrete derivatives by commutators we can express discrete physics in many variables in a context of non-commutative algebra (non-commutative worlds as in [20]).

In the next section we use the temporal nexus (the square root of minus one as a clock) and rewrite these commutators to match quantum mechanics.

8. Epilogue and Simplicity

Finally, we arrive at the simplest place. Time and the square root of minus one are inseparable in the temporal nexus. The square root of minus one is a symbol and algebraic operator for the simplest oscillatory process. As a symbolic form, i is an eigenform satisfying the equation

$$i = \ 1/i.$$

One does not have an increment of time all alone as in classical t. One has it, a combination of an interval and the elemental dynamic that is time. With this understanding, we can return to the commutator for a discrete process and use it for the temporal increment.

We found that discrete observation led to the commutator equation

$$[x, Dx] = J(\Delta x)^2/\Delta t$$

which we will simplify to

$$[q, p/m] = (\Delta x)^2/\Delta t.$$

taking q for the position x and p/m for velocity, the time derivative of position and ignoring the time shifting operator on the right hand side of the equation.

Understanding that Δt should be replaced by$i\Delta t$, and that, by comparison with the physics of a process at the Planck scale one can take

$$(\Delta x)^2/\Delta t = \hbar/m,$$

we have

$$[q, p/m] = (\Delta x)^2/i\Delta t = -i\hbar/m,$$

whence

$$[p, q] = i\hbar,$$

and we have arrived at Heisenberg's fundamental relationship between position and momentum. This mode of arrival is predicated on the recognition that only it represents a true interval of time. In the notion of time there is an inherent clock or an inherent shift of phase that is making a synchrony in our ability to observe, a precise dynamic beneath the apparent dynamic of the observed process. Once this substitution is made, once the correct imaginary value is placed in the temporal circuit, the patterns of quantum mechanics appear. In this way, quantum mechanics can be seen to emerge from the discrete.

The problem that we have examined in this paper is the problem to understand the nature of quantum mechanics. In fact, we hope that the problem is seen to disappear the more we enter into the present viewpoint. A viewpoint is only on the periphery. The iterant from which the viewpoint emerges is in a superposition of indistinguishables, and can only be approached by varying the viewpoint until one is released from the particularities that a point of view contains.

References

[1] H. P. Barendregt, The Lambda Calculus - Its Syntax and Semantics, North Holland Pub. (1981,1985).
[2] H. Bortoft, The Whole - Counterfeit and Authentic, Systematics , vol. 9, No. 2, Sept. (1971), 43-73.
[3] G. Spencer-Brown, Laws of Form, George Allen and Unwin Ltd. (1969).
[4] Heinz von Foerster, Observing Systems, The Systems Inquiry Series, Intersystems Publications (1981).
[5] Heinz von Foerster, Objects: tokens for (eigen-) behaviors, in "Observing Systems," The Systems Inquiry Series, Intersystems Publications (1981), pp. 274 - 285.
[6] Heinz von Foerster, Notes on an epistemology for living things, in: Observing Systems, The Systems Inquiry Series, Intersystems Publications (1981), pp. 258 - 271.
[7] Heinz von Foerster, On constructing a reality, in: Observing Systems, The Systems Inquiry Series, Intersystems Publications (1981), pp. 288 - 309.
[8] L. H. Kauffman, Self-reference and recursive forms, Journal of Social and Biological Structures (1987), 53-72.
[9] L. H. Kauffman, Knot logic, in: Knots and Applications, ed. by L. H. Kauffman, World Scientific Pub. Co. (1995), pp. 1-110.
[10] L. H. Kauffman, The mathematics of Charles Sanders Peirce, in Cybernetics and Human Knowing, Volume 8, No. 1-2, (2001), pp. 79-110.
[11] Louis H. Kauffman, Eigenform, Kybernetes - The Intl J. of Systems and Cybernetics, Vol. 34, No. 1/2 (2005), Emerald Group Publishing Ltd, p. 129-150.
[12] Louis H. Kauffman, Eigenforms - Objects as Tokens for Eigenbehaviors, Cybernetics and Human Knowing, Vol. 10, No. 3-4, 2003, pp. 73-90.

[13] F.W. Lawvere, Introduction to "Toposes, Algebraic Geometry and Logic", Springer Lecture Notes on Mathematics Vol. 274 (1970) , pp. 1-12.
[14] B. B. Mandelbrot, The Fractal Geometry of Nature, W. H. Freeman and Company (1977, 1982).
[15] W. S. McCulloch, What is a number that a man may know it, and a man, that he may know a number?, in: Embodiments of Mind, MIT Press (1965), pp. 1-18.
[16] B. Piechocinska, Physics from Wholeness, PhD. Thesis, Uppsala Universitet (2005).
[17] J. J. Sakurai, Modern Quantum Mechanics, Benjamin/Cummings Publishing Company, Inc. (1985).
[18] D. Scott, Relating theories of the lambda calculus, in: To H. B. Curry: Essays on Combinatory Logic, Lambda Calculus and Formalism, (P. Seldin and R. Hindley eds.), Academic Press (1980), pp. 403-450.
[19] L. Wittgenstein, Tractatus Logico-Philosophicus, Routledge and Kegan Paul Ltd, London and New York (1922).
[20] Louis H. Kauffman, Non-Commutative Worlds, New Journal of Physics, Vol. 6, (2004), 173 (47 pages).
[21] Patrick DeHornoy, Braids and Self-Distributivity, Birkhauser (2000).
[22] Louis H. Kauffman, Reflexivity and Eigenform - The Shape of Process, Constructivist Foundations, Vol. 4, No. 3, July 2009, pp. 121-137.
[23] Louis H. Kauffman, Reflexivity and Foundations of Physics, In: Search for Fundamental Theory - The VIIth International Symposium Honoring French Mathematical Physicist Jean-Pierre Vigier, Imperial College, London, UK, 12-14 July 2010, AIP - American Institute of Physics Pub., Melville, N.Y., pp.48-89.

EmerQuM 11: Emergent Quantum Mechanics 2011 IOP Publishing
Journal of Physics: Conference Series **361** (2012) 012035 doi:10.1088/1742-6596/361/1/012035

Emergence of classical theories from quantum mechanics

P Hájíček

Institute for Theoretical Physics, University of Bern, Sidlerstrasse 5, CH-3012 Bern, Switzerland

E-mail: hajicek@itp.unibe.ch

Abstract. Three problems stand in the way of deriving classical theories from quantum mechanics: those of realist interpretation, of classical properties and of quantum measurement. Recently, we have identified some tacit assumptions that lie at the roots of these problems. Thus, a realist interpretation is hindered by the assumption that the only properties of quantum systems are values of observables. If one simply postulates the properties to be objective that are uniquely defined by preparation then all difficulties disappear. As for classical properties, the wrong assumption is that there are arbitrarily sharp classical trajectories. It turns out that fuzzy classical trajectories can be obtained from quantum mechanics by taking the limit of high entropy. Finally, standard quantum mechanics implies that any registration on a quantum system is disturbed by all quantum systems of the same kind existing somewhere in the universe. If one works out systematically how quantum mechanics must be corrected so that there is no such disturbance, one finds a new interpretation of von Neumann's "first kind of dynamics", and so a new way to a solution of the quantum measurement problem. The present paper gives a very short review of this work.

1. Introduction

All systems investigated by classical physics seem to consist of quantum particles and their dependent fields. Then, they seem to be also quantum systems and their classical properties ought to be derivable from quantum mechanics. More precisely, we expect that one can find quantum models of all classical properties. There are three independent problems that stand in the way of a satisfactory completion of this project.

First, classical systems and their properties are objective at least in the sense that the assumption of their real existence, independent of measurements, does not lead to any contradictions. Exactly this does not seem to hold for quantum systems and their properties, which is called *Problem of Realist Interpretation* of quantum mechanics. Second, classical properties such as a position do not allow linear superposition. Nobody has ever seen a table to be in a linear superposition of being simultaneously in the kitchen as well as in the bedroom. Moreover, observation of the table does not shift it while quantum registration changes the state of the registered system. Let us call this *Problem of Classical Properties*. Finally, evidence suggests the assumption that the registration apparatus always is in a well-defined classical state at the end of any quantum measurement indicating just one value of the registered observable. This is called *objectification requirement* [1]. However, if the registered system is in a linear superposition of different eigenvectors of the observable, then the linearity of Schrödinger equation implies that the apparatus is also in a linear superposition eigenstates of its pointer observable. Let us call this *Problem of Quantum Measurement.*

Our work during the recent five years has lead to a surprising discovery: The problems are considerably aggravated by some inveterate tacit assumptions about quantum mechanics that are clearly wrong if one states them explicitly [2, 3, 4, 5, 6, 7]. Quantum theory purified from these errors has been called *The Reformed Quantum Mechanics*. The present paper gives a very short review of our main ideas. The paper starts with the Problem of Quantum Measurement because only this requires changes in the conceptual and mathematical structure of quantum mechanics, proceeds to the Problem of Realist Interpretation and finishes by the Problem of Classical Properties.

2. Problem of Quantum Measurement

In classical physics, any measurement can be done, at least in principle, so that the system on which we measure is not disturbed by the measurement. We assume, as did the co-called Copenhagen interpretation, that this is not true for quantum measurement. A *registration* of any observable on quantum system $\mathcal{S}$ needs a special device $\mathcal{A}$, the so-called *registration apparatus*, and the registration is always an interaction between $\mathcal{S}$ and $\mathcal{A}$ that changes the state of the $\mathcal{S}$ except for some special cases. Moreover, to speak of the state of $\mathcal{S}$, there must be a *preparation*. Thus each measurement on microsystems can be split into preparation and registration.

The errors that cause the confusion in the theory of quantum measurement are two: 1) neglecting the disturbance of registration due to identical particles and 2) disregarding the role and structure of detectors.

2.1. Disturbance of registration due to identical particles

Let us consider two experiments.

Experiment I: State $\psi(\vec{x}_1)$ of particle $\mathcal{S}_1$ is prepared in our laboratory.

Experiment II: State $\psi(\vec{x}_1)$ is prepared as in Experiment I and state $\phi(\vec{x}_2)$ of particle $\mathcal{S}_2$ of the same type is prepared simultaneously in a remote laboratory.

If our laboratory does not know about the second one, it believes that the state of $\mathcal{S}_1$ is $\psi(\vec{x}_1)$. If it does then it believes that the state is

$$2^{-1/2}\big(\psi(\vec{x}_1)\phi(\vec{x}_2) \pm \phi(\vec{x}_1)\psi(\vec{x}_2)\big) . \tag{1}$$

Thus, it seems that the notions of preparation and of state are ambiguous. Has this ambiguity any observable consequences? To answer this question, let us first consider Experiment I supplemented by a registration corresponding to the observable of $\mathcal{S}_1$ with kernel $a(\vec{x}_1;\vec{x}'_1)$, and let the registration be made in our laboratory. Measurements of this kind lead to average value

$$\int d^3x_1 d^3x'_1\, a(\vec{x}_1;\vec{x}'_1)\psi^*(\vec{x}_1)\psi(\vec{x}'_1) . \tag{2}$$

Second, perform Experiment II supplemented by the registration by the same apparatus in our laboratory as above. Because the apparatus cannot distinguish between the contributions by the two particles, the correct observable corresponding to this registration now is:

$$a(\vec{x}_1;\vec{x}'_1)\delta(\vec{x}_2-\vec{x}'_2) + \delta(\vec{x}_1-\vec{x}'_1)a(\vec{x}_2;\vec{x}'_2) . \tag{3}$$

Such measurements lead to the average value defined by Eqs. (1) and (3):

$$\int d^3x_1 d^3x'_1\, a(\vec{x}_1;\vec{x}'_1)\psi^*(\vec{x}_1)\psi(\vec{x}'_1) + \int d^3x_1 d^3x'_1\, a(\vec{x}_1;\vec{x}'_1)\phi^*(\vec{x}_1)\phi(\vec{x}'_1)\ . \tag{4}$$

Expression (4) appreciably differs from (2) for all standard observables such as position, momentum, spin, angular momentum energy etc.

We conclude that the registrations of the standard quantum observables on a system is always disturbed by all other systems of the same kind existing somewhere in the universe.

2.2. Separation status

Cluster separability principle is a kind of locality assumption that has been fruitful in several branches of quantum physics, see e.g. [8] and Chap. 4 of [9]. An application of the principle to identical particles can be found for instance in [10] p. 128. We formulate it as follows:

Cluster Separability Principle *No quantum experiment with a system in a local laboratory is affected by the mere presence of an identical system in remote parts of the universe.*

This is violated by results similar to those of the previous section but confirmed by experience. We must therefore correct quantum mechanics so that it satisfies the Cluster Separability Principle.

We introduce an important locality property of observables [4] (for generalisation to composite systems, see [5]; a similar local condition on observables has been introduced in [11, 12]):

Definition 1 *Let $D \subset \mathbb{R}^3$ be open. Operator with kernel $a(\vec{x}_1;\vec{x}'_1)$ is D-local if*

$$\int d^3x'_1\, a(\vec{x}_1;\vec{x}'_1)f(\vec{x}'_1) = \int d^3x_1\, a(\vec{x}_1;\vec{x}'_1)f(\vec{x}_1) = 0\ ,$$

for any test function f that vanishes in D.

Now assume for Experiment II, to keep everything simple, that

(i) our laboratory is inside open set $D \subset \mathbb{R}^3$,
(ii) $\mathrm{supp}\,\phi \cap D = \emptyset$.

Then, the second term in (4) vanishes for all D-local observables and Eqs. (2) and (4) agree in this case (for a more general theorem see [5]). This suggests the following approach.

(i) We introduce the key notion of our theory:

Definition 2 *Let $\mathcal{S}$ be a particle and $D \subset \mathbb{R}^3$ an open set satisfying the conditions:*

- *Registrations of any D-local observable A of $\mathcal{S}$ lead to average $\langle\psi(\vec{x})|\mathsf{A}\psi(\vec{x})\rangle$ for all states $\psi(\vec{x})$ of $\mathcal{S}$.*
- *$\mathcal{S}$ is prepared in state $\phi(\vec{x})$ such that $\mathrm{supp}\,\phi \cap D \neq \emptyset$.*

In such a case, we say that $\mathcal{S}$ has separation status D.

For generalisation to composite systems and non-vector states, see [5].[1]

(ii) We assume: Any preparation of $\mathcal{S}$ must give it a non-trivial separation status $D \neq \emptyset$. Then D-local observables are individually registrable on $\mathcal{S}$ but only these are.

[1] Further generalisation is possible. We are working on two: First, to replace the condition $\mathrm{supp}\,\phi \cap D \neq \emptyset$ by $\int_D d^3x|\phi|^2 < \epsilon$ and that on the registered averages to the averages lying in intervals $(\langle\psi(\vec{x})|\mathsf{A}\psi(\vec{x})\rangle - \epsilon', \langle\psi(\vec{x})|\mathsf{A}\psi(\vec{x})\rangle + \epsilon')$, the epsilons being related to the finite sensitivity of registration apparatuses. Second, to impose conditions on the ranges of momenta defined by the sensitivity of detectors in addition to those on coordinates.

2.3. Registration

Preparation transfers a trivial into a non-trivial separation status. Thus, the separation status changes during a preparation. What is the relation of registrations to separation status change?

An important assumption of our theory of measurement is:

Pointer hypothesis *Any registration apparatus for microsystems must contain at least one detector and every 'reading of a pointer value' (see e.g. [1]) is a macroscopic signal from a detector.*

We take the definition of detector from experimentalists [13, 14]. Then, during registration, the system must enter the sensitive matter of a detector and its non-trivial separation status changes into a trivial one [4, 5].

In this way, preparation and registration have even more importance in our theory than in Copenhagen interpretation: they must include changes of separation status. Clearly, any change of separation status of a system $\mathcal{S}$ is also a change of its observable algebra, i.e., change of the kinematics of $\mathcal{S}$ [6].

What has been said up to now shows that textbook quantum mechanics is incomplete in the following sense:

(i) It accepts and knows only two separation statuses:
  (a) that of isolated systems, $D = \mathbb{R}^3$, with all self-adjoint operators as observables, and
  (b) that of a member of a system of identical particles, $D = \emptyset$, with no individual observables of its own.

(ii) It provides no rules for changes of separation status.

Hence, quantum mechanics must be supplemented by a theory of general separation status [4], a correct theory of observables [4] and by new rules that govern processes in which separation status changes [7].

In [7], a general rule for interaction between microsystem $\mathcal{S}$ and macrosystem $\mathcal{A}$ has been motivated and discussed. Formally, the process is decomposed into three steps. First, there is a change of kinematics and second, the unitary evolution of the state of $\mathcal{S}+\mathcal{A}$ in the framework of the new kinematics. Finally, the state of $\mathcal{S}+\mathcal{A}$ given by the unitary evolution must be corrected by a state reduction at the time of any detector signal in a way that is uniquely determined by the signals. The decomposition into the three steps is not a description of how things proceed in time: the rule gives only the end state.

The state reduction itself consists of two changes. First, some non-diagonal elements of the state operator of the total isolated system are erased. Second, the resulting mixture is postulated to be a proper one (a gemenge, see [4]). Something similar to the first change can also be achieved for an open subsystem of the total isolated system within standard quantum mechanics by methods such as quantum decoherence [1]. However, the second change can only be accomplished (or circumvented) with help of new assumptions that must be added to quantum mechanics such as the Everett interpretation in the case of decoherence, see [1, 15]. For a discussion of super-selection methods such as Wan's [11], see [7].

The Reformed Quantum Mechanics thus returns to von Neumann's "two kinds of dynamics" (see also [4]) but its notion of state reduction differs from von Neumann's in two points. First, it is less ad hoc because it is justified by the argument of separation status change, which is logically independent from the proper quantum measurement problem and, second, it is more specific because it happens only in a detector and its form is determined by objective processes inside the detector sensitive matter including approach to thermal equilibrium.

3. The problem of realist interpretation

The very subject of quantum mechanics is usually formulated very cautiously, e.g., [10], p. 13:

> ...quantum theory is a set of rules allowing the computation of *probabilities* for the outcomes of tests [registrations] which follow specific preparations.

This can be contrasted with naive expectation that

> quantum theory studies properties of existing microscopic objects.

The cause of such reserve is an apparent lack of objective properties of quantum systems. For example, one cannot ascribe values of observables to quantum systems prior to registrations, or else there will be nasty contradictions such as those due to contextuality [10], p. 187. An existing object must have enough objective properties. Thus, the very existence of quantum systems is uncertain, see, e.g., [10], p. 13.

The origin of this difficulty is the surprisingly inveterate tacit assumption that the values of observables of a quantum system $\mathcal{S}$ are a kind of basic properties of $\mathcal{S}$ so that no other properties can be accepted in their place.

In our opinion, every value a of observable A is only created during registration of A by an apparatus $\mathcal{A}$ and it is an objective property of composite $\mathcal{S} + \mathcal{A}$, not $\mathcal{S}$. It contains only some indirect information about $\mathcal{S}$. Then, all the enormous effort to justify values of observables of $\mathcal{A}$ as properties of $\mathcal{S}$ is futile and must be rejected. Instead, we postulate [2, 5]:

Basic Ontological Hypothesis of Quantum Mechanics *A property is objective if its value is uniquely determined by a preparation according to the rules of quantum mechanics. The "value" is the value of the mathematical expression that describes the property and it may be more general than just a real number. No registration is necessary to establish such a property but correct registrations cannot disprove its value; in most cases, registrations can confirm the value.*

Thus, we associate objective properties with preparations instead of registrations. Let us give some obvious examples of objective properties. First, the so-called *structural properties*: mass, charge, spin, structure of Hamiltonian for isolated systems, etc. Second, the so-called *dynamical properties*: state operator (not just wave function), average value and variance of an observable, etc.

It may be helpful to recall that preparations include more general processes than just manipulations in a human laboratory. For example, solar neutrinos are prepared at the centre of the Sun. But even if a property is defined by such a manipulation, the assumption that it can be ascribed to the prepared system is non-trivial and helpful. In fact, this is also assumed in classical physics for systems created in laboratory. In any case, a preparation is defined by objective (classical) conditions on $\mathcal{S}$, which determine the statistics of subsequent registrations on $\mathcal{S}$. More about the relation between objective and subjective aspects of probability theory can be found in [5].

We have shown in [5] that all properties determined by preparations form a Boolean lattice similarly to all objective properties of a classical system. Hence, there is no need for an extra logic in quantum mechanics.

Moreover, one can completely describe the dynamical situation of any quantum system by these objective properties. Hence, the condition on a system to be a physical object are satisfied. Such an object—a real quantum system—is theoretically modelled by an *occupied quantum state* [5]. In all considerations about quantum systems, it is then possible to refer directly to their objective properties, e.g., to speak of registration devices instead of observers. "Unspeakable" becomes "speakable" in quantum mechanics and it is a surprising liberation.

4. The problem of classical properties

Quantum mechanics leads to linear superposition of different states of macroscopic systems but classical theories such as Newtonian mechanics and Maxwellian electrodynamics do not allow such superpositions. How can this feature of the theories be derived from quantum mechanics?

The origin of all difficulties to obtain such derivation is a common inveterate assumption that a sharp trajectory is the key objective property of classical systems. Hence, quantum models of classical systems usually start from states with minimum uncertainty. These are pure and can easily be linearly superposed.

However, a completely sharp classical trajectory is a figment of imagination. *Real* states of all classical theories always have a non-zero variance, which is always much larger than the minimum quantum uncertainty. This is an old idea, cf. [16, 17]. It is, therefore, sufficient that quantum models of classical systems explain such fuzzy states. Instead of sharp trajectories, we assume:

Basic classicality hypothesis *To construct models of classical systems, high-entropy states of macroscopic quantum systems must be used.*

This seems to be in accord with state reduction being associated with approach to the thermal equilibrium of a detector (Sec. 2.3). Moreover, high-entropy states are mixed and cannot be linearly superposed. Of course, it is possible (but difficult) to prepare other states of macroscopic systems: strong laser beams, large EBC, etc. This simply means that not every state of a macroscopic system must be classical.

An example of suitable states modelling Newton mechanics of mass centre are the so-called *maximum entropy (ME) packets*: they maximise entropy for fixed averages and variances of coordinates and momenta [3]:

Theorem 1 *Let $\mathcal{S}$ be a quantum system with coordinates q_k and momenta p_k, $k = 1, \cdots, n$. Then, the state operator of the ME packet with averages and variances Q_k, ΔQ_k of coordinates and P_k, ΔP_k of momenta is*

$$\mathsf{T}(Q, P, \Delta Q, \Delta P) = \prod_{k=1}^{n} \left[\frac{2}{\nu_k^2 - 1} \exp\left(-\frac{1}{\hbar} \ln \frac{\nu_k + 1}{\nu_k - 1} \mathsf{K}_k \right) \right] , \tag{5}$$

where

$$\mathsf{K}_k = \frac{1}{2} \frac{\Delta P_k}{\Delta Q_k} (\mathsf{q}_k - Q_k)^2 + \frac{1}{2} \frac{\Delta Q_k}{\Delta P_k} (\mathsf{p}_k - P_k)^2 \tag{6}$$

and

$$\nu_k = \frac{2 \Delta P_k \Delta Q_k}{\hbar} \in (1, \infty) . \tag{7}$$

Here, K_k is *not* the Hamiltonian of $\mathcal{S}$: the Hamiltonian can be arbitrary. A similar theorem can be shown for classical EM packets [3].

Von Neumann entropy of $\mathsf{T}(Q, P, \Delta Q, \Delta P)$ is

$$S(\mathsf{T}(Q, P, \Delta Q, \Delta P)) = tr[\mathsf{T}(Q, P, \Delta Q, \Delta P) \ln \mathsf{T}(Q, P, \Delta Q, \Delta P)] =$$
$$\sum_{k=1}^{n} \left(\frac{\nu_k + 1}{2} \ln(\nu_k + 1) - \frac{\nu_k - 1}{2} \ln(\nu_k - 1) - \ln 2 \right)$$

and can be shown to be an increasing function of ν_k. Limit

$$\nu_k \to 1$$

for all k gives

$$S(\mathsf{T}(Q, P, \Delta Q, \Delta P)) \to 0 \ , \quad \mathsf{T}(Q, P, \Delta Q, \Delta P) \to |Q, P, \Delta Q, \Delta P\rangle\langle Q, P, \Delta Q, \Delta P| \ ,$$

where $|Q, P, \Delta Q, \Delta P\rangle$ is the Gaussian wave packet for the averages and variances $Q, P, \Delta Q, \Delta P$.

The time dependence of the averages and variances can be calculated from ME packets as initial states using the Hamiltonian of $\mathcal{S}$ in both Newtonian and quantum mechanics. Comparison of function $\mathsf{Q}_k(t)$, $\mathsf{P}_k(t)$, $\Delta\mathsf{Q}_k(t)$ and $\Delta\mathsf{P}_k(t)$ in the two theories shows full agreement for all Hamiltonians (potential functions) if

$$\nu_k \to \infty$$

for all k. Thus, classical limit is just the opposite of the textbook one: it is not the Gaussian wave packet but a high entropy limit.

Observe that classical properties $\mathsf{Q}_k(t)$, $\mathsf{P}_k(t)$, $\Delta\mathsf{Q}_k(t)$ and $\Delta\mathsf{P}_k(t)$ are uniquely determined by the preparation process and are therefore objective. More discussion can be found in [2] and [5].

We close this section by a few words on semi-classical (or WKB) approximation. It is an approximation method within quantum mechanics, usually defined as the limit $\hbar \to 0$ in some quantum expressions [10]. The resulting equations may be similar to the corresponding classical equations. Limit $\nu \to \infty$ also results from $\hbar \to 0$, if the variances are kept constant. However, the semi-classical approximation alone is not sufficient for a derivation of complete classical theories [10].

5. Conclusion

(i) Quantum mechanics is as objective as any other theory of physics.

(ii) Classical objects can be modelled by high-entropy states of macroscopic quantum systems. Classical limit is a suitably taken high-entropy limit. Coherent states have nothing to do with classical limit.

(iii) The theory of observables and of quantum measurement must be corrected to remove the disturbance due to identical particles. This leads to a solution of quantum measurement problem based on a state reduction. Our state reduction, however, is less ad hoc than von Neumann's because it is justified by a change of separation status, and it is more specific because it is associated with objective processes in the sensitive matter of a detector.

Acknowledgments
The author is indebted to Jiří Tolar for discussion.

References

[1] Bush P, Lahti P J and Mittelstaedt P 1996 *The Quantum Theory of Measurement* (Berlin: Springer)
[2] Hájíček P and Tolar J 2009 Intrinsic properties of quantum systems *Found. Phys.* **39** 411
[3] Hájíček P 2009 Quantum models of classical mechanics: maximum entropy principle *Found. Phys.* **39** 1072
[4] Hájíček P 2011 Quantum measurement problem and cluster separability *Found. Phys.* **41** 640
[5] Hájíček P and Tolar J 2010 Survey of an approach to quantum measurement, classical properties and realist interpretation problems *Acta Phys. Slovaca* **60** 613
[6] Hájíček P 2011 *J. Phys.: Conf. Ser.* **306** 012035
[7] Hájíček P 2011 *preprint* arXiv:1108.0053
[8] Keister B D and Polyzou W N 2002 Relativistic Hamiltonian dynamics in nuclear and particle physics *Advances in Nuclear Physics* vol 20, ed J W Negele and E Vogt (New York: Plenum) chapter 4 p 225
[9] Weinberg S 1995 *The Quantum Theory of Fields* vol I (Cambridge: Cambridge University Press)
[10] Peres A 1995 *Quantum Theory: Concepts and Methods* (Dordrecht: Kluwer)
[11] Wan K K and McLean R G D 1984 *J. Phys. A: Math. Gen.* **17** 837

[12] Wan K K 2006 *From Micro to Macro Quantum Systems. A Unified Formalism with Superselection rules and its Applications* (London: Imperial College Press) p 185
[13] Leo W R 1987 *Techniques for Nuclear and Particle Physics Experiments* (Berlin: Springer)
[14] Twerenbold D 1996 *Rep. Progr. Phys.* **59** 239
[15] Bassi A and Ghirardi G 2000 *Phys. Letters* A **275** 373
[16] Exner F 1922 *Vorlesungen über die physikalischen Grundlagen der Naturwissenschaften* (Leipzig: Deuticke)
[17] Born M 1955 *Phys. Blätter* **11** 49

EmerQuM 11: Emergent Quantum Mechanics 2011 IOP Publishing
Journal of Physics: Conference Series **361** (2012) 012036 doi:10.1088/1742-6596/361/1/012036

Towards Einstein's dream of a unified field theory: Reports from a journey on a long and winding road

Theo M. Nieuwenhuizen

Institute for Theoretical Physics, University of Amsterdam, Science Park 904, P.O. Box 94485, 1090 GL Amsterdam, The Netherlands

E-mail: `t.m.nieuwenhuizen@uva.nl`

Abstract. With general arguments it is motivated that we are still in search of a hidden-variables theory or sub-quantum mechanics. Aspects of a solvable model for quantum measurements are discussed, and a solution of the quantum measurement problem within ordinary quantum theory. The related statistical interpretation of quantum mechanics is advocated. Soliton approaches for elementary particles are proposed and some classical looking aspects of the relativistic hydrogen atom are mentioned. The integration of these solitons with stochastic electrodynamics is sketched.

1. Introduction

The long and winding road
that leads to your door
The Beatles

Quantum mechanics (QM) is our most successful theory ever. Whoever doubts it will rightfully face an uphill battle to call interest in his argumentation. Still, its most famous founding father, Albert Einstein, remained unsatisfied with its logical position. This led to the Einstein-Bohr debate on ontology versus epistemology, the question between "what is" and "what can we say about". For several decades the debate was thought to be won by Bohr, but the present opinion is that both were right from their perspectives. An every day analog concerns the weather in our own town. "Does it rain now?" is an ontological question, and "what are the chances for rain today?" an epistological one. Both questions are scientifically valid, but different.

In QM the situation leads to surprises. The question through which of the slits a particle went in a double slit experiment, is "forbidden", in the sense that QM cannot give an answer. If the setup is changed such that it can be monitored through which of the slits a particle goes, the outcomes of the set of experiments are different and do not exhibit interference effect. But this does not mean that the which-slit question is "wrong". Even though QM does not answer it, there should exist the physical answer: "this subset of measured particles went through slit A, and that subset through slit B". While this thesis may not convince the reader if the particles are photons or electrons, the case of double slit experiments with bucky balls (C_{60} molecules) and large organic molecules [1] makes it hard to believe that naive intuition is wrong: they, at least, went each through one slit, and nevertheless interference effects are observed. Having assured us that our naive intuition is right, we are left with the next question: How does Nature get this done?

Though the greatest minds of last century payed attention to it, the prevailing opinions nowadays are of "party line" type, namely "there is no problem to solve", "everything is explained by decoherence", or "the many worlds interpretation solves it all". We call such approaches "talking dynamics", proper physical questions are denied, trivialized or "solved" by linguistics instead of convincing physical modeling. Painful as it may be, we believe that the right approach is to do a better physics.

Here we face the dilemma between "party line consensus" and "free thinking." Though many new – or not so new – ideas are indeed too silly to be considered, the standard consensus, be it the Copenhagen interpretation of quantum mechanics or the Cold Dark Matter hypothesis in cosmology, deserves to be questioned as long as it lacks a convincing physical basis.

The mind set for investigating such fundamental questions should be ripe enough. Teenage girls singing about lost lovers may sound well, but cannot impress. The Dutch director Bernard Haitink admitted that it had not been right to have the Concertgebouw Orkest play Mahler in his early career. My own teacher at the University of Utrecht, Prof. Ben Nijboer, would not resist to stress that the study of foundations of QM needs the ripe mind of a researcher who already demonstrated the capability to solve various ordinary applications of QM.

If QM provides us with statistical answers, the question comes to the mind: *the statistics of what?* Probably like many others before, I have envisioned more than a decade ago the picture of QM being a statistical description *of an underlying reality* that is intuitively better understandable than QM itself. QM would then just be the statistical physics of this underlying dynamics, like classical statistical physics describes the statistics of the underlying classical mechanics. In this view a particle, like an electron, a photon or a neutrino, is some type of soliton that creates waves on its way, with which it interferes later so that correlations are present that are each time random but statistically correlated.

The plausibility of such a picture has now been beautifully demonstrated by the Couder experiments, where, in a suitable parameter domain, fluid drops move on top of the same fluid, and interference effects occur due to bow waves emitted by the moving drop and later coming back to it. In each trajectory this leads to a chaotic effect, but there is an interference effect at the statistical level. The observed interference can be viewed as a "particle-wave duality". Interestingly, the same setup allows for a different regime in the phase diagram where these interference effects do not occur, the quantum-like effects can be turned off [2].

If QM is emergent and can be explained from a deeper level theory, the integration with gravity might be achievable only at that level. That gives an a posteriori a reason why Quantum Gravity approaches starting from a quantized theory may not involve the right physics and be doomed, as they seem to be. This view is nowadays shared by Gerard 't Hooft as he exposed in his keynote lecture of the Heinz von Förster Congress on Emergent Quantum Mechanics in Vienna, November 2011. He considers that nature satisfies some classical-type dynamics, and that quantum techniques are effective tools to describe statistical properties thereof.

For the semi-classical approach to gravity the problems of blindly applying quantum rules are evident already: when one inserts in the classical Einstein equations the expectation value of the operator-valued energy-momentum tensor for quantum matter, this can only make physical sense for averaging over processes that have time scales shorter than those that characterize the evolution of the metric; those that are slower should not be averaged over, but taken instantaneously. This situation is no different from glasses, that have fast and very long time scales. They may satisfy a thermodynamics, provided thermal averages are done only over the fast modes, while the slow modes may setup their own, sliding equilibrium at an effective temperature that is much higher than the bath temperature. These insights already have helped to overcome the problems connected to the Ehrenfest relations and the Prigogine-Defay ratio in glass physics [3], but are more general.

2. Models for quantum measurement, the quantum measurement problem and the statistical interpretation

It is not by chance that the comparison with probabilities in everyday life makes sense in a discussion on foundations of QM. Indeed, through the Born rule QM provides us with probabilities for outcomes in experiments. But there is more to that. The ongoing debate on the interpretation QM comes to the mind. An important point to make here is the following: The only point of contact between quantum theory and the reality in the laboratory lies in quantum measurements. It is therefore likely that understanding the measurement process should guide us towards the minimalist interpretation of the theory. However, most approaches involve measurements only at the consistency level of incorporating the Born rule. For this reason, it seems difficult to trust the anyhow intricate logics of, e.g., the Copenhagen interpretation, the mind-body interpretation, consistent histories, the many worlds interpretation, the real ensemble interpretation, not to speak about multiverse interpretations. They are based on imaginative thinking, but not on what goes on during a real measurement process.

Clearly the cause of the trouble is the lack of understanding realistic quantum measurements. This can only be done by studying models that describe the process within some degree of precision. A model accounting for the various properties of ideal quantum measurements should in principle satisfy the following requirements [4]:

R1: simulate as much as possible real experiments;

R2: ensure unbiased, robust and permanent registration by the pointer of A, which should therefore be macroscopic; the pointer should give at each run a well-defined indication, which requires sufficiently complex interactions within the apparatus (dynamical stability and hierarchic structure of subensembles);

R3: involve an apparatus initially in a metastable state and evolving towards one or another stable state under the influence of S, so as to amplify this signal; the transition of A, instead of occurring spontaneously, is triggered by S;

R4: include a bath where the free energy released because of the irreversibility of the process may be dumped;

R5: be solvable so as to provide a complete scenario of the joint evolution of S + A and to exhibit the characteristic times;

R6: conserve the tested observable;

R7: lead to a final state devoid of "Schrödinger cats"; for the whole set of runs (truncation), and to a von Neumann reduced state for each individual run;

R8: satisfy Born's rule for the registered results;

R9: produce, for ideal measurements or preparations, the required diagonal correlations between the tested system S and the indication of the pointer, as coded in the expression (1) for the final state of S + A;

R10: be sufficiently flexible to allow discussing processes that are not perfect measurements.

These features need not be fulfilled with mathematical rigor. A physical scope is sufficient, where violations may occur over unreachable time scales or with a negligible probability.

Though a few dozen models have been proposed, they mostly fail on one or more of these criteria. With Armen Allahverdyan and Roger Balian, I have written an Opus Magnum on models for quantum measurement that starts with giving an extensive overview of the literature on models [4]. Next we focus on a rather realistic, but still solvable model, termed the Curie-Weiss model [5]. Of the tested system S, a spin $\frac{1}{2}$, the z-component is measured by an apparatus A that consists of a magnet M coupled to a bath B. The magnet is of Ising-type consisting of $N \gg 1$ spins $\frac{1}{2}$ and B is a bosonic bath. All steps of the measurement process can be analyzed analytically. The initial state of the apparatus is characterized as a metastable paramagnet. Such a metastability allows an amplification of the quantum signal, because it can trigger the phase transition towards the stable state. Remember that a photographic plate is also in a metastable state, as are the vision cells in our retina. The magnet starting in a mixed state (paramagnet) as well as the bath (Gibbs state) implies that there is no unitarity paradox in arriving from a mixed initial state of the total system (tested system plus apparatus) at a mixed final state, also not when the tested system starts in a pure state. This solves the "unitarity paradox", one of the mysteries of quantum measurements.

The measurement proceeds in two steps. First there is a quick decay of the Schrödinger cat terms, the off-diagonal terms of the density matrix in the diagonal basis of the measured operator. They first vanish due to a dephasing by creating a cascade of very many, very small multi-particle correlations, that all decay in a short time. Next the bath starts to act, and washes out the phase-coherent, dephased terms. Interestingly, this decoherence is not once-and-for-all, but goes on continuously. This process is called *truncation* of the density matrix. The term *reduction*, on the other hand, is reserved for an individual run of the measurement, and the assignment of a state to S+A at the end of the process [4].

As a second step, the registration takes place. This process is slower, because excess free energy of the initial metastable state has to be dumped in the bath. In case of measuring the z-component of a spin $\frac{1}{2}$, the final shape of the density matrix of S+A is

$$\hat{\mathcal{D}}(t_{\rm f}) = p_\uparrow |\uparrow\rangle\langle\uparrow| \otimes \hat{\mathcal{R}}_\Uparrow + p_\downarrow |\downarrow\rangle\langle\downarrow| \otimes \hat{\mathcal{R}}_\Downarrow. \tag{1}$$

where $p_\uparrow = r_{\uparrow\uparrow}(0)$ is probability for finding the spin "up" according to the Born rule (for a pure state: the square of the amplitude in the wave function) in terms of the initial density matrix $\hat{r}(0)$ of the spin, and, likewise, $p_\downarrow = r_{\downarrow\downarrow}(0)$. $\hat{\mathcal{R}}_\Uparrow$ and $\hat{\mathcal{R}}_\Downarrow$ denote the final density matrix of the apparatus with magnetization "up" and "down", respectively. More generally, this expression reads

$$\hat{\mathcal{D}}\left(t_{\rm f}\right) = \sum_i \left(\hat{\Pi}_i \hat{r}\left(0\right) \hat{\Pi}_i\right) \otimes \hat{\mathcal{R}}_i = \sum_i p_i \hat{r}_i \otimes \hat{\mathcal{R}}_i, \tag{2}$$

where $\hat{r}_i = \hat{\Pi}_i \hat{r}(0) \hat{\Pi}_i / p_i$ is one of the final shapes of density matrix of the tested system, with $\hat{\Pi}_i$ the projector on the Hilbert space of the measured eigenvalue; if the latter is non-degenerate, one just has $\hat{\Pi}_i = |\psi_i\rangle\langle\psi_i|$. The Born probability is generally $p_i = \mathrm{tr}\hat{\Pi}_i \hat{r}(0)$. Finally, $\hat{\mathcal{R}}_i$ is the final state of the apparatus when its pointer indicates the value labeled by i.

2.1. The quantum measurement problem

We observe a definite a outcome in each and every single experiment. *The quantum measurement problem* is name given to the task to explain this, preferably within QM. But since a convincing explanation within QM was never given, people gave up the hope and started to look into extensions of QM, like spontaneous collapse models of Ghirardi-Rimini-Weber and connections to gravity of Penrose.

However, we came to the conclusion that an answer is possible within QM. First of all, we have to face the fact that QM does not deal with individual systems. However, here the notion of pure subensembles comes to help: if we can find a pure subensemble on the diagonal basis of the measured operator, every one of its members will exhibit the same value for its measured observable.

It is tempting to adopt the naive explanation of (2) as describing one subensemble of spins being up, and A having magnetization up, and another subensemble describing spins down, characterized by the magnetization being down. After all, we know from experience that the apparatus will stay in its stable state i for a very long time. However, one has to face the problem that in QM a density matrix allows an infinity of decompositions. Now it can be argued that all other decompositions are dynamically unstable with respect to very small perturbations of the Hamiltonian. Therefore the only stable decomposition is (indeed) the naive one, which means that this is the only physical one. This solves the quantum measurement problem within ordinary QM and connects outcomes of ideal experiments to a classical probability theory [4].

2.2. The statistical interpretation, a minimalist interpretation of quantum mechanics

Many interpretations of QM have been proposed, but none has been generally accepted. However, with the only point of contact between quantum theory and reality lying in experiments, interpretation(s) should be based on analyzing quantum measurements. The literature on models has been reviewed in [4]. In absence of rich enough models this has neither led to a convincing situation. But a rather realistic but still solvable model was proposed by us nearly a decade ago [5].

So it is the description of the measurement process that should guide us. In the solution of the Curie-Weiss model we have heavily relied on quantum statistical physics: the initial state of the magnet is a paramagnet; of the bath it is a Gibbs state. Hence there emerges the statistical interpretation. Shortly, it reads: for an ensemble of identically prepared quantum systems (this may be an ensemble of real systems or a thought ensemble of supposedly similar systems), our knowledge is coded in the density matrix. A "quantum measurement" deals with an ensemble of measurements on this ensemble of systems and produces statistical statements. Many subtleties of this interpretation are discussed in [4].

3. Emergent QM: On sub-quantum mechanics and hidden variables theories

The Heinz Förster conference in Vienna 2011 was dedicated to emergent QM. This is just Einstein's dream, and the dream of many people after him. Views on this matter are expressed by: Adler; Hiley and others on extending on Bohm; Cetto and de la Pena; Couder; Everitt; Groessing and Schwabl; Hajicek; 't Hooft; Hu; Isidro; Kauffman; Khrennikov; Nelson; Vitiello. Perhaps several of these colleagues arrived at their view before realizing that they were following the footsteps of Einstein; at least, for me this was the case.

I have had many discussions with colleagues on the question whether we should expect an underlying "hidden variables" description of Nature. I could convince few. There were several groups (camps): those who were convinced already, those who remained unconvinced, those who were agnostic but open minded. For me the case is clear: in an individual measurement something is going on in the apparatus. QM allows us only to understand this statistically for a large number of repeated measurements. But each time something very definite goes on, and it leads to a definite outcome. *There must be a theory for individual events, since Nature it using it everyday the whole day.* The experiments of Couder on oil drops moving on the surface of the oil from which they are made, which lead to quantum-like (i.e., interference) phenomena in classical physics give hope that quantum behavior can be derived from such an underlying reality.

It is often said that Bell inequalities prevent the existence of hidden variable models. I have put forward that this is based on a mathematical assumption that Bell made when he took into account the hidden variables of the apparatuses [6]. Related views are expressed by several people, see the references in that paper, and I also mention the recent view of Cetto and de la

Peña. Which ever reason is the definite one, I consider Bell inequalities as too simple by their nature to rule out local realism. Moreover, quantum field theories would be subject to the same problems, but they do not expose this in the shape of their perfectly local Lagrangians.

4. On models for the elementary particles

Cosmic ray particles are known to have traveled millions of years; cosmic microwave background photons over thirteen billion years. We have asked ourselves the question: what happened in the mean time? And the only answer we could arrive at was as deep as trivial: they traveled as certain packets, that remained intact in the course of time. In certain one dimensional problems such solutions are known as *solitons*, a well known case is the Korteweg-de Vries equation.

Which structure could be needed to describe the solitons? There are two options: the presently known theories are sufficient, or they are not. In the latter case, our task seems hopeless. So let us investigate how far we can get with the present theories. Classical electrodynamics alone will probably be too simple, so we may have to add: general relativity, the weak force and the strong force. Given that there is good hope that the Higgs particle will be discovered in 2012, we may add the Higgs field as well. And in the 5d Kaluza-Klein theory also the inflaton would appear; the benefit of this approach would be that the $U(1)$ group would be naturally combined with gravity and not just stand next to it.

4.1. The electron

Since Faraday's insights in electromagnetic fields were put in mathematical form by Maxwell, nature is described by two antagonistic entities: matter and fields. A triumph for this split was the discovery of the electron by J.J. Thomson in 1897. Still, in those days it was now and then supposed that matter was itself some manifestation of fields, in the very same way that a vortex (a whirl of fluid or vapor), a hurricane and a tsunami are 'solitons' in hydrodynamics.

Models for the electron with extended charge distributions were considered by Abraham in 1903 and Lorentz in 1904. Though this led to a finite energy, they could not satisfactorily explain the criticism that Coulomb repulsion between different parts would prevent stability. Another twist came from the postulate of the electron spin by Goudsmit and Uhlenbeck, which would imply in these models also a surface rotation speed that exceeds the speed of light by a factor of order $1/\alpha$. In those days, classical intuition still held such a firm position that de Kronig refrained from publishing his idea after a discussion with Pauli, while Uhlenbeck begged Ehrenfest to withdraw the manuscript [7], Ehrenfest replying: "You are young enough to make an error"[1]. Only two years later the essential correctness of the idea was demonstrated by Dirac: the spin of the electron is identical to that of a spinning top, but it is an "intrinsically quantum" phenomenon. Nevertheless, here we shall discuss spin on a classical basis.

Our argument starts with considering the Kerr-Newman black hole. It is a solution in electrogravity for a massive, spinning, charged black hole. Based on the observation that is has a g-factor equal to 2, Carter proposed to consider it as a model for the electron [8]. For this application, it has a naked singularity with charge located on a ring of diameter equal to the Compton length $\hbar/mc$. If one goes through the ring, one ends up in another R^3 space, on going through it a second time, one is back in the original space. This strange behavior may be eliminated by assuming a singular disk bordered by the singular ring, or by assuming a certain elongated ball shaped region filled with a Higgs vacuum [9].

[1] Goudsmit recalls the following: Now it is being told that Uhlenbeck got frightened, went to Ehrenfest and said: "Don't send it off, because it probably is wrong; it is impossible, one cannot have an electron that rotates at such high speed and has the right moment". And Ehrenfest replied: "It is too late, I have sent it off already". But I do not remember the event, I never had the idea that is was wrong because I did not know enough. The one thing I remember is that Ehrenfest said to me: "Well, that is a nice idea, though it may be wrong. But you don't yet have a reputation, so you have nothing to lose". That is the only thing I remember [7].

A drawback of the Kerr-Newman model is that the parameters of the solution, mass, charge and spin, are not fixed. Other ingredients are needed to achieve this. One may search for Kerr-Newman type solutions in the Weinberg-Salam model. The hope is that the nonlinearities of the model put some constraints on the allowable parameters.

4.2. Application to the neutrino

One would expect for the neutrino soliton solutions appear that are similar to those for the electron. The electric charge would be zero, but not the hypercharge. Moreover, it is likely that, next too the usual "active" neutrinos, also sterile (right handed) neutrinos and (left handed) antineutrinos exist. They probably are Majorana fermions, rather than the Dirac fermions of the electron-positron pair.

5. Classical view on the relativistic hydrogen atom

When wishing to describe quantum physics from a classical starting point, it is a motivation if QM itself presents structures that invite for such a view point. Such a case has been found in the description of the relativistic hydrogen atom [10]. The eigenstates have energies that are certain square roots of square roots.

By considering a classical reality, the stationary classical phase space density will be a function of the conserved quantities, energy and angular momentum. This has been worked out for the Yrast states (n,l)=$(n,n-1)$; for $n=1$ this is the $1s$ ground state, for $n=2$ the $2p$ excited state, etc. In doing so, the Rydberg value for the energy is reproduced, but the fine structure correction, of order $\alpha=1/137$, is not correct. However, if the average of a certain product is replaced by the product of the averages, the proper correction is derived, and when this is done at arbitrary order, the exact formula for the Yrast energies is reproduced. The omission of correlations can be interpreted as a form of time scale separation. For details, see [10].

6. Stochastic Electrodynamics and Stochastic Soliton Dynamics

Stochastic electrodynamics (SED) was introduced by Nernst as an attempt to derive quantum theory from a classical basis. It is assumed that in classical electrodynamics particles such as the electron and the proton are classical entities, that are subject to the Coulomb force and a fluctuating electrodynamic field. The power spectrum of the field should be Planck's zero point spectrum, by which Planck's constant enters as a physical constant, the strength of the power. To mention one success of the approach, the leading logarithm of the Lamb shift is derived in a few lines [11]. Recent progress in this field is discussed by Cetto and de la Peña in these proceedings.

One would expect that the above picture achieves to produce a soliton description of elementary particles, as well as for their compositions like the proton and nuclei. The strength of the spin should then somehow be connected to the zero point field, since both involve $\hbar$. Combined with stochastic electrodynamics this would lead to a picture that we have termed *Stochastic Soliton Mechanics* [10]. At the statistical level it should almost everywhere reproduce QM, but in its corners it may offer an integration with gravity. What it could offer on top of results known from QM, is a description of transient effects before the quantum regime is reached.

6.1. On the cosmological constant problem

In quantum theory it is known that adding up the zero point energy of all electromagnetic modes up to the Planck energy brings a result that exceeds the observed cosmological constant by some 123 orders of magnitude. The very same would hold in SED. It is said that the related pressure equals minus this large value, so that the whole acts as a cosmological constant. This seems to be plainly wrong, for radiation the pressure is one-third of the energy density. But how then can this even more difficult cosmological constant problem be solved?

We believe that the answer lies in gravitation. The huge zero point energy cannot be created for free, but only at the expense of a nearly equally large negative gravitational energy. Likewise, in a given time-slice (3d space at a given time) the positive zero point pressure will be compensated by a nearly as large negative pressure from gravitation. Their combined effect, however, must have the form of the observed cosmological constant, which assures coordinate invariance. There may exist though a very early phase of the Universe where the zero-point spectrum, and hence a quantum structure of the dynamics, was not established.

Clearly, one needs a theory of gravitation that has a well defined energy momentum tensor. Such is possible if it is defined as a field in Minkowski space. There exist approaches in that direction, for instance by Baryshev and Teerikopri [12], and one may also consider the ghost-free massive gravity of de Rham, Tolley and Gabadadze [13].

7. Outlook

With Einstein's dream still being no more than a dream, a great challenge still lies ahead of us: to find a new theory of Nature that at the statistical level reproduces quantum theory. Quantum mechanics would then appear to be emergent. A new application could be the description of transient effects, before the quantum regime is reached. Deviations may be found too, and one of them may be a working integration of sub-quantum mechanics and gravitation.

References

[1] S. Gerlich, S. Eibenberger, M. Tomandl et al., Nature Comm. **2**, 263 (2011).
[2] Y. Couder and E. Fort, Phys. Rev. Lett. **97** 154101 (2006).
[3] L. Leuzzi and Th. M. Nieuwenhuizen, *Thermodynamics of the glassy state.* (Taylor and Francis, New York, 2008).
[4] A. E. Allahverdyan, R. Balian and Th. M. Nieuwenhuizen, arXiv:1107.2138; submitted to Physics Reports.
[5] A. E. Allahverdyan, R. Balian and Th. M. Nieuwenhuizen, Europhys. Lett. **61**, 453 (2003).
[6] Th. M. Nieuwenhuizen, Found. Phys. **41**, 580 (2010).
[7] S.A. Goudsmit, Ned. Tijdschr. Nat. **37**, 386 (1971). For an English translation, see: http://www.lorentz.leidenuniv.nl/history/spin/goudsmit.html.
[8] B. Carter, Phys. Rev. **174**, 1559 (1968).
[9] A. Burinskiy, Class. Quant. Grav. **16**, 3497 (1999); Grav. Cosmol. **10**, 50 (2004).
[10] Th. M. Nieuwenhuizen, AIP Conf. Proc. **810**, *Quantum Theory: Reconsideration of Foundations – 3*, G. Adenier, A. Yu. Khrennikov and Th. M. Nieuwenhuizen, eds. (AIP, Melville, NY, 2006), pp 198.
[11] L. de la Pena and A.M. Cetto, *The Quantum Dice: An Introduction to Stochastic Electrodynamics*, (Kluwer, Dordrecht, 1996).
[12] Y. Baryshev and P. Teerikorpi, *Fundamental questions of practical cosmology*, (Springer, Berlin, 2012).
[13] C. de Rham, G. Gabadadze and A. J. Tolley, Phys. Rev. Lett. **106**, 231101 (2011).

No Drama Quantum Theory?

Andrey Akhmeteli
LTASolid Inc., 10616 Meadowglen Ln 2708, Houston, TX 77042, USA
E-mail: akhmeteli@ltasolid.com

Abstract. Is it possible to offer a "no drama" quantum theory? Something as simple (in principle) as classical electrodynamics - a theory described by a system of partial differential equations in 3+1 dimensions, but reproducing unitary evolution of a quantum field theory in the configuration space? The following results suggest an affirmative answer:

1. The scalar field can be algebraically eliminated from scalar electrodynamics; the resulting equations describe independent evolution of the electromagnetic field.
2. After introduction of a complex 4-potential (producing the same electromagnetic field as the standard real 4-potential), the spinor field can be algebraically eliminated from spinor electrodynamics; the resulting equations describe independent evolution of the electromagnetic field.
3. The resulting theories for the electromagnetic field can be embedded into quantum field theories.

Another fundamental result: in a general case, the Dirac equation is equivalent to a 4th order partial differential equations for just one component, which can be made real by a gauge transform.

Issues related to the Bell theorem are discussed.

Is it possible to offer a "no drama" quantum theory? Something as simple (in principle) as classical electrodynamics - a local realistic theory described by a system of partial differential equations in 3+1 dimensions, but reproducing unitary evolution of quantum theory in the configuration space (see details and references in Refs. [1, 2, 3])? Of course, the Bell inequalities cannot be violated in such a theory. But there are some reasons to believe these inequalities cannot be violated either in experiments or in quantum theory. Indeed, there seems to be a consensus among experts that "a conclusive experiment falsifying in an absolutely uncontroversial way local realism is still missing" [4]. On the other hand, to prove theoretically that the inequalities can be violated in quantum theory, one needs to use the projection postulate (loosely speaking, the postulate states that if some value of an observable is measured, the resulting state is an eigenstate of the relevant operator with the relevant eigenvalue). However, such postulate, strictly speaking, is in contradiction with the standard unitary evolution of the larger quantum system that includes the measured system and the measurement device, as such postulate introduces irreversibility and turns a superposition of states into their mixture. Therefore, mutually contradictory assumptions are required to prove the Bell theorem, so it is on shaky grounds both theoretically and experimentally and can be circumvented if, for instance, the projection postulate is rejected.

For example, we can start with a system of interacting scalar charged field ψ and electromagnetic field A^μ with the Klein-Gordon-Maxwell equations of motion (scalar electrodynamics). For each solution A^μ, ψ of these equations there is a physically equivalent (i.e. coinciding with it up to a gauge transform) solution B^μ, φ, where φ is real [5]. Surprisingly, the resulting equations for B^μ and φ describe independent dynamics of electromagnetic field in the following sense: if components B^μ of the potential and their first derivatives with respect to x^0, $\dot{B}^\mu$, are known in the entire space at some time point ($x^0 = \text{const}$), the equations yield the values of their second derivatives, $\ddot{B}^\mu$, for the same value of x^0, so integration yields B^μ for any value of x^0. In particular, φ can be naturally eliminated from the equations.

The result remains valid if conserved external currents are added to the right-hand side of the Maxwell equations. This is important for description of the hydrogen atom, the Aharonov-Bohm effect, and other quantum phenomena in terms of electromagnetic field only.

This result can also be relevant to interpretation of quantum theory. For example, it allows an interesting modification of the de Broglie - Bohm interpretation. The results of this work suggest that electromagnetic field, rather than the matter field wave function, can be regarded as the guiding field, and in each point the particles move along the potential B^μ.

A second-quantized theory describing unitary evolution can be obtained from the resulting equations by a generalization of the Carleman linearization procedure (Ref. [6]). For a system of nonlinear partial differential equations, this procedure generates a system of linear equations in the Hilbert space, which looks like a second-quantized theory and is equivalent to the original nonlinear system on the set of solutions of the latter.

Similar results can be obtained if we start with the (more realistic) equations of the Dirac-Maxwell electrodynamics (spinor electrodynamics). After introduction of a complex 4-potential (producing the same electromagnetic field as the standard real 4-potential), the spinor field can be algebraically eliminated from spinor electrodynamics; the resulting equations describe independent evolution of electromagnetic field. Again, the resulting theory for electromagnetic field can be embedded into a quantum field theory.

As a spin-off, this work produced another fundamental result: in a general case, the Dirac equation is equivalent to a 4th order PDE for just one component, which can be made real by a gauge transform. As the Dirac equation is one of the most fundamental equations, this result both belongs in textbooks and can be used for development of new efficient methods and algorithms of quantum chemistry.

If, for the sake of simplicity, a system of units is used where $\hbar = c = m = 1$, and the electric charge e is included in A_μ for spinor electrodynamics ($eA_\mu \to A_\mu$), then the Dirac equation has the following form:

$$(i\not{\partial} - \not{A})\psi = \psi, \tag{1}$$

where, e.g., $\not{A} = A_\mu \gamma^\mu$ (the Feynman slash notation). In the chiral representation of γ-matrices (Ref. [7]) this equation is generally equivalent to the following equation for the first component ψ_1 of the Dirac spinor ψ:

$$\left((\Box' - iF^3) \left(iF^1 + F^2 \right)^{-1} (\Box' + iF^3) - iF^1 + F^2 \right) \psi_1 = 0, \tag{2}$$

where $F^i = E^i + iH^i$, electric field E^i and magnetic field H^i are defined by the standard formulas, and

$$\Box' = \partial^\mu \partial_\mu + 2iA^\mu \partial_\mu + iA^\mu_{,\mu} - A^\mu A_\mu + 1. \tag{3}$$

[1] Akhmeteli A quant-ph/1111.4630
[2] Akhmeteli A 2011 *J. Math. Phys.* **52** 082303
[3] Akhmeteli A 2011 *Int. J. Quantum Inf.* **9** 17
[4] Genovese M 2005 *Phys. Rep.* **413** 319
[5] Schrödinger E 1952 *Nature* **169** 538
[6] Kowalski K and Steeb W H 1991 *Nonlinear dynamical systems and Carleman linearization* : (World Scientific, Singapore)
[7] Itzykson C and Zuber J B 1980 *Quantum field theory* (McGraw-Hill)

Journal of Physics: Conference Series **361** (2012) 012038 doi:10.1088/1742-6596/361/1/012038

Application of the Doubly Special Relativity to the Dirac equation formalism

A Bérard[1], Z Belhadi[2], F Ménas[2], P Gosselin[3], Grandati[1] and H Mohrbach[1]

[1] Institut de Physique, Equipe "BioPhyStat", ICPMB, IF CNRS 2843, Université Paul Verlaine-Metz, 1 Bd Arago, 57078 Metz, Cedex 3, France

[2] Laboratoire de Physique et de Chimie Quantique, Faculté des Sciences, Université Mouloud Mammeri, BP 17 Tizi Ouzou, Algerie

[3] Institut Fourier, UMR 5582 CNRS-UJF, UFR de Mathématiques, Université de Grenoble I, BP74, 38402 Saint Martin d'Hères, Cedex, France

Abstract. In this letter we generalize the Dirac equation in the context of the Magueijo-Smolin (MS) [1] approach to the Doubly Special Relativity. This recent theory introduced by Amelino-Camelia [2] is based on the addition to the speed of light of a second relativistic invariant κ related to the Planck energy that implies a noncommutative space-time structure, the $\kappa-$ Minkowski space-time. In a recent paper [3] we deduced the "Dirac equation" in momentum space in MS base (see also [4]). Interesting studies with this equation have also been conducted, in particular in [5] where it was applied to a calculation of the hydrogen atom spectrum, or in [6] where the discreteness of the "generalized uncertainty principle" was used. Here we will establish the Dirac equation in position space.

Starting from the deformed dispersion relation of MS

$$E^2 = p^2c^2 + m^2c^4\left(1 - \frac{E}{E_p}\right)^2, \tag{1}$$

with $E_p = \sqrt{c^5\hbar G^{-1}}$ the Planck energy and using the equivalence principle that was proven to be valid in [8] , we readily arrive at the following Dirac equation in position space and MS base [8] :

$$\left(\left(1 + \frac{mc}{E_p}\beta\right) i\frac{\partial}{\partial t} - i\overrightarrow{\alpha}.\overrightarrow{\nabla} + \frac{mc}{\hbar}\beta\right)\widetilde{\psi}(\overrightarrow{r}, t) = 0.$$

Discussion of the associated deformed current can be found in [8] .

The prospects of this approach could be of at least be of two types, firstly toward a direct study of another equations as Kemmer-Duffin-Petiau or Feschbach-Villars, secondly toward a generalisation of the type "Gravity's rainbow" of J. Magueijo and L. Smolin [7] . We wish to insist on the simplicity of the deformed energy dispersion relation which has enabled us to continue the calculus so easily. It is obvious that the generalisation of this kind of results in the case of a general non-deformed energy dispersion is certainly not trivial.

References

[1] J. Magueijo and L. Smolin, Phys. Rev. Lett. 88(2002) 190403; Phys. Rev. D67 (2003) 044017.
[2] G. Amelino-Camelia, Nature 418 34 (2002); Phys. Lett. B 510 (2001) 255; Int. J. Mod. Phys. A11 (2002) 35.
[3] P. Gosselin, A. Bérard, H. Mohrbach and S. Ghosh, Phys.Lett.B660 (2008) 267-274.
[4] A. Agostini, G. Amelino-Camelia and M. Arzano, Class. Quant. Grav. 21 (2004) 2179.
[5] E. Harikumar and M. Sivakumar, arXiv:0910.5778.
[6] S. Das, E. C. Vagenas and A. Farag Ali, Phys.Lett.B690 (2010) 407-412.
[7] J. Magueijo and L. Smolin, Class. Quantum Grav. 21 (2004) 1725.
[8] Z. Belhadi, F. Ménas, A. Bérard, P. Gosselin and H. Mohrbach, arXiv:1106.1302 [hep-th].

Journal of Physics: Conference Series **361** (2012) 012039 doi:10.1088/1742-6596/361/1/012039

From Maximal Entropy Random Walk to quantum thermodynamics

Jarek Duda

Smoluchowski Institute of Physics, Jagiellonian University, Cracow, Poland

E-mail: `dudaj@interia.pl`

Abstract. Surprisingly, the natural looking random walk leading to Brownian motion occurs to be often biased in a very subtle way: emphasizing some possibilities by only approximating maximal uncertainty principle. A new philosophy of stochastic modelling has been recently introduced, in which we use the only maximizing entropy choice of transition probabilities instead. Local behaviour of both approaches is similar, but they usually lead to completely different global situations. In contrast to Brownian motion leading to nearly uniform stationary density, this recent approach turns out in agreement with having strong localization properties, thermodynamical predictions of quantum mechanics, like thermalizing to dynamical equilibrium state of probability density as the quantum ground state: squares of coordinates of the lowest energy eigenvector of the Bose-Hubbard Hamiltonian for single particle in discrete case, or of the standard Schrödinger operator while including potential and making infinitesimal limit.

While constructing probabilistic models, the choice of probabilities is often made arbitrarily. However, universal maximum uncertainty principle strictly imposes the optimal choice of probabilities as maximizing entropy for the knowledge we posses. For example, if among 2^n length n sequences of $0/1$, we restrict to sequences in which $p \in [0,1]$ of symbols are "1", their number is approximately $\binom{n}{pn} \approx 2^{nh(p)}$, where $h(p) := -p\lg(p) - (1-p)\lg(1-p)$ is Shannon's entropy, which has single maximum for $p = 0.5$. The entropy is coefficient in the exponent, so in $n \to \infty$ limit, the maximizing entropy cases completely dominates all possibilities - without additional relations, a long random $0/1$ sequence almost surely contains nearly half of "1" symbols.

Random walk is natural situation in physics of choosing probabilities - we discretize the space of possibilities and assume that because of complexity or our lack of information, the system makes transitions between these chosen states in practically random way. According to maximum uncertainty principle, as long as there are no additional clues about the transition probabilities, there should be used the maximizing entropy choice. The space of possibilities for any other choice becomes asymptotically negligible.

Specifically, for given graph defined by its adjacency matrix ($M_{ij} \in \{0,1\}$), we would like to choose transition probabilities ($0 \le S_{ij} \le M_{ij}$, $\sum_j S_{ij} = 1$), such that average entropy production ($H(S)$) is the largest possible: $H(S) := -\sum_i \pi_i \sum_j S_{ij}\lg(S_{ij})$, where $\pi_i \in [0,1]$ is stationary probability distribution for given S: $\sum_i \pi_i S_{ij} = \pi_j$, $\sum_i \pi_i = 1$. The standard choice leading to Brownian motion is that for each vertex, each outgoing edge is equally probable ($S_{ij}^{\mathrm{GRW}} = 1/\sum_j M_{ij} \;\Rightarrow\; \pi_i^{\mathrm{GRW}} \propto \sum_j M_{ij}$) - we will call such choice a Generic Random Walk. It can be seen as local approximation of entropy maximization. However, it usually is not

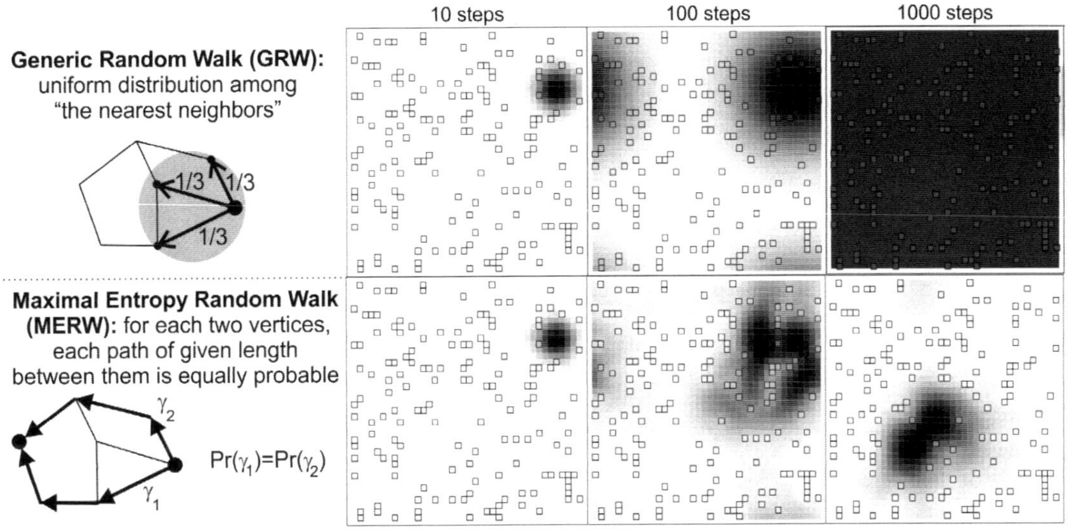

Figure 1. Schematic explanation of GRW and MERW choice of transition probabilities and simulation of their probability distribution evolution on 40×40 lattice graph with cyclic boundary conditions, in which all vertices but marked ones have self-loops (edge to itself).

the real maximum, which can be obtained by taking scale-invariant limit of GRW: instead of choosing uniform distribution among possible length 1 path, assume uniform probability among length k paths and take $k \to \infty$ limit. Such Maximum Entropy Random Walk is given by

$$S_{ij}^{\mathrm{MERW}} = \frac{M_{ij}}{\lambda}\frac{\varphi_j}{\varphi_i} \qquad \pi_i^{\mathrm{MERW}} \propto \varphi_i^2 \qquad \text{where} \qquad \sum_j M_{ij}\varphi_j = \lambda\varphi_i$$

is the dominant eigenvector. Local behaviour of both GRW and MERW looks similar, but from Fig. 1 we see that on defected lattice they lead to completely different stationary distributions - GRW to nearly uniform one, while MERW has strong localization properties ([1], [2]) - in this case practically completely localizes in the largest defect-free region (called Lifshitz sphere).

The adjacency matrix can be seen as minus Hamiltonian of Bose-Hubbard model for single particle - the dominant eigenvector corresponds to quantum ground state and so in opposite to GRW, MERW stationary probability distribution agrees with thermodynamical prediction of quantum mechanics: deexcitation to the ground state density. This localization effect can be seen in recent STM measurements of electron density on defected lattice of semiconductor [3].

The defects from Fig. 1 can be seen as locally larger potential - after generalization into continuous potential and taking infinitesimal limit of such lattice, the Hamiltonian becomes Schrödinger operator. It also leads to thermodynamical analogues of Ehrenfest equation, Heisenberg uncertainty principle and Pauli exclusion principle [4].

[1] Burda Z, Duda J, Luck J M, Waclaw B 2009, *Phys. Rev. Lett.* **102** 160602
[2] Burda Z, Duda J, Luck J M, Waclaw B 2010 *Acta Phys. Pol. B* **41** 949
[3] Richardella A, Roushan P, Mack S, Zhou B, Huse D A, Awschalom D D, Yazdani A 2010, *Science* **327** 5966
[4] Duda J 2011 (*Preprint* arXiv:1111.2253)

Quantum mechanical motion of classical particles

V D Rusov[1], K Lukin[2]

[1] Department of Theoretical and Experimental Nuclear Physics, Odessa National Polytechnic University, 65044 Odessa, Ukraine
[2] Institute for Radiophysics and Electronics NAS of Ukraine, 61085 Kharkiv, Ukraine.

E-mail: lukin.konstantin@gmail.com

Abstract. The Schrödinger equation for the particle wave function introduced via its action was derived from Newton equation for a point-like particle moving under effect of both potential force and fluctuation-dissipative environment, provided considering only stable motion of the particle. The model considered assumes existence of wave function as a physical field rather than just a mathematical abstraction.

1. Introduction

Copenhagen (orthodox) interpretation of Quantum Mechanics (QM) rejects existence of real wave function associated with a QM particle and suggests dealing with an "empty" QM wave function, the squared modulus of which gives probability of finding the particle at a given point of space at certain time (Max Born). At the same time, this interpretation leaves outside the QM theory a consideration of possible causes and origin of such probabilistic behaviour of the particles with mass. Nevertheless, many attempts to clarify "physical nature" in interpretations of QM have never stopped. Moreover, the interest in explanations of the physical nature of wave-particle dualism and indeterminism in QM has grown significantly in the last decade. There are several interpretations of QM, which in the authors opinion explain the essence of the things from the viewpoints much closer to the classical physics rather than to the abstract mathematical constructions of QM. It is accepted that a QM particle is governed by Schrödinger equation which describes its wave properties and enables to evaluate probability of its finding at a given coordinate at the given time instance.

In this short paper we show the way of derivation of Schrödinger equation from Newton equation for a point-like particle moving in both a potential force and fluctuation-dissipative environment. Application of the Chetaev motion stability theorem for a particle moving under dissipative forces in a potential field is an essential point of this derivation [1]. In this way we have shown that quantum potential which provides compensation of the environment action onto the particle motion has a fluctuation-dissipative nature and defines stable motion of the particles. The second objective of this short paper is verification of the fluctuation-dissipative interpretation of a quantum potential.

2. Schrödinger equation as a condition for the stability of the particle motion in the presence of dissipative forces

Consider a non-relativistic particle motion under both a potential force and fluctuation-dissipative environment, including the forces generated by the quantum potential.

The main assumptions of the developed theory are as follows: (1) non-relativistic particle with mass m governed by the Newton motion equation, in the right side of which the forces generated by external potentials, including the fluctuation-dissipation potential due to the fluctuations of the vacuum are taken into account; (2) in addition, the particle is affected by the forces produced by the quantum potential, whose existence is postulated; (3) an essentially new proposition of the theory is the assumption on the dissipative nature of particle interaction with the environment (fluctuating vacuum), on the one hand, and assumption of the existence of stable trajectories of the particle due to exact compensation of its fluctuations by the quantum potential (QP). Equation of one-dimensional motion of a classical particle under the action of potential and dissipative forces, characterized by the potentials $U(x,t)$ and $Q(x,t)$, respectively, may be written in terms of its action and energy as follows:

$$\frac{\partial S(x,\nu,t)}{\partial t} = T(\nu,t) + U(x,t) + Q(x,t) \tag{1}$$

where $T(\nu,t) = \frac{1}{2}m\nu^2$ is the particle kinetic energy, $S(x,\nu,t) = \int m\nu dx$ is the particle action.

The dissipative potential force, which leads to the destruction of the motion integral of the corresponding conservative system has been introduced in Newton equation and as a potential $Q(x,t)$ in Eq.(1) in contrast to the Jacobi equation, which relates the kinetic energy of a particle with its action and potential for *conservative* systems. If we neglect dissipation term in Eq. (1) the latter will be identical to the Jacobi equation. On the other hand, for only stable trajectories of the particle under consideration, the equations describing such motions turn out to be *conservative* again, i.e. the system under consideration is Hamiltonian one! Now, if we introduce wave function ψ in standard way $\psi = Aexp(i\hbar S)$ and apply Chetaev condition for stable motion [1] we may readily derive Schrödinger equation with respect to the function ψ :

$$i\hbar\frac{\partial\psi}{\partial t} = -\frac{\hbar^2}{2m}\Delta\psi + U\psi \tag{2}$$

along with probability continuity equation and relation between quantum potential and wave function amplitude, respectively:

$$\frac{\partial A}{\partial t} = -\frac{A}{2m}\sum\frac{\partial^2 S}{\partial x_i^2} - \sum\frac{\partial A}{\partial x_i}\frac{p_i}{m} \quad and \quad Q = -\frac{\hbar^2}{2m}\frac{\Delta A}{A} \tag{3}$$

We have shown that QP which provides compensation of the environment action onto the particle motion has a fluctuation-dissipative nature and defines stable motion of the particles [2]. QP calculated with the help of wave function for the stochastic Schrödinger equation describing quantum oscillator with dissipation, produces the dissipative forces and fluctuations with zero mean value, which leads us to quantum Langevin equation. The particle wave properties may be tied with its local complicated (but periodical) motion around a mass center and may be described in terms of known "zitterbewegung" models [3]. Deriving the Fischer information for the QP and using Cramer-Rao inequality we directly arrive at the Heisenberg uncertainty principle for moving particles. In this way description of QM particle is derived from classical equations of motion for a point-like particle in potential field and fluctuation-dissipative environment as its stable motion provided compensation of the particle fluctuations by the QP.

References

[1] Rusov V D, Vlasenko D S and Mavrodiev S Cht 2011 *Annals of Physics* **326** 1807-15
[2] Rusov V D, Lukin K A and Vlasenko D S 2011 Schrödinger Equation as Equation for Stable Motion of Classical Particles in Fluctuation-Dissipative Environment, *Proc. Third Int. Conf. on Quantum Electrodynamics and Statistical Physics (Kharkov, Ukraine, 29 August - 2 September)*, p.127
[3] Salesi G 1997 Int. J. Mod. Phys. **A12**, 5103

Classical Simulation of Double Slit Interference via Ballistic Diffusion

J Mesa Pascasio[1,2], S Fussy[1], H Schwabl[1], G Grössing[1]

[1] Austrian Institute for Nonlinear Studies, Akademiehof, Friedrichstr. 10, 1010 Vienna, Austria
[2] Institute for Atomic and Subatomic Physics, Vienna University of Technology, Operng. 9, 1040 Vienna, Austria

E-mail: ains@chello.at

Abstract. Based on a proposed classical explanation, the quantum mechanical "decay of the wave packet" is shown to simply result from sub-quantum diffusion with a specific diffusivity varying in time due to a particle's changing thermal environment. The exact quantum mechanical intensity distribution, as well as the corresponding trajectory distribution and the velocity field of a Gaussian wave packet are therewith computed. We utilize no quantum mechanics, but only familiar simulation techniques for diffusion, e.g., finite differences or coupled map lattices (CML).

We derive a solution for an anomalous diffusion equation with a time-dependent diffusion coefficient $D(t) = kt^\alpha$ of the general form

$$\frac{\partial P}{\partial t} = kt^\alpha \frac{\partial^2 P}{\partial x^2}, \quad \alpha > 0. \tag{1}$$

Here, t and k denote the time and a constant factor, respectively. $P(x,t)$ is the solution of Eq. (1). We assume a Gaussian distribution function as ansatz for P,

$$P(x,t) = \frac{1}{\sqrt{2\pi}\sigma} \mathrm{e}^{-\dfrac{(x-x_0)^2}{2\sigma^2}}, \tag{2}$$

where $\sigma(t)$ denotes the standard deviation, i.e. the distance from the mean value to the left or right interception point of the distribution function P. For further considerations we put the initial location x_0 in Eq. (2) into the coordinate's origin, i.e., $x_0 = 0$. Then we calculate the partial derivatives of P using the short form $\dot{\sigma}$ for the time-derivative of $\sigma(t)$,

$$\frac{\partial P}{\partial t} = -P\frac{\dot\sigma}{\sigma} + P\frac{x^2\dot\sigma}{\sigma^3}, \qquad \frac{\partial P}{\partial x} = -P\frac{x}{\sigma^2}, \qquad \frac{\partial^2 P}{\partial x^2} = P\left(-\frac{x}{\sigma^2}\right)^2 + P\left(-\frac{1}{\sigma^2}\right), \tag{3}$$

then substitute these results into Eq. (1),

$$\frac{P\dot\sigma}{\sigma}\left(\frac{x^2}{\sigma^2} - 1\right) = kt^\alpha \frac{P}{\sigma^2}\left(\frac{x^2}{\sigma^2} - 1\right), \tag{4}$$

and obtain

$$\frac{\sigma^2}{2} = k\frac{t^{\alpha+1}}{\alpha+1} + \frac{c_0}{2} \,. \tag{5}$$

At this stage we introduce a result of [1] which is an expression for the spreading of the Gaussian,

$$\sigma^2 = \sigma_0^2 \left(1 + \frac{D^2 t^2}{\sigma_0^4}\right) , \tag{6}$$

with σ_0 as initial standard deviation at time $t = 0$. Note that the diffusivity $D = \frac{\hbar}{2m}$ is constant for all times t and has to be distinguished from the diffusion coefficient $D(t)$. Substitution of Eq. (6) into (5) yields $c_0 = \sigma_0^2$ and

$$k\,\frac{2t^{\alpha+1}}{\alpha+1} = \frac{D^2 t^2}{\sigma_0^2} \,. \tag{7}$$

Eq. (7) can only be fulfilled by $\alpha = 1$, so that $k = \frac{D^2}{\sigma_0^2}$. Hence, our time-dependent diffusion coefficient becomes

$$D(t) = \frac{D^2 t}{\sigma_0^2} \,. \tag{8}$$

Finally, Eq. (1) becomes

$$\frac{\partial P}{\partial t} = D(t)\,\frac{\partial^2 P}{\partial x^2} \tag{9}$$

and turns out to be a ballistic diffusion equation because $\alpha = 1$.

It is straightforward to simulate the diffusion process of Eq. (9) in a computer model. One approximates the diffusion equation by

$$P[x, t+1] = P[x, t] + \frac{D[t+1]\Delta t}{\Delta x^2} \left\{P[x+1, t] - 2P[x, t] + P[x-1, t]\right\} \tag{10}$$

with space and time grid indices x and t, respectively, and an initial Gaussian $P(x, 0)$ at $t = 0$ with initial standard deviation σ_0.

The simulation of interference of two beams emerging from Gaussian slits is established with the aid of a simulation as in Refs. [1, 2, 3]. To account for interference, we simply follow the classical rule for the intensities $P_{\rm tot} := R^2 = |R_1\mathbf{k}_1 + R_2\mathbf{k}_2|^2 = R_1^2 + R_2^2 + 2R_1R_2\cos\varphi = P_1 + P_2 + 2\sqrt{P_1P_2}\cos\varphi$, with $R\mathbf{k} = R_1\mathbf{k}_1 + R_2\mathbf{k}_2$ and $\varphi = m\Delta v_x\, x/\hbar$. The trajectories are the flux lines obtained by choosing a set of initial points at $y = 0$. Two adjacent flux lines thereby define regions of constant flux, i.e., $\int_A P\,\mathrm{d}A = $ const., with A being the cross section of a flux tube.

We have shown in [3] that it is possible to simulate a particle's trajectories behind a double slit with usual classical simulation tools. The key is the ballistic diffusion equation (9) which describes the underlying physics by the time-dependent diffusion coefficient (8).

References

[1] G. Grössing, S. Fussy, J. Mesa Pascasio, H. Schwabl, Emergence and collapse of quantum mechanical superposition: Orthogonality of reversible dynamics and irreversible diffusion, Physica A 389 (2010) 4473–4484. `arXiv:1004.4596v1`.

[2] G. Grössing, S. Fussy, J. Mesa Pascasio, H. Schwabl, Elements of sub-quantum thermodynamics: quantum motion as ballistic diffusion, J. Phys.: Conf. Ser. 306 (2011) 012046. `arXiv:1005.1058v2`.

[3] G. Grössing, S. Fussy, J. Mesa Pascasio, H. Schwabl, An explanation of interference effects in the double slit experiment: Classical trajectories plus ballistic diffusion caused by zero-point fluctuations, Ann. Phys. 327 (2012) 421–437. `arXiv:1106.5994v3`.

Institute of Physics
Dirac House, Temple Back
Bristol BS1 6BE UK

CURRAN ASSOCIATES INC
proceedings.com